DATE DUE

CRC Handbook of

ELECTRICAL FILTERS

CRC Handbook of

ELECTRICAL FILTERS

Edited by

John T. Taylor
Qiuting Huang

CRC Press
Boca Raton New York

Acquiring Editor:	Navin Sullivan
Associate Editor:	Felicia Shapiro
Project Editor:	Jennifer Richardson
Marketing Manager:	Susie Carlisle
Direct Marketing Manager:	Becky McEldowney
Cover Design:	Dawn Boyd
PrePress:	Kevin Luong
Manufacturing:	Sheri Schwartz

Library of Congress Cataloging-in-Publication Data

CRC handbook of electrical filters / edited by John T. Taylor and
Qiuting Huang.
p. cm.
Includes bibliographical references and index.
ISBN 0-8493-8951-8 (alk. paper)
1. Electric filters. I. Taylor, John T., Ph.D. II. Huang,
Qiuting.
TK7872.F5C73 1996
621.3815′324--dc20

96-26723
CIP

This book contains information obtained from authentic and highly regarded sources. Reprinted material is quoted with permission, and sources are indicated. A wide variety of references are listed. Reasonable efforts have been made to publish reliable data and information, but the author and the publisher cannot assume responsibility for the validity of all materials or for the consequences of their use.

Neither this book nor any part may be reproduced or transmitted in any form or by any means, electronic or mechanical, including photocopying, microfilming, and recording, or by any information storage or retrieval system, without prior permission in writing from the publisher.

All rights reserved. Authorization to photocopy items for internal or personal use, or the personal or internal use of specific clients, may be granted by CRC Press LLC, provided that $.50 per page photocopied is paid directly to Copyright Clearance Center, 27 Congress Street, Salem, MA 01970 USA. The fee code for users of the Transactional Reporting Service is ISBN 0-8493-8951-8/97/$0.00+$.50. The fee is subject to change without notice. For organizations that have been granted a photocopy license by the CCC, a separate system of payment has been arranged.

The consent of CRC Press LLC does not extend to copying for general distribution, for promotion, for creating new works, or for resale. Specific permission must be obtained in writing from CRC Press LLC for such copying.

Direct all inquiries to CRC Press LLC, 2000 Corporate Blvd., N.W., Boca Raton, Florida 33431.

© 1997 by CRC Press LLC

No claim to original U.S. Government works
International Standard Book Number 0-8493-8951-8
Library of Congress Card Number 96-26723
Printed in the United States of America 1 2 3 4 5 6 7 8 9 0
Printed on acid-free paper

Preface

In essence, an electric filter is a device whose response differs in some specified way from the applied stimulus, such that the relationship between response and excitation can be designed to fit a specification. The precise relationship between response and stimulus is often very complex and this may be characterized in the frequency or time domains, depending on the specification.

Although the requirement for filter systems originated in the early telephone networks which were being developed from the end of the nineteenth century, it was not long before many more applications appeared. In fact, filters have become essential components of a very wide variety of modern telecommunications, instrumentation, and signal processing systems. This increasing pressure from such a diverse spectrum of application areas has had a dramatic effect on the field of filter design. Firstly, the meaning of the term has been broadened and modified. From its origins as a passive device supplying approximate frequency domain amplitude shaping in the audio band, modern filter characteristics can be specified in a variety of ways and implemented in a variety of technologies, both active and passive, over a very wide frequency range. The intervening stages in this evolutionary process have involved the development of specialized mathematical tools, new technologies, and computer software packages for analysis and synthesis.

This handbook is not only intended to be a practical guide for experienced engineers who are nonspecialists in filter theory and design but also to serve as an introduction for newcomers in the field. Some general information and basic theory is therefore provided for each filter type, but without entering into detailed analysis. The frequency range approximately DC-1GHz is covered, and so, e.g., microwave filters are not included.

The Editors

John T. Taylor is Reader in Electronic Engineering at University College, London, U.K. Dr. Taylor received the Ph.D. degree from Imperial College, London, in 1984, his subject having been exact design methods for switched-capacitor ladder filters. His current research interests, in addition to a continuing interest in analogue filters, include the use of optoelectronic methods to enhance the performance of electronic circuits and the realization of artificial neural networks using analogue circuit techniques.

Qiuting Huang is an assistant professor at the Swiss Federal Institute of Technology, Zurich, Switzerland. Professor Huang received his Ph.D. degree from the ESAT Laboratories, Katholieke Universiteit Leuven, Belgium, in 1987. The subject of his dissertation was optimal design of switched-capacitor circuits and oscillators. His current research activities are in RF receivers for wireless communications, integrated filters, and data converters, as well as high-speed digital integrated circuits.

Contributors

P. Agathoklis
Professor
Department of Electrical and Computer Engineering
University of Victoria
Victoria, Canada

Andreas Antoniou
Professor
Department of Electrical and Computer Engineering
University of Victoria
Victoria, Canada

Majid Ahmadi
Professor
Department of Electrical Engineering
University of Windsor
Windsor, Canada

Marcello Luiz Rodrigues de Campos
Associate Professor
Department of Electrical Engineering
Military Institute of Engineering
Rio de Janeiro, Brazil

Satoru Fujishima
Corporate Advisor, Technology
Murata Manufacturing Co., Ltd.
Nagaokakyoshi, Japan

Philip R. Geffe
Independent Consultant
Bethlehem, Pennsylvania

Jacqueline H. Hines
Manager, Research and Development
Sawbek, Inc.
Orlando, Florida

Peter Horn
Horsch Elektronik AG
Gams, Switzerland

Qiuting Huang
Professor
Integrated Systems Laboratory
Swiss Federal Institute of Technology
Zurich, Switzerland

Robert A. Johnson
Consultant
Filter Products
Rockwell
Costa Mesa, California

Robert Kinsman
Retired
Naperville, Illinois

Jürgen Kintscher
LS Allegem. u. Theoret. E-technik
University Erlangen
Erlangen, Germany

William B. Kuhn
Assistant Professor
Electrical and Computer Engineering
Kansas State University
Manhattan, Kansas

Stuart Lawson
Reader
Department of Engineering
University of Warwick
Coventry, United Kingdom

L.F. Lind
Department of Electronic Systems Engineering
University of Essex
Essex, England

Wu-Sheng Lu
Professor
Department of Electrical and Computer Engineering
University of Victoria
Victoria, Canada

E.I. El-Masry
Department of Electrical Engineering
Technical University of Nova Scotia
Nova Scotia, Canada

F. William Stephenson
Dean of Engineering
College of Engineering
Virginia Polytechnic Institute and State University
Backsburg, Virginia

George Szentirmai
DGS Associates, Inc.
Menlo Park, California

Ayman Tawfik
Department of Electrical Engineering
Technical University of Nova Scotia
Halifax, Canada

John T. Taylor
Department of Electronic and Electrical Engineering
University College London
London, United Kingdom

R. Unbehauen
LS Allegem. u. Theoret. E-technik
University Erlangen
Erlangen, Germany

Jiri Vlach
Professor
Department of Electrical and Computer Engineering
University of Waterloo
Waterloo, Canada

Contents

Chapter 1

Introduction

John T. Taylor and Qiuting Huang

CONTENTS

1. BRIEF HISTORY

Interest in filter theory and design began with the development of the early telecommunications industry in Europe and the U.S., beginning in the late 19th century. Specifically, the need for frequency division multiplexing (FDM) systems to extend the scope of voice telephone networks was a key factor in the stimulation of early research in the area. This led to the discovery of the first image-parameter electric wave filters (independently by Wagner in Germany and Campbell in the U.S.) in 1915. These filters were passive, lumped-element systems which provided an approximate solution to the problem of filtering in the frequency domain. Subsequently, the definition of what a filter is was substantially enlarged to include the time-domain as well as the frequency-domain response.

During the 1920s and 1930s much effort went into developing the theory of the filter design problem, that is, the problem of designing a network with real components to provide a prescribed response to a particular stimulus. This problem divides naturally into two parts: *approximation* and *synthesis*. Approximation is concerned with choosing a mathematical function which (1) describes the problem to be solved with sufficient accuracy and (2) is of a form which is known to be realizable using real components (during the period under discussion the components available were passive, consisting of inductors, capacitors, resistors, and transformers). Synthesis identifies the topology and component values of an electrical network (which may or may not be unique) which realizes the approximating function.

In addition to theoretical research, the practical requirement for high-quality filters stimulated the development of components for use in passive LCR filters. Most voice-frequency filters ($f < 4$ kHz) were realized using these techniques up to the 1960s, and passive LCR filters are still extensively used today in specialized applications where, e.g., wide dynamic range and low noise are required.

In the period following World War II, stimulated by the practical difficulties associated with inductors suitable for low-frequency filtering applications, an interest in *inductorless* filters began to develop. These active RC filters employed active components (initially thermionic valves, later

0-8493-8951-8/97/$0.00+$.50
© 1997 by CRC Press LLC

semiconductor devices and subsystems) in combination with resistors and capacitors to simulate the electrical properties of inductors without incurring the penalties of the use of physical inductors. However, since the time constants of these active RC filters depended on RC products, each unit had to be tuned individually, making the approach fundamentally incompatible with the emerging integrated circuit technologies. This situation was drastically changed in the 1970s by the appearance of *switched-capacitor* filters. In this approach, which employs both sampled data and analogue techniques, the time constants are defined by capacitor ratios and by the frequency of an externally generated clock, both of which can be specified to high levels of accuracy. In addition, more recently, methods have been described which enable fully integrated continuous-time active-RC filters to be fabricated.

In very demanding applications where bandpass characteristics are required with very sharp attenuation characteristics which cannot be realized using passive LCR filters, *mechanical filters* are frequently employed. These devices make use of acoustic (mechanical) resonances to perform the filtering of electrical signals. In addition, mechanical filters are highly stable to temperature changes and ageing and are readily manufacturable in miniaturized form at low frequencies. Other types of discrete mechanical filters, employing different types of electromechanical resonance, have also been evolved which are suitable for use in certain important application areas. The main types include crystal filters and ceramic filters, both of which use the piezoelectric effect, but with fundamentally different materials and surface acoustic wave (SAW) filters. All these devices are extensively used in radio communication transceivers, where the noise and dynamic range requirements preclude active filters. Because of the rapid growth of wireless communications services currently, miniaturized passive filters have been the focus of research and development by major manufacturers and their performance improves continually. The chapter on mechanical, crystal, ceramic, and SAW filters will have particular relevance to RF designers.

In complete contrast to the types of filters mentioned above, *digital filters* have become extremely important in recent years. These filters process binary-coded versions of samples obtained by uniformly and linearly sampling an analogue signal and are therefore entirely digital in operation, unlike switched-capacitor filters, which retain many aspects of analogue signal processing. Digital filters have gained popularity steadily since the 1980s when the first inexpensive single-chip digital signal processors became available.

At the present time, filter research continues at a great pace in industrial companies and academic institutions around the world. For example, much work is being done in the design and development of advanced digital filters, mirroring developments in the design of integrated digital signal processors. Integrated continuous-time filters have also made tremendous progress. They can be found in products such as a disk drive, where the linearity and noise requirements are not so stringent as in other, more traditional applications. As the design of such filters involves a great deal of transistor circuit-level details, it is left out of this handbook.

2. PURPOSE AND SCOPE OF THIS HANDBOOK

This handbook was written as a practical guide for electronic and electrical engineers who are nonspecialists in the areas of filter theory and design. The intention was therefore to provide enough general theory and general information about the various types to be useful, without entering into detailed analysis of the various filter types. Such detailed material can be found in the very many specialist textbooks which already exist on the subject (an exception to this occurs in Chapter 2 where it was felt appropriate to include a section giving some detailed explanation of approximation theory to back up the CAD packages described later on). The handbook covers the frequency range approximately DC-1 GHz and so, e.g., microwave filters are not included, since these are a separate study in themselves.

Chapter 2

Approximation Methods

Jürgen Kintscher and R. Unbehauen

CONTENTS

1. SURVEY

1.1 PROBLEM

The design of linear, time-invariant, and stable lumped-element filters is mostly based on the evaluation of real, rational transfer functions assuming that the specifications are given in the form of prescriptions for the transfer behavior of the network. This is typical for the synthesis of continuous as well as for the synthesis of digital filters.[1-4] In the case of continuous-time filters the transfer function $H(s)$ must be free of poles in the half-plane Re $s \geq 0$, normally excluding "indifferent (simple) poles" on the imaginary axis of the s-plane; in the case of digital filters the

0-8493-8951-8/97/$0.00+$.50
© 1997 by CRC Press LLC

transfer function must be free of poles outside of the closed unit disk. This condition guarantees the realizability. In continuous-time filters, resistors (R), inductors (L), capacitors (C), transformers (T), gyrators (Γ), and controlled sources are the admissible elements; in digital filters, adders, delay elements, and constant multipliers are the admissible elements.

If only a part of the above-mentioned kinds of elements for continuous-time networks is admitted (for instance, only C and R), the transfer functions must satisfy additional restrictions. It should be mentioned that in the case of continuous-time filters, it is essential whether the input (cause) and the output (effect), which are related to each other by the transfer function, occur at the same point of the circuit (one-port) or at different points of the circuit (two-port). In the case of an RLCT-one-port, that transfer function (referred to as immittance) $Z(s)$ must be a positive real function; i.e., Re $Z(s) > 0$ must hold for all Re $s > 0$. In this case the class of admissible transfer functions is considerably restricted. Instead of the transfer function representation the state space description of the network can also be used.

The design specification can be given in the time or in the frequency domain. A typical problem in the time domain is, for instance, the design of a two-port with specified rectangular response; a classic design problem in the frequency domain is the synthesis of filters with desired attenuation tolerance plots. Also, mixed design problems (in the time and frequency domain) appear. The parameters of the transfer function (polynomial coefficients, or poles and zeros, or residues and poles, or Laurent series coefficients, or poles and coefficients of an orthogonal representation) serve, for fixed degree, as design parameters to meet the specifications.

Normally, the design specification is described in the form of a function in the approximation interval; it is to be approximated according to an appropriate measure. The design specification may also be given in the form of a tolerance scheme. One of the following cirteria is used as a standard approximation measure:

1. The Chebychev criterion — The absolute maximum of the modulus of the error function (difference between approximating function and specification) in the approximation interval is to be minimized (to the so-called mini-max error).
2. The criterion of the least mean squared error — The integral of the squared error function in the approximation interval is to be minimized (to the so-called least square error).
3. The criterion of maximum flatness — According to the number of available approximation parameters one must take care that all derivatives of the error function (including that of order zero) vanish up to an order as high as possible at a specific point in the approximation interval.
4. Interpolation — It is required that the error function vanishes at specified discrete points of the approximation interval. The number of the interpolation points corresponds to the number of the approximation parameters. Specifications on derivatives in some of the interpolation points can accordingly be included.

While applying those approximation criteria, one has always to consider that the solution fulfils the above-mentioned realizability conditions.

The degree of the approximating function is an essential parameter. It must be chosen sufficiently high especially to fulfil tolerance specifications. On the other hand, economy considerations dictate that the degree must be as low as possible.

1.2 SPECIFICATIONS

An essential aspect in the design problem is the type of specification, which always must be seen in view of the realizability.

In the case of the two-port design the following categories of specifications are of high importance:

1. Specification of an input and a corresponding output signal: As far as both signals are compatibly prescribed, the corresponding impulse response can be specified by applying deconvolution. The approximation can be achieved up to any accuracy. In many cases, however, it is more convenient to approximate directly. Such problems occur in pulse shaping, data transmission, and television measurements, in processing of read tape signals, in shortening the response time of measuring systems, and in system or signal identification.

2. Specification of the amplitude function (or, equivalently, of the corresponding attenuation) in the form of a continuous, bounded and non-negative function in a real frequency interval that may be the whole frequency domain: In filter design the approximating problem is often such that only unilateral stop-and-pass tolerance prescriptions exist. Such approximation can be accomplished up to any accuracy.
3. Specification of the phase function or the group delay in the real frequency domain: Such problems occur in data transmission, in processing of video signals, and in measurements. It is known that all-pass transfer functions can be determined, the phase of which approximates a specified real function up to an additive linear part (constant group delay part which is often immaterial) in a given real frequency range to any desired accuracy, for the discrete-time case as well as for the continuous-time case.[5]
4. Specification of the real or the imaginary part of the transfer function in the domain of real frequencies.
5. Simultaneous specification of the attenuation and the phase (or the group delay) or simultaneous specification of the real and the imaginary part in a segment of the real frequency domain: A solution is possible up to any accuracy. In the case of such specification in the total frequency interval, the coupling between the real and imaginary part of the prescription via the Hilbert transform must be claimed (at least approximately).

In the one-port design the same classes of specifications are practically important. As far as the transfer function must become positive real, one has to be aware of the fact that the specification cannot always be approximated to any accuracy.

1.3 METHODS

The approximation problems can be solved analytically only for special characteristics;[4] the use of tables is also limited to relatively simple specifications. For nonstandard, sophisticated cases numerical methods are required. In principle, this means that the parameter values are varied iteratively in order to optimize some quality measure.

An obvious possibility to achieve this is the *direct circuit optimization* which consists of adjusting the component values of a given circuit by keeping in view the physical realizability. But this method requires great know-how in order to select a suitable network topology and initial element values in advance. And, more disadvantageously, the result of the optimization gives only a solution out of a subclass of filters, and there may exist better results which cannot be realized by the chosen topology.

To satisfy higher demands it is advantageous to separate the design process into several parts. First of all, in the *characterization* step, a mathematical description is constructed which comprehends the whole class of the circuits allowed. In the second, the proper *approximation* step, out of all admissible functions one is ascertained which approximates the given specifications sufficiently well in some sense. Starting from this solution, in the following *realization* step, usually analytical methods are available to get the circuit topology and element values. This step often contains some degrees of freedom which lead to several equivalent circuits, out of which the best appropriate, e.g., the one with the least element spread, can be taken. Finally, in the *verification* step the calculated circuit is built up, and its performance is tested.

For numerical solution of the approximation task, various proposals have been published. Some of the existing methods use mathematical transformations to convert the original task to simpler standard approximation problems.[6] To this class belongs, for example, the well-known characteristic function for solely approximating the attenuation of doubly terminated reactance two-ports. Besides, there are methods using state space techniques,[7] and further methods which are based on the modeling theory.[1,8] All these methods reach the limits of their practical applicability when sophisticated filters, e.g., those with group delay equalizations or special properties in the time domain, are to be designed. However, such sophisticated filters play an increasingly important role today.

The subject of this chapter is a method which avoids the above-mentioned difficulties by approximating directly the underlying specifications by using the transfer function parameters as design parameters. The universal technique presented can be applied to RLCT, LCT, active RC, digital, and switched-capacitor circuits to mention the most important. It allows several specifications to be processed simultaneously, especially mixed in the frequency and time domain. In addition, special features, e.g., allpass sections or variable basic group delays, can be taken into account. Even for standard filter types, which are covered as special cases, a novel numerical approximation with high-speed modern-day computers may sometimes be more comfortable than deriving the filters from tabulated prototypes.

2. FUNCTION PARAMETRIZATIONS

An appropriate selection of the parametrization for the functions of interest ensures that such important realizability aspects as stability and causality can be controlled easily. Furthermore, the functions themselves, as well as their partial derivatives which are needed by the subsequent optimization procedure, should be computable with little expense and high numerical precision.

2.1 CONTINUOUS-TIME SYSTEMS

For numerical reasons it is advantageous to express the numerator and denominator polynomials of the transfer function $H(s)$ in factorized forms. With regard to approximation purposes the zeros and poles of $H(s)$ can be considered uniformly by the parametrization

$$H(s) = K\prod_{\lambda=1}^{l}\left(s - s_\lambda\right)^{\upsilon_\lambda} = K\prod_{\lambda=1}^{l}\left(s - \left(\sigma_\lambda + j\omega_\lambda\right)\right)^{\upsilon_\lambda} \tag{1}$$

The integer exponents $\upsilon_\lambda \neq 0$ are regarded as fixed, the real variables K; σ_λ, ω_λ ($\lambda = 1,\ldots,l$) serve as the design parameters for the approximation. s_λ stands for a zero if $\upsilon_\lambda > 0$ and for a pole if $\upsilon_\lambda < 0$, respectively. Using this notation the stability conditions read

$$\sigma_\lambda < 0 \text{ for } \upsilon_\lambda < 0(\lambda = 1,\ldots,l) \quad \text{and} \quad \sum_{\lambda=1}^{l}\upsilon_\lambda \leq 0 \tag{2}$$

Methods to satisfy the constraints $\sigma_\lambda < 0$ as well as to guarantee real transfer functions are discussed in Section 4.

In the area of filter design the circuit behavior in the frequency domain is the most important. The attenuation $\alpha(\omega) = -20\log_{10}|H(j\omega)|\text{dB}$, the phase $\varphi(\omega) = \arg H(j\omega)$, and the group delay $\tau(\omega) = -d\varphi(\omega)/d\omega$ can be calculated easily by summing the contributions of all zeros and poles in Equation 1. This leads to

$$\alpha(\omega) = -20\left(\log_{10}|K| + \frac{1}{2}\sum_{\lambda=1}^{l}\upsilon_\lambda\log_{10}\left(\sigma_\lambda^2 + \left(\omega - \omega_\lambda\right)^2\right)\right)\text{dB} \tag{3}$$

$$\varphi(\omega) = \pi u(-K) + \sum_{\substack{\lambda=1\\ \sigma_\lambda\neq 0}}^{l}\upsilon_\lambda\left(\pi u\left(\sigma_\lambda\right) + \arctan\frac{\omega-\omega_\lambda}{-\sigma_\lambda}\right) - \frac{\pi}{2}\sum_{\substack{\lambda=1\\ \sigma_\lambda=0}}^{l}\upsilon_\lambda + \pi\sum_{\substack{\lambda=1\\ \sigma_\lambda=0\\ \upsilon_\lambda \text{ odd}}}^{l}u\left(\omega-\omega_\lambda\right) \tag{4}$$

$$\tau(\omega) = \sum_{\lambda=1}^{l}\upsilon_\lambda\frac{\sigma_\lambda}{\sigma_\lambda^2 + \left(\omega-\omega_\lambda\right)^2} \tag{5}$$

where $u(t)$ denotes the unit step function. It should be pointed out that the angle of $H(j\omega)$ is originally ambiguous with integer multiples of 2π. Choosing the principal value for each arctan function in Equation 4 prevents steps in the phase function. This special definition of $\varphi(\omega)$ helps to evade later difficulties, especially when approximating linear phase specifications.

Based on the mentioned frequency characteristics, further functions of interest may be calculated. These are, namely, the magnitude $M(\omega) = 10^{-\alpha(\omega)/20\text{dB}}$, the real part $R(\omega) = M(\omega)\cos\varphi(\omega)$, and the imaginary part $X(\omega) = M(\omega)\sin\varphi(\omega)$.

In the time domain the filter behavior is often characterized by its impulse response. This can be calculated by means of the partial fraction expansion

$$H(s) = c_0 + \sum_{\substack{\lambda=1 \\ \upsilon_\lambda<0}}^{l} \sum_{\kappa=1}^{-\upsilon_\lambda} \frac{c_{\lambda\kappa}}{\left(s - s_\lambda\right)^{\kappa}} \tag{6}$$

For the coefficients in this expression, we get

$$c_0 = \lim_{s\to\infty} H(s), \quad c_{\lambda,-\upsilon_\lambda} = K \prod_{\substack{\mu=1 \\ \mu\neq\lambda}}^{l} \left(s_\lambda - s_\mu\right)^{\upsilon_\mu} \tag{7}$$

In the case of multiple poles ($\upsilon_\lambda < -1$) the auxiliary quantities

$$d_{\lambda\rho} = (-1)^{\rho} \sum_{\substack{\mu=1 \\ \mu\neq\lambda}}^{l} \frac{\upsilon_\mu}{\left(s_\lambda - s_\mu\right)^{\rho+1}} \quad \left(\rho = 0, 1, \ldots, -\upsilon_\lambda - 2\right) \tag{8}$$

can be introduced.[9] They allow the recursive computation of the remaining coefficients

$$c_{\lambda\kappa} = \frac{1}{-\upsilon_\lambda - \kappa} \sum_{\rho=0}^{-\upsilon_\lambda-\kappa-1} d_{\lambda,-\upsilon_\lambda-\kappa-1-\rho}\, c_{\lambda,-\upsilon_\lambda-\rho} \quad \left(\kappa = -\upsilon_\lambda - 1, \ldots, 2, 1\right) \tag{9}$$

This way of deriving the coefficients c_0 and $c_{\lambda\kappa}$ directly from the pole-zero representation (Equation 1) is important for mixed approximations in the frequency and time domain. Applying the inverse Laplace transform to Equation 6 yields the impulse response

$$h(t) = c_0\delta(t) + \sum_{\substack{\lambda=1 \\ \upsilon_\lambda<0}}^{l} e^{s_\lambda t} \sum_{\kappa=1}^{-\upsilon_\lambda} c_{\lambda\kappa} \frac{t^{\kappa-1}}{(\kappa-1)!} u(t) \tag{10}$$

with $\delta(t)$ denoting the unit impulse function. Another frequently used time-domain characteristic is the step response. This can be calculated in just the same way via a partial fraction expansion of $H(s)/s$.

In the important case of the continuous-time one-port design, reactance parts with all poles and zeros lying on the imaginary axis can often be excluded or considered in advance. If the constraint $\operatorname{Re} Z(j\omega) \geq 0$, which must hold for all frequencies, is taken into account by an additional real part specification, the approximation of immittances $Z(s)$ can be done similarly to the described approximation of two-port transfer functions. Furthermore, a special parametrization for immittances $Z(s)$ is known which guarantees the driving point function properties of $Z(s)$ for *any* parameter values.[10] Additionally, reactance parts require no special treatment when using this parametrization. In this way the original approximation task is converted to an *unconstrained* optimization problem which can also be solved with the algorithm described in Section 5. This method has been successfully used; further details cannot be discussed here for lack of space.

2.2 DISCRETE-TIME SYSTEMS

Similar to the previous section, the pole–zero representation (for discrete-time transfer functions)

$$H(z) = K\prod_{\lambda=1}^{l}\left(z - z_\lambda\right)^{\upsilon_\lambda} = K\prod_{\lambda=1}^{l}\left(z - \beta_\lambda e^{j\gamma_\lambda}\right)^{\upsilon_\lambda} \tag{11}$$

can be introduced. Notice that here it is advantageous to use the polar form parameters β_λ and γ_λ for the poles and zeros z_λ. The realizability conditions can then be formulated as

$$\left|\beta_\lambda\right| < 1 \quad \text{for} \quad \upsilon_\lambda < 0\left(\lambda = 1,\ldots,l\right) \quad \text{and} \quad \sum_{\lambda=1}^{1}\upsilon_\lambda \le 0 \tag{12}$$

The frequency characteristics are obtained for $z = e^{j\omega T}$, where T denotes the clock period. One gets the attenuation $\alpha(\omega)$, the phase $\varphi(\omega)$, and the group delay $\tau(\omega)$ from Equation 11 as

$$\alpha(\omega) = -20\left(\log_{10}|K| + \frac{1}{2}\sum_{\lambda=1}^{l}\upsilon_\lambda \log_{10}\left(1 + \beta_\lambda^2 - 2\beta_\lambda \cos\left(\omega T - \gamma_\lambda\right)\right)\right)\text{dB} \tag{13}$$

$$\varphi(\omega) = \pi u(-K) + \sum_{\lambda=1}^{l}\upsilon_\lambda \arg\left(e^{j\omega T} - \beta_\lambda e^{j\gamma_\lambda}\right) \tag{14}$$

$$\tau(\omega) = T\sum_{\lambda=1}^{l}\upsilon_\lambda \frac{\beta_\lambda \cos\left(\omega T - \gamma_\lambda\right) - 1}{1 + \beta_\lambda^2 - 2\beta_\lambda \cos\left(\omega T - \gamma_\lambda\right)} \tag{15}$$

For the phase function $\varphi(\omega)$, remarks corresponding to those for Equation 4 can be made. Based on these functions, the magnitude, the real part, and the imaginary part of $H(e^{j\omega T})$ are evaluated with no formal difference from the continuous-time case.

The time-domain behavior can be calculated via a partial fraction expansion of $H(z)$ which is given by Equation 6 when substituting s by z. By applying the inverse Z-transform one obtains the impulse response

$$h(k) = c_0\delta(k) + \sum_{\substack{\lambda=1 \\ \upsilon_\lambda<0}}^{l}\sum_{\kappa=1}^{-\upsilon_\lambda} c_{\lambda\kappa}\binom{k-1}{\kappa-1} z_\lambda^{k-\kappa} u(k-\kappa) \tag{16}$$

with $\delta(k)$ and $u(k)$ denoting the unit impulse sequence and the unit step sequence, respectively. In the same way the step response can be calculated by replacing $H(s)$ in Equation 6 with $H(z)\, z/(z-1)$.

3. ERROR CALCULATION

In order to judge which result from a class of functions is the best, it is necessary to define a suitable mathematical error criterion which allows the deviations between actual functions $f(\xi)$ and the specifications to be measured. In this section $f(\xi)$ is a real valued function and stands for any of the frequency- or time-domain characteristics $\alpha(\omega)$, $\tau(\omega)$, $h(t)$, $h(k)$, etc., introduced above, ξ denotes the independent variable ω, t, or k, respectively. For the purpose of error calculation it makes no difference which function is considered actually, but it is essential to distinguish between several important types of specifications.

3.1 ERROR FUNCTIONS

First of all, as the basis for the error criterion, error functions have to be introduced. In the simple case of a *desired function* $f_D(\xi)$ which is to be approximated with maximal deviations of $\pm\,\varepsilon(\xi)$, the normalized difference

$$e(\xi) = \left[f(\xi) - f_D(\xi)\right] / \varepsilon(\xi) \tag{17}$$

suggests itself as an error function. But filter specifications are rather given in terms of tolerance schemes for the function $f(\xi)$. If only a *lower bound* $f_L(\xi)$ is required (e.g., for the attenuation in stopbands), one can get an expression similar to Equation 17 by introducing an arbitrary tolerance $\varepsilon(\xi) > 0$ and approximating the shifted specification

$$f_D(\xi) = f_L(\xi) + \varepsilon(\xi), \quad e(\xi) = \min\left[f(\xi) - f_D(\xi), 0\right] / \varepsilon(\xi) \tag{18}$$

$\varepsilon(\xi)$ has the meaning of a ξ-dependent weighting function; in normal cases a constant ε is sufficient anyway. In just the same way for a sole *upper bound* $f_U(\xi)$ one chooses a tolerance $\varepsilon(\xi) > 0$ and defines

$$f_D(\xi) = f_U(\xi) - \varepsilon(\xi), \quad e(\xi) = \max\left[f(\xi) - f_D(\xi), 0\right] / \varepsilon(\xi) \tag{19}$$

A *two-sided demand* $f_L(\xi) \le f(\xi) \le f_U(\xi)$ can be treated as a simultaneous upper and lower bound. But in this special case using Equation 17 with the tolerance $\varepsilon(\xi) = [f_U(\xi) - f_L(\xi)]/2$ and the mean value of both bounds as the desired function, $f_D(\xi) = [f_U(\xi) + f_L(\xi)]/2$ has proved to be better.

Equations 17 to 19 have the important property that $|e(\xi)| \le 1$ holds if $f(\xi)$ fulfils the tolerance demands. Another advantage of the introduced errors is their similar calculation for all important specification types and their at least piecewise linear dependency on $f(\xi)$ and $f_D(\xi)$.

3.2 GENERALIZED SPECIFICATIONS

For distortionless systems the approximation of a linear phase, i.e., a constant group delay, at least within the passbands is of great interest. Also frequency-independent, additive terms for the attenuation or linear terms for the phase are often irrelevant to filter circuits. The best values of these terms, with respect to optimum performance, are normally not known *a priori*. Thus, it is an advantage when these degrees of freedom are included in the approximation process as additional parameters.

The discussed variations can be taken into consideration by modifying the respective desired functions to

$$\bar{f}_D(\xi) = f_D(\xi) + r\xi + q \tag{20}$$

The quantities r and q are real variables describing the additional linear and constant terms in ξ (r or q may be set to zero if desired). However, the tolerable deviations $\varepsilon(\xi)$ in Equations 17 to 19 are assumed to be fixed. Any other extensions of $f_D(\xi)$ are possible but have no significance for filter applications. Using Equation 20, one obtains the error functions in the general form

$$\bar{e}(\xi) = \left[f(\xi) - \bar{f}_D(\xi)\right] / \varepsilon(\xi) \tag{21}$$

For single-sided tolerance plots the error $\bar{e}(\xi)$ additionally must be limited to nonpositive or non-negative values, respectively, as in Equations 18 and 19.

If several specifications are given, extra parameters r_λ and q_λ can be introduced for each approximation interval. In this way, it is, for instance, possible to design filters with different variable basic group delays in several passbands. On the other hand, by applying parameter couplings from Section 4 one can also force equal delays if it is necessary. This shows the flexibility of the suggested method in adapting to specific requirements.

3.3 ERROR CRITERIA

How good a function $f(\xi)$ approximates a specification given in an interval $\xi_A \leq \xi \leq \xi_B$ can be measured by the L_p-norm (p may be real, $p \geq 1$) of the error function*

$$\|\bar{e}\|_p = \left[\frac{1}{\xi_B - \xi_A} \int_{\xi_A}^{\xi_B} |\bar{e}(\xi)|^p \, d\xi\right]^{1/p} \tag{22}$$

In order to reduce the number of error function evaluations, k mesh points ξ_κ ($\kappa = 1,\ldots,k$) are defined and the integral in Equation 22 is calculated numerically. This leads to the discretized version

$$E = \left[\sum_{\kappa=1}^{k} b_\kappa |\bar{e}(\xi_\kappa)|^p\right]^{1/p} \tag{23}$$

The quantities b_κ are real positive weighting coefficients. For the sake of simplicity often the mesh points are regularly spaced and all b_κ are chosen to be equal. But depending on the slope of $|\bar{e}(\xi)|^p$ a nonuniform mesh and proper values of b_κ (e.g., by applying the trapezoidal rule) are recommended.

Now let us look at the important case where several specifications for potentially distinct functions $f_\lambda(\xi)$ ($\lambda = 1,\ldots,L$) are given simultaneously. The error criteria E_λ for each approximation interval can be calculated by Equation 23 and may be combined in the way

$$Q = \left[\sum_{\lambda=1}^{L} a_\lambda E_\lambda^p\right]^{1/p} \tag{24}$$

The real positive coefficients a_λ may be used for weighting purposes; in most cases, however, the setting of all $a_\lambda = 1$ will be sufficient. The value Q represents a suitable overall quality measure and will be used as the *objective function* which is to be minimized by the optimum solution. For simplification, the index transformation

* For discrete-time sequences such as $h(k)$ the measure in Equation 23 can be directly applied.

$$\mu = \sum_{\nu=1}^{\lambda-1} k_\nu + \kappa \tag{25}$$

and the abbreviations

$$\tilde{e}_\mu = \bar{e}_\lambda(\xi_{\lambda\kappa}), \quad \tilde{\varepsilon}_\mu = \varepsilon_\lambda(\xi_{\lambda\kappa}), \quad g_\mu = \sqrt[p]{a_\lambda b_{\lambda\kappa}} \tag{26}$$

are introduced where $\xi_{\lambda\kappa}$ denotes the κth mesh point in the λth approximation interval which contains k_λ mesh points. Then Equation 24 can be rewritten as the single sum over the total number of $m = \Sigma_{\lambda=1}^{L} k_\lambda$ mesh points

$$Q = \left[\sum_{\mu=1}^{m} \left|\hat{e}_\mu\right|^p\right]^{1/p} \quad \text{with} \quad \hat{e}_\mu = g_\mu \tilde{e}_\mu \tag{27}$$

The minimization of Q in Equation 27 is called the least pth approximation. The case $p = 2$ yields a weighted mean square error. For high values of p the contributions of large errors $\tilde{e}_\mu$ are emphasized. For $p \to \infty$ the resulting measure $Q = \max_{1 \le \mu \le m} |\tilde{e}_\mu|$ represents a discrete Chebychev criterion. In this special case the values of the coefficients a_λ and $b_{\lambda\kappa}$ are immaterial.

4. ACCESSORY CONDITIONS

For practical applications usually real, stable transfer functions are needed. To achieve these properties, the parameters must satisfy additional conditions.

4.1 PARAMETER FIXINGS AND COUPLINGS

In general, the universal parametrizations in Section 2 describe complex transfer functions. Real functions are included in Equations 1 and 11 as special cases where both poles and zeros are either real or occur as conjugate complex pairs.

A real pole or zero s_λ is obtained by setting ω_λ to zero; a real z_λ is obtained by setting γ_λ to an integer multiple of π. Thus, first of all, it is important to introduce *fixed parameters* in order to get real transfer functions. This fixing to constant values is useful for other parameters, too. For example, to force finite attenuation poles, zeros on the imaginary axis of the s-plane ($\sigma_\lambda = 0$) or on the unit circle of the z-plane ($\beta_\lambda = 1$) are needed.

The coupling between conjugate complex pairs of poles or zeros can be achieved by introducing *dependent parameters*. Other important properties of the transfer functions can also be described by appropriate parameter dependencies. Therefore, all these possibilities are discussed here at the same time.

As can be seen from Equations 3 to 5 and 13 to 15, there exist special pole–zero combinations which yield no essential contributions to the attenuation, to the phase, or to the group delay, respectively. Such pairs are relevant for approximation purposes as well as in the following realization step. For example, all-pass sections allow phase corrections without changing the attenuation and can often be realized by preferable circuit topologies. In the continuous-time case, dependencies of interest are given by

$$\begin{aligned}
&(a) \ \sigma_\mu = \sigma_\nu, \quad && \omega_\mu = -\omega_\nu, \quad && \upsilon_\mu = \upsilon_\nu; \\
&(b) \ \sigma_\mu = -\sigma_\nu, \quad && \omega_\mu = \omega_\nu, \quad && \upsilon_\mu = \upsilon_\nu; \\
&(c) \ \sigma_\mu = -\sigma_\nu, \quad && \omega_\mu = \omega_\nu, \quad && \upsilon_\mu = -\upsilon_\nu.
\end{aligned} \tag{28}$$

For the discrete-time case, adequate couplings read

$$\begin{aligned}
&\text{(a)}\ \beta_\mu = \beta_\nu, && \gamma_\mu = -\gamma_\nu, && \upsilon_\mu = \upsilon_\nu;\\
&\text{(b)}\ \beta_\mu = 1/\beta_\nu, && \gamma_\mu = \gamma_\nu, && \upsilon_\mu = \upsilon_\nu;\\
&\text{(c)}\ \beta_\mu = 1/\beta_\nu, && \gamma_\mu = \gamma_\nu, && \upsilon_\mu = -\upsilon_\nu.
\end{aligned} \tag{29}$$

A coupling of type (a) generates a conjugate complex pair, type (b) offers a constant contribution (continuous-time systems) or a linear contribution (discrete-time systems) to the phase function, and type (c) describes a (complex) all-pass section. The couplings (a) to (c) can be applied simultaneously to several poles or zeros. In this way, for example, real transfer functions with all-pass sections of degree two or phase-invariant quadruples of zeros can be constructed.

The above parameter dependencies can be treated in a uniform manner. For that purpose each of the $M = 1 + 2l + 2L$ available real parameters K, σ_λ, ω_λ $(\lambda = 1,\ldots,l)$ and q_λ, r_λ $(\lambda = 1,\ldots,L)$ is referenced as a quantity x_μ $(\mu = 1,\ldots,M)$ in the following. Instead of $(\sigma_\lambda, \omega_\lambda)$ in the continuous-time case, $(\beta_\lambda, \gamma_\lambda)$ are used in the discrete-time case. The dependency between several parameters x_μ is generated by deriving these parameters from other real variables $\tilde{x}_\nu(v = 1,\ldots,n)$. The number n of these *free parameters* $\tilde{x}_\nu$ is smaller than the number M of the parameters x_μ, if some values are fixed or coupled. With the simple functions

$$\begin{aligned}
&\text{(a)}\ x_\mu = \tilde{x}_\nu\\
&\text{(b)}\ x_\mu = -\tilde{x}_\nu\\
&\text{(c)}\ x_\mu = 1/\tilde{x}_\nu
\end{aligned} \tag{30}$$

the above-discussed parameter couplings are covered. Other invertible functions are possible, but are normally not necessary in filter design.

4.2 PARAMETER CONSTRAINTS

In order to guarantee stable transfer functions, the constraints $\sigma_\lambda < 0$ or $|\beta_\lambda| < 1$ from Equations 2 and 12 for the poles $(\upsilon_\lambda < 0)$ are still to be satisfied.

A widely used method to take parameter constraints into account is the introduction of so-called *penalty functions*. This means a modification of the objective function Q in such a manner that Q is extremely increased if any of the constraints is violated. On the other hand, the original error criterion Q should not be influenced if all parameter constraints are fulfilled. It turned out that the construction of an appropriate extension of Q strongly depends on the specific problem and often causes difficulties in the optimization process. Therefore, the use of penalty functions is discouraged in this context.

Simple parameter transformations are the better way to take the present constraints into consideration. This is possible because the selected parametrizations yield no mixed constrints that contain more than one variable, e.g., $(\text{Re } z)^2 + (\text{Im } z)^2 < 1$ is avoided by using polar coordinates. To convert the constrained approximation problem, each free parameter $\tilde{x}_\nu$ is derived from an *unconstrained parameter* $\hat{x}_\nu(v = 1,\ldots,n)$. For the parameters in question the following transformations are sufficient:

$$\begin{aligned}
&\text{(a) unconstraint:} && \tilde{x}_\nu = \hat{x}_\nu\\
&\text{(b)}\ \tilde{x}_\nu < 0: && \tilde{x}_\nu = -\exp(\hat{x}_\nu)\\
&\text{(c)}\ |\tilde{x}_\nu| < 1: && \tilde{x}_\nu = \frac{2}{\pi}\arctan \tilde{x}_\nu
\end{aligned} \tag{31}$$

If these transformations are applied to all parameters $\tilde{x}_\nu$, not only can the realizability be guaranteed. For example, in this way it is also possible to construct minimum phase transfer functions. With similar transformations other constraints for each $\tilde{x}_\nu$ can be implemented if needed.

5. OPTIMIZATION PROCEDURE

At this point, the original approximation task is reduced to that of minimizing the objective function Q in Equation 27 with respect to the real parameters $\hat{x}_\nu$ from Section 4. Several classic methods for treating such nonlinear unconstraint optimization problems are known.[11] Out of this variety, a modified Gauss–Newton method will be described briefly. This method yields good results with regard to computational time, memory requirements, and programming efforts.

5.1 FUNDAMENTAL PROCEEDING

The discrete errors $\hat{e}_\mu$ ($\mu = 1,\ldots,m$) forming Q in Equation 27 are put together in an error vector $\boldsymbol{e}$ containing the elements

$$[\boldsymbol{e}]_\mu = \left(\frac{|\hat{e}_\mu|}{\hat{e}_\mathrm{M}}\right)^{p/2-1} \frac{\hat{e}_\mu}{\hat{e}_\mathrm{M}}, \qquad \hat{e}_\mathrm{M} = \max_{1\le\mu\le n} |\hat{e}_\mu| \tag{32}$$

This definition ensures that the sign of each component $[\boldsymbol{e}]_\mu$ coincides with the sign of the error $\hat{e}_\mu$. Furthermore $|[\boldsymbol{e}]_\mu| \le 1$ always holds, which avoids numerical overflows in the computation. Using this notation the objective function reads

$$Q = \hat{e}_\mathrm{M} \sqrt[p]{\boldsymbol{e}^\mathrm{T}\boldsymbol{e}} \tag{33}$$

Therefore, minimizing Q is equivalent to minimizing $\boldsymbol{e}^\mathrm{T}\boldsymbol{e} = \|\boldsymbol{e}\|_2^2$ for any chosen L_p-norm. The variables $\hat{x}_\nu$ ($\nu = 1,\ldots,n$) introduced in Section 4 can be comprehended to the parameter vector $\boldsymbol{x}$ which simply consists of the elements $[\boldsymbol{x}]_\nu = \hat{x}_\nu$. For further considerations the error vector $\boldsymbol{e}$ is expanded into a Taylor series at a particular parameter vector $\boldsymbol{x}_k$ in the form

$$\boldsymbol{e}(\boldsymbol{x}_k + \Delta\boldsymbol{x}_k) = \boldsymbol{e}(\boldsymbol{x}_k) + \boldsymbol{J}(\boldsymbol{x}_k)\Delta\boldsymbol{x}_k + \ldots \tag{34}$$

where $\boldsymbol{J}(\boldsymbol{x}_k)$ denotes the Jacobian matrix with the elements

$$[\boldsymbol{J}(\boldsymbol{x}_k)]_{\mu\nu} = \left.\frac{\partial[\boldsymbol{e}]_\mu}{\partial[\boldsymbol{x}]_\nu}\right|_{\boldsymbol{x}=\boldsymbol{x}_k} = \frac{p}{2}\frac{1}{\hat{e}_\mathrm{M}}\left(\frac{|\hat{e}_\mu|}{\hat{e}_\mathrm{M}}\right)^{p/2-1} \left.\frac{\partial\hat{e}_\mu}{\partial\hat{x}_\nu}\right|_{\hat{x}_\nu=[\boldsymbol{x}_k]_\nu(\nu=1,\ldots,n)} \tag{35}$$

By using Equations 21, 25, 26, 30, and 31, the needed partial derivatives can be traced back to

$$\frac{\partial\hat{e}_\mu}{\partial\hat{x}_\nu} = \frac{g_\mu}{\tilde{\varepsilon}_\mu}\frac{\partial\tilde{x}_\nu}{\partial\hat{x}_\nu}\sum_{\iota=1}^{M}\frac{\partial x_\iota}{\partial\tilde{x}_\nu}\left(\frac{\partial f_\lambda(\xi_{\lambda\kappa})}{\partial x_\iota} - \frac{\partial\bar{f}_{D\lambda}(\xi_{\lambda\kappa})}{\partial x_\iota}\right) \tag{36}$$

The calculation of $\partial f_\lambda/\partial x_\iota$ depends on which type of function f_λ and on which type of parameter x_ι stands for concretely. Notice that the partial derivatives of the frequency characteristics are very easy to compute because the derivatives of all but one term of the sums in Equations 3 to 5 and 13 to 15 vanish. Therefore, an analytical calculation should be favored compared with an approximative differentiation by difference quotients.

Truncating the series expansion in Equation 34 after the linear term leads to an approach for $\|\boldsymbol{e}(\boldsymbol{x}_k + \Delta\boldsymbol{x}_k)\|_2$ which represents a quadratic form in $\Delta\boldsymbol{x}_k$. Provided that the matrix $\boldsymbol{J}(\boldsymbol{x}_k)$ is nonsingular, $\boldsymbol{J}^T(\boldsymbol{x}_k)\boldsymbol{J}(\boldsymbol{x}_k)$ will be positive definite, and the solution $\Delta\boldsymbol{x}_k$, for which the absolute minimum of the said approach for $\|\boldsymbol{e}(\boldsymbol{x}_k + \Delta\boldsymbol{x}_k)\|_2$ is reached, can be calculated analytically:

$$\Delta\boldsymbol{x}_k = -\left[\boldsymbol{J}^{\mathrm{T}}(\boldsymbol{x}_k)\boldsymbol{J}(\boldsymbol{x}_k)\right]^{-1}\boldsymbol{J}^{\mathrm{T}}(\boldsymbol{x}_k)\boldsymbol{e}(\boldsymbol{x}_k) \tag{37}$$

However, for numerical reasons, Equation 37 is not suitable for ill-conditioned matrices $\boldsymbol{J}(\boldsymbol{x}_k)$ which are unfortunately rather the usual case in the course of optimization. For practical computations it is much better to inspect the equivalent problem

$$\boldsymbol{J}(\boldsymbol{x}_k)\Delta\boldsymbol{x}_k + \boldsymbol{e}(\boldsymbol{x}_k) = \boldsymbol{r}, \quad \|\boldsymbol{r}\|_2^2 \overset{!}{=} \min \tag{38}$$

wherein the system of linear equations is overdetermined if the total number m of mesh points exceeds the number n of available free unconstraint parameters. The Householder transformation has proved to be a well-suited numerical method for solving such equation systems with the least residual norm $\|\boldsymbol{r}\|_2$.[11]

The displacement vector $\Delta\boldsymbol{x}_k$ calculated in this way is based on a linearization of the error vector $\boldsymbol{e}$. In the general case of nonlinear functions the Jacobian matrix $\boldsymbol{J}(\boldsymbol{x}_k)$ depends on the particular vector $\boldsymbol{x}_k$ and the described process is to be repeated iteratively for the parameter vectors $\boldsymbol{x}_{k+1} = \boldsymbol{x}_k + \Delta\boldsymbol{x}_k$.

5.2 STEP DAMPING

Especially for large displacements $\Delta\boldsymbol{x}_k$ it often occurs that although the linearized problem is minimized, of course, the original nonlinear problem is not. In this case there is a risk of obtaining an objective function value $Q(\boldsymbol{x}_k + \Delta\boldsymbol{x}_k) > Q(\boldsymbol{x}_k)$, i.e., a risk that the calculated displacement actually causes a worsening of the approximation. But in order to guarantee the convergence of the iteration process in any case, monotonically decreasing values of Q are required.

A simple method to ensure a decreasing Q is to reduce the actual length of the displacement vector while retaining its direction. This can be done by introducing a scalar damping factor δ in order to satisfy the condition

$$Q(\boldsymbol{x}_k + \delta\,\Delta\boldsymbol{x}_k) < Q(\boldsymbol{x}_k) \quad \text{with} \quad 0 < \delta \le 1 \tag{39}$$

The process of finding an appropriate value for δ is called *one-dimensional* or *line search*. The numerical expense to find the exact minimum of $Q(\boldsymbol{x}_k + \delta\Delta\boldsymbol{x}_k)$ is normally too great; therefore, often a parabolic shape with respect to δ is assumed. This approach leads approximately to an estimated value $\delta_m = 1/\left(1 + \sqrt{Q(\boldsymbol{x}_k + \Delta\boldsymbol{x}_k)/Q(\boldsymbol{x}_k)}\right)$. If δ_m does not yet fulfil Equation 39, the line search process has to be repeated iteratively.

Another possibility is to bound $\|\Delta\boldsymbol{x}_k\|_2$ when solving Equation 38. This can be achieved by the modified equation system

$$\begin{bmatrix} J(x_k) \\ \lambda I \end{bmatrix} \Delta x_k + \begin{bmatrix} e(x_k) \\ \mathbf{0} \end{bmatrix} = \begin{bmatrix} r \\ \lambda \Delta x_k \end{bmatrix}, \quad \|r\|_2^2 + \lambda^2 \|\Delta x_k\|_2^2 \overset{!}{=} \min \tag{40}$$

where $\boldsymbol{I}$ denotes an $n \times n$ unity matrix and $\boldsymbol{0}$ a zero vector of dimension n. The real quantity λ is known as the Levenberg parameter and represents a weighting between $\|\boldsymbol{r}\|_2$ and $\|\Delta \boldsymbol{x}_k\|_2$ with regard to the minimization. Solutions of Equation 40 have great significance for ill-conditioned matrices $\boldsymbol{J}$; even singular matrices $\boldsymbol{J}$ are no longer an obstacle. Again, the Householder transformation can be applied to solve this equation with high numerical precision. The overall advantage of the method using the Levenberg parameter λ in comparison with the one using the damping factor δ is that the direction of $\Delta \boldsymbol{x}_k$ is also corrected. The computational expense for solving the already known linear equation system of Equation 40 again with another value of λ can be lower than evaluating the objective function Q once more.

For the investigated approximation problems a combination of the two methods discussed above for step damping has proved to be the most efficient. Here the successive parameter vectors are calculated as

$$\boldsymbol{x}_{k+1} = \boldsymbol{x}_k + \delta \, \Delta \boldsymbol{x}_k(\lambda) \tag{41}$$

where λ and δ have to be chosen so that $Q(\boldsymbol{x}_{k+1}) < Q(\boldsymbol{x}_k)$ holds. Unfortunately, appropriate values are strongly dependent on the particular problem. It should be clear that the calculation of the optimum pair (λ, δ) which leads to the minimum of $Q(\boldsymbol{x}_{k+1})$ is too expensive. But as an experimental result, it was found that the best values in successive iteration steps differ only a little. Therefore, it is possible to contruct a heuristic adaptation process for λ and δ in the course of the optimization. The best results are obtained when $\delta = 1$ is taken and λ is chosen as low as possible. Only if $\delta < 1$ is necessary, λ might be increased by a few percent for the next iteration step.

6. EXAMPLES

Based on the preceeding considerations a computer program package has been developed to carry out the numerical calculations. With an interactive implementation the parameter settings, couplings, and constraints, as well as all mentioned types of specifications, can be managed comfortably. Out of the variety of treated examples two typical filter design problems are chosen to demonstrate how efficiently the method presented here works in practice.

6.1 ANALOG LOWPASS WITH PRESCRIBED STEP RESPONSE

This example shows a mixed approximation in the frequency and time domain. For control engineering applications or antialiasing filters in digital signal-processing systems, special analog lowpass filters are required. Besides the spectral bounding of the signal, here it is often important that little or no overshoot occurs in the step response

In the concrete case, a stopband attenuation of at least 60 dB for normalized frequencies $\omega \geq 1$ is demanded. In addition, the step response shall reach its final value of 1 after a normalized time of $t = 15$ at the latest. An overshoot of maximal 0.1% has been considered to be low enough.

The parameter values for a transfer function of degree seven resulting from an L_{100}-approximation are listed in Table 1. The corresponding pole–zero plot, the attenuation characteristic, and the step response are depicted in Figure 1.

Table 1 Parameter Values of the Continuous-Time Lowpass

K	= 0.006108778,					
σ_1	= –0.471905319,	ω_1	= 0.000000000,	υ_1	= –1
$\sigma_{2,3}$	= –0.458893006,	$\omega_{2,3}$	= ±0.282865151,	$\upsilon_{2,3}$	= –1
$\sigma_{4,5}$	= –0.398302042,	$\omega_{4,5}$	= ±0.562614615,	$\upsilon_{4,5}$	= –1
$\sigma_{6,7}$	= –0.310765050,	$\omega_{6,7}$	= ±0.830113507,	$\upsilon_{6,7}$	= –1
$\sigma_{8,9}$	= 0.000000000,	$\omega_{8,9}$	= ±1.025492227,	$\upsilon_{8,9}$	= 1
$\sigma_{10,11}$	= 0.000000000,	$\omega_{10,11}$	= ±1.268135999	$\upsilon_{10,11}$	= 1
$\sigma_{12,13}$	= 0.000000000,	$\omega_{12,13}$	= ±2.226103176	$\upsilon_{12,13}$	= 1

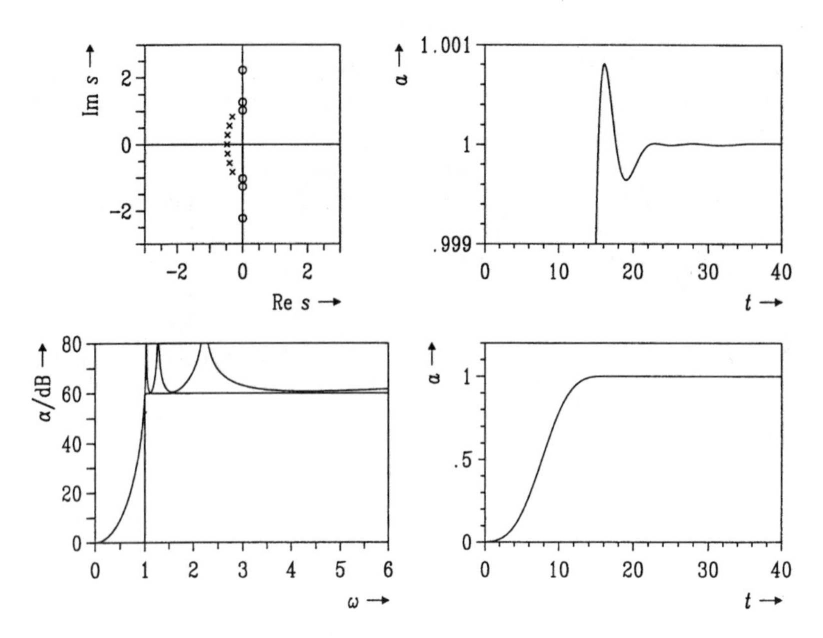

Figure 1 Pole–zero plot, attenuation α(ω), and step response *a*(*t*) of the analog lowpass.

Finally, a hint for carrying out the approximation is given. Often the convergence can be speeded by starting with $p = 2$ and increasing the value of p subsequently. For values of p greater than about 100, the differences between and L_p-solution and a Chebychev solution are almost imperceptable in practice. But notice that the *discrete* norm used requires mesh points lying close enough to get the extremal points of the error function with sufficient accuracy.

6.2 DIGITAL BANDPASS WITH GROUP DELAY EQUALIZATION

A digital filter with a clock period of $T = 1/(44.1\text{ kHz})$ is to be designed. The stopband edges for this bandpass example are 0.44 and 1.56 kHz, respectively; a minimum stopband attenuation 81 dB is demanded. In the passband interval $0.96\text{ kHz} \leq f \leq 1.04\text{ kHz}$, an attenuation less than 0.5 dB is required; within the interval $0.92\text{ kHz} \leq f \leq 1.08\text{ kHz}$, the attenuation may rise to 1.5 dB. The shape of the tolerance scheme used can be seen in Figure 2. In addition, the group delay is desired to be constant up to a maximum deviation of $\pm\, T/2$ between 0.76 and 1.24 kHz. This frequency interval is wider than the passband and covers part of both transition bands.

A L_{1000}-solution of degree 12 with all transmission zeros lying on the unit circle (one zero fixed at $z = 1$ and one at $z = -1$) was calculated. The resulting transfer function parameters are listed in Table 2; the corresponding pole–zero plot, the attenuation, and the group delay characteristics are shown in Figure 2. It is to be emphasized that the basic group delay was not specified in advance; the best value turned out to be $142.587T$.

Notice that neither the attenuation nor the group delay shows an *equirripple* behavior. This is due to the known fact that, in general, equiripple results differ from Chebychev solutions and, moreover, lead to greater maximum deviations.[12]

Compared with special nonrecursive digital filters with strictly linear phase functions even in the stopbands, the result presented is of much lower degree. Therefore, its implementation on a signal processor needs less computational time, which is important for real-time applications.

7. CONCLUSIONS

In the following list the highlights of the previously described function parametrization, the error criterion, and the optimization procedure are summarized. The essential advantages for the designer are:

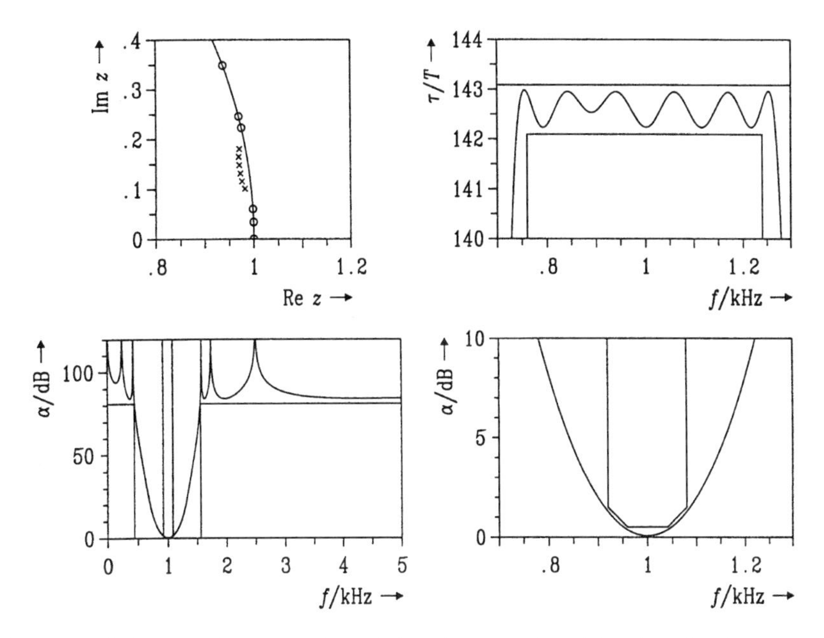

Figure 2 Pole–zero plot, attenuation $\alpha(f)$, and group delay $\tau(f)$ of the digital bandpass.

Table 2 Parameter Values of the Discrete-Time Bandpass

K	=	0.000027568							
$\beta_{1,2}$	=	0.987174529,	$\gamma_{1,2}$	=	±0.032655648,	=	$\upsilon_{1,2}$	=	−1
$\beta_{3,4}$	=	0.982807524,	$\gamma_{3,4}$	=	±0.037620873,	=	$\upsilon_{3,4}$	=	−1
$\beta_{5,6}$	=	0.981557791,	$\gamma_{5,6}$	=	±0.042799351,	=	$\upsilon_{5,6}$	=	−1
$\beta_{7,8}$	=	0.981766279,	$\gamma_{7,8}$	=	±0.048200549,	=	$\upsilon_{7,8}$	=	−1
$\beta_{9,10}$	=	0.983103963,	$\gamma_{9,10}$	=	±0.053495270,	=	$\upsilon_{9,10}$	=	−1
$\beta_{11,12}$	=	0.987301843,	$\gamma_{11,12}$	=	±0.058448449,	=	$\upsilon_{11,12}$	=	−1
β_{13}	=	1.000000000,	γ_{13}	=	0.000000000,	=	υ_{13}	=	1
β_{14}	=	−1.000000000,	γ_{14}	=	0.000000000,	=	υ_{14}	=	1
$\beta_{15,16}$	=	1.000000000,	$\gamma_{15,16}$	=	±0.010824750,	=	$\upsilon_{15,16}$	=	1
$\beta_{17,18}$	=	1.000000000,	$\gamma_{17,18}$	=	±0.019213711,	=	$\upsilon_{17,18}$	=	1
$\beta_{19,20}$	=	1.000000000,	$\gamma_{19,20}$	=	±0.071483264,	=	$\upsilon_{19,20}$	=	1
$\beta_{21,22}$	=	1.000000000,	$\gamma_{21,22}$	=	±0.079038235,	=	$\upsilon_{21,22}$	=	1
$\beta_{23,24}$	=	1.000000000,	$\gamma_{23,24}$	=	±0.113388888,	=	$\upsilon_{23,24}$	=	1

- No restriction to special circuit topologies;
- Direct processing of the underlying tolerance schemes;
- Simultaneous processing of an arbitrary number of specifications, potentially mixed in the frequency and time domain;
- Consideration of generalized specifications, e.g., for treating variable basic group delays;
- Approximation in an arbitrary L_p-sense with any real $p \geq 1$, practically enabling Chebychev approximations;
- Direct control of the poles and zeros as important design parameters allowing special transfer function properties, e.g., finite transmission zeros and allpass sections; and
- Guaranteed realizability of the resulting transfer functions.

The described computer-aided filter design technique has stood the test for inumerable examples in practice. As an illustration, both examples in Section 6 may suffice.

REFERENCES

1. **Unbehauen, R.,** *Netzwerk- und Filtersynthese* (in German), 4th ed., R. Oldenbourg Verlag, Munich, 1993.
2. **Parks, T. W. and Burrus, C. S.,** *Digital Filter Design,* John Wiley & Sons, New York, 1987.

3. **Temes, G. C. and Mitra, S. K.,** *Modern Filter Theory and Design,* John Wiley & Sons, New York, 1973.
4. **Daniels, R. W.,** *Approximation Methods for Electronic Filter Design,* McGraw-Hill, New York, 1974.
5. **Unbehauen, R. and Weinzierl, K.,** On the approximation of phase specifications by discrete-time or continuous-time allpass transfer functions, *Int. J. Circuit Theory Appl.,* 19, 211, 1991.
6. **Hohneker, W., Kicherer, H., Unbehauen, R., and Wüpper, A.,** Analog circuit design in the time and frequency domains — approximation problems, *ntz-Archiv,* 6, 53 (part 1) and 71 (part 2), 1984.
7. **Thiele, L.,** An analytic approach to curve fitting, *Arch. Elektron. Übertragungstechnik,* 40, 51, 1986.
8. **Jackson, L. B.,** *Digital Filters and Signal Processing,* 2nd ed., Kluwer Academic Publishers, Boston, 1986.
9. **Brugia, O.,** A noniterative method for the partial fraction expansion of a rational function with high order poles, *Soc. Ind. Appl. Math. Rev.,* 7, 381, 1965.
10. **Forster, U. and Unbehauen, R.,** Comments on parametric representations of Brune functions, *Int. J. Circuit Theory Appl.,* 8, 179, 1980.
11. **Lawson, C. L. and Hanson, R. J.,** *Solving Least Squares Problems,* Prentice-Hall, Englewood Cliffs, NJ, 1974.
12. **Deczky, A. G.,** Equiripple and minimax (Chebychev) approximations for recursive digital filters, *IEEE Trans. Acoust., Speech Signal Process.,* 22, 98, 1974.

Chapter 3

Computer-Aided Design Methods in Filter Design: S/FILSYN and Other Packages

George Szentirmai

CONTENTS

0-8493-8951-8/97/$0.00+$.50
© 1997 by CRC Press LLC

1. S/FILSYN PROGRAM

1.1 INTRODUCTION

As outlined in previous chapters, the modern (insertion-loss) method of filter synthesis and design involves a very large amount of numerical computations, as well as, in most cases, the need to make choices that are anything but clear or simple. Furthermore, the numerical computations are nearly always very ill conditioned, necessitating the use of either a large number of decimal places or esoteric procedures to overcome.

Hence, it is not surprising to find that the use of computers in the field of filter synthesis is just about as old as the computers themselves. In fact, even the name computer-aided design (CAD) was originally nearly synonymous with filter design.

One of the earliest programs developed for this purpose in the mid 1960s at Bell Telephone Laboratories was the progenitor of what is now known as S/FILSYN, at that time called the General Filter Synthesis Program, which was written for the design of passive LC filters only. In the early 1970s that program was completely rewritten to be used for educational purposes at Cornell University and was renamed FILSYN. The first commercial version of this program made its appearance in 1975, now handling passive LC filters as well as active RC and digital filters. Now called S/FILSYN, it handles the widest set of filter types with the most options of any of the commercial programs.

While the first version of the program was naturally written for mainframe computers to be used in batch mode only, later versions were fully interactive and ran on anything from personal computers to workstations to mainframes (if they are still around).

In addition to relieving the user from performing tedious numerical computations, the program also offers several levels of automation in making the choices referred to above. If a user does not know how to select, or wishes not to be involved in selecting, the proper options, the program quietly takes over. At the same time all these options can also be made visible and available to any user for the purpose of experimentation or learning.

In the last 5 years, a large number of other filter design programs have made their appearances. These range from simple table lookup programs to quite sophisticated ones, covering selected filter types. We shall mention some of these latter ones, listing their main features.

1.2 SYNTHESIS PROCEDURE

Without going into the details, covered in other places in this handbook (see also Reference 1), the general description of a filter in the (analog) frequency domain is through the transfer function:

$$\text{Output/input} = H(s) = Q(s)/E(s)$$

where E(s) is a strict Hurwitz polynomial of the complex frequency variable "s" and whose zeros are the natural modes of the network and Q(s) is a pure even or pure odd polynomial of "s" and whose zeros are the transmission zeros of the filter. Both are real polynomials; i.e., their coefficients are purely real.

A bit more convenient to use is the "characteristic function":

$$K(s) = F(s)/Q(s)$$

where Q(s) is the same as before and F(s) is a completely arbitrary real polynomial in "s." The relationship between these two functions is the celebrated Feldtkeller equation:

$$H(s)H(-s) = 1/[1 + K(s)K(-s)]$$

or, expressed in term of the polynomials,

$$E(s)E(-s) = F(s)F(-s) + Q(s)Q(-s)$$

The characteristic function is very useful, because the filter insertion loss can be expressed very simply as

$$a \text{ [in dB]} = 10 \log_{10} [1 + |K(j\omega)|^2]$$

and, apart from the fact that the denominator of K(s) should be pure even or odd, it is a completely arbitrary real, rational function.

Once these polynomials are known, we can generate the filter input and output impedances and admittances by generating the four polynomials:

$$\text{Even part of } E(s) \pm \text{Even part of } F(s)$$

and

$$\text{Odd part of } E(s) \pm \text{Odd part of } F(s)$$

Once these functions are known, the actual synthesis is conceptually very simple, consisting of simple manipulations of these polynomials, like evaluating, subtracting, etc.

Performing this synthesis, hence, consists of three basic steps.

1. First, we must find a characteristic function K(s) that meets the loss requirements or possibly the transfer function H(s) that meets both the loss and delay requirements. This step also includes the calculation of all these additional polynomials.
2. The second step is to find a configuration and calculate the corresponding element values.
3. The third and last step consists of possible modification of the circuit and rearranging the structure to provide a more easily realizable set of elements.

The first step, approximation, is reasonably simple, since for maximally flat or equal ripple passbands, the F(s) polynomial can be expressed in terms of the Q(s) polynomial (Reference 20, p. 49 ff.); hence, the approximation is reduced to an essentially polynomial type. For Chebyshev and elliptic filters the unknown transmission zeros can be found by explicit expressions, while in the most general stopband case a very rugged approximation procedure is available,[6] which is guaranteed to converge and converge very fast.

If the passband delay is specified, we have explicit expressions for the E(s) polynomial for the maximally flat case (the familiar Bessel polynomials), while still another approximation procedure is available for the equal ripple case. This procedure starts by generating a phase function that interpolates a linear function at equidistant frequencies,[17] which is followed by a modified Remez method, and produces excellent results but a bit slowly. Finally, if we need to combine this with an equal-minima-type stopband, we have another explicit procedure[10] to obtain the corresponding transmission zeros.

If the passband has a prescribed, nonconstant shape, we need to consider a more general approximation procedure that works with the transfer function, thereby permitting the possible specification of the delay as well. This procedure[22] is somewhat slower, and global convergence is not guaranteed, but was found to yield satisfactory solutions in all cases tried.

The second step, synthesis, is much more complex with problems classified into two groups. One is the sequence of operations, that is to say, the selection of one of usually very many equivalent configurations; the other is numerical, and the more insidious.

The first problem is solved by a two-step procedure. While the program permits users to specify their own configuration, the better option is to let the program select one for us. Then, during the final stages of the design we can rearrange the components and modify their values in any way we wish, using the large number of equivalency transformations available. These include the familiar tee-to-pi conversion (including Brune's equivalence), Norton (impedance) transformations, three-element branch conversions, interchanging the locations of resonant branches, etc. In fact, we can safely say that the available operations permit us to convert *any* ladder structure to *any* other equivalent one.

The source of the numerical problems is the fact that the zeros of all the polynomials are grouped close together, near the edges of the filter passband, making these polynomials very ill conditioned. This means that small changes in the coefficients lead to large changes in the zero locations, which are the crucial parameters. As a consequence, the actual synthesis cannot, for all practical purposes, be performed if we use the polynomial coefficients of these functions.

One might ask at this stage, why our insistence in using this design method at all? It turns out that the other side of this same coin, large sensitivity of the roots to coefficients, is a *very small* sensitivity of these same roots to the final element values! That is to say, our filters will be very stable and relatively simple to implement accurately.

Historically, two methods have been developed to overcome this problem and they seem to be about equally efficient. One is to represent these polynomials by their roots (and a multiplier) at all times,[5] without ever expanding them into polynomial forms. The other is to use a transformation of the independent (frequency) variable in order to spread the locations of these zeros. The proper transformation is[7]

$$z^2 = [s^2 + \omega_A^2]/[s^2 + \omega_B^2]$$

where ω_A and ω_B are the lower and upper passband edge frequencies, respectively. Finally, of course, we may also combine the two methods, i.e., use this transformed variable *and* represent all polynomials by their roots at all times, yielding the most accurate method known at this time.[9]

The remaining problem is simply the fact that these methods need computational procedures that are impossible to perform either by hand or by hand-held calculators and need efficient computer programs to perform, which explains why these modern synthesis techniques took their time in taking their place in the armory of engineers.

The third step, the post-synthesis operations, includes the equivalency transformations already discussed above, the further inclusion of delay equalizers and, possibly, loss equalizers, analysis in the time and frequency domains, and others.

Next we shall describe in some detail the features of S/FILSYN, a general-purpose filter synthesis and design program, that uses the combination method just described for the numerical calculations.

We shall not consider problems arising out of using transmission line segments as components (microwave filters) or those associated with the design of digital filters. Suffice it to say that passive LC filter design is found very useful as a starting point even for the design of active RC, switched-capacitor, and digital filters, as well.

1.3 FEATURES OF S/FILSYN

1.3.1 Filter Kinds and Configurations

1.3.1.1 Passive LC

Lumped Design — Ladder, single lattice, ladder with embedded lattices.
Microwave Design — Ladder using commensurate length stubs and unit elements.

1.3.1.2 Active RC

Cascaded biquads, biquads with leapfrog feedback,[8] biquads with follow-the-leader feedback.[18]
Active elements may be operational amplifiers or transconductance amplifiers.

Switched-Capacitor — Cascaded biquads, biquads with leapfrog feedback, biquads with follow-the-leader feedback.

1.3.1.3 Digital

IIR — IIR (infinite impulse response) filters can be designed directly as such, or can be obtained from an analog transfer function using any of the following transformation methods:
- Impulse-invariant transform
- Matched transform
- Bilinear transform

Methods of implementation include the direct, cascaded biquads, parallel biquads, Gray–Markel lattice,[4] differential allpass,[11] and many ladder and lattice wave-digital forms.[2]

FIR (finite impulse response)
- Equal ripple design (up to 10 bands)
- Windowed designs (20+ windows)

Implementation is in a single segment or up to five cascaded segments.

1.3.2 Filter Types

Lowpass
Highpass
Bandpass
Band-reject
Delay equalizers
Delay lines
Matching networks (passive/microwave)
Interstage networks (passive/microwave)

1.3.3 Characteristics

1.3.3.1 Passband

Maximally flat loss
Maximally flat delay (lowpass and bandpass)
Equal ripple loss
Equal ripple delay (lowpass and bandpass)
Sloping loss (bandpass)
Specified shape (using a stand-alone optimizing preprocessor. Works for analog as well as digital, and microwave, frequencies)

1.3.3.2 Stopband

Monotonic loss
Equal minima loss (either closed-form or PLACER preprocessor)
Specified transmission zeros (including complex ones)
Specified loss shape (either PLACER or the stand-alone optimizing preprocessor)

1.3.3.3 Others

Arbitrary (including extreme) terminations (passive/microwave)
Parametric bandpass
Arithmetically symmetrical bandpass
Predistortion for dissipation (passive)
Contributing unit elements (microwave)

The combinations of these characteristics include, among many others, the following filter families:

Butterworth
Chebyshev
Inverted Chebyshev

Elliptic (Cauer)
Bessel
Generalized Bessel (with equal-minima stopband)
Ulbrich–Piloty (equal-ripple delay)

Through the use of script files that either contain or generate the natural modes of the filters, we may also design filters of the following families:

Gaussian (either maximally flat or equal-ripple approximations)
Legendre (Papoulis-type steepest monotonic filter)
Transitional Butterworth–Thomson
Raised-cosine lowpass
Equal-ripple pulse response
Equal-ripple step response

The program capabilities include generating all known closed-form solutions, while using iterative approximation procedures wherever closed-form solutions do not exist. However, as explained above, most of these approximation procedures are restricted ones applicable only to a very limited set of circumstances, thereby insuring their robustness.

In particular, we have four such approximation procedures in the program:

1. PLACER stopband loss approximation[6] — This is used when the stopband loss requirements are not uniform, or the bandpass is a parametric one, or not exactly geometrically symmetrical. The process is guaranteed to converge and converge very fast.
2. Equal-ripple delay lowpass — This is a little less stable and much slower approximation, and its success depends very much on the numbers specified. To help this step, a set of curves is provided to find a good starting point. However, it yields very accurate results.
3. Stand-alone optimizing preprocessor — This is used when passband loss (and delay) and stopband loss are of specified (nonconstant) shape. It works with the poles and zeros of the transfer function and, so far, has been found to yield acceptable results in all cases tried. The process is interactive, uses three methods of optimization, and the results can be observed, interrupted, and restarted at will.
4. Delay equalizer design — To equalize a given filter delay, we need still another optimization procedure. This one adds a specified number of (first- and) second-order delay equalizer sections and determines their paramaters such that the overall delay approximates a constant in the least-squares sense over the indicated band. The procedure is fast and always converges.

All other computations performed in the synthesis and design steps are exact. In the analysis stages, computations are exact except the computation of the delay in certain instances, like the analysis of the parallel implementation of IIR filters or the analysis of passive LC/microwave structures, where a divided difference method is used instead of exact differentiation.

1.3.4 Program Limits

Maximum filter degree:	50
Maximum FIR filter length:	512
Maximum number of delay sections:	25

Transfer data files are available to the following programs:

Touchstone (passive/microwave)
SuperCompact (passive/microwave)
SPICE (passive LC/active RC)
SuperStar (passive LC/active RC)
SWITCAP (switched-capacitor)
SCASY (switched-capacitor)
ASCII files to other programs and documentation

1.4 CONFIGURATION

1.4.1 Passive/Microwave

Finding a configuration and computing the corresponding element values of passive LC (and microwave) ladder filters are the most difficult steps in filter design.

For experts or students of filter synthesis, S/FILSYN offers several levels of hands-on synthesis capabilities that allow control of the process at the lowest, innermost level.

For day-to-day operation, the program generates a ladder structure that has the minimum number of inductors and the minimum number of total components. While this structure is usually acceptable, quite often, especially in bandpass cases, additional modification steps are needed to obtain a satisfactory set of element values. For this purpose, S/FILSYN offers a full suite of tools that permits the conversion of any ladder into any other equivalent ladder structure.

As already mentioned above, this suite includes Norton (impedance) transformations (including the extended set of Kuroda's identities for microwave filters), exchange of neighboring resonant circuits, dual circuits, tee–pi conversions, adding, deleting, splitting, combining, and interchanging branches, etc.

For specific circuits, like capacitively coupled resonator filters, we have developed script files (using the supplied EBL [extended batch language] utility) to perform these steps automatically until the structure has all inductances, say, equal. Dozens of these script files are supplied with the program, and others are constantly being developed.

1.4.2 Active RC

The selection of configuration for active RC filters is much simpler. If transconductance amplifiers are used as active elements, the circuit selected is unique. In case operational amplifiers are used there are options, but the computer-selected sequence and configuration is usually acceptable. In any case, the choices are much more limited than in the passive LC case.

1.4.3 IIR Digital

With the exception of wave-digital structures, selecting an implementation for IIR filters is also simple. The direct implementation is hardly ever used; the usual choice is that among the cascade, the Gray–Markel lattice, and differential allpass (if it exists).

To design a wave-digital filter, the user should be somewhat familiar with the design of microwave filters. One first designs a microwave filter with the "quarter-wave frequency" requested by the program being equal to the Nyquist frequency (half the sampling rate). When the design is complete, a single command converts the structure to a wave-digital filter.

1.5 ANALYSIS AND MODIFICATIONS

Once a satisfactory circuit is obtained, it can be analyzed, stored in a data file, or printed for documentation purposes. Data transfer files are also available to a number of other CAD programs for further processing. Filters can be analyzed both in the time and frequency domains and the results tabulated or displayed graphically.

Other significant steps are the capability of delay equalization and the inclusion of dissipation into the analysis for passive/microwave circuits.

Digital filters can additionally be optimally scaled, their coefficients truncated and modified in various other ways.

The program has a full, context-sensitive help system available a keystroke away. The large number (120 plus) of script files is integrated into a menutype front end, which simplifies program usage substantially. The script files can be activated from the menu, but a bit of time may be saved by calling them directly from the DOS prompt. The menu system can also generate new batch files, which can be saved, edited, and reused.

2. EXAMPLES

A few representative sample runs are shown here to demonstrate how simple it is nowadays to obtain efficient designs to satisfy complex requirements.

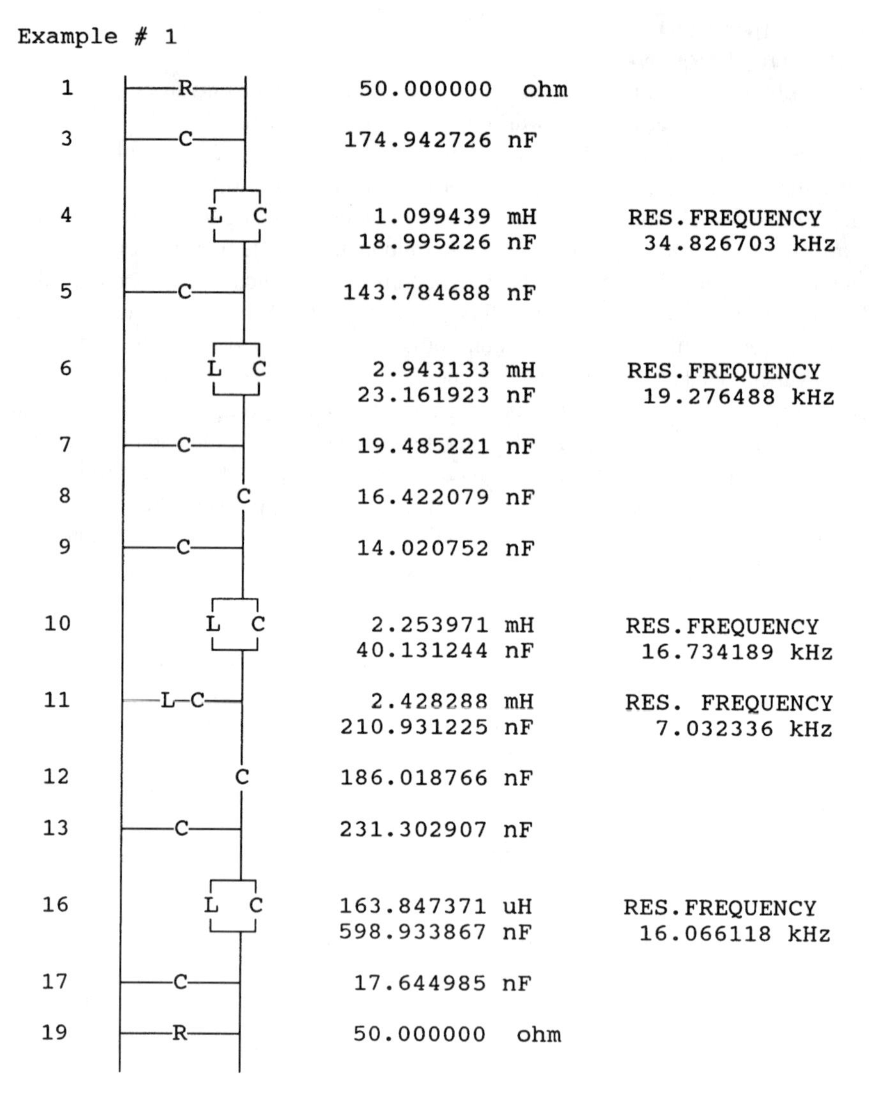

Figure 1 Passive LC filter (Example 1).

EXAMPLE 1

This is a bandpass filter with passband from 10 to 15 kHz with 0.1 dB passband ripple. We need 35 dB loss below 7.5 kHz and 60 dB loss above 16 kHz; terminations are 50 Ω at both ends. This is just about all the information the program needs, and the results are shown in Figure 1. The circuit could be modified in many ways, but that is mainly for expert users.

Naturally, we need confirmation, and the computed response of this circuit is consequently shown in Figure 2.

EXAMPLE 2

This is a quasi-elliptic microwave lowpass filter with a passband up to 200 MHz and passband ripple of 0.1 dB, again. Stopband begins at 220 MHz, terminations are 50 Ω, and the quarter-wave frequency is 2 GHz. We include contributing unit elements to simplify the physical implementation. Again, nothing else is needed and the program generates the configuration shown in Figure 3. Note that the L and C symbols designate shorted and open-circuited quarter-wave line segments, respectively, while UE designates a unit element (a cascaded segment of a quarter-wave long line) and the values printed next to them are their respective characteristic impedances. The series connection of an L and C stub, that implements a transmission zero, may also be replaced by the cascade

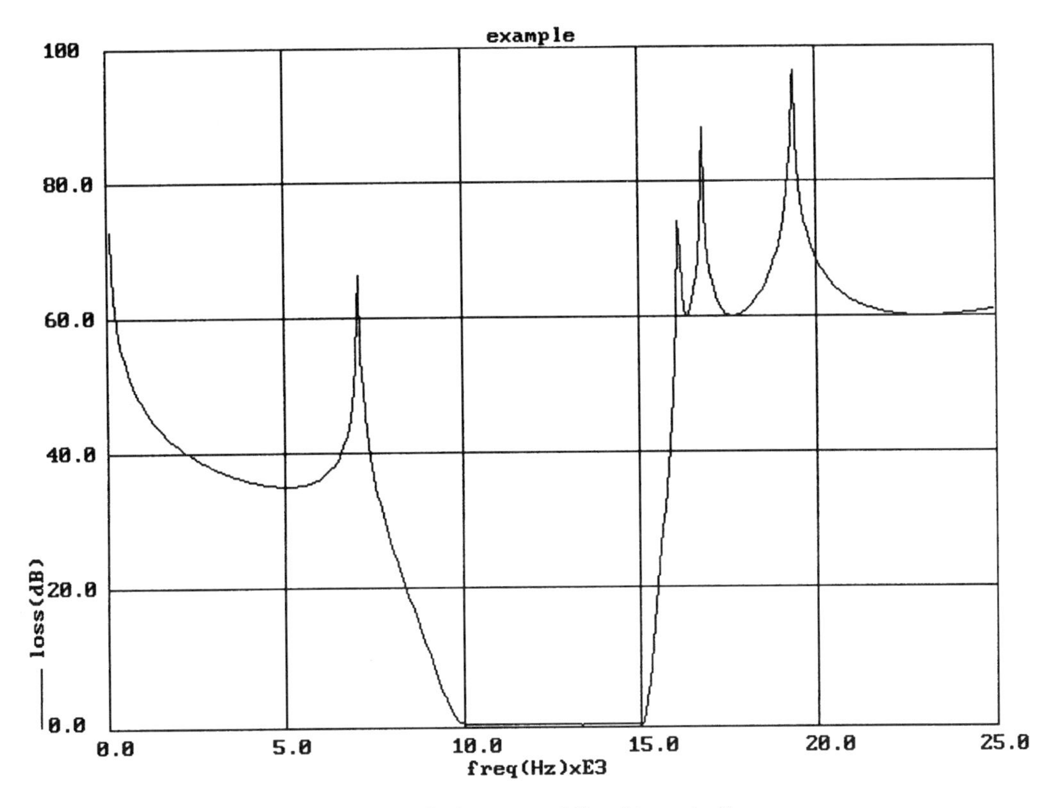

Figure 2 Performance of filter (Example 1).

connection of two quarter-wave line segments, and their respective characteristic impedances are given by the Z1 and Z2 values printed.

Again, computed performance is shown in Figure 4 and indicates a minimum stopband loss of 47 dB. Note that the true elliptic design with noncontributing unit elements would only provide a 20 dB stopband loss.

EXAMPLE 3

We can use the same example to illustrate the design of wave-digital filters of the ladder type. Note that all the pass- and stopband frequencies are relative to the quarter-wave frequency in this case, while, similarly, in an IIR digital filter all frequencies are relative to the Nyquist frequency (half the sampling frequency). So this filter, with 200 MHz passband edge and 2 GHz quarter-wave frequency, could equally well be a digital filter with 100 kHz Nyquist rate and 10 kHz passband edge.

Invoking the WAVE command converts this filter to a wave-digital form[2] that is very convenient for pipelined implementation; see Figure 5. Naturally, this printout needs a bit more explanation. The actual structure, using two- and three-port adaptors and shift registers (unit delay elements) is shown in the Figure 6. The implementation of the two- and three-port adaptors is shown in Figures 7 and 8. The computed performance of this filter is naturally identical to the previous one and is not shown.

EXAMPLE 4

The next example is an 11th-order elliptic lowpass IIR digital filter, with 20 kHz passband edge, 0.1 dB passband ripple, 200 kHz sampling frequency, and a stopband that starts at 22 kHz. This is again all the information necessary, and we show in Figure 9 the cascaded second-order, the two-multiplier Gray–Markel lattice,[4] and the differential allpass[11] implementations. The computed performance is shown in Figure 10. The Gray–Markel lattice structure is shown in Figure 11, where the boxes contain the lattice sections shown in Figure 12. Finally, the differential allpass structure

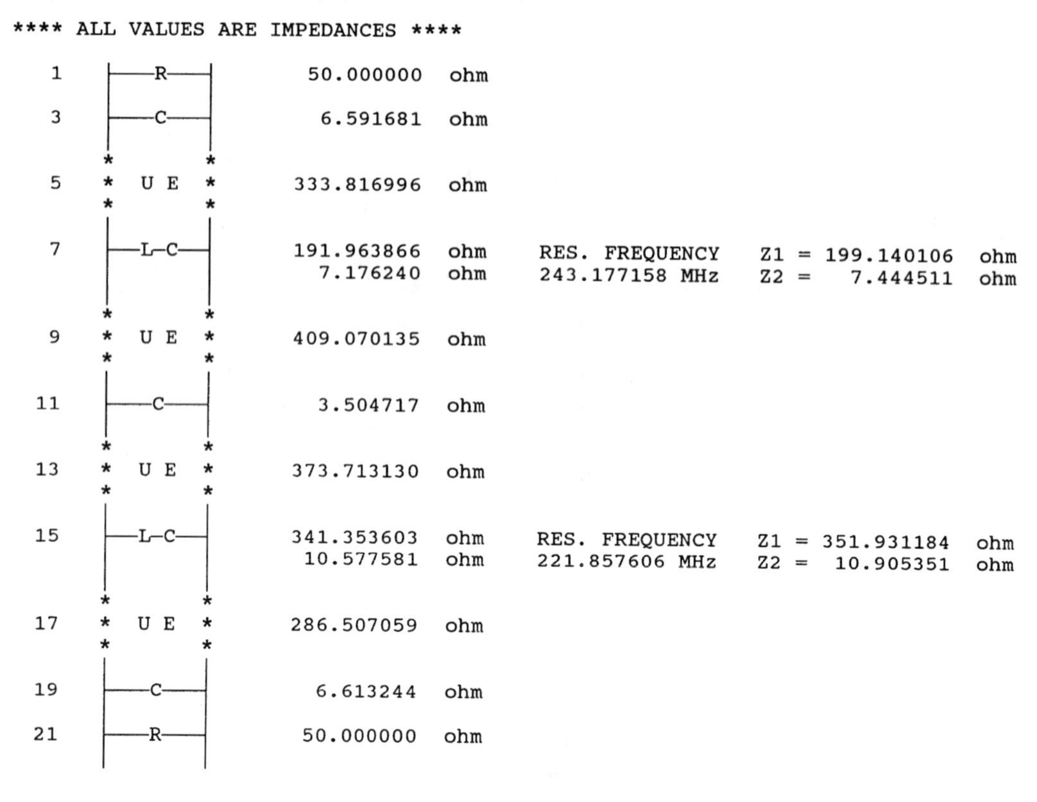

Figure 3 Example 2.

is shown in Figure 13, where $A_0(z)$ and $A_1(z)$ represent two allpass structures containing cascaded (first- and) second-order sections. The structure, in fact, generates two transfer functions, H(z) and G(z), that are complementary to each other, one of them being the filter we are synthesizing. Usually, a frequency-domain analysis is needed to decide which is which.

EXAMPLE 5

Our next example is an elliptic highpass filter implemented as an active RC filter using operational amplifiers as active elements. The filter is of degree 6, the stopband ends at 10 kHz, the passband begins at 12 kHz with 0.1 dB passband ripple.

Again, this is just about all the data input the program needs, and one possible implementation of this filter is shown in Figure 14. We specified here that all capacitors have a value of 10 nF, while the resistor values are computer generated. Again, this is only one of the many possible equivalent implementations, but we leave the investigations of other equivalents for another time and place. The computed performance of this structure is shown in Figure 15.

EXAMPLE 6

The last example is more interesting and unusual. It is an arithmetically symmetrical bandpass, with passband centered at 150 kHz and having an equal-ripple delay of 40 μsec over a 30 kHz band. The stopband is of the equal-minima type outside of a 40 kHz band (i.e., below 130 kHz and above 170 kHz). Finally, we would like to have as many of the inductors in the circuit as possible to have equal values.

The final circuit is shown in Figure 16, with the computed performance displayed in Figure 17. Note that all but one inductor are indeed equal, at the cost of some extra capacitors. All the inductors *can* be made equal, but then the terminations will not be.

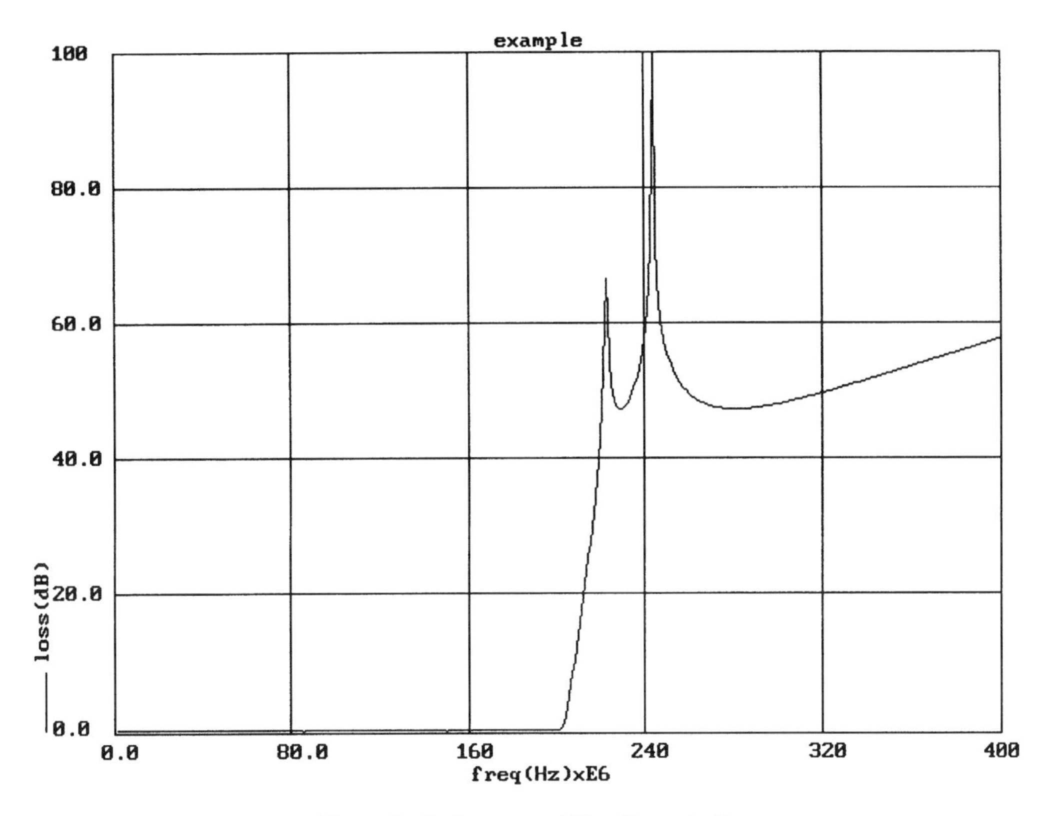

Figure 4 Performance of filter (Example 2).

3. OTHER SYNTHESIS PROGRAMS

Interestingly enough, we have found only five other programs, of widely different nature, that actually perform true synthesis of passive ladder and other filters.

3.1 FILTORX PROGRAM

Department of Electrical Engineering
University of Toronto
10 King's College Road
Toronto, Canada M5S 1A4

While this program is available for SUN workstation systems only, it has features that may be of interest to filter experts. Basically, this program is developed into a nearly full-fledged programming language by combining the basics of a C-like language with some specific constructs and functions applicable to filters. It contains most of the familiar filter functions, like Butterworth, Chebyshev, and elliptic, and other filter shapes can be arrived at by a built-in optimizer. The program can handle passive LC, active RC (including and, in fact, emphasizing switched-capacitor filters), and IIR digital filters. Passive LC filters are deemphasized and used only as a precursor for some advanced active RC design techniques.

For the actual synthesis, one must write procedures in this language, but, presumably, the most significant and often-used such procedures have already been developed and are available in the library. One of the unique features of this program is its ability to handle complex filters and multirate digital filters.

Although we have not been able to try it, the program seems to be extremely flexible, but also very complex and intimidating to the uninitiated. To alleviate this problem, a menu system has

```
Example # 3

 *** LADDER WAVE-DIGITAL IMPLEMENTATION ***

   *T* INDICATES UNIT DELAY

3-PORT PAR. AD.
A= 2.289614D-01 *T*
B= 3.995728D-02

*T*

3-PORT PAR. AD.       2-PORT ADAPTOR
A= 7.749654D-01 *T* A= 9.398881D-01 *T*
B= 5.941286D-01

*T*

3-PORT PAR. AD.
A= 1.842557D-02 *T*
B= 1.683239D-02

*T*

3-PORT PAR. AD.       2-PORT ADAPTOR
A= 4.672748D-01 *T* A= 9.279207D-01 *T*
B= 5.726018D-01

*T*

3-PORT PAR. AD.
A= 3.429250D-02 *T*
B= 2.289614D-01


COMMAND:
>
```

Figure 5 Wave-digital filter (Example 3).

been wrapped around the program to make it more user-friendly, but, just as with S/FILSYN, do not expect to become an expert user on short notice, unless you know the basics of filter design.

3.2 FILTERMASTER PROGRAMS

Intusoft
P.O. Box 710
San Pedro, CA 90733
(310)833-0710

There are actually three programs available under this name, FilterMaster, FilterMaster Professional, and FilterMaster Active.

The first two are for passive LC filters, the last one for active RC ones. The first one is basically a "calculator" type program (see below), while the other two may be considered to represent a subset of the S/FILSYN program. In fact, we are told that the developers of the program were previous users of S/FILSYN when they started developing their product.

The programs are basically menu driven and, as such, are considered to be easier to use, but many features of S/FILSYN are missing from them. These include many of the circuit transformation steps, arithmetically symmetrical bandpasses, equal-ripple delay approximations, microwave filters, and others. No optimizer is offered for general passband shapes either.

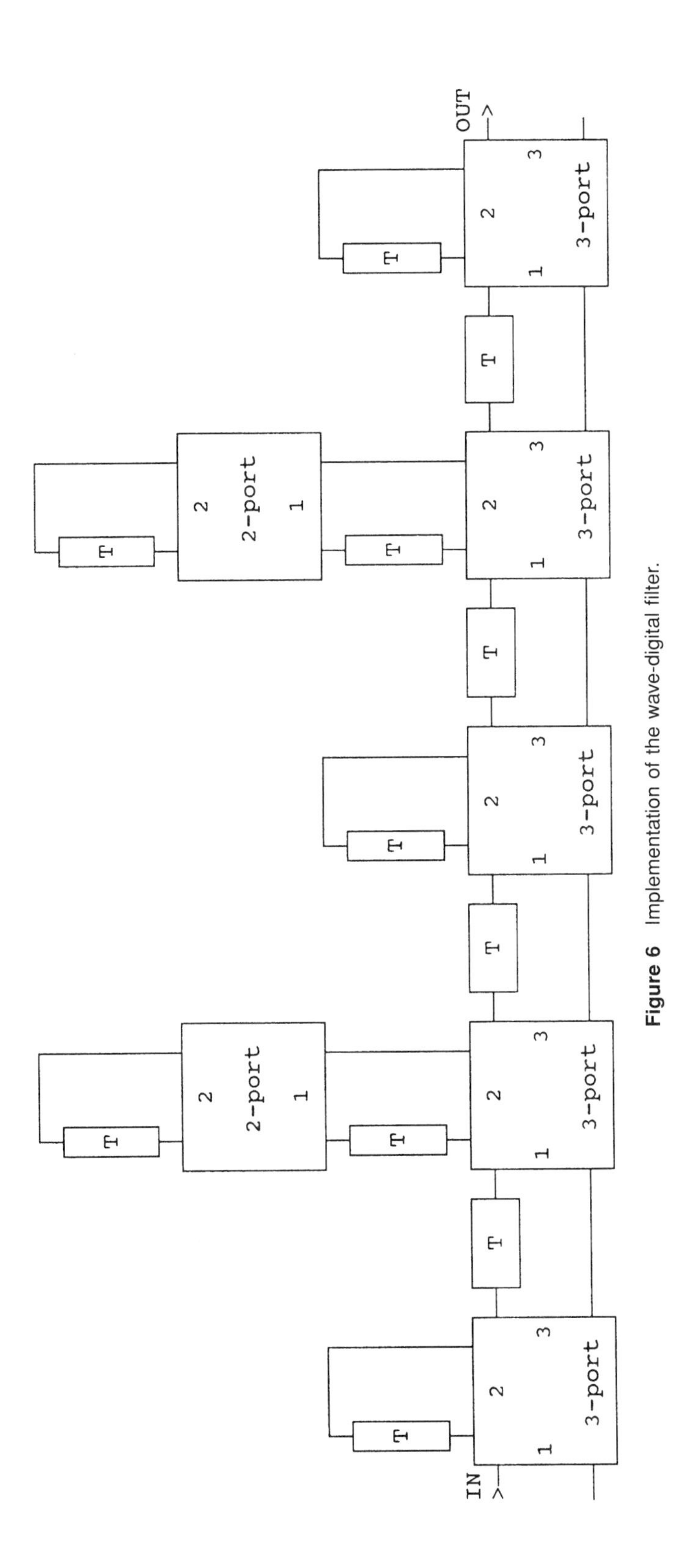

Figure 6 Implementation of the wave-digital filter.

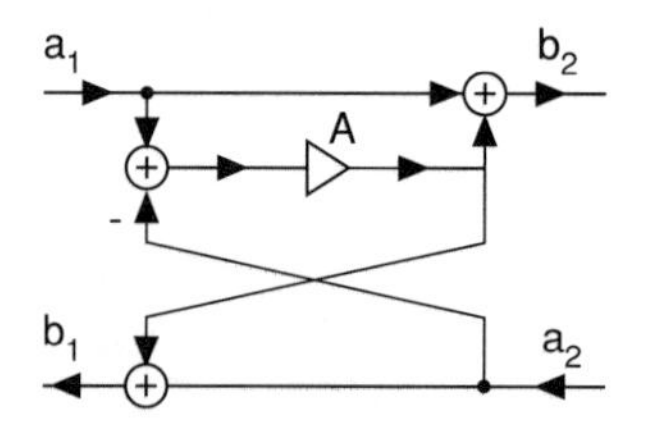

Figure 7 Two-port adaptor.

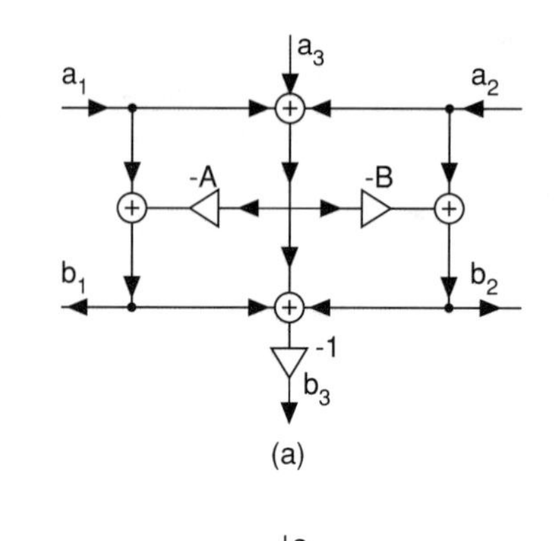

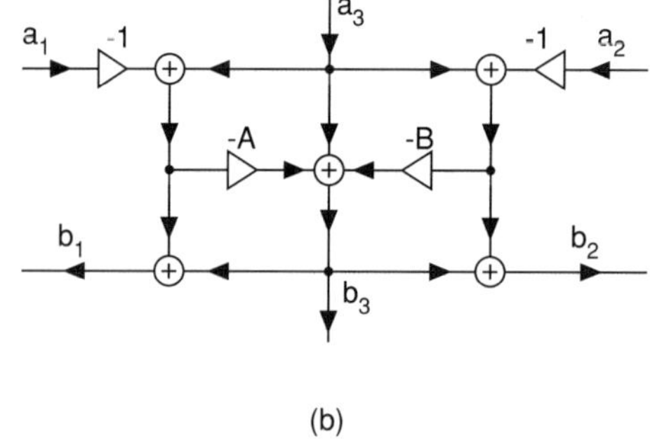

(b)

Figure 8 Series and parallel three-port adaptors.

3.3 PCFILT

ALK Engineering
1310 Emerson Avenue
Salisbury, MD 21801
(410)546-5573

This is a program similar to FilterMaster above, but with a bit more capabilities, strictly for passive LC filters.

3.4 PROMETHEUS PROGRAM

Scientific Computing Services
14 Magnolia Street
Margate, Queensland
Australia 4019

This is a strict synthesis program for passive LC filters, available from its manufacturer from Australia. The program is a bit dated and of limited capabilities.

```
*** S/FILSYN ***   FILTER PROGRAM

Example # 4
   LOW-PASS FILTER
      DIGITAL FILTER
      SAMPLING FREQUENCY                              = 200.000000 kHz
      EQUAL-RIPPLE PASS BAND
         BANDEDGE LOSS                                =     .100000 DB.
         UPPER PASSBAND EDGE FREQUENCY                =  20.000000 kHz
      EQUAL-MINIMA STOPBAND TYPE
         UPPER STOPBAND EDGE FREQUENCY                =  22.000000 kHz
         MULTIPLICITY OF ZERO AT NYQUIST FREQUEN.     =   1
         NUMBER OF FINITE TRANSMISSION ZEROS          =   5
         OVERALL FILTER DEGREE                        =  11

         TRANSMISSION ZEROS

               REAL PART          IMAGINARY PART

            0.0000000D+00         2.2082872D+04
            0.0000000D+00         2.2839169D+04
            0.0000000D+00         2.4963323D+04
            0.0000000D+00         3.0496313D+04
            0.0000000D+00         4.6968785D+04

 **IIR** DIGITAL FILTER TRANSFER FUNCTION
    Z TRANSFORM USED : BILINEAR WITH PREWARP
    FILTER TYPE      :  LOWPASS

       FRONT-END MULTIPLIER       =  4.11857482329D-04

                        **** CASCADE FORM ****

      NUMERATOR COEFFICIENTS

          Z**( 0)                Z**(-1)                Z**(-2)

     1.000000000D+00     1.000000000D+00      0.000000000D+00
     1.000000000D+00    -1.901691363D-01      1.000000000D+00
     1.000000000D+00    -1.150200022D+00      1.000000000D+00
     1.000000000D+00    -1.415842129D+00      1.000000000D+00
     1.000000000D+00    -1.506885711D+00      1.000000000D+00
     1.000000000D+00    -1.537702195D+00      1.000000000D+00

      DENOMINATOR COEFFICIENTS

          Z**( 0)                Z**(-1)                Z**(-2)

     1.000000000D+00    -7.933654456D-01      0.000000000D+00
     1.000000000D+00    -1.586187754D+00      6.821591164D-01
     1.000000000D+00    -1.585473685D+00      7.914675022D-01
     1.000000000D+00    -1.586286259D+00      8.869718051D-01
     1.000000000D+00    -1.590430691D+00      9.479930499D-01
     1.000000000D+00    -1.601061122D+00      9.851822306D-01
```

Figure 9(a) Digital Example 4.

3.5 FILTER DESIGN AND ANALYSIS SYSTEM

Mr. William B. Lurie
8503 Heather Place
Boynton Beach, FL 33437
(407)369-3218

This is a package of programs available from its author, written in APL, a programming language one either loves or hates. Since this language is an interpreter, a potential user must also purchase the language itself, before the programs can be used.

```
                **** GRAY-MARKEL LATTICE ****

              MULTIPLIER              TAP-WEIGHT

           -8.155328836D-01        -1.575244853D-06
            9.986819572D-01        -2.787422046D-07
           -8.640230853D-01        -1.744862495D-05
            9.806850798D-01         1.353272856D-04
           -9.290387178D-01         7.985372745D-04
            9.490211441D-01         2.135171529D-03
           -9.401579802D-01         4.385143972D-03
            9.331759446D-01         5.916150631D-03
           -9.078686179D-01         5.973180537D-03
            7.864565680D-01         3.916990724D-03
           -3.548326659D-01         1.623544569D-03
            0.000000000D+00         4.118574823D-04


                 **** DIFFERENTIAL ALLPASS FORM ****

ALLPASS NO. 1

     NUMERATOR COEFFICIENTS

          Z**( 0)              Z**(-1)              Z**(-2)

     1.000000000D+00     -1.585473685D+00     7.914675022D-01
     1.000000000D+00     -1.590430691D+00     9.479930499D-01
     1.000000000D+00     -7.933654456D-01

     DENOMINATOR COEFFICIENTS

     7.914675022D-01     -1.585473685D+00     1.000000000D+00
     9.479930499D-01     -1.590430691D+00     1.000000000D+00
    -7.933654456D-01      1.000000000D+00

ALLPASS NO. 2

     NUMERATOR COEFFICIENTS

          Z**( 0)              Z**(-1)              Z**(-2)

     1.000000000D+00     -1.586187754D+00     6.821591164D-01
     1.000000000D+00     -1.586286259D+00     8.869718051D-01
     1.000000000D+00     -1.601061122D+00     9.851822306D-01

     DENOMINATOR COEFFICIENTS

     6.821591164D-01     -1.586187754D+00     1.000000000D+00
     8.869718051D-01     -1.586286259D+00     1.000000000D+00
     9.851822306D-01     -1.601061122D+00     1.000000000D+00
```

Figure 9(b) (continued)

The programs themselves are equally idiosyncratic. They concentrate on passive LC filters, offer some of the standard filter functions, generate and analyze ladder implementations, and offer some manipulation and optimization capabilities and special configurations.

4. "CALCULATOR" PROGRAMS FOR FILTER DESIGN

As far as we can determine, all other programs on the market are what we call calculator programs. By this we mean programs that use closed-form expressions and algorithms that are of nearly closed form, that were developed for simple filter kinds, like Butterworth, Chebyshev, Bessel, and elliptic lowpass filters. These were simple enough to be entered into hand-held programmable calculators first; they then later migrated to desktop computers and were spruced up with fancy

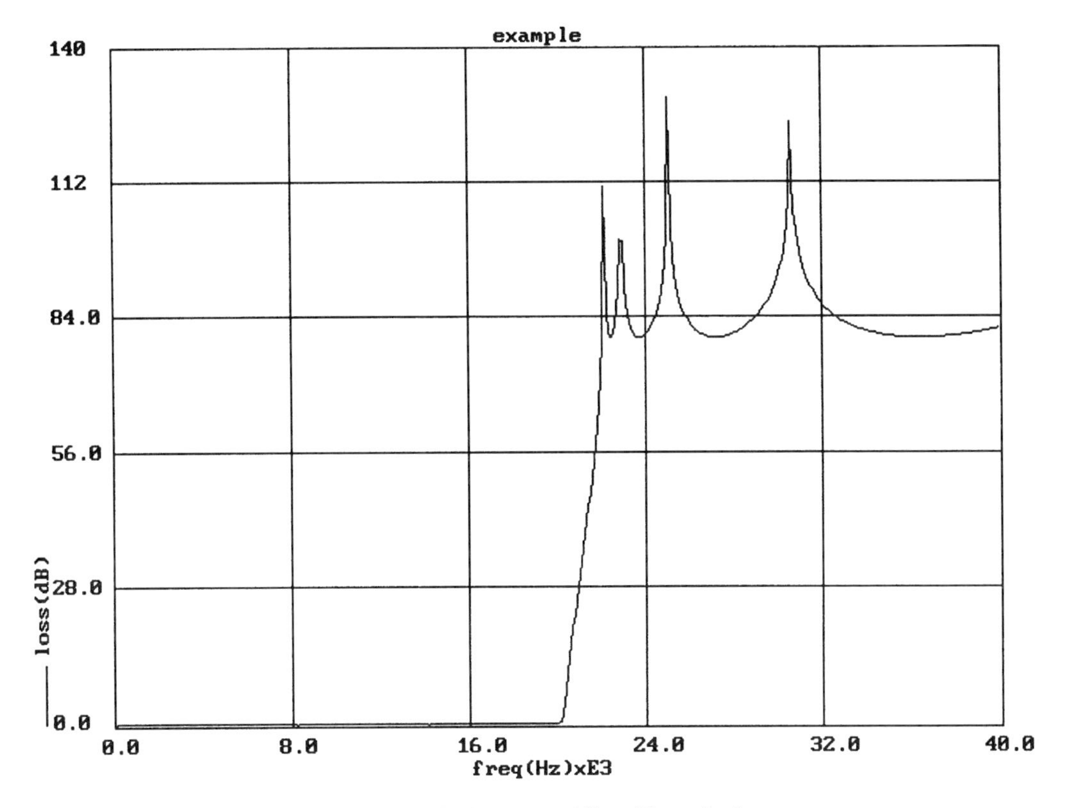

Figure 10 Performance of filter (Example 4).

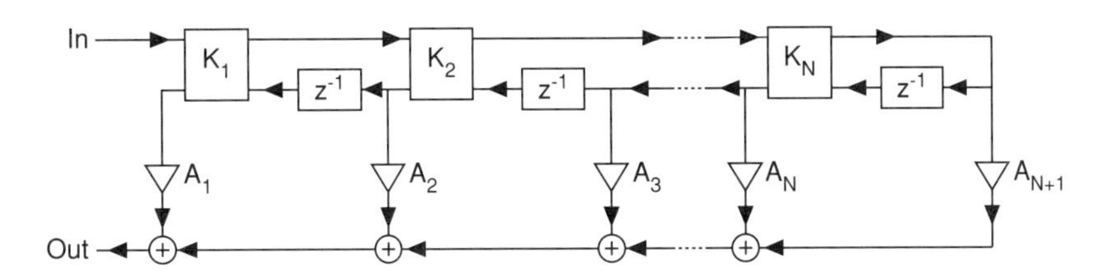

Figure 11 Gray–Markel lattice structure.

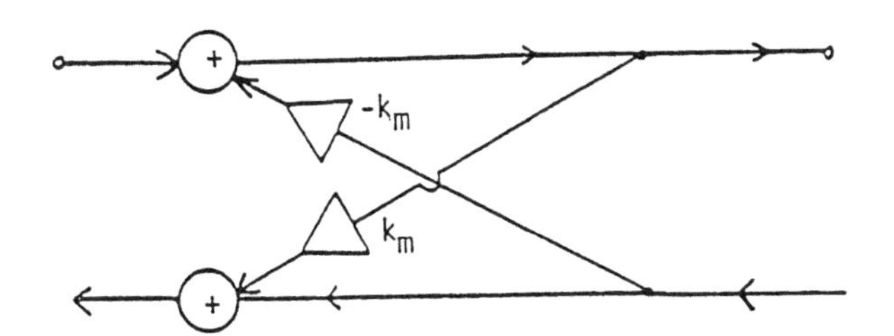

Figure 12 The lattice section in the Gray–Markel structure.

graphics and menu systems. Basically, however, their nature is unchanged and they are very simple to distinguish by their way of generating bandpass filters and filter functions. They can only generate bandpass filters by the standard lowpass-to-bandpass transformation, thereby eliminating the more general bandpass filters which have lower and upper stopbands that are independent of each other.

Most of these programs are also unable to handle different and, specifically, open- or short-circuited terminating impedances.

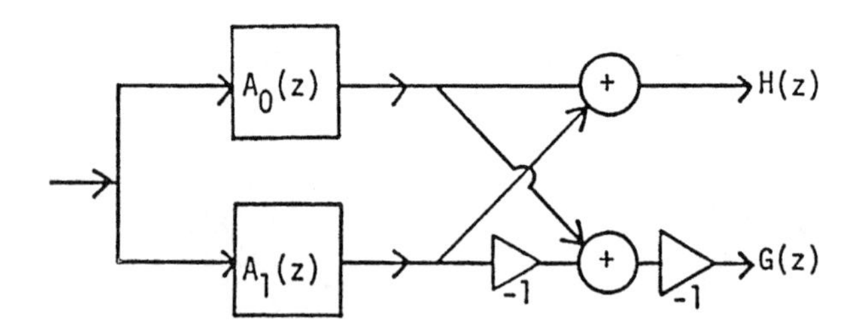

Figure 13 Differential allpass implementation.

We shall simply list all of the commercial programs of this type we know of, with a brief mention of what they do. We will specifically exclude those applicable only to microwave (and waveguide) filters.

1. *Filter* — Eagleware Corp., 1750 Mountain Glen, Stone Mountain, GA 30087 (404)939-0156; passive LC and active-RC filters
2. *QEDesign* — Momentum Data Systems, 1520 Nutmeg Place, Suite 108, Costa Mesa, CA 92626 (714)557-6884; both IIR and FIR digital filters
3. *DADiSP/Filters* — DSP Development Corp., 1 Kendall Square, Cambridge, MA 02139 (617)577-1133; both IIR and FIR digital filters
4. *Digital Filter Design System* — Hanson Engineering, 708 Bettyhill Avenue, Duarte, CA 91010; both IIR and FIR digital filters; built-in optimizer for magnitude and delay
5. *DISPRO* — Bridgenorth Signal Processing, Inc., P.O. Box 469, Custer, WA 98240 (604)538-0003; both IIR and FIR digital filters
6. *FilSolv* — Webb Laboratories, 139 E. Capitol Drive, Suite 4, Hartland, WI 53029 (414)367-6823; passive LC filters
7. *Active* — Tatum Labs, Inc., 3917 Research Park Drive, B-1, Ann Arbor, MI 48108 (313)663-8810; active RC filters
8. *DSP Designer* — Zola Technologies, Inc., 6195 Heards Creek Drive, N.W., Atlanta, GA 30328 (404)843-2972; both IIR and FIR digital filters, plus some other signal processing functions
9. *PC-OPT* — Electrical Engineering Software, Inc., 4675 Stevens Creek Blvd., Suite 200, Santa Clara, CA 95051 (408)296-8151; a pure optimizer that can design filters, among other circuits
10. *AFDPLUS* — RLM Research, P.O. Box 3630, Boulder, CO 80307 (303)499-7566; active RC filters
11. *Filter Designer* — MicroSim Corp., 20 Fairbanks, Irvine, CA 92718 (714)770-3022; active RC filters, with optimizer
12. *Digital Filter Design* — The Athena Group, Inc., 3424 N.W. 31st Street, Gainesville, FL 32605 (904)371-2567; both IIR and FIR filters, among other signal processing features
13. *FDS* — Comdisco Systems Inc., 919 E. Hillsdale Blvd., Foster City, CA 94404 (415)574-5800; both FIR and IIR digital filters on workstation systems only
14. *DFDP* — Atlanta Signal Processors, Inc., 770 Spring Street N.W., Suite 208, Atlanta, GA 30308 (404)892-7265; both IIR and FIR digital filters
15. *Filtoid* — Geesaman Software Inc., P.O. Box 219, Sebago Lake, ME 04075 (207)642-2728; passive LC filters
16. *Parfil* — WaveCon, P.O. Box 2697, Escondido, CA 92033 (619)747-6922; various microwave filters (subsets of the capabilities are also available under different names)
17. *FilterCAD* — LinearX Systems, Inc., 7556 S.W. Brideport Road, Portland, OR 97224 (503)620-3044; passive LC and active RC filters

In addition, many, if not all, general-purpose signal processing software packages include a module for the design of FIR and/or IIR digital filters, since these are going to be needed as building blocks.

5. OTHER PROGRAMS

There are many other programs, some of them very interesting and exciting, either under development in universities or in use as in-house programs developed for special filter kinds, most often

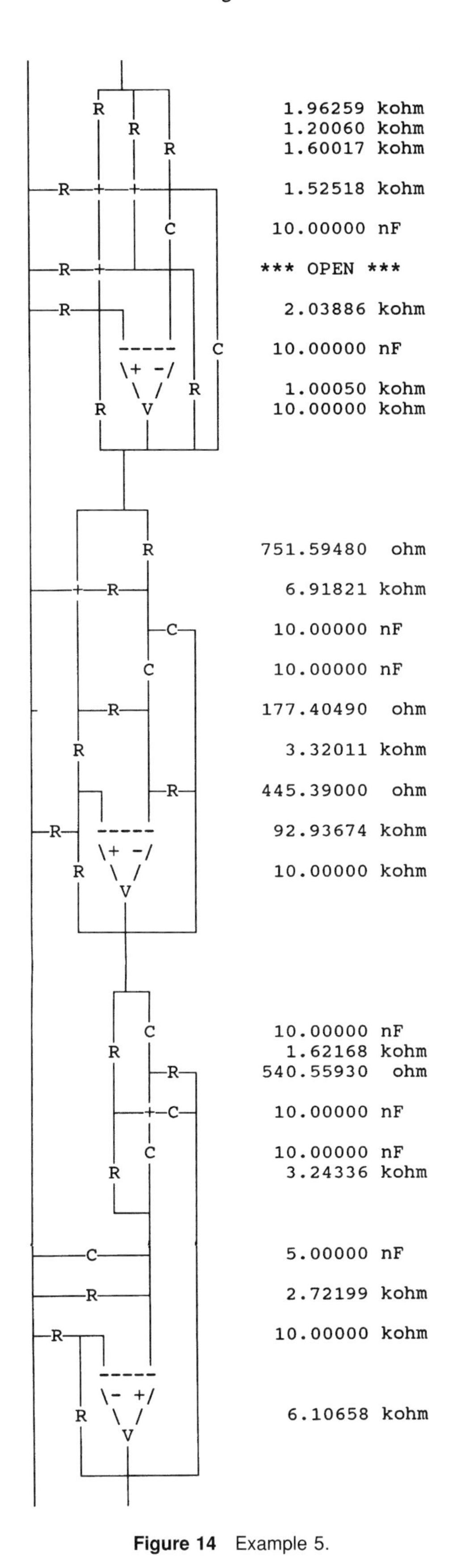

Figure 14 Example 5.

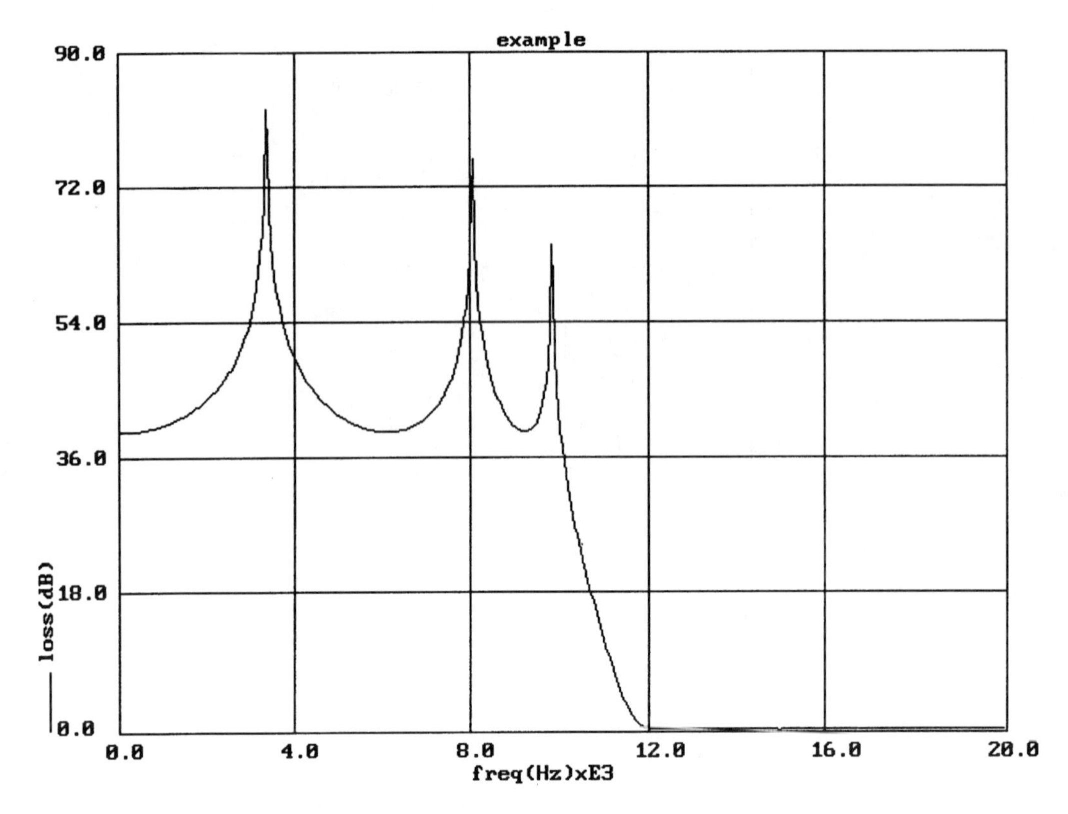

Figure 15 Performance of filter (Example 5).

switched-capacitor ones. Industry-developed in-house programs are naturally not generally available, while those coming from universities are changing too fast to report on and often are not very well supported. Here we mention only a few, which have reasonably good track records.

1. The PANDDA program for switched-capacitor, active RC, and digital filters, available from the University of Glasgow, Glasgow, England. This program can handle some unusual implementations (R.K. Henderson, L. Ping, J.I. Sewell: A design program for digital and analogue filters: *PANDDA Proc. ECCTD,* pp. 289–293, 1989).
2. The DIGICAP program for digital filters from the University of Calgary, Calgary, Alberta, Canada. The program can optimize word length, compare sensitivities, and generate DSP code (L.E. Turner, D.A. Graham, P.B. Denyer: The analysis and implementation of digital filters using a special purpose CAD Tool. *IEEE Trans. Educ.,* 32, 287–297, 1989).

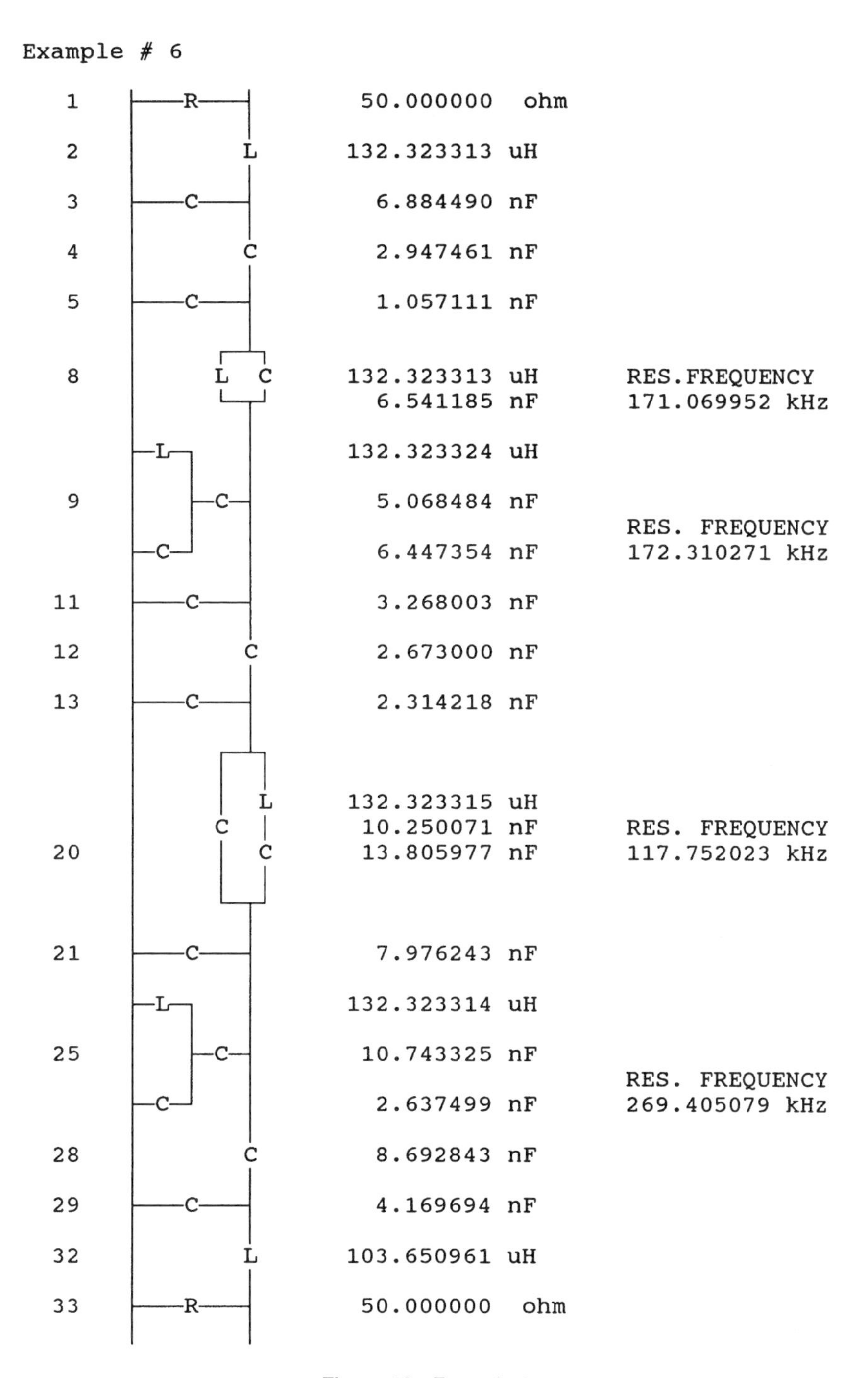

Figure 16 Example 6.

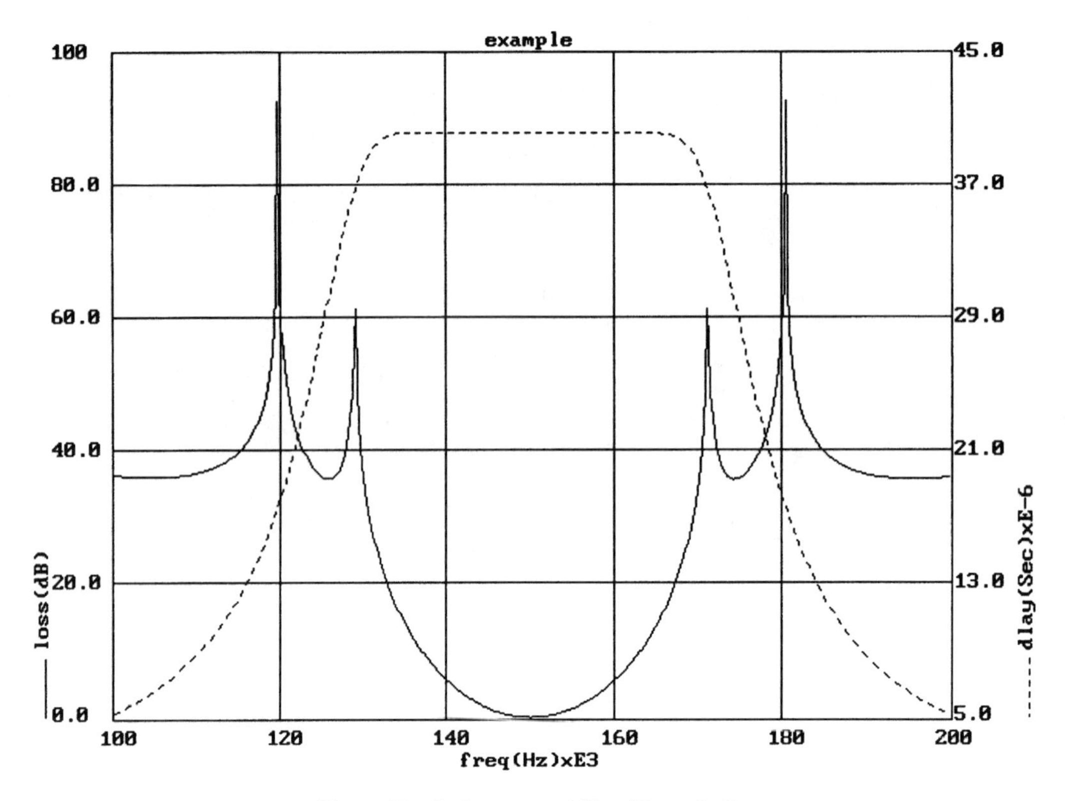

Figure 17 Performance of filter (Example 6).

REFERENCES

We list below just a few references out of an enormous literature, those that have a direct bearing on what has been mentioned in the chapter.

1. **Darlington, S.**: Synthesis of reactance fourpoles which produce prescribed insertion loss characteristics, *J. Math. Phys.*, 18, 257–353, 1939.
2. **Fettweis, A.**: Wave digital filters: theory and practice, *Proc. IEEE,* 74, 270-327, 1986.
3. **Fleischer, P.E., Laker, K.R.**: A family of active switched capacitor biquad building blocks, *Bell Syst. Tech. J.*, 58, 2235–2269, 1979.
4. **Gray, A.H., Jr., Markel, J.D.**: Digital lattice and ladder filter structures, *IEEE Trans. Audio Electroacoust.*, AU-21, 491–500, 1973.
5. **Norek, C.**: Product method for the calculation of the effective loss LC filters, *Belgrade Symp. on Network Theory,* 1968, 353–365.
6. **Smith, B.R., Temes, G.C.**: An iterative approximation procedure for automatic filter synthesis, *IRE Trans. Circuit Theory,* CT-12, 107–112, 1965.
7. **Szentirmai, G.**: Theoretical basis of a digital computer program package for filter synthesis, *Proc. 1st Annual Allerton Conf. on Circuit and System Theory,* 1963, 37–49.
8. **Szentirmai, G.**: Synthesis of multiple feedback active filters, *Bell Syst. Tech. J.*, 52, 527-555, 1973.
9. **Szentirmai, G.**: FILSYN — a general purpose filter synthesis program, *Proc. IEEE,* 65, 1443–1458, 1977.
10. **Temes, G.C., Gyi, M.**: Design of filters with arbitrary passband and Chebyshev stopband attenuation, *IEEE Int. Conv. Rec.*, 15(5), 2–12, 1967.
11. **Vaidyanathan, P.O., Mitra, S.K., Neuvo, Y.**: A new approach to the realization of low sensitivity IIR digital filters, *IEEE Trans. Acoust. Speech Signal Process,* ASSP-34, 1986.

A few more general references are given below:

12. **Cuthbert, T.R., Jr.,**: *Circuit Design Using Personal Computers*, John Wiley & Sons, New York, 1983.
13. **Elliott, D.F.** (Ed.): *Handbook of Digital Signal Processing*, Academic Press, San Diego, CA, 1987.
14. **Guillemin, E.A.**: *Synthesis of Passive Networks*, John Wiley & Sons, New York, 1957.
15. **Moschytz, G.S., Horn, P.**: *Active Filter Design Handbook*, John Wiley & Sons, Chichester, England, 1981.
16. **Rabiner, L.R., Gold, B.**: *Theory and Applications of Digital Signal Processing*, Prentice Hall, Englewood Cliffs, NJ, 1975.

17. **Rhodes, J.D.**: *Theory of Electrical Filters*, John Wiley & Sons, London, 1976.
18. **Schaumann, R., Ghausi, M.S., Laker, K.R.**: *Design of Analog Filters*, Prentice Hall, Englewood Cliffs, NJ, 1990.
19. **Szentirmai, G.** (Ed.): *Computer Aided Filter Design*, IEEE Press, New York, 1973.
20. **Temes, G.C., Mitra, S.K.** (Eds.): *Modern Filter Theory and Design*, John Wiley & Sons, New York, 1973.
21. **Vlach, J.**: *Computerized Approximation and Synthesis of Linear Networks*, John Wiley & Sons, New York, 1969.
22. *S/FILSYN User Manual*, DGS Associates, Inc., 1992, Sec. 14.

Chapter 4

LCR Filters

LC Filter Design

Phillip R. Geffe

CONTENTS

0-8493-8951-8/97/$0.00+$.50
© 1997 by CRC Press LLC

1. FREQUENCY RESPONSE

1.1 RESPONSE, INSERTION LOSS

The frequency response of a filter is usually measured by making a calculation on voltage measurements. In Figure 1, the filter is driven by a variable frequency voltage source, through the source resistance, R_s, to the output load, R_l. The ratio $H = e_3/e_1$ is called the *voltage ratio*. The insertion loss is obtained at any frequency, f, by measuring $H_1 = e_3/e_1$ without the filter connected, and then measuring $H_2 = e_3/e_1$ with the filter inserted. The ratio of these is the *insertion ratio*: when expressed in decibels, it becomes *insertion loss*. In other words, insertion loss is the loss which results from inserting the filter into the circuit. Thus, we have

$$\text{insertion loss} = 20\log_{10}\left(\frac{H_2}{H_1}\right) \quad [\text{dB}] \tag{1}$$

Relative attenuation is obtained from the insertion loss by subtracting the minimum passband insertion loss from the insertion loss at all frequencies. Thus, the maximum gain point of the passband is at 0 dB relative loss. Alternatively, some users may specify the 0 dB frequency arbitrarily.

Important note: The ratio e_3/e_2 is not filter loss of any kind, and has no particular significance in filter theory or practice.

1.2 LOWPASS, HIGHPASS, BANDPASS, AND BANDSTOP FILTERS

Filters have one or more passbands (frequency intervals of low attenuation), and one or more stopbands with substantial attenuation. Figure 2 shows attenuation plots of (a) a lowpass filter, (b) a highpass filter, (c) a bandpass filter, (d) an ideal bandstop filter, and (d) a bandstop filter with finite passbands.

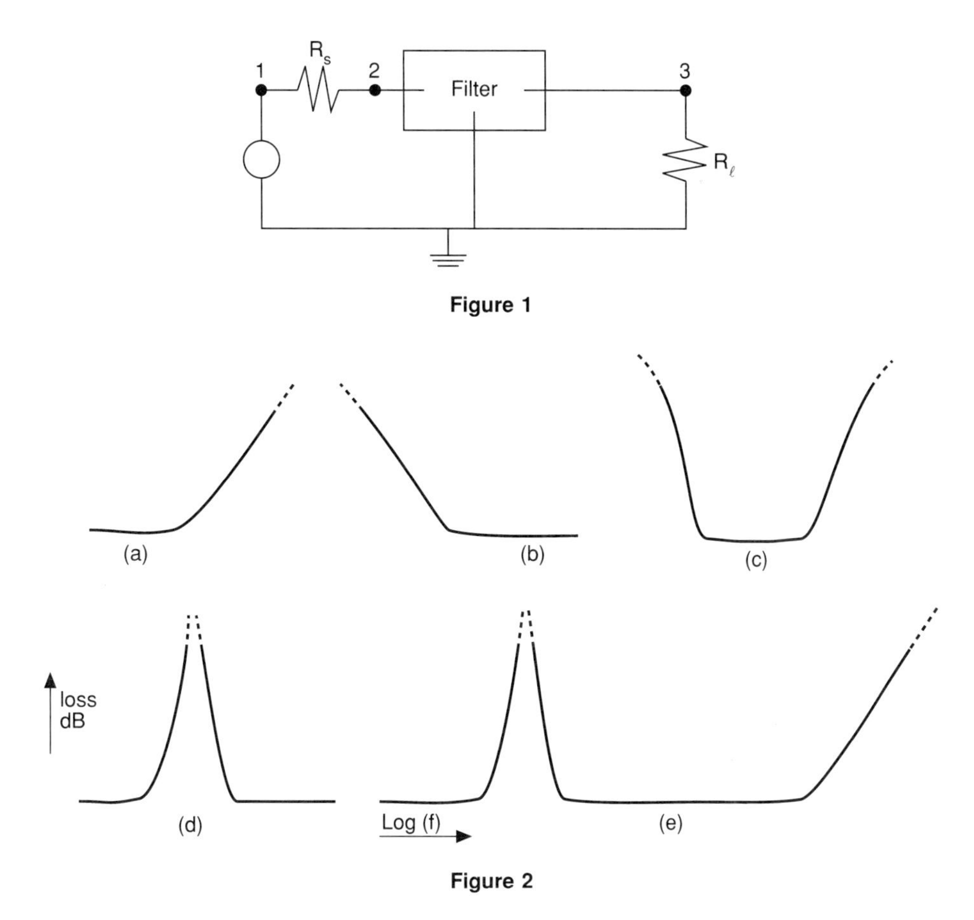

Figure 1

Figure 2

1.3 PASSBAND TYPES

Figure 3 shows (a) a lowpass filter and (b) a bandpass filter, both with equiripple passbands. In these cases, the ripple value is usually called either A_{max} or A_p. Filters with equiripple passbands and monotonic stopbands are *Chebyshev* filters. The filters with smooth (actually, *maximally flat*) passbands and monotonic stopbands are *Butterworth* filters, and they are limiting cases of Chebyshev filters. When normalized for the half power band edge at 1 rad/s, the Butterworth filters have the loss function

$$a(\omega) = 10 \log_{10}(1 + \omega^{2n}) \tag{2}$$

The loss of normalized Chebyshev filters is

$$a(\omega) = 10 \log_{10}(1 + (eCn(\omega))^2) \tag{3}$$

where

$$A_{max} = 10 \log_{10}(1 + e^2) \tag{4}$$

Here, $C_n(\omega)$ is the n–th degree Chebyshev polynomial. (The coefficients of the polynomials may be found in the references.) A convenient way of calculating the $C_n(\omega)$ functions, for the stopband only, is:

$$C_n(\omega) = \cosh\left(\operatorname{arccosh}\frac{(\omega)}{n}\right), \quad \omega > 1 \tag{5}$$

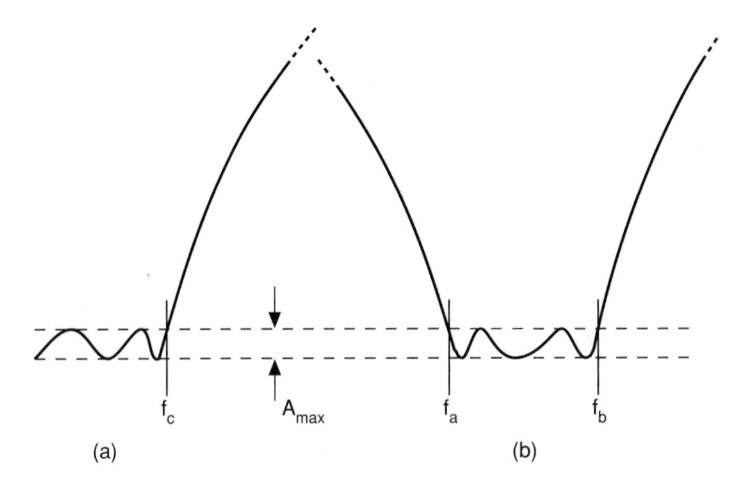

Figure 3

and, for the passband only, is

$$C_n(\omega) = C_p\left(\arccos\frac{(\omega)}{n}\right), \quad \omega < 1 \tag{6}$$

Note that, for the same degree, the Chebyshev filters have the greater stopband loss.

1.4 RETURN LOSS AND VOLTAGE STANDING WAVE RATIO

Return loss is a measure of how closely the internal impedance of a filter matches its terminations. The higher the loss, the closer the match. Thus, in a line of (say) 50 Ω, we would like to see a 50 Ω impedance in all the 50-Ω devices that are inserted into the line, since this would cause all the devices to be properly terminated, and, thus, they could all be expected to work as they were designed to do. In this situation, it is common to specify a minimum value for return loss. If Z is the internal impedance of the terminated filter and R is the line impedance, then the return loss is

$$A_e = 20\log_{10}\left(\text{abs}\left(\frac{R+z}{R-z}\right)\right) \tag{7}$$

When the filter has an equiripple passband, then the minimum value assumed by the return loss in the passband is related to the ripple value:

$$x = \frac{-A_p}{10} \tag{8}$$

$$A_e = -10\log_{10}\left(1-10^x\right) \quad [\text{dB}] \tag{9}$$

Alternatively, the filter impedance can be specified with a quantity known as VSWR (voltage standing wave ratio). This is the ratio of the magnitude of the internal impedance to the line impedance (or vice versa), and is taken to be greater than unity. In an equiripple passband filter, it can be calculated from the reflection coefficient, ρ:

$$\text{VSWR} = \frac{(1+\rho)}{(1-\rho)} \tag{10}$$

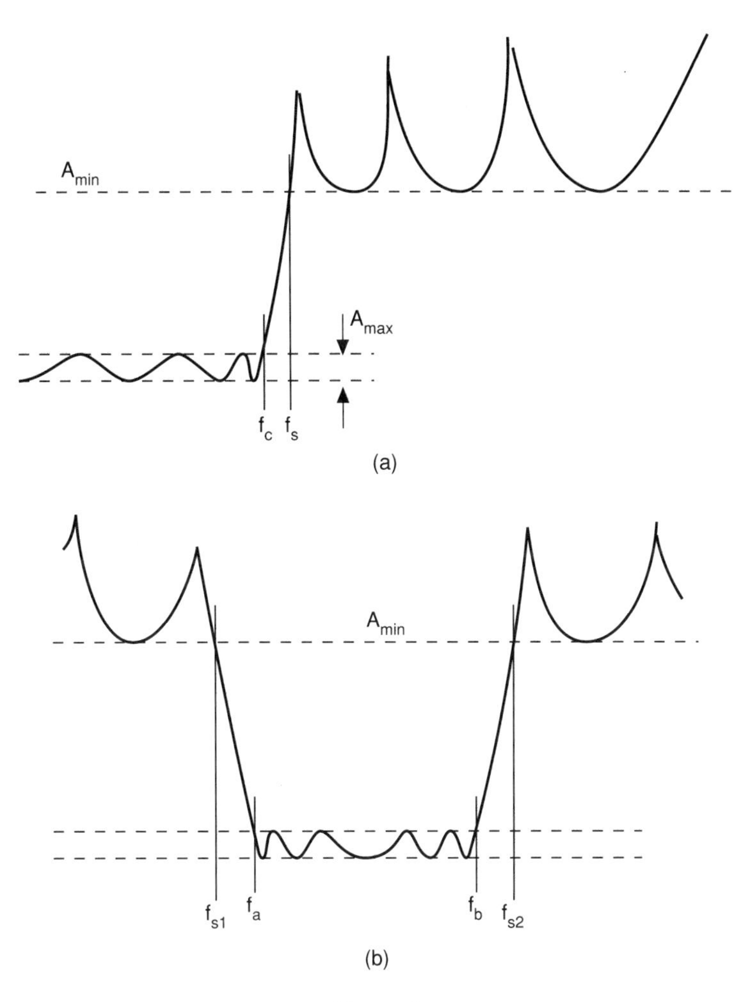

Figure 4

where

$$\rho = \sqrt{1 - 10^{-A_p/10}} \tag{11}$$

1.5 ELLIPTIC PARAMETERS

Figure 4 shows the normalized loss of (a) lowpass elliptic filters and (b) bandpass elliptic filters. These responses are rational functions which are rather complicated, and the coefficients are almost never needed in practical design. Instead, we use the elliptic parameters shown. These are

A_{max} or A_p	= the guaranteed maximum loss in the passband
A_{min} or A_s	= the guaranteed minimum loss in the stopband
f_c or ω_c	= the edge of the ripple band (lowpass)
f_a, f_b or ω_a, ω_b	= the edges of the ripple band (bandpass)
f_s or ω_s	= the edge of the stopband
n	= the degree of the filter (this equals the number of ladder branches see Figures 5a and 5b)

(In the above, "f" or "ω" indicate hertz or radians per second.)

Note that elliptic filters have the most selectivity; i.e., the stopband is closest to the passband.

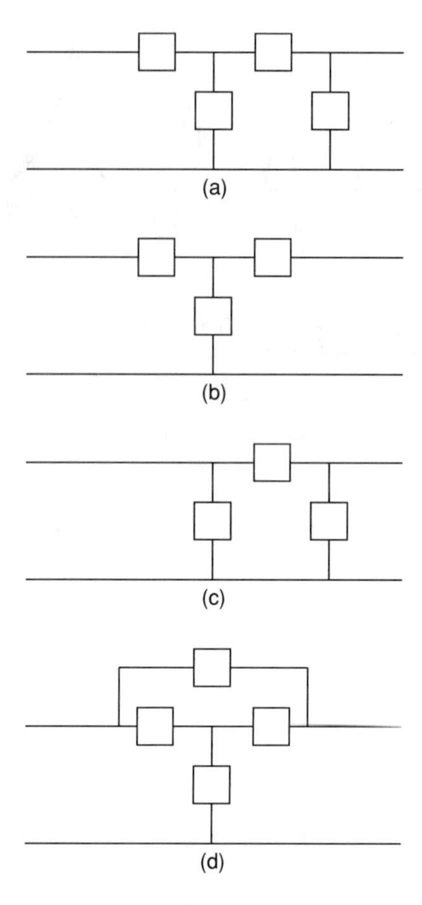

Figure 5

1.6 BESSEL AND MISCELLANEOUS TYPES

Bessel filters (not shown here) are similar to the Butterworth, but have constant group delay and very little selectivity. They see frequent application and are highly prized for their excellent time response, which is a consequence of both attenuation response and group delay. Note that this very useful property of the Bessel filters is found in only two cases:

1. Lowpass filters (ω)
2. Narrowband ($b_\omega < 10\%$) bandpass filters

Bessel filters of all degrees approximate the same transfer function (one that gives ideal time response). As a result, increasing the degree tends to give no increase in attenuation, except deep in the stopband. The higher degrees also push the constant delay bandwidth out into the stopband.

Many other types of frequency response families have been defined. These may be found in the references, but, since they have few applications, they are not discussed here.

2. BASIC CIRCUIT ALGORITHMS

2.1 LADDER AND BRIDGED-T

A *ladder* consists of alternating series and shunt branches, as shown in Figure 5a. A *branch* is simply any two-terminal passive network. Figure 5b shows a T network, and Figure 5c shows a π network. Figure 5d shows a T network with a bridging branch and is called a bridged-T.

2.2 DUALITY

To say that one network is the *dual* of another is to say that the impedance of one is the admittance of the other. Very simple rules may be used to transform a ladder network into its dual, as follows:

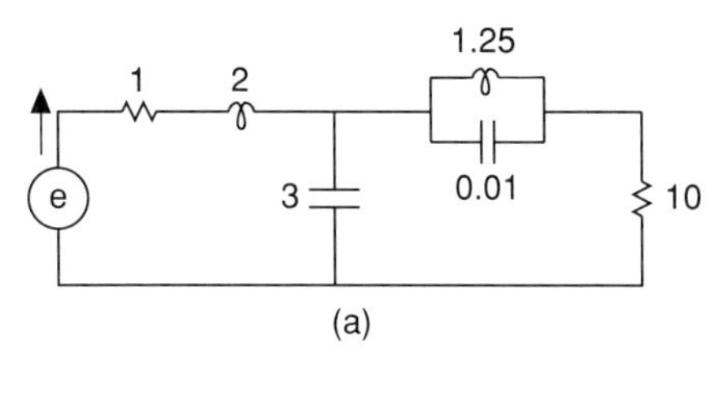

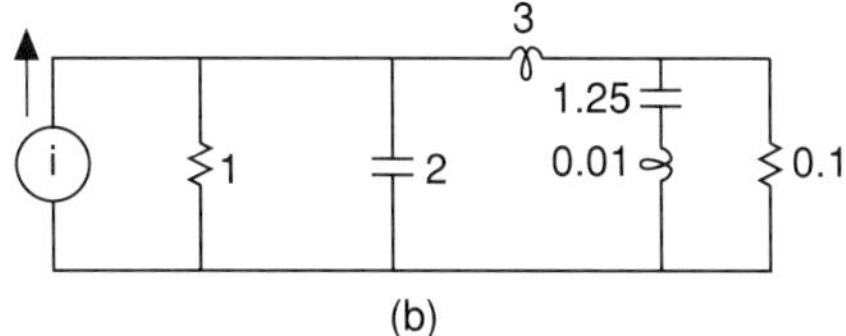

Figure 6

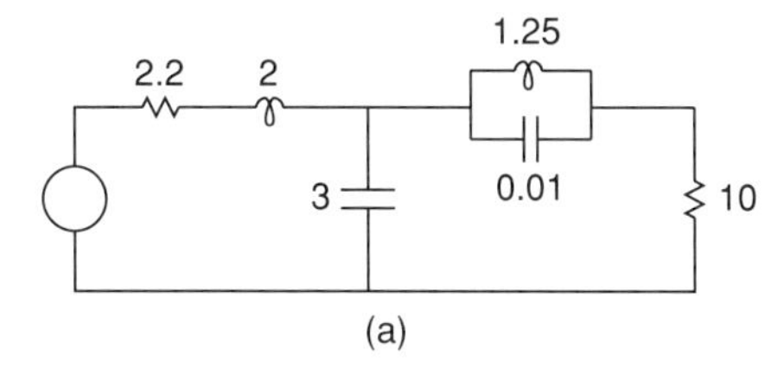

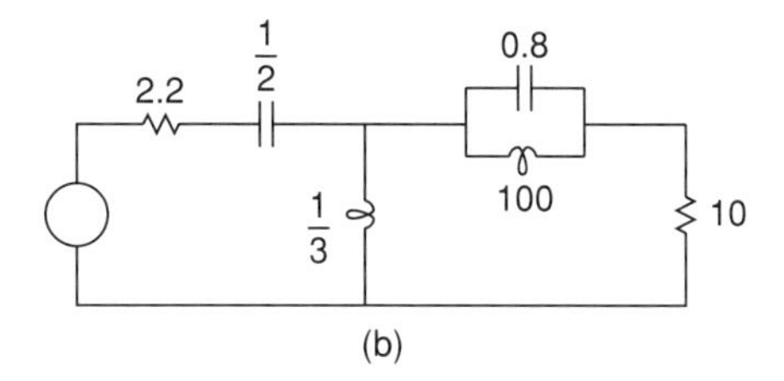

Figure 7

1. Inductors are transformed into capacitors of the same value, and vice versa.
2. Resistors are reciprocated.
3. Current sources become voltage sources, and vice versa.
4. Series branches become shunt branches, and vice versa.
5. The duals of elements in parallel are connected in series, and vice versa.

Figure 6 shows a pair of mutually dual networks.

2.3 RECIPROCATION

A network is transformed into its *reciprocal* by applying the following rules:

1. Coils become capacitors of the reciprocal value, and vice versa.
2. Resistors are unaffected.

Figure 7 shows a pair of reciprocal networks.

2.4 T–TO–π AND π–TO–T

Figure 8 shows a pair of T and π networks. The branches are all resistors, all capacitors, or all inductors. If the two are equivalent, then the following relations hold:

Case 1: Resistors or inductors

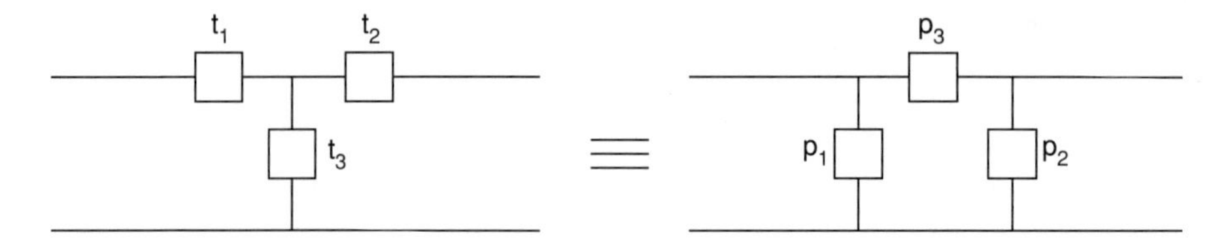

Figure 8

$$p_0 = p_1 + p_2 + p_3 \qquad t_0 = t_1t_2 + t_2t_3 + t_1t_3$$

$$t_1 = \frac{p_1p_3}{p_0} \qquad p_3 = \frac{t_0}{t_3}$$

$$t_2 = \frac{p_2p_3}{p_0} \qquad p_1 = \frac{t_0}{t_2}$$

$$t_3 = \frac{p_1p_2}{p_0} \qquad p_2 = \frac{t_0}{t_1} \tag{12}$$

Case 2: Capacitors

$$p_0 = p_1p_2 + p_1p_3 + p_2p_3 \qquad t_0 = t_1 + t_2 + t_3$$

$$t_1 = \frac{p_0}{p_2} \qquad p_3 = \frac{(t_1t_2)}{t_0}$$

$$t_2 = \frac{p_0}{p_1} \qquad p_1 = \frac{(t_1t_3)}{t_0}$$

$$t_3 = \frac{p_0}{p_3} \qquad p_2 = \frac{(t_2t_3)}{t_0} \tag{13}$$

2.5 NORTON TRANSFORMERS

This algorithm gives true transformer action without using a coiled wire transformer. Figure 9 shows an equivalence in which a transformer, of turns ratio, t, is preceded by a series impedance, z. This is equivalent to the π shown, wherein

$$Z_1 = \frac{tz}{(t-1)}$$

$$Z_2 = tz$$

$$Z_3 = \frac{-t^2z}{(t-1)}$$

Figure 10 shows a circuit in which the element preceding the transformer is a shunt element. The conditions for equivalence are

$$Z_1 = -(t-1)z$$

$$Z_2 = tz$$

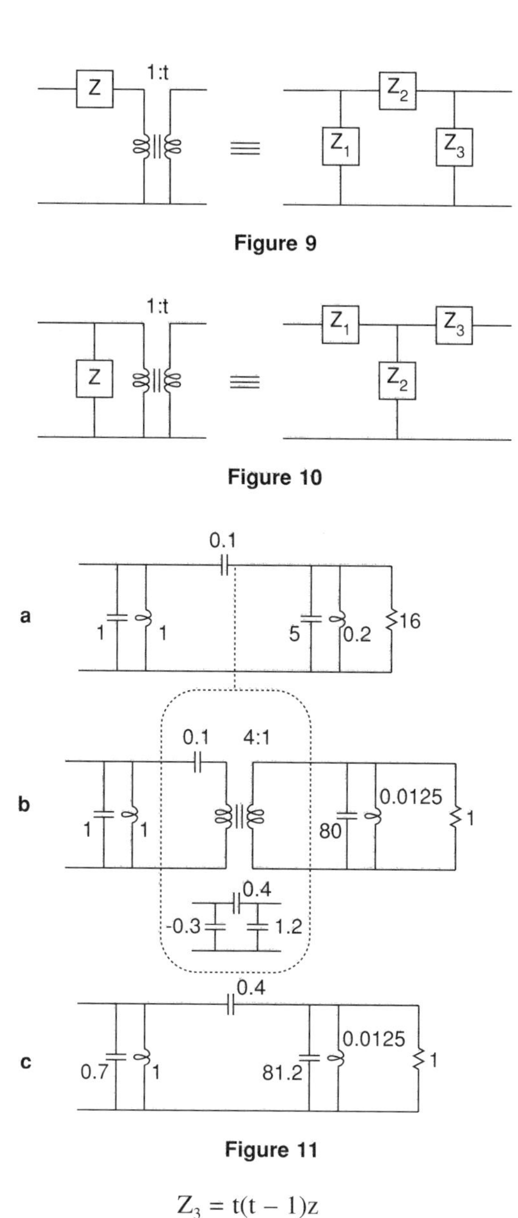

Figure 9

Figure 10

Figure 11

$$Z_3 = t(t - 1)z$$

A numerical example is shown in Figure 11, where (a) is the original circuit, (b) is the Norton equivalence, and (c) is the final circuit, in which we have substituted the Norton circuit into the original circuit. The transformer, of course, is inserted into the original circuit at the dashed line. Because the Norton equivalent has both series and shunt elements of the same kind, it is realizable only in bandpass circuits.

Important note: Norton transformers are vital in transforming bandpass filters with infeasible element values into practical circuits.

An important application of a Norton transformation is shown in Figure 12, in which an embarrassingly large series coil is reduced to a low value. The reader should note that one of the commonest uses of Norton transformers is to reduce the inductance spread.

2.6 APPROXIMATE END MATCHING

It often happens that a bandpass filter will have its impedance level (say, 50 Ω) raised, or lowered, for the purpose of using reasonable element values. We will then need to find a really practical transformer that will match 50 Ω at the ends, and simultaneously keep the return loss

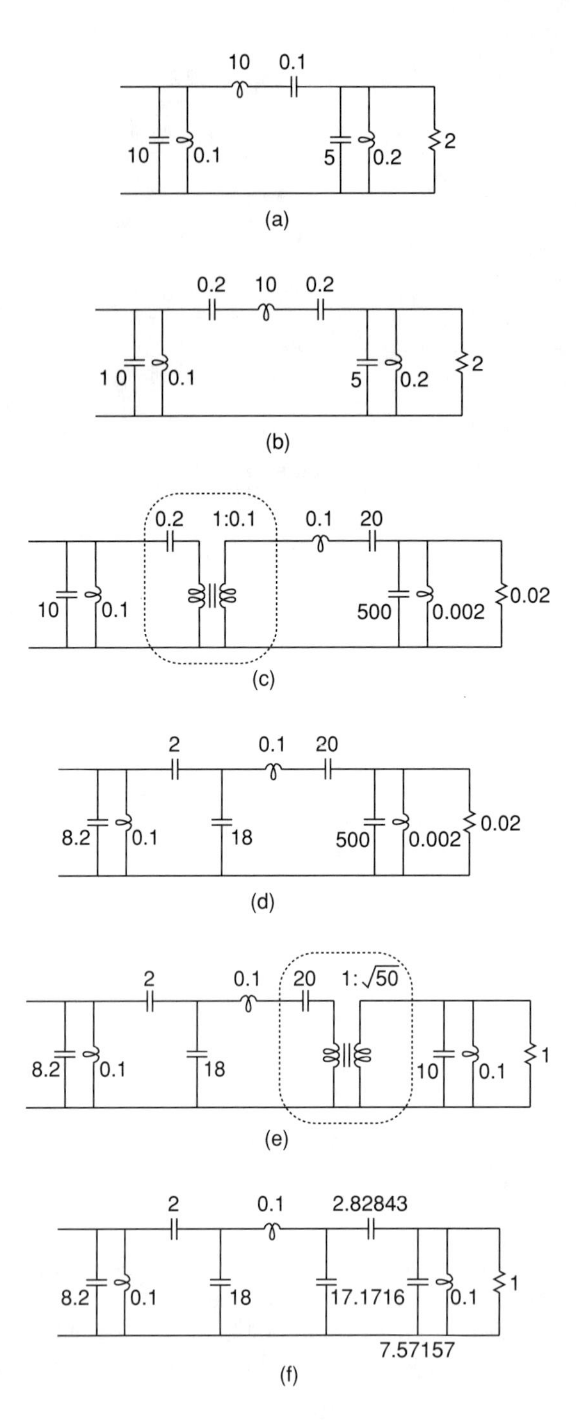

Figure 12

high in the passband. There are various schemes for using "tapped" capacitors to do the matching, but they have a bad effect on the return loss, and tapped coils are infeasible at carrier frequencies.

If the band is not too wide, we may change part of an end capacitor from shunt to series, or vice versa. This results from setting the impedance of a series RC equal to the impedance of a parallel RC. In Figure 13, the condition for equivalence is

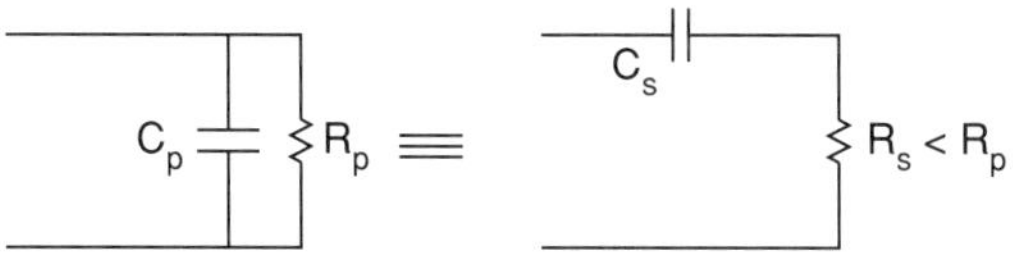

Figure 13

$$\frac{(1 + j\omega r_s c_s)}{j\omega c_s} = \frac{r_p}{(1 + j\omega r_p c_p)} \tag{14}$$

Application of this algorithm requires manipulation of Equation 14, depending on which of the quantities are given. For example, suppose we are given f_0, r_s, and $r_p > r_s$. If we need to calculate c_p and c_s, we easily obtain:

$$\omega = 2\pi f_0$$

$$c_p = \frac{\sqrt{\frac{r_p}{r_s - 1}}}{r_p \omega}$$

$$c_s = \frac{\left(1 + (r_p c_p \omega)^2\right)}{(c_p r_p^2 \omega_p^2)} \tag{15}$$

Notice that the resistors change here, so the algorithm can be used to match ends to terminations, which makes the circuit an approximate transformer. This algorithm is an example of a narrowband transformation. The equivalence is exact only at f_0, but is approximately true at nearby frequencies.

Keep in mind that this transformer preserves return loss only over a narrow range of frequencies and impedances. To extend its usefulness to wider ranges of impedances, say, from 600 to 50 Ω, we may use the following strategy.

First, we step the impedance down by only a small amount — say, 5% (from 600 to 570 Ω). This gives an extra capacitor at the ends and causes only a small disturbance in the return loss. The circuit is now suitable for a Norton transformation, which will take us down the rest of the way to 50 Ω and will cause no further change in the return loss. The cost here is a mere pair of capacitors at each end, and yet it has the very important merit, in many practical cases, of letting us use the highest Q coils that are available to us.

Note that this narrowband transformation gives us an end match which transforms the impedance down from an end shunt capacitor, or up from an end series capacitor.

3. REACTANCE TRANSFORMATIONS

3.1 INTRODUCTORY NOTE

Reactance transformations are used to transform a circuit so that its frequency response (phase, attenuation) will be transformed in a specified way. Proofs of the properties stated below consist merely of elementary circuit theory.

3.2 BASIC LOWPASS FILTERS

Consider a lowpass filter which is normalized for a band edge at 1 rad/s, and working between 1-Ω terminations (we call this a "prototype" filter). By scaling it in both frequency and impedance, we can obtain an unlimited number of practical lowpass filters from it.

Frequency scaling: To multiply the band edge frequency by any number, K we divide all reactances by K and leave the resistances unaltered.

Impedance scaling: To multiply all impedances of the circuit by K and leave the frequency response unchanged, we multiply all resistance and inductance values by K and divide all capacitances by K.

Thus, one lowpass prototype can lead to many lowpass filters simply by choosing the appropriate scale factors for frequency and impedance. It should be noted that scaling applies to any filter type and not merely to lowpass filters. For other types, the rule for frequency scaling should be rephrased:

Frequency scaling: If we divide all filter reactances by K, then any plot of the filter response will have the frequency axis numbers all multiplied by K.

3.3 HIGHPASS FILTERS

Any lowpass prototype can be transformed into a normalized highpass filter by reciprocation. The precise effect on the frequency response is

The highpass response at ω is equal to the lowpass response at $1/\omega$. Thus, the stopband lies between $\omega = 0$ and $\omega = 1$ and the passband goes from $\omega = 1$ to $w = \infty$.

Of course, the normalized highpass may then be made into a practical filter by using impedance and frequency scaling. We will consider design examples in Section 5.

3.4 BANDPASS FILTERS

If we use p as the complex frequency variable, then the circuit impedances, z, and admittances, y, can be expressed in terms of p:

$$\text{an inductor of L henrys} \rightarrow z = Lp$$

$$\text{a capacitor of C farads} \rightarrow y = \frac{1}{Cp}$$

By combining these in the familiar series, parallel manner, we can get the impedance of very large filter networks, since the branch impedances of ladder filters are all series/parallel circuits.

Now, suppose that, in the above expressions, we transform the frequency variable:

$$p \rightarrow Q_c\left(\frac{p+1}{p}\right) \tag{16}$$

We will easily find that the following transform pairs result:

.....L..... → .....L′...C′.....

where $L' = Q_c L$ and $C' = 1/L'$.

.....C..... → ....C′........
. .
.....L′....

where $C' = Q_c C$ and $L' = 1/C'$.

To find the effect of this transformation on the frequency response, we can simply make the same substitution into the transfer function of the lowpass prototype. It is easy to show that the results can be described as follows:

1. All the lowpass elements yield LC pairs that resonate at $\omega = 1$.

2. Any point of the lowpass response is transformed into a pair of points of the bandpass filter. The frequencies of the pair of points are reciprocals. This means that, after frequency scaling we can write

$$f_0 = \sqrt{f_1 f_2}$$

where f_1 and f_2 are the scaled frequencies of the transforms of a single LP point, and f_0 was scaled from $\omega = 1$. This effect tells us that the BP filter has geometric-mean symmetry, which is a very useful thing to know. Thus, let us plot the BP response on log-linear graph paper and draw a vertical line at f_0. Now, we can fold the paper along the f_0 line, and the two halves of the frequency response will coincide.

3. The bandwidth of the LP prototype is unity (by definition). Then the bandwidth of the BP filter is the distance between the two frequencies that were transformed from $\omega = 1$. This bandwidth is given by

$$B_\omega = \frac{f_0}{Q_c} \tag{17}$$

Alternatively, we can write Q_c (which we call the "circuit Q") as

$$Q_c = \frac{f_0}{(f_2 - f_1)} \tag{18}$$

4. The band edges of the bandpass filter will have the same attenuation that the lowpass filter has at $\omega = 1$. The band edge frequencies (before scaling) are given by

$$b = \frac{1}{Q_c}$$

$$\omega_1, \omega_2 = \frac{\left(\sqrt{4 + b^2} \pm b\right)}{2} \tag{19}$$

Example: Let the prototype have 3 dB at $\omega = 1$, and let the 3 dB bandwidth of the bandpass filter be 20% of the center frequency, i.e., let $Q_c = 5$. Then the normalized BP has band edges at

$$\omega_1, \omega_2 = \frac{\left(\sqrt{4.04} - 0.2\right)}{2}, \frac{\sqrt{4.04} + 0.2}{2}$$

$$= .904988,\ 1.104988$$

Checking this calculation, we find that $\omega_1\omega_2 = 1$ and $\omega_2 - \omega_1 = 0.2$.

5. Suppose that the LP elements are lossy. More specifically, let inductors have a resistor in series and capacitors have a resistor in parallel. The element Qs are then the ratio of reactance to resistance, i.e.,

$$Q_{ind} = \frac{\omega L}{r} \tag{20}$$

$$Q_{cap} = \frac{1}{\omega Cr} \tag{21}$$

Now, let the element Qs of the prototype all have the same value, Q_p, and let the branch Q of the BP filter be Q_b. If the capacitor Qs are very high and the coil Qs are low, we usually neglect the capacitor Q, so, in this case, Q_b would be the Q of the inductors. When we perform the LP-to-BP transformation on this lossy prototype, the resulting network is not the true transform of the prototype unless it meets the Q condition:

$$Q_b = Q_p Q_c \tag{22}$$

This important relation makes it possible for us to exactly predict the performance of a bandpass filter when it is built with practical elements.

6. *Important note:* The LP-to-BP transformation permits the exact BP phase to be calculated from the phase of the LP filter, but the stretching of the frequency axis causes the group delay to be distorted. Consequently, a constant delay prototype yields a BP filter having constant delay only over a narrow band centered at f_0.

3.5 BANDSTOP FILTERS

One way of looking at the LP-to-BP transformation is to consider that whatever happens in the low-frequency range of [0,1] is made to occur in the range $[f_1,f_2]$. If we start with a passband in [0,1], then we get a passband in $[f_1,f_2]$. But if we start with a stopband in [0,1], then we will get a stopband in $[f_1,f_2]$. So, if we do the LP-to-BP transformation on a highpass filter, the result will be a bandstop filter.

4. PROTOTYPE ELEMENT VALUES

4.1 ABOUT THE TABLES

Filter tables give element values for prototype filters of various types. For many years, we have resorted to them for design of Chebyshev, elliptic, and Bessel filters. Today, we mostly use computer programs running on PCs to get these numbers. In this handbook we give you:

1. Formulas for Butterworth prototypes that can be used on any scientific calculator, and the short Table 1 for use in the design examples.
2. Table 2 of Bessel prototypes.
3. Formulas for Chebyshev prototypes. These can be used on a programmable calculator. We also give the short Table 3 for degrees 2 through 5.
4. For elliptic prototypes, we give two short tables of degrees 3 (Table 4) and 4 (Table 5), to use for examples. We also refer you to a most excellent paper by Amstutz.[7] Amstutz gives listings of computer programs that design both even-degree and odd-degree prototypes.
5. In the Filter Utilities diskette[8] that you can buy to do the calculations in this chapter (see references), we include MS-DOS computer programs that calculate Butterworth/Chebyshev and elliptic prototypes to high degrees.

Be sure to remember that every prototype has a dual, which will serve as another prototype. If one of them has more coils than capacitors, then its dual has fewer coils.

4.2 BUTTERWORTH

For a prototype of degree n, the i-th element is

Table 1 Butterworth Prototypes

n	×1	×2	×3	×4	×5
2	1.4142	1.4142			
3	1.0000	2.0000	1.0000		
4	0.7654	1.8478	1.8478	0.7654	
5	0.6180	1.6180	2.0000	1.6180	0.6180

Table 2 Bessel Prototypes

n	×1	×2	×3	×4	×5	×6	×7	×8
2	2.1478	0.5755						
3	2.2034	0.9705	0.3374					
4	2.2404	1.0815	0.6725	0.2334				
5	2.2582	1.1110	0.8040	0.5072	0.1743			
6	2.2645	1.1126	0.8538	0.6392	0.4002	0.1365		
7	2.2659	1.1052	0.8690	0.7020	0.5249	0.3259	0.1106	
8	2.2656	1.0956	0.8695	0.7303	0.5936	0.4409	0.2719	0.0919

Table 3 Chebyshev Prototypes for Ap = 0.1, w3 = 1

N	C1	L2	C3	L4	C5	RL
2	0.7378	1.2086				0.7378
3	1.4328	1.5937	1.4328			1
4	1.3451	1.5845	2.1476	0.9924		0.7378
5	1.3013	1.5559	2.2411	1.5559	1.3013	1

Table 4 Elliptic Prototypes for n = 3, Ap = 0.1

As	Ws	C1	C2	L2	C3
30	2.4550	0.9439	0.1257	1.0122	0.9439
35	2.9331	0.9712	0.08395	1.0543	0.9712
40	3.5195	0.9901	0.05647	1.0835	0.9901
45	4.2359	1.0032	0.03814	1.1036	1.0032
50	5.1087	1.0122	0.02584	1.1175	1.0122

Table 5 Elliptic Prototypes for n = 4, Ap = 0.1, Type C (RL = 1)

As	Ws	C1	C2	L2	C3	L4
30	1.7178	0.7149	0.2647	1.0778	1.2977	0.9348
35	1.9498	0.7666	0.1916	1.1617	1.3292	0.9340
40	2.2123	0.8062	0.1402	1.2270	1.3540	0.9332
45	2.5209	0.8364	0.1033	1.2772	1.3733	0.9324
50	2.8818	0.8593	0.07651	1.3156	1.3881	0.9318

$$X_i = 2\sin\left(\frac{(2i-1)\pi}{2n}\right) \tag{23}$$

The element values are symmetrical, as can be seen from an inspection of Table 1. For all degrees, the attenuation at $\omega = 1$ is 3.0103 dB, so the power transmitted at $\omega = 1$ is exactly half the power transmitted at 0 frequency.

4.3 CHEBYSHEV

Let the prototype element values be $x_1, x_2, \ldots x_n$. Choose the degree, n, and the passband ripple in decibels, A_p. Then the following pseudocode, based on the Orchard formulas,[6] gives an efficient calculation of the element values:

$$\exp = \frac{A_p}{10}$$

$$e = \sqrt{10^{\exp} - 1}$$

$$z = \frac{1}{e}$$

$$p = z + \sqrt{(z^2 + 1)}$$

$$s = \frac{(p^{1/n} - p^{-1/n})}{2}$$

$$\text{Initial values: } a = 1,\ b = s,\ y = 2,\ t = \frac{\pi}{(2n)},\ d = \frac{-\pi}{n},\ d_1 = \frac{d}{2}$$

for $k = 1$ to n

$$t = t + d$$

$$f = t + d_1$$

$$a_1 = \sin(t)$$

$$x(k) = \frac{4a^2{}_1}{by}$$

$$y = x(k)$$

$$a = a_1$$

$$b = s^2 + \sin^2(f)$$

next k

For even n, the output load is RL:

$$h = \sqrt{e^2 + 1}$$

$$RL = 2h^2 - 2h\sqrt{h^2 - 1} - 1$$

x(k) is a shunt capacitor for k odd, and a series inductor for k even. The above formulas give designs normalized for the edge of the ripple band at $\omega = 1$. To normalize for 3 dB at $\omega = 1$, multiply all element values by ω_3:

$$z = \frac{1}{e}$$

$$t = \frac{\log\left(2 + \sqrt{(2^2 - 1)}\right)}{n}$$

$$\omega^3 = \frac{(e^{+} + e^{-+})}{2}$$

These formulas give designs for terminations having a minimum mismatch loss: they are 1:1 for odd n, and as near 1:1 as possible for even n.

The Filter Utilities diskette[8] contains a program that calculates Butterworth and Chebyshev prototypes using the Takahasi algorithms, which give the element values for any load-to-source ratio that is realizable.

4.4 BESSEL

Table 2 gives Bessel prototypes having 3 dB at w = 1, for degrees 2 to 8. Higher degrees have little utility.

4.5 ELLIPTIC

Table 4 gives prototypes for n = 3, $A_p = 0.1$ dB. Here, A_s is in decibels and ω_s is the edge of the stopband in radians per second. Even degree prototypes are given in Table 5, for n = 4. Elliptic filters may be of types A, B, or C. Type A is not ladder realizable, type B has unequal terminations, and type C has equal terminations, but is less selective than type B; i.e., the ω_s values are smaller.

5. EXAMPLES

5.1 AN ELLIPTIC LOWPASS

Let it be required to design a 50-Ω lowpass filter with the edge of the 0.1-dB ripple band at 800 MHz, and 35 dB attenuation at 1600 MHz. Looking over the elliptic designs available (Table 5), we see that a four–pole gives 35 dB at 1.9498 times the band edge. This gives a margin of 2/1.9498 = 1.0257, or about 2.6%. We prefer to divide the margin equally between pass- and stopbands, so we set the cutoff frequency to 800 MHz * 1.013 = 810.4 MHz. The denormalizing factors are then calculated:

$$w_c = 2\pi\left(810.4 \times 10^6\right) = 5.09189e^9$$

$$r = 50$$

$$L' = \left(\frac{r}{\omega_c}\right)L = \left(9.81954 \times 10^{-9}\right)L$$

$$C' = \frac{C}{r\omega_c} = \left(3.92781 \times 10^{-12}\right)C$$

The result is shown in Figure 14 with the normalized values (from the table) shown in parentheses.

5.2 AN ELLIPTIC HIGHPASS

A high-power public address system needs to limit the low-frequency audio power below 250 Hz, so we specify a three–pole elliptic highpass with a 40-dB stopband, and the edge of the 0.1-dB ripple band at 250 Hz. We take the prototype from Table 4. After denormalization for a 600-Ω line (Figure 15a), we notice the large tuning capacitor (for the response notch near 61 Hz), and so we perform a T–to–π transformation on the capacitors. This gives Figure 15b.

5.3 AN ELLIPTIC BANDPASS

We design a 47 to 53 kHz bandpass filter based on the three-pole elliptic prototype with 0.1-dB passband ripples, and a 50-dB stopband, in Table 4.

$$b = 6.0 \text{ kHz}$$

$$f_0 = \sqrt{47 \times 10^3 \times 53 \times 10^3} = 49.9099 \text{ kHz}$$

$$Q_c = \frac{f_0}{b} = 8.31832$$

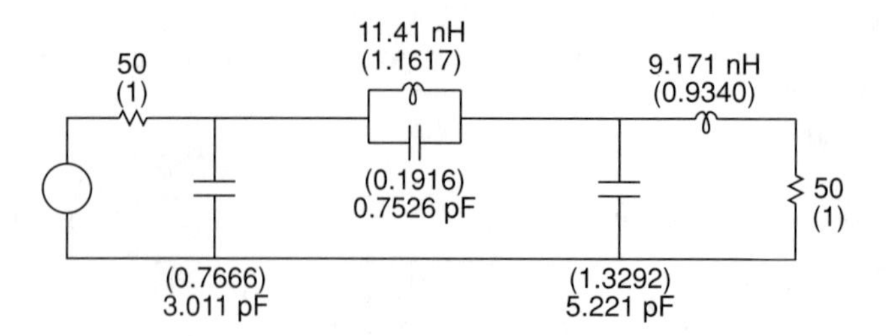

Figure 14

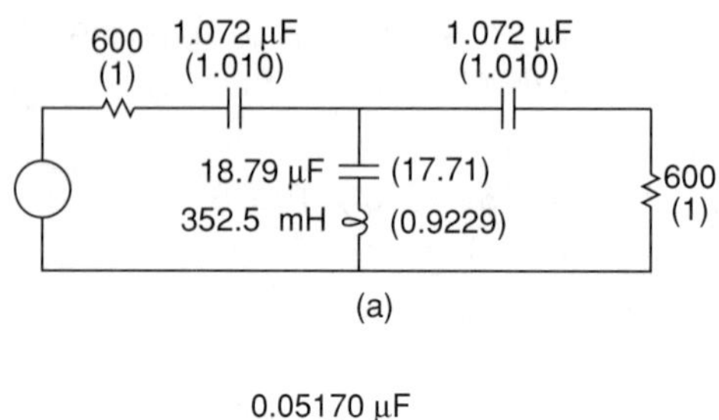

(a)

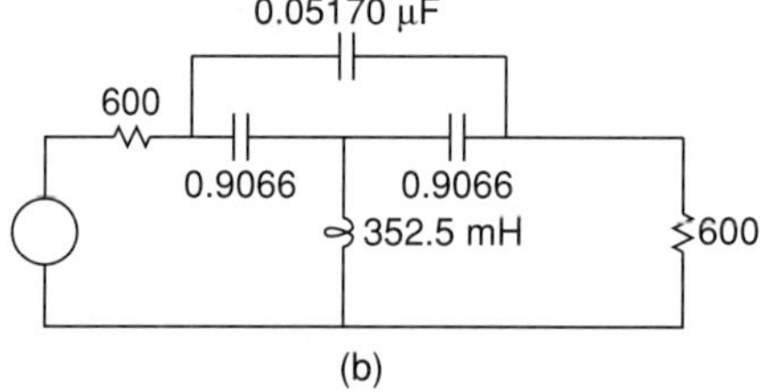

(b)

Figure 15

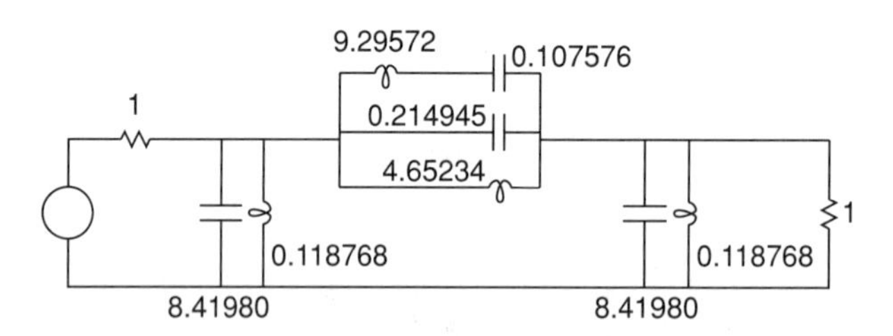

Figure 16

The direct result (normalized) of the lowpass–to–bandpass transformation is shown in Figure 16. Note that the series branch has an inductance of 9.3 H, while the shunt branches have only 119 mH. This branch can be improved with a dipole transformation, leading to a pair of antiresonances in series. We prefer to show you a modified LP–to–BP transformation which takes you directly to the better branch configuration from the prototype. This modified branch becomes mandatory for narrow bands (but gives worse element spread for wide bands).

Figure 17 shows two branch configurations that may be found in a lowpass prototype. Whichever one is used, the formulas to be applied are the same, as shown in the figure. Using these formulas here gives the normalized bandpass schematic of Figure 18.

5.4 A 70-MHz IF FILTER

While Butterworth filters have less selectivity than Chebyshev filters, they also have less delay distortion, and so may sometimes be chosen when economical delay equalization is needed. Let the band edges be at 65 and 75 MHz, and design a three-pole Butterworth for 75 Ω.

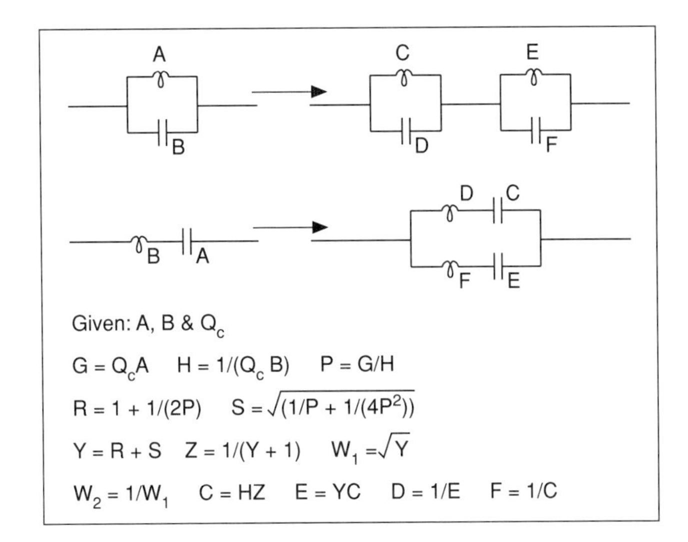

Figure 17

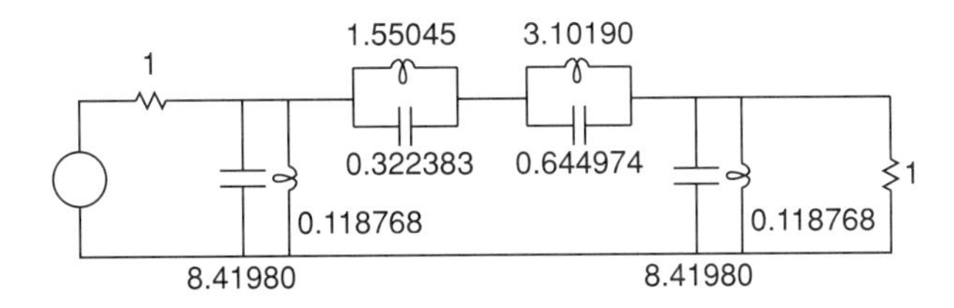

Figure 18

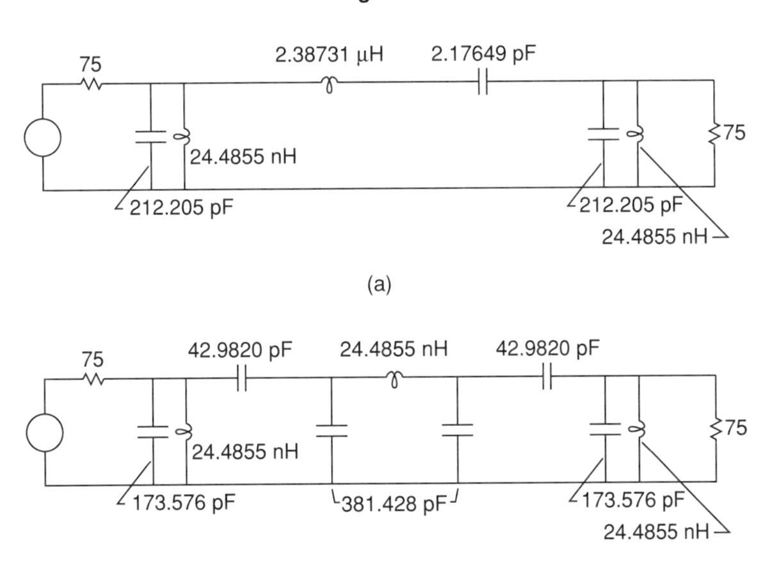

Figure 19

Applying the LP–to–BP transformation to the prototype, we have the denormalized design shown in Figure 19a. Using the technique shown in Figure 12, we may make all the coils equal. The result is displayed in Figure 19b. The coil values here are rather small, but can be improved by a narrowband transformation followed by a Norton transformation, as described in Section 2.6, and applied in the next section below.

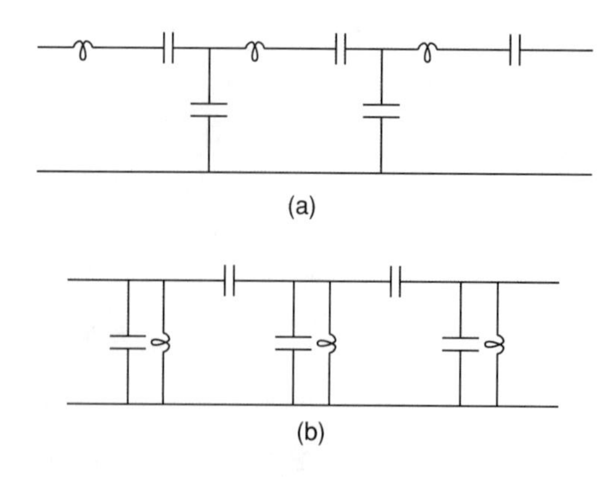

Figure 20

The rather tedious calculations of this section can all be done very easily by computer programs included in the Filter Utilities diskette described in the references.

6. COUPLED RESONATOR FILTERS

6.1 INTRODUCTION

A coupled resonator filter consists of resonant or antiresonant circuits coupled with single elements, as shown in Figure 20. This type of structure makes it possible for us to avoid one of the worst consequences of the LP-to-BP transformation:

> In bandpass filters, the series coils become large as the passband narrows, and the shunt coils become small. The inductance spread increases inversely as the square of the bandwidth.

This effect is especially difficult to accommodate at high RF frequencies, as the air core coils that are used (above about 100 MHz) have good Q only over a narrow range of inductance. This implies, of course, that the coupling elements should be capacitors only. In principle, design of a coupled resonator filter proceeds in two steps:

1. Design a resonant branch BP filter with the LP-to-BP transformation.
2. Use immittance inverters to transform the design into a coupled resonator structure.

6.2 IMMITTANCE INVERTERS

Figure 21 shows that an impedance, z, as seen through an inverter having a conversion constant of i, appears as its dual, multiplied by i. Numerical examples are shown in Figure 21b and c.

Now without considering at the moment how an inverter may be realized, we show a three-pole Butterworth prototype in Figure 22a and a normalized resonant branch BP filter in Figure 22b. This filter was obtained by using the LP-to-BP transformation on the prototype.

Note that the 2% bandwidth results in an inductance spread of 5000. In Figure 22c, we use inverters with inversion constants of unity to get a filter with an inductance spread of 2.0. In Figure 22d, we see that a constant of 2.0 makes all the coils equal!

6.3 IDEAL INVERTERS

In Figure 23, we show pure reactances, j, permitting them to be constants that do not vary with frequency. Of course, such constants do not exist in nature, but we regard them as useful mathematical abstractions. If we substitute any of these inverters in Figure 21a and apply some network analysis, we will find that they do, indeed, have the inversion property. Now, in Figure 24, we see two practical inverters which are approximations to those in Figure 23. These have

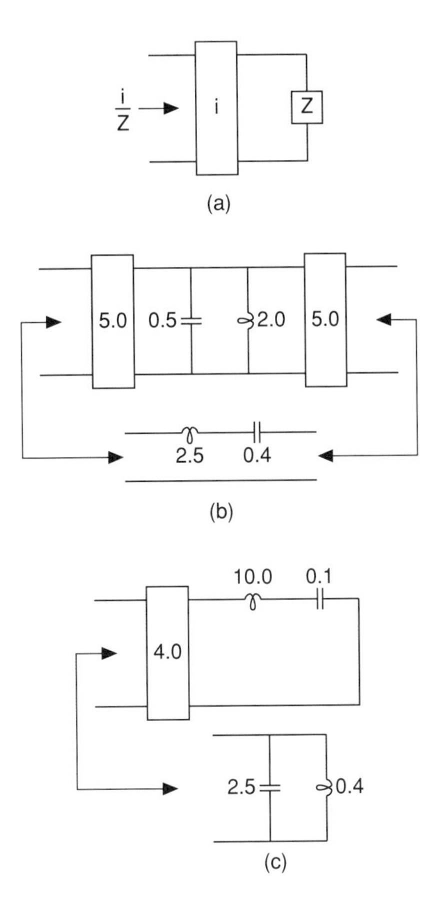

Figure 21

the exact inversion property only at ω = 1. However, since they change reactance slowly with frequency, they will do approximate immittance conversion over a narrow, but useful, bandwidth. The conversion constant is calculated from the inverter element values and is unity in these examples.

The negative capacitors must, of course, be inserted into a physical circuit that has adjacent positive capacitors which are large enough to absorb the negative capacitance.

6.4 PRACTICAL DESIGN

The concepts described above are sufficient to do rigorous design; so, without going into all the theoretical details, we now describe a straightforward design procedure for practical filters. The procedure is detailed in Figure 25.

In Figure 25a, we see the LP prototype, which may be of either even or odd degree, as shown. Now we draw the schematic with ideal inverters, as in Figure 25b. After this, we do the work in Figure 25c. One value of CS and one value of LS will do for all the shunt resonators. (The resonators are all identical.) Then calculate RL, and CC(k) for k = 1…n − 1.

Next, we decide whether we want all the coils equal or all the shunt capacitors equal. Usually, if the coils are adjustable, then we would want the capacitors (which may be purchased parts) to have as few values as possible, so we choose equal shunt capacitors. Similarly, the use of trim capacitors would motivate us to choose equal coils. This decision tells us which set of inverters to take in Figure 25. Of course, the shunt elements of the inverters must then be combined with the adjacent elements of the resonators.

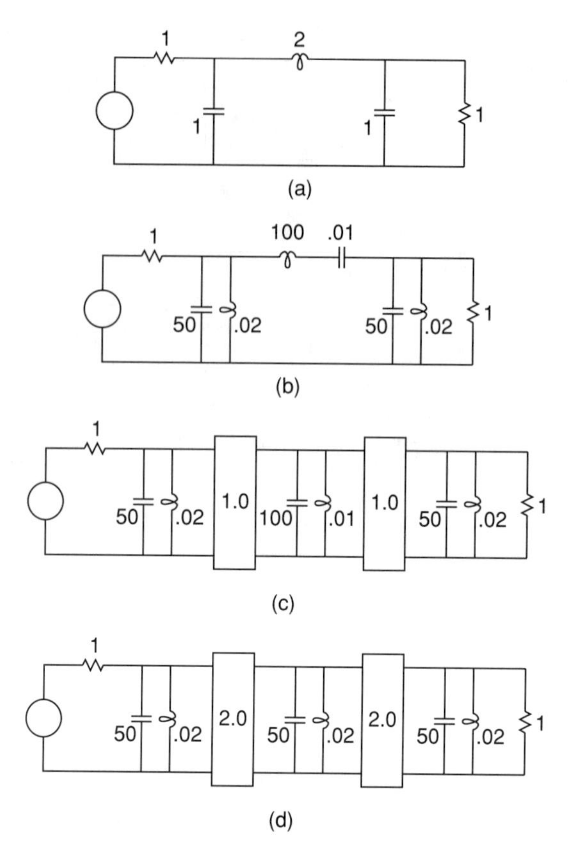

Figure 22

6.5 CIRCUIT EQUIVALENCES

It is worth noting that the two normalized resonator filters of Figure 20 can be obtained from one another by taking the dual of the reciprocal network. If one takes only the dual, or only the reciprocal, then the result is a valid design for inductive coupling.

6.6 EXAMPLE

Let us design a BP filter with three resonators, top-c coupled, at 5 MHz with $Q_c = 10$, working between 50-Ω terminations, with all coils equal. The prototype we use here is the three-pole Butterworth. Referring to Figure 25, the normalized design has CS = $Q_c * x_1 = 10$, LS = 0.1, and RS = RL = 1. Also CC(1) = CC(2) = 0.707107. The normalized BP filter is shown in Figure 26. Scaling to 5 MHz and 50 Ω, we have the denormalizing factors

$$
\begin{aligned}
L' &= \left(\frac{R_S}{\omega_0}\right)L = \left(1.59155 \times 10^{-6}\right)L \\
C' &= \frac{C}{\left(R_S\omega_0\right)} = \left(636.620 \times 10^{-6}\right)C
\end{aligned}
\tag{24}
$$

which give us the final design of Figure 27.

6.7 USING OPTIMUM INDUCTORS

One important problem in designing practical filters is to make use of those inductance values that have the best Q. Since the Q of physical coils varies considerably with inductance (especially at carrier frequencies), we want to make use of a narrow range of values, and it often happens that

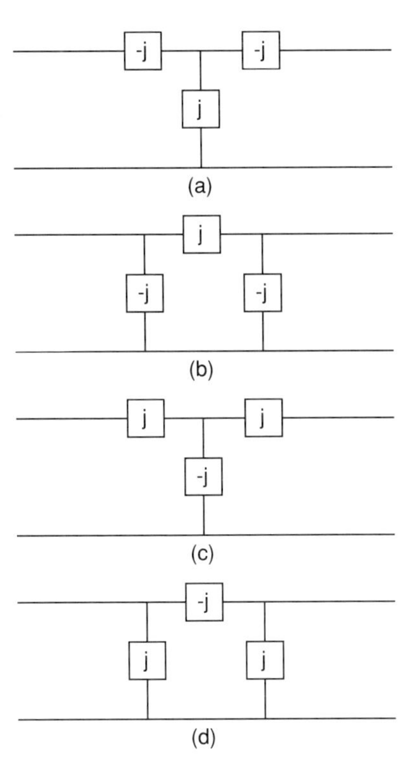

Figure 23

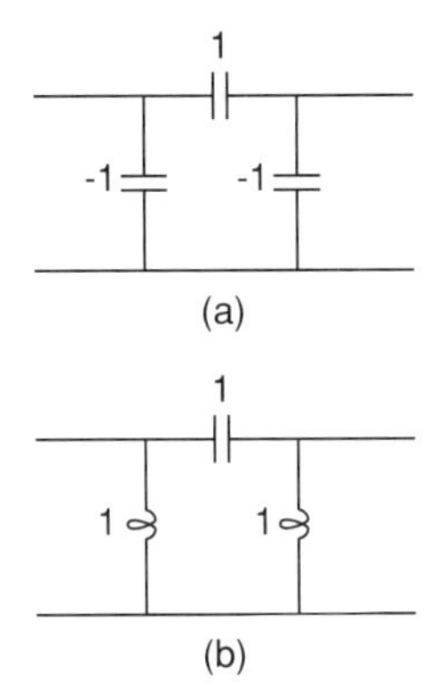

Figure 24

the numerical design is uncooperative. In the case of coupled resonators, we have a very useful technique to deal with this problem.

First, take note of the fact that top-c-coupled filters tend to have very small inductance values, so, basically, we need a design artifice that will increase the values required by the design. We begin by choosing a coil value, L_0, that has good Q, and we scale up the impedance of the filter so as to use this value. Now the problem is to step down the filter impedance at the ends to the original resistance termination — say, 50 Ω. Since the customer usually has some specification on VSWR or return loss, the step-down method has to meet that requirement.

This is accomplished by using the RC equivalent circuit of Figure 19.

6.8 EXAMPLE

Turning again to the example of Figure 27, let us transform this circuit so that it will use coil values of 2.0 μH (i.e., more than 12 times as large as the coil in the figure). Scaling the filter impedance,

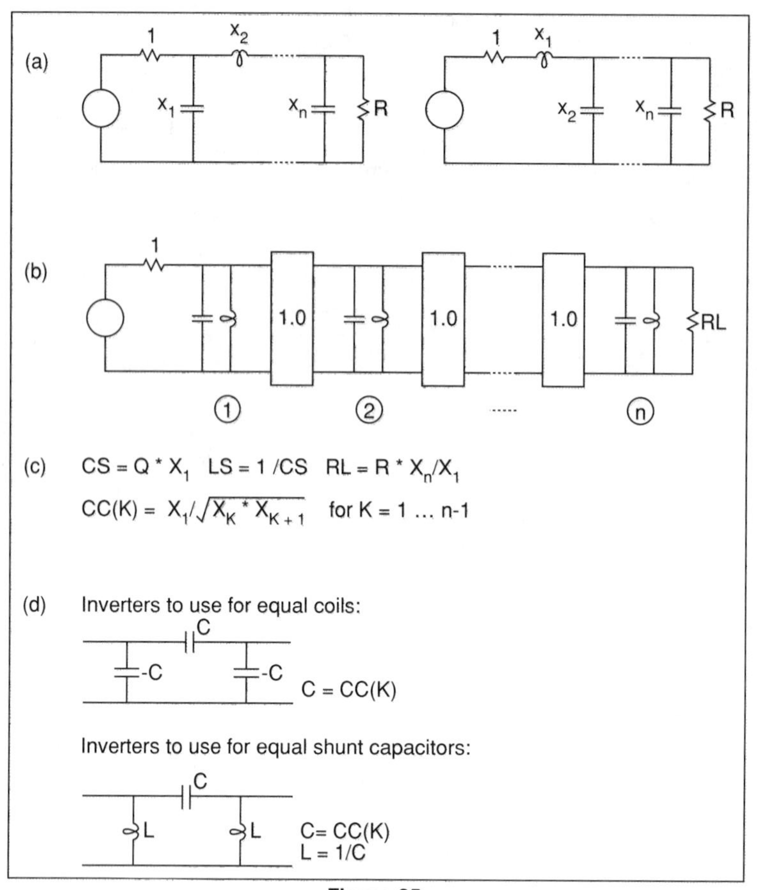

Figure 25

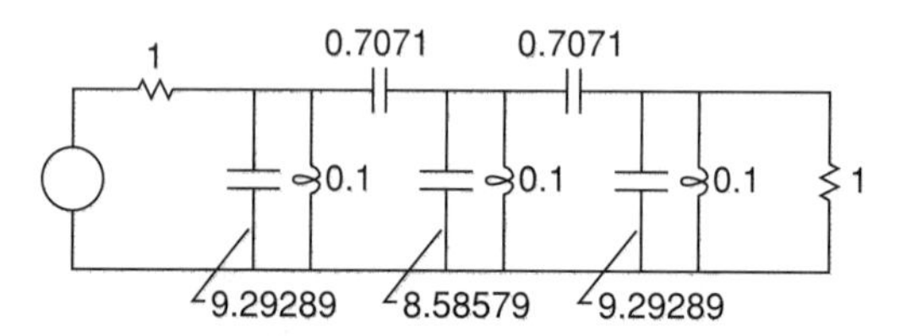

Figure 26

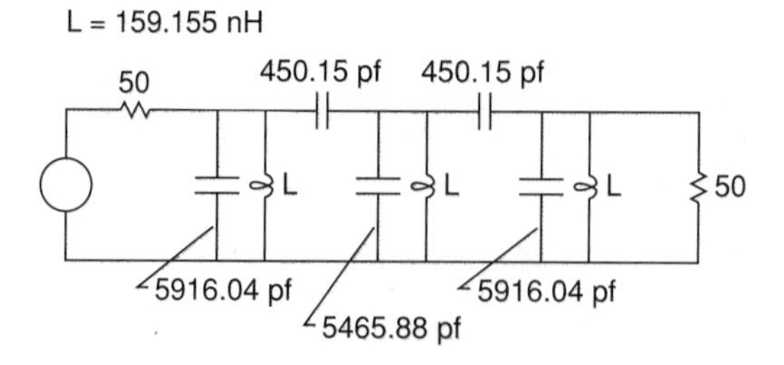

Figure 27

we obtain the schematic of Figure 28, in which the load resistors are 628.32 Ω. Now, we use the equivalence of Equation 15 to obtain the equivalent circuits of Figure 29, transforming the end down about 10% to $r_p = 565\ \Omega$. Thus

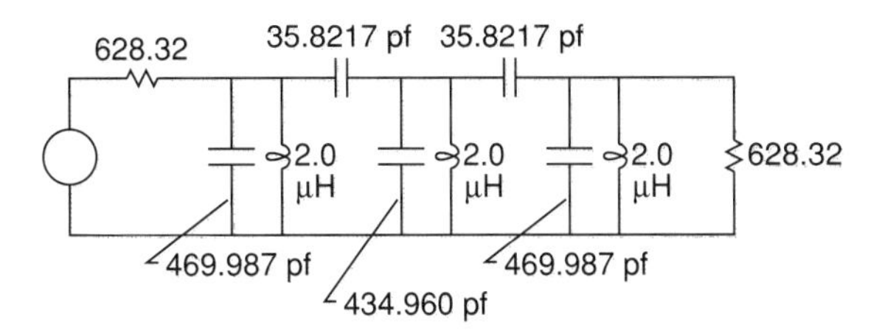

Figure 28

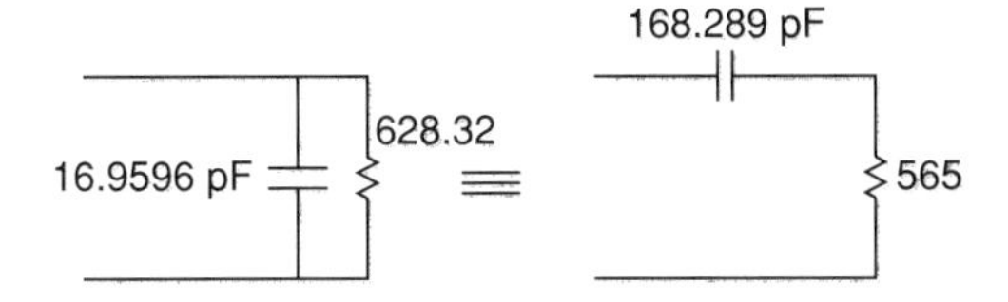

Figure 29

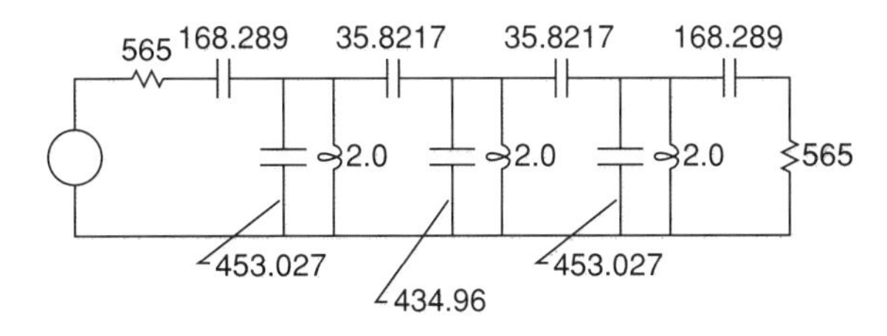

Figure 30

$$w = 2\pi \times 5 \times 10^6 = 31415926$$

$$c_p = 11.6223 \text{ pF}$$

$$c_s = 232.446 \text{ pF}$$

We subtract c_p from the output shunt capacitor and insert the values c_s and r_s into the circuit to get the schematic shown in Figure 30. By symmetry, we can change the input part of the filter in the same way, without making any more calculations.

Finally, we insert transformers next to the end capacitors to bring the load resistors all the way down to 50 Ω. The turns ratio is given by

$$t = \sqrt{\frac{50}{565}} = 0.297482$$

Replacing the series cap and transformer with the Norton equivalent gives the final circuit of Figure 31.

6.9 BOTTOM-C-COUPLED FILTERS

The configuration shown in Figure 20a has advantages which will bear some discussion. To begin, we note that system engineers usually want RF bandpass filters to have arithmetic symmetry. While this is not exactly achievable, a well-known rule of thumb is quite useful:

> In an all-pole BP filter, good arithmetic symmetry is obtained when one quarter of the zeros lie at zero frequency, with the rest at infinity.

Now, when we make a bandpass filter from a lowpass prototype of degree n, the resonant branch form has n zeros at zero, and n zeros at infinity. The top-c-coupled filter is even worse, with 1 zero at

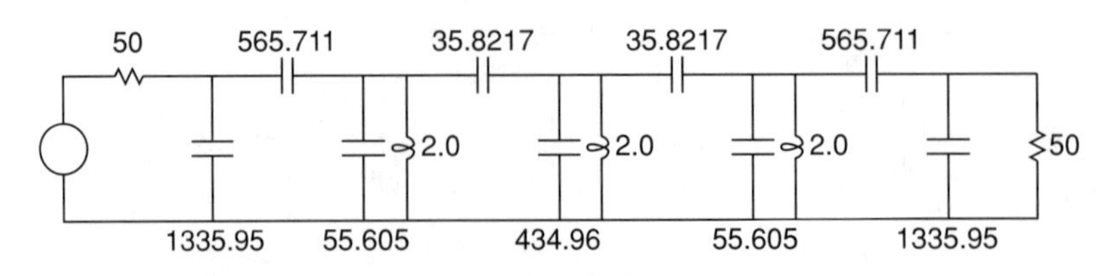

Figure 31

infinity and n – 1 zeros at the origin. On the other hand, the bottom-c-coupled filter has 1 zero at the origin and n – 1 at infinity. Thus, bottom-c coupling with n = 4 meets the condition for good symmetry. As n increases, the response departs from symmetry rather slowly, and so comes closer to symmetry than the other types. This property makes bottom-c coupling a good choice for many applications.

Another advantage of the bottom-c-coupled configuration is that it is readily transformed into the tubular configuration. This is accomplished by partitioning series capacitors, C, into the series connection of two capacitors of value, 2C. Then, by placing one of these on each side of the series coils, we find that the bandpass filter now consists of series coils coupled by T's of capacitors. We then obtain better capacitor values by transforming the T's into π's. This topology can then be realized, physically, with a cylindrical metal tube which is electrically grounded, having concentric capacitors connecting the series branches to ground. Of course, the series branches are inside the tube.

Even without the physical tube, the tubular topology is a good one for realizing miniature RF filters and so sees frequent application.

6.10 DESIGN PROCEDURE

The design calculations are closely similar to those we used for designing a top-c-coupled filter. First, we draw the schematic with ideal inverters, as in Figure 32a. Then calculating the quantities in Figure 32b, we choose inverters in Figure 32c and conclude by scaling to the user's frequency and impedance level. One important difference here is that the coils tend to be too large, so we want to scale the impedance down to a suitable low level, and then match up at the ends.

6.11 BOTTOM-C-COUPLED EXAMPLE

Again we design for 5 MHz, with 10% bandwidth and a three–pole Butterworth prototype. Doing the calculations in Figure 32, we get CS = 0.1, LS = 10, RL = 1, CC(1) = CC(2) = 1.41421, as shown in Figure 33. Scaling to 5.0 MHz and 50 Ω gives Figure 34, in which we see that we have very large coil values. A simple ratio calculation tells that we can use 2.0 μH coils if we scale the filter to 6.28318 Ω, as shown in Figure 35. The problem now is to step up to 50 Ω at the ends.

Figure 36 shows the narrowband equivalent circuits which will step up the impedance about 11%. Then Figure 37a shows the capacitance calculation needed, and Figure 37b shows the new capacitors installed in the 7 Ω filter. It is now only necessary to perform a Norton transformation to bring the terminations to 50 Ω, as in Figure 38.

The attenuation and group delay responses of this filter are shown in Figure 39.

7. ATTENUATION EQUALIZERS

7.1 INTRODUCTION

Attenuation equalizers are RLC circuits that make relatively small adjustments to a given frequency response and are mostly used for correcting errors caused by parasitic components or other effects considered undesirable.

Figure 40 shows the response of a lowpass equalizer. The attenuation is constant for low frequencies. It shows more attenuation at high frequencies, and equals a_0, the "total equalization" at infinity. In the neighborhood of f_0, we see a smooth transition between the two values of attenuation. The dotted line shows a highpass equalizer response.

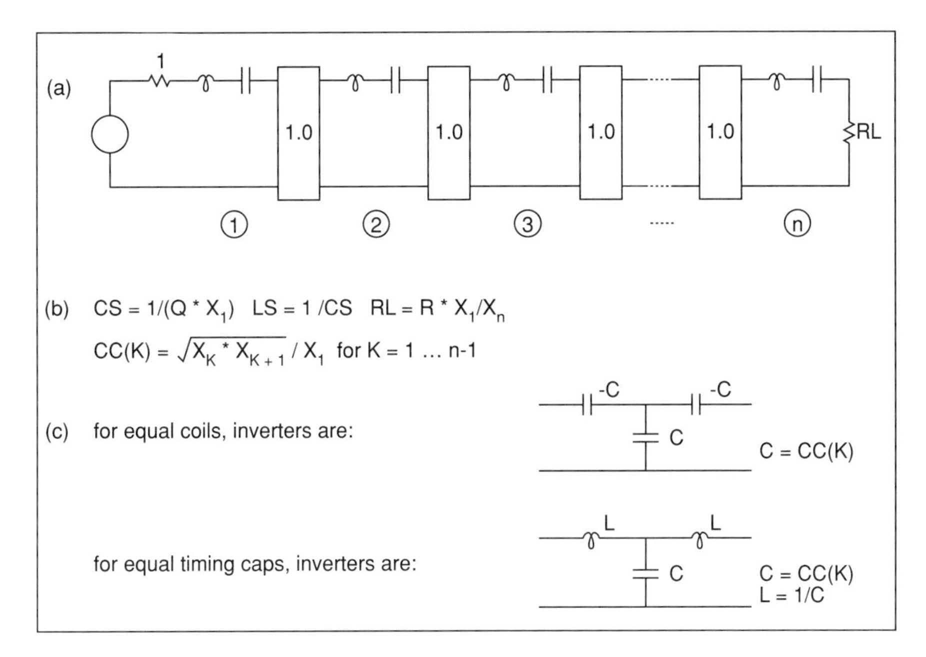

Figure 32

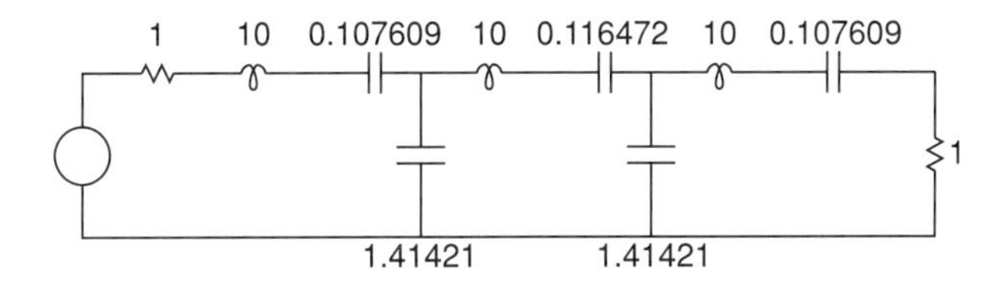

Figure 33

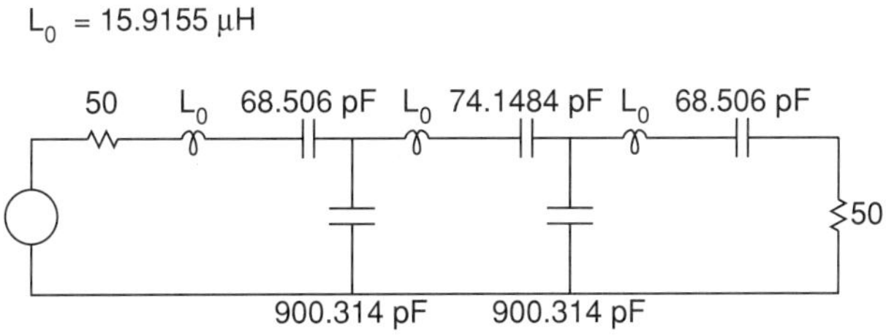

L_0 = 15.9155 µH

50 L_0 68.506 pF L_0 74.1484 pF L_0 68.506 pF

50

900.314 pF 900.314 pF

Figure 34

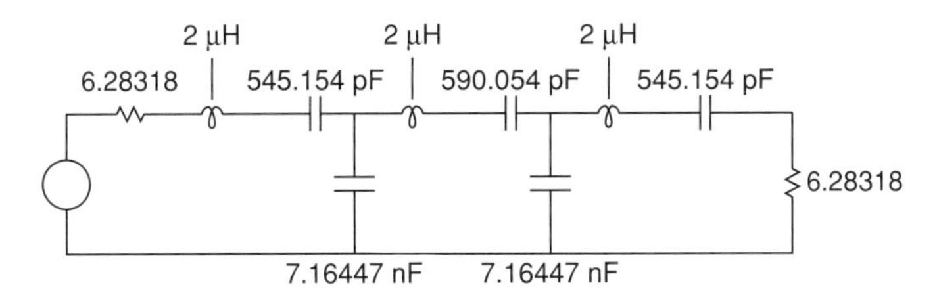

Figure 35

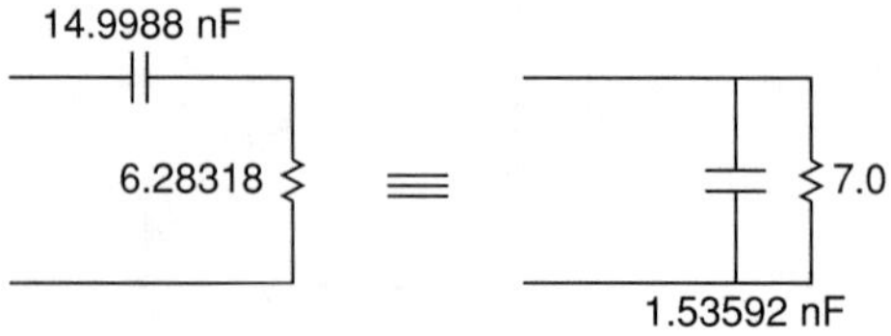

14.9988 nF

6.28318 ≡ 7.0

1.53592 nF

Figure 36

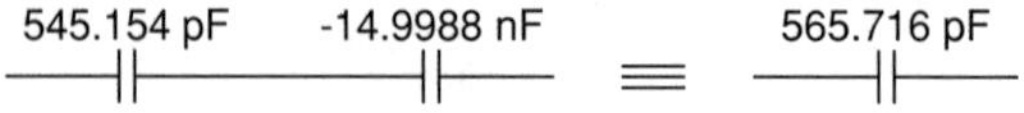

545.154 pF -14.9988 nF ≡ 565.716 pF

(a)

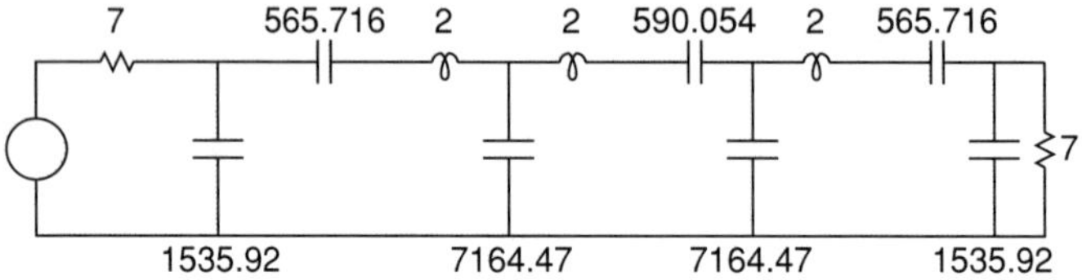

7 565.716 2 2 590.054 2 565.716

7

1535.92 7164.47 7164.47 1535.92

(b)

Figure 37

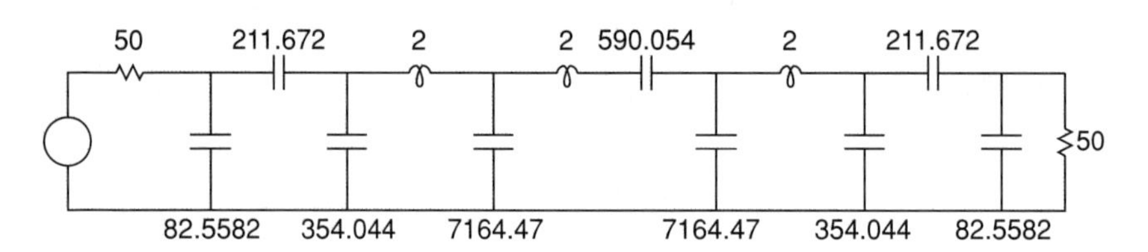

Figure 38

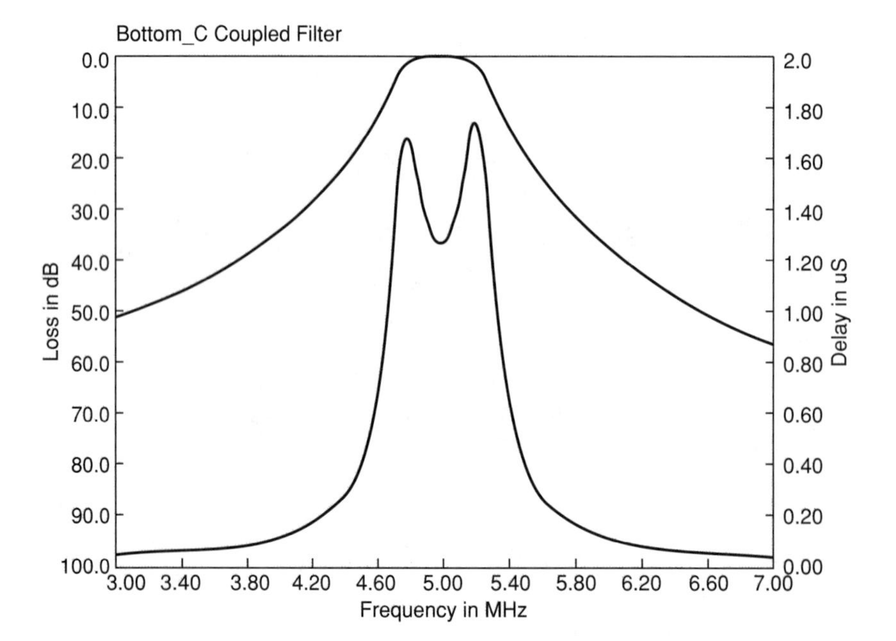

Figure 39

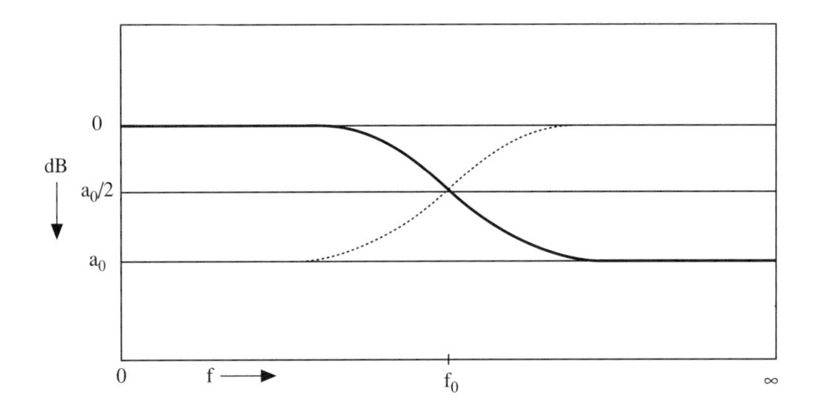

Figure 40

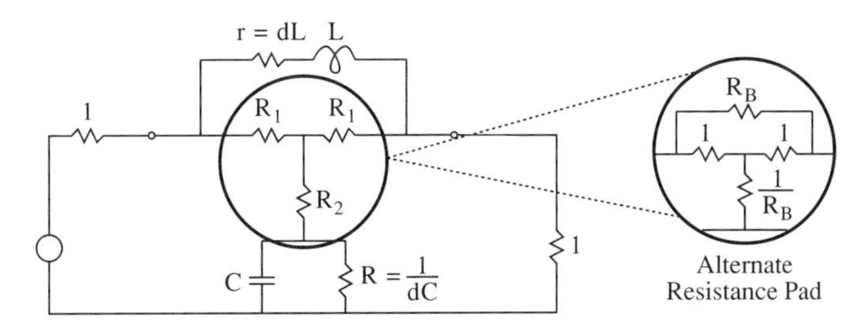

Figure 41

A normalized lowpass equalizer circuit is shown in Figure 41. The center frequency is at $\omega = 1$, between 1 Ω terminations. For lossless elements, the design equations are:

$$k = 10^{a_0/20}$$

$$L = C = \frac{K-1}{\sqrt{K}}$$

$$R1 = \frac{K-1}{K+1}$$

$$R2 = \frac{2K}{K^2-1}$$

$$Rb = K-1$$

An outstanding property of this circuit is that the terminal impedance is a constant resistance (equal to 1 Ω) at all frequencies if the circuit is properly terminated at the other end. This means that it can correct the frequency response of a filter while terminating it with the required resistance and without suffering any interaction between the two circuits. Additional equalizer sections can be cascaded without deleterious effects.

7.2 USING LOSSY ELEMENTS

Circuit analysis shows that if a constant resistance equalizer is made with all components having the same Q then it is still a constant resistance circuit. This is useful in making practical lowpass and bandpass equalizers. The losses in highpass and notch equalizers can be allowed for by removing some resistance from the bridged-T pad.

A predistorted lowpass equalizer is made using the following design formulas, in which $d = 1/Q$, and Q is the same for all L and C elements.

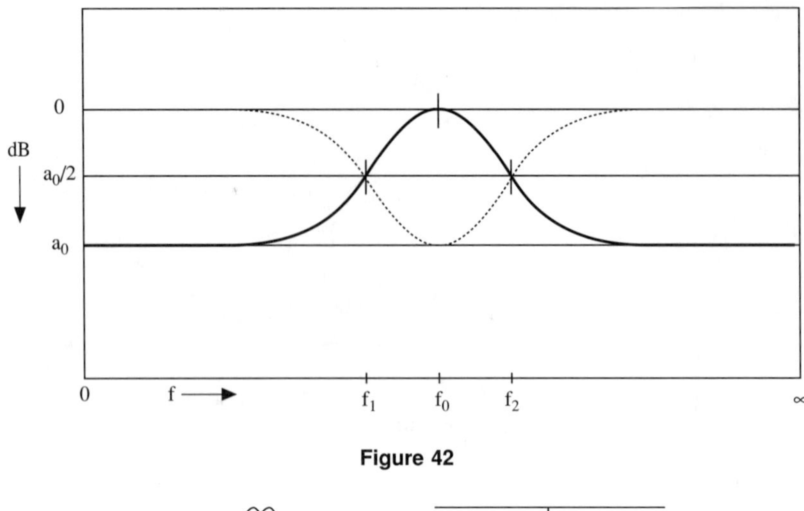

Figure 42

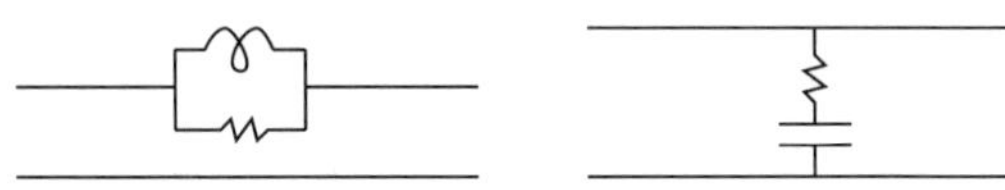

Figure 43

$$k = 10^{(a_0/20)} \quad k_1 = \sqrt{(k)} \quad k_2 = \frac{1}{k_1} \quad a = k_1 - k_2 \quad b = k_1 - d \quad e = k_2 - d$$

$$L = C = \frac{1}{be} \quad R_1 = \frac{a}{b+e} \quad R_2 = \frac{2be}{b^2 - e^2} \quad RB = \frac{a}{e}$$

$$A_f = \text{flat loss} = 20\log_{10}\left(\frac{b}{k_1 - dk}\right)$$

The predistorted design is realizable if and only if $d < k_2$. It is worth remarking that predistorted equalizers do not show the high sensitivity that we find in predistorted filters.

7.3 REACTANCE TRANSFORMATIONS

The same techniques that transformed prototype filters into highpass, bandpass, and bandstop filters can be applied to the lowpass equalizer to obtain highpass, bump, and notch equalizers. The response shown by the dotted line in Figure 40 is obtained from a circuit similar to that of Figure 41, but the elements are reciprocated.

Figure 42 shows the response of a bump section (solid line) and of a notch section (dotted line). The bump circuit is designed in the following steps:

1. Define $Q_c = f_0/(f_2 - f_1)$ where f_1 and f_2 and the half-attenuation frequencies.
2. Replace L by $L_1 = LQ_c$, and C by $C_2 = CQ_c$.
3. Connect a capcitor in series with L_1, $C_1 = 1/L_1$.
4. Connect a coil in parallel with C_2, $L_2 = 1/C_2$.
5. Scale this design, in frequency and impedance, for the desired application.

Similarly for the notch design:

1. Replace L by $C_1 = Q_c/L$, and C by $L_1 = Q_c/C$.
2. Connect an inductor across C_1, $L_1 = 1/C_1$, and connect a capacitor in series with L_2, $C_2 = 1/L_2$.
3. Scale for the application.

In the notch design, one must replace the T-pad with a bridged-T pad so that resistance will be able to accomodate the Q of the bridging elements.

7.4 NONCONSTANT RESISTANCE EQUALIZERS

Equalizers that do not have the constant resistance property are easily obtained from the equalizer that uses a bridged-T pad. Double the impedance of the bridging elements, and connect them as a ladder series branch, or halve the impedance of the shunt branch and connect the circuit as a ladder shunt branch. These circuits are shown in Figure 43.

7.5 DETERMINING Q_C

The requirements for the bump equalizer may be such that the half attenuation points are not readily known. If other points are accurately known, say F and A, we may calculate Q_c as follows:

1. $x = F/f_0$
2. $c = 10^{(A - a_0)/10}$
3. $e = \text{abs}(x - 1/x)$
4. $\omega = \sqrt{(ck - 1/k)/(1 - c)}$
5. $Q_c = \omega/e$

7.6 ATTENUATION

The attenuation of a lowpass equalizer section is given by

$$a(\omega) = a_0 - 10\log_{10}\frac{(\omega^2 + k)}{(\omega^2 + 1/k)} \quad [\text{dB}]$$

Attenuation of a bump equalizer at f is given by

1. $x = f/f_0$
2. $\omega = Q_c \, \text{abs}(x - 1/x)$
3. $b = \dfrac{\omega^2 + 1/k}{\omega^2 + k} \quad [\text{dB}]$
4. $a = 20 \log10(k\sqrt{b})$

Example: $f_0 = 10$, $f = 6$, $Q_c = 2$, $a_0 = 8$ dB (hence $k = 2.5118864$)

1. $x = 0.6$
2. $\omega = 2.1333333$
3. $b = 0.70072491$
4. $a = 6.455$ dB

The responses of highpass and notch equalizers are simply the inverted response of the corresponding lowpass and bump equalizers.

8. FILTER SYNTHESIS VS. APPROXIMATE METHODS

8.1 SYNTHETIC TECHNIQUES

In References 3, 4, and 5 the reader will find full statements of the complicated mathematical techniques which design filters for precise transfer functions. They are exact (except for the floating point representations of the numbers). For example, equiripple passbands obtained by "conventional" synthesis have ripples that are exactly equal. In the case of "parametric" synthesis, we wish to enable certain circuit configurations, so the ripples are reduced in the middle of the passband by enough to accomplish this (roughly 0.1%).

The principle advantage here is not that the ripples are equal, which is frequently unimportant, but rather that the internal filter impedance is kept within a certain tolerance, i.e., the minima of the return loss are equal.

In the above design algorithms, we have used approximate immittance inverters and approximate transformers (as in Figure 36). These have the effect of producing small distortions in the passband

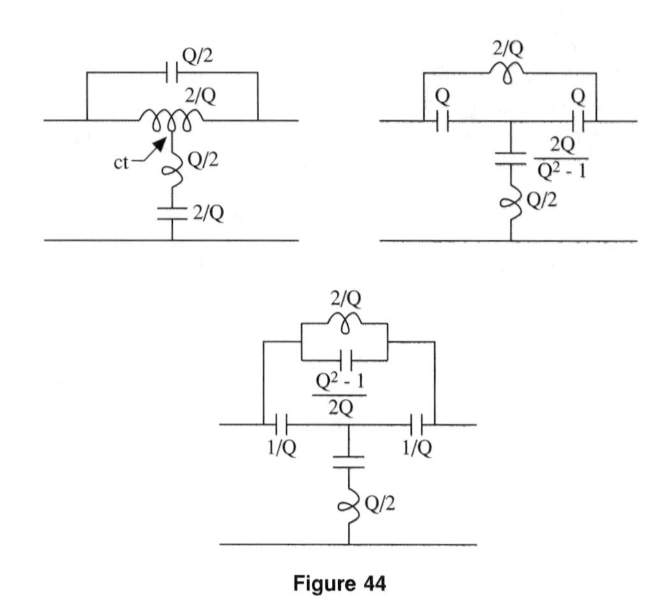

Figure 44

ripples so that they are unequal. In order to meet common requirements on the return loss, it will sometimes be necessary to reduce the design value of the ripple below what would be used in a synthetic design. In this case, the advantage (usually quite small) of exact synthesis would appear in the stopband.

By far the most important advantage of synthesis, however, is that it can give us designs having general parameter stopbands. By this we mean that the stopband attenuation is tailored to a given specification. If the system engineers can define the shape of the interfering spectrum, then the most economical design would attenuate that spectrum by the desired amount throughout the band. An elliptic function filter, on the other hand, would have to supply the desired maximum attenuation throughout the entire band.

In addition, synthesis can make it possible to produce designs that use fewer inductors and also produce designs that realize arbitrary transfer functions. This latter capability is enormously useful when linear phase filters are needed.

Unfortunately, computer programs to do analog filter synthesis are very complicated and, therefore, expensive. But, if the reader wants to cope with the complications, the references show how to do it.

9. ALLPASS CIRCUITS: DELAY EQUALIZERS

9.1 ALLPASS CIRCUITS

Figure 44 shows bridged-T allpass constant resistance circuits normalized for 1 Ω, with unit pole frequency and prescribed pole Q. The pole-zero geometry and delay response of these circuits is shown in Figure 45.

If the pole-zero coordinates are x and jy, then the transfer function is

$$t(\omega) = \frac{4x\left(\omega^2 + x^2 + y^2\right)}{\omega^4 + 2\left(x^2 - y^2\right)\omega^2 + \left(x^2 + y^2\right)^2}$$

$$= \frac{2\omega_0 Q\left(\omega^2 + \omega_0^2\right)}{Q^2\left(\omega^2 - \omega_0^2\right)^2 + \omega_0^2\omega^2}$$

where the pole frequency is $\omega_0 = \sqrt{x^2 + y^2}$ and the pole Q is $Q = \omega_0/(2x)$

The delay at ω_0 is

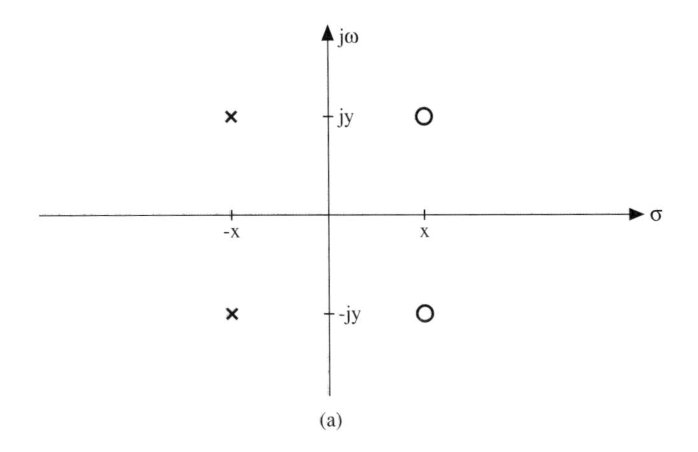

(a)

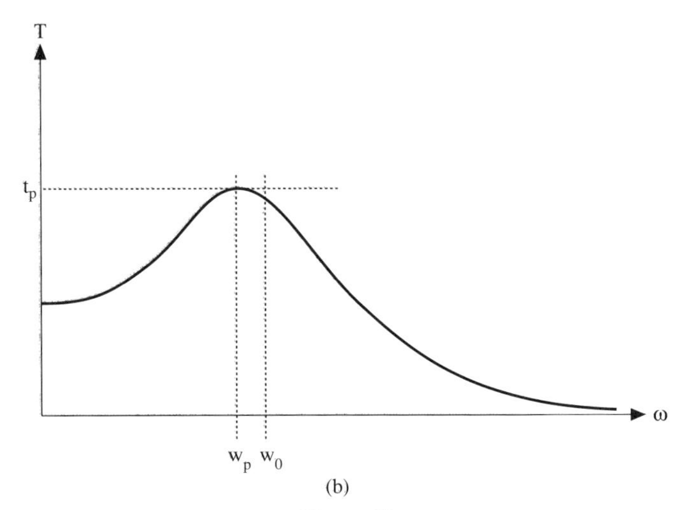

(b)

Figure 45

$$t(\omega_0) = \frac{4Q}{\omega_0}$$

The conditions for a delay peak, ω_p, are:

$$\omega_p = 0 \quad \text{unless} \quad d < \sqrt{3}$$

then we have

$$\omega_p = u\omega_0 \quad \text{for} \quad u = \sqrt{\sqrt{4 - d^2} - 1}$$

and the peak delay is

$$tp = \frac{2d}{\omega_0(1 - u^4)}$$

It is also possible to use a first degree constant resistance allpass circuit, shown in Figure 46, whose delay response, peaked at zero frequency, is

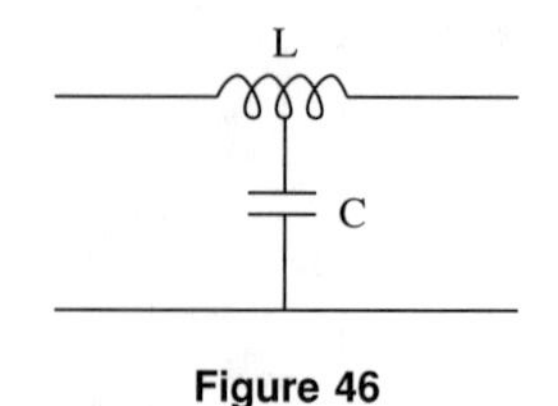

Figure 46

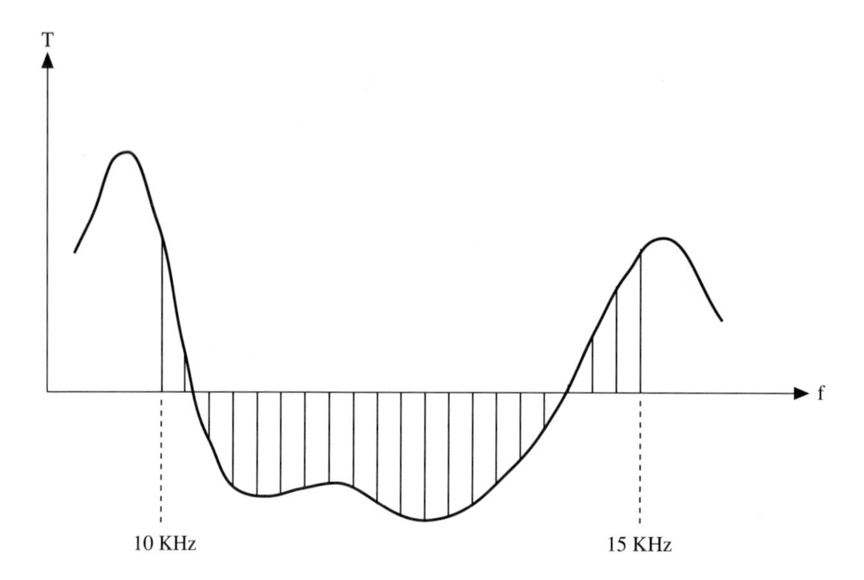

Figure 47

$$t(\omega) = \frac{2x}{x^2 + \omega^2}$$

The appearance of mutual inductance here, in the center-tapped coil, makes it difficult or impossible to use the circuit at high frequencies.

9.2 OUTLINE OF A DELAY EQUALIZATION ALGORITHM

Let us suppose that we wish to equalize the delay of a filter whose delay response is like that shown in Figure 47. The delay after equalization should be close to some constant delay value, t_0, over the band from f_1 = 10 kHz to f_2 = 15 kHz (or as close as two allpass sections can make it).

We arbitrarily divide the equalization band into 20 equal parts (6 or more per equalizer section). This gives 21 frequency-delay points of the filter response, in frequency steps of 250 Hz. Making a forthright choice of pole numbers for each allpass section, we evaluate the delay of the equalizer (i.e., of the cascade of allpass sections) at all 21 frequencies. Then at each frequency we add the filter delay to the equalizer delay, and subtract the nominal target delay, t_0. Thus, we have calculated 21 deviations from t_0 of the circuit delay. These are called the residuals. (The initial allpass parameters might be chosen so that each section has its ω_0 delay equal to t_0. The poles should be equally spaced across the band.)

Now, the sum of the squares of the residuals is called the objective function, and our problem is solved by minimizing this function. Of course, the "solution" obtained this way is not always good enough, but we can usually improve it by using more equalizer sections, up to the practical limit.

Systematic minimization of such a function involves very complicated mathematics, and has been a prime target of research for the last 40 years or more. However, making use of a minimization routine that has already been worked out and embodied in a computer program is not difficult. The filter utilities diskette (see references) contains such a program and is easier to use than most because it does not require the user to program the partial derivatives of the objective function.

After becoming familar with the algorithmic details, above, the reader should be able to achieve useful and practical delay equalizers without a great deal of trouble.

REFERENCES

1. **G. Szentirmai (Ed.),** *Computer-Aided Filter Design,* IEEE Press, 1973. Reprints some of the papers referenced below and many others.
2. **DeVerl S. Humpherys,** *The Analysis, Design, and Synthesis of Electrical Filters,* Prentice-Hall, 1970. Although written before age of the PC, this is an excellent book on filter theory.
3. **R. Saal and E. Ulbrich**, On the design of filters by synthesis, *IRE Trans. Circuit Theory,* December, 1958.
4. **H.J. Orchard and G.C. Temes**, Filter design using transformed variables, *IEEE Trans. Circuit Theory,* December, 1968.
5. **J.K. Skwirzynski,** On synthesis of filters, *IEEE Trans. Circuit Theory,* January, 1971.
6. **H.J. Orchard,** Formulae for ladder filters, *Wireless Eng.,* January, 1953, pp. 3–5.
7. **Pierre Amstutz,** Elliptic approximation and elliptic filter design on small computers, *IEEE Trans. Circ. Systems,* vol. 25, no. 12, December, 1978, pp. 1001–1011.
8. **P.R. Geffe,** *Geffe's Filter Utilities,* a set of filter design programs on diskette. Requires an IBM or compatible computer, MS-DOS 3.3 or later, 80 × 86 or better; numeric coprocessor optional, but recommended. The programs do Butterworth, Chebyshev, elliptic prototypes; elliptic parameters, top__c and bottom__c bandpass filters; lossless or predistorted attenuation equalizers; lowpass to bandpass transformation; tee-to-pi and endcap transformations. Filter network simulation (analysis of filter response) is included, and an optimization program using the Nelder–Mead algorithm with documentation. All source code is included. The diskette is $69.00 (US) prepaid, and is shipped by mail on a 3.5″ diskette only. Send your order with payment (no COD or credit cards) to Filterware, 5095 Country Top Trail, Bethlehem, PA 18017–8814, USA.

Data Transmission Filters

L. F. Lind

CONTENTS

1. IDEAL FILTERS

Assume there is a stream of binary data to be transmitted, in electronic form. An impulse can be used to send each bit. This representation is pleasing, because the bits are cleanly separated in time (impulses have zero width). A portion of the data and the resulting signal are shown in Figure 1. A data bit rate of 1 s is assumed for simplicity.

A receiver would have no difficulty in decoding this signal back into data. The problem with this signal is that the channel needs infinite bandwidth to send it without distortion. The requirement of infinite bandwidth is obviously inefficient, and impractical.

The main question is, how little bandwidth is needed to send the data without distortion. This question was answered theoretically by Nyquist in 1928.[1] The absolute minimum bandwidth is $^1/_2$ Hz. If the signal impulse train is applied to a filter having this bandwidth and a rectangular amplitude frequency characteristic, the filter output can still be resolved exactly into bit-by-bit detection, without interference from other bits.

To see that this is true, the ideal filter response to just one impulse is considered, as shown in Figure 2. Even though the input impulse has been considerably smeared out in time, the output y(t) still has zero value at all integers (except t = 0). These zero crossings are at the peaks of the other impulse responses. Thus, if the composite output is sampled at integer time points, each sample will contain information solely due to its corresponding input. The intersymbol interference (isi) for such a system is zero.

0-8493-8951-8/97/$0.00+$.50
© 1997 by CRC Press LLC

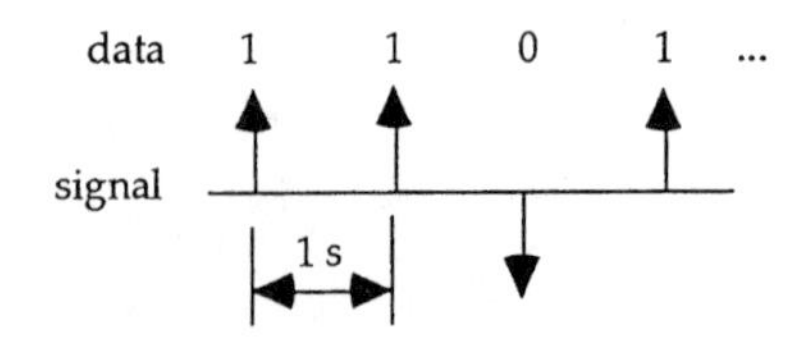

Figure 1 A stream of data and its encoding into electronic impulses.

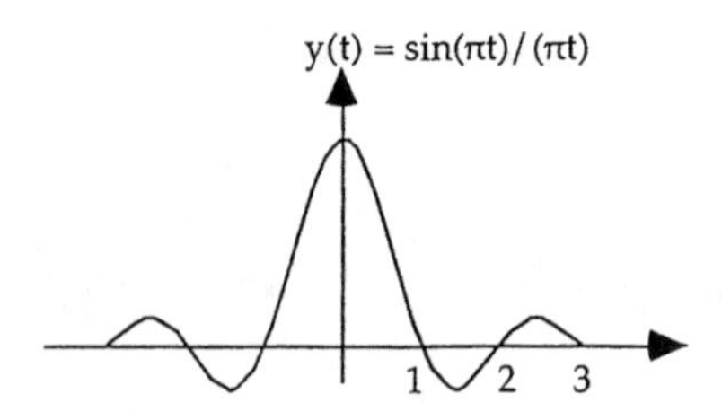

Figure 2 Impulse response of an ideal Nyquist filter.

The ideal Nyquist filter cannot be built. It has an anticipatory impulse response (the output starts oscillating before the input is applied), and the frequency response has a discontinuity at $f = \pm^1/_2$ Hz, where it abruptly jumps from 1 to 0.

2. REALIZABLE FILTERS

The filters described next are realizable and closely approach the Nyquist ideal filter. The result of the design is a set of poles and zeros for the transfer function of a composite filter. Its poles and zeros can be split rather arbitrarily between the transmitter and receiver filters. If approximately equal frequency selectivity is desired in the two filters, the poles and zeros of the composite filter should be divided into two nearly equal sets, with one set of poles and zeros interlacing those of the other set. It should be noted that this method does *not* result in a pair of identical filters, contrary to results from ideal communication theory. It does however realise the composite filter, which has extremely good performance in both time and frequency domains.

These filters have been specially created for data transmission. Every degree of freedom has been used to maximal effect for this purpose. Because of their optimal properties, these filters require a more elaborate realization procedure than classical filters. But the extra effort soon pays for itself in lower overall complexity. Indeed, it has been found that filters with only nine poles are within a few percent of ideal performance in both the time and frequency domains. Curiously, the performance of these filters is also excellent in realizing simultaneous brick wall amplitude and linear phase performance, which are targets from classical filter theory.

As the filters are self-equalized, no separate group delay equalizer is required. The filter design has its data transmission performance placed at a global minimum. Thus, it has zero differential sensitivity to changes in element values, which makes the design relatively insensitive to component tolerances.

A primary requirement of the realizable filter is to provide near zero isi when subjected to a synchronous stream of input impulses, with each impulse carrying a bit of data information. Since the filter has a finite number of poles and zeros, it is impossible for it to have exactly zero isi at the infinity of integer sample points (a data rate of 1 baud being assumed). The net isi can be controlled, however, to lie at a very low level.

Figure 3 shows the desired impulse response. For the realizable filter, the main sample point is not at t = 0, but at $t = t_0$, to get over the problem of an anticipatory response. The other primary requirement is that the realizable filter must have the minimum possible bandwidth.

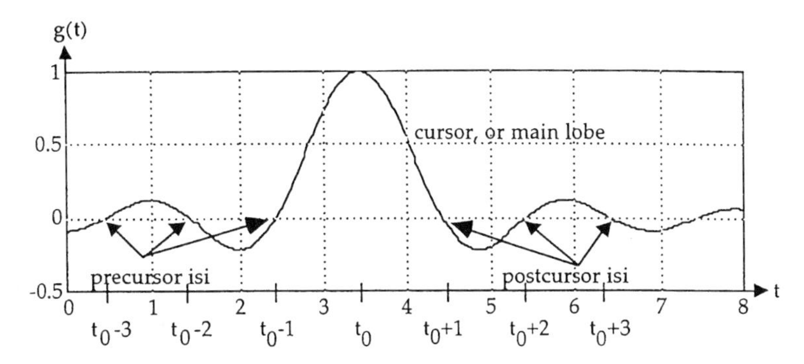

Figure 3 The impulse response of a realizable data transmission filter, having near zero isi at the sample points $t_0 \pm 1$, $t_0 \pm 2$, ...

3. SOLVING THE APPROXIMATION PROBLEM

The goal is to find a set of poles and zeros which provide a preset low isi specification in the time domain and, also, minimal bandwidth in the frequency domain. The solution is not easy to derive. It involves pages of algebra and a special method of numerical solution.[2] Some key points will be given here for an overall understanding of the method.

1. A realizable all-pole filter impulse response is assumed first, which is written in the form

$$g(t) = \sum_{n=1}^{N} r_{ne} P_{nt} \tag{1}$$

where r_n and p_n are the unknown residues and poles of the filter. Later, finite transmission zeros will be introduced, to cancel the precursor isi.

2. Because any realizable filter will have some postcursor isi, an isi squared error function E is formed:

$$E_1 = \sum_{k=1}^{\infty} g^2(t_0 + k) \tag{2}$$

In this equation the value of t_0, the main sample point, is also unknown. Observe that the error is monitored over an infinity of sample points. In addition to being satisfied that all postcursor isi is being measured, the construction of E also has another important use. When Equation 1 is put into Equation 2, one gets some infinite geometric series, which can be summed and put into a closed form.

3. To this specified error the constraint $g(t_0) = 1$ is added. If this constraint were not used then a trivial result $r_n = 0$ in Equation 1 would give zero isi (and zero output for any input!). The constraint is added to E via a Lagrange multiplier.[2]

4. The resulting functional is differentiated with respect to each r_n residue and differentiated with respect to the Lagrange multiplier, all differentials being set to zero. This is where the method acquires its optimality. There is an enforced zero differential sensitivity with respect to changes in element values. A set of nonlinear equations is found from this step, containing the unknown poles, unknown residues, and unknown t_0.

5. Another set of nonlinear equations, giving each residue as a function of the poles, can be found by the normal pole-residue results for an all-pole transfer function.[2]

6. When the results of step 4 are equated to step 5, a set of N equations is found for the poles:

$$\prod_{\substack{i=1\\i\neq n}}^{N}\frac{\sinh(p_n-p_i)/2}{(p_n-p_i)/2}=\frac{2^{N-1}g(t_0)\exp(p_n t_n)}{\left[\prod_{i=1}^{N}(p_i)\right]\sinh\left(\sum_{n=1}^{N}p_n\right)}\prod_{i=1}^{N}\sinh\frac{(p_i+p_n)}{2} \tag{3}$$

$$n=1,2,\ldots,N$$

7. The equations in Equation 3 are nonlinear and have many solutions! It turns out that the solutions have vertical periodicity in the s-plane. For minimal bandwidth, all poles should have an imaginary part less than π, the Nyquist frequency. Imposing this condition, there is then just one solution. The equations are used in two different ways. First, they are simplified through approximation to provide a closed form starting point. Then the exact equations are used in a first-order Taylor series expansion to find the solution.

8. These equations contain the free parameter t_0, the location of the main sample point. In solving these equations, an initial value of t_0 is guessed, and the pole locations found. Then the impulse response is formed, and checked to see if t_0 occurs at the peak of the main lobe. If not, t_0 is changed, and the process repeated until t_0 is at the peak.

9. Once a set of exact solutions was found for various E and N values, the pole locations and t_0 were fitted by empirical formulas. These gave even better starting points. It has been found that the nonlinear Equation set 3 has many local minima, and so it is important to have good starting values which are within a few percent of the global solution. In solving the Equation set 3 there are two iterative loops. The inner one finds the poles for a given t_0. The outer loop adjusts t_0 until it is at the peak of the main lobe of the impulse response. Because of the good starting point, the process takes only seconds to run on a personal computer.

10. The result of step 8 is an optimal all-pole data transmission filter, and probably has $t_0 > 1$ second. If this is the case, there will be some precursor isi at $t_0 - 1$, $t_0 - 2$, and so on. This precursor isi can be zeroed by using an even numerator polynomial in the transfer function. Each coefficient (except the zeroth order one) can be adjusted to eliminate isi at one of the precursor points. For example, the rational transfer function

$$\frac{(a_0+a_2s^2)}{D(s)} \tag{4}$$

has an impulse response

$$h(t)=a_0g(t)+a_2g''(t) \tag{5}$$

where g(t) is the all-pole impulse response and g″(t) is the second derivative of g with respect to t. A value of a_2 can be found which sets $h(t_0 - 1) = 0$. The numerator polynomial coefficients can also be used to provide stopband transmission zeros, to increase stopband selectivity.

11. When the numerator polynomial has been found and included in the transfer function, then the postcursor isi and t_0 will be slightly distorted. These need to be reoptimised. Once this is done, then the numerator might need a small readjustment, and so on, until all requirements are met. In practice, the numerator/denominator interaction has been found to be fairly small, resulting in quick overall convergence.

4. RESULTS

The results of this process are given in Table 1. The poles and zeros are given for an input data rate of one impulse per second. For an input rate of k bit/s, all poles and zeros are multiplied by k.

Table 1 Poles, Zeros, and t_0 Values for Data Transmission Filters

E		10^{-4}	10^{-5}	10^{-6}
N-4	t_M	1.1619	1.0721	—
	$P_1 \cdot P_4$	−0.860491 ± j2.88317	−1.08185 ±j2.93160	—
	$P_2 \cdot P_3$	−1.44212 ± j1.08148	−1.79638 ±j1.12002	—
	$z_1 \cdot z_2$	±13.4677	±32.0892	—
N-5	t_M	1.5902	1.4784	1.3833
	$P_1 \cdot P_5$	−0.647102 ± j2.94584	−0.800952 ± j2.98658	1.11408 ± j3.05228
	$P_2 \cdot P_4$	−1.07447 ± j1.60246	−1.34119 ±j1.65715	−1.60751 ± j1.71245
	P_3	−1.16207	1.47217	−1.79951
	$z_1 \cdot z_2$	±4.08939	±5.31036	±6.94645
N-6	t_M	2.0412	1.9143	1.8024
	$P_1 \cdot P_6$	−0.515430 ± j2.98676	−0.638429 ± j3.01828	−0.760566
	$P_2 \cdot P_5$	−0.864420 ± j1.91054	−1.06956 ± j1.95663	−1.27391 ± j2.00592
	$P_3 \cdot P_4$	-0.922758 ± j0.62775	−1.41940 ± j0.669077	−1.66897 ± 0.691227
	$z_1 \cdot z_2$	±2.40180	±2.87699	±3.43732
	$z_3 \cdot z_4$	±57.8628	—	—
N-7	t_M	2.4941	2.3572	2.2357
	$P_1 \cdot P_7$	−0.423595 ± j3.01469	−0.522301 ± j3.04191	−0.622050 ± j3.06532
	$P_2 \cdot P_6$	−0.716328 ± j2.13020	0.888251 ± j2.17168	−1.05827 ± j2.21285
	$P_3 \cdot P_5$	−0.778452 ± j1.05600	−0.981718 ± j1.07240	−1.18410 ± j1.09064
	P_4	−0.768469	−0.971926	±2.34591
	$z_1 \cdot z_2$	±2.12453	±2.17563	±2.34591
	$z_3 \cdot z_4$	±4.74193	±7.52895	±12.5457
N-8	t_M	2.9575	2.8085	2.6754
	$P_1 \cdot P_8$	−0.361188 ± j3.03327	−0.443228 ± j3.05676	−0.525115 ± j3.07719
	$P_2 \cdot P_7$	0.609120 ± j2.27952	−0.751889 ± j2.31979	−0.894080 ± j2.35981
	$P_3 \cdot P_6$	−0.663671 ± j1.36900	−0.843875 ± j0.459921	−1.01461 ± j1.41115
	$P_4 \cdot P_5$	−.668631 ± j0.458181	−0.843875 ± j0.459921	−1.02007 ± j0.460979
	$z_1 \cdot z_2$	2.00632 ±j0.914105	2.47340 ± j0.601014	±2.29201
	$z_3 \cdot z_4$	2.00632 ±j0.914105	2.47340 ±j0.601014	±3.82521
N-9	t_M	3.4242	3.2676	3.1257
	$P_1 \cdot P_9$	−0.312902 ± j3.04802	−0.383609 ± j3.06853	−0.454166 ± j3.08611
	$P_2 \cdot P_8$	−0.530569 ± j2.39388	−0.653755 ± j2.42934	−0.774990 ± j2.4678
	$P_3 \cdot P_7$	−0.577983 ± j1.59897	−0.727637 ± j1.62132	−0.877830 ± j1.64397
	$P_4 \cdot P_6$	−.583166 ± j0.809606	−0.737692 ± j0.818444	−.894414 ± 0.824327
	P_5	-0.595979	−0.751082	−0.904954
	$z_1 \cdot z_2$	1.59810 ±j1.05998	1.75501 ±j1.00883	1.98964 ± j0.934250
	$z_3 \cdot z_4$	−1.59810 ±j1.05998	−1.75501 ± j1.00883	−1.98964 ± j0.934250
	$z_5 \cdot z_6$	±6.02630	±10.9307	±26.6207
N-10	t_M	3.9851	3.7299	3.5801
	$P_1 \cdot P_{10}$	−0.276498 ± j3.05881	−0.337635 ± j3.07725	−0.398573 ± j3.09304
	$P_2 \cdot P_9$	−0.468058 ± j1.77994	−0.575586 ± j2.51527	−0.681889 ± j2.54819
	$P_3 \cdot P_8$	−0.511854 ± j1.77994	−0.643983 ± j1.30256	−0.775573 ± j1.82394
	$P_4 \cdot P_7$	−0.517323 ± j1.07821	−0.652125 ± j1.09044	−0.788589 ± j1.10060
	$P_5 \cdot P_6$	−0.528878 ± j0.363506	−0.668904 ± j0.370437	−0.809254 ± j0.375688
	$z_1 \cdot z_2$	1.54307 ± j1.32152	−1.61780 ± j1.13597	−1.68059 ± j1.05732
	$z_3 \cdot z_4$	−1.54307 ± j1.32152	−1.61780 ± j1.13597	−1.68059 ± j1.13597
	$z_9 \cdot z_6$	±2.19181	±3.41705	±5.14895

Note: E is the total squared isi, and N the number of poles in the filters

The poles are found to lie approximately on three sides of a rectangle in the left half of the s-plane, with imaginary parts roughly equally spaced. The outermost two poles have only half the real part of the other poles and are fairly close to $\pm j\pi$, the Nyquist frequency. This is to be expected, for this pair is mainly responsible for the impulse response ringing at the correct rate.

The transmission zeros are generally in both the left half and the right half of the s-plane. They can be at infinity, on the real axis at $s = \pm a$, on the imaginary axis at $\pm jb$, or in a complex quartet at $\pm a \pm jb$. The transfer function is nonminimum phase, and therefore not limited by the Hilbert transform pair. This property results in the good phase as well as amplitude response, but also implies that a more complicated structure than a ladder network will be needed for realization.

Intermediate points in Table 1 can be found by using either linear or quadratic interpolation, and so a great range of designs can be derived from the table.

Knowing the poles and zeros, there are many possible realizations. For example, an active filter based on a cascade of biquadratic pairs could be designed directly from the information in Table 1. In some commercial filter design software packages, the only input required is pole and zero information, which could be supplied from Table 1.

5. PASSIVE FILTER EXAMPLE

Passive lossless filters are more difficult to design, because of the complex transmission zeros. The details are beyond the scope of this chapter. To illustrate the process however, the main steps of the $N = 9$, $E = 10^{-4}$ case will be given. The pole and zero factors, when combined, give the transfer function

$$S_{21}(s) = \frac{N(s)}{D(s)} \tag{6}$$

where

$$N(s) = -0.0020s^6 + 0.0798s^4 - 0.2390s^2 + 1.0000$$

$$D(s) = 0.0105s^9 + 0.0482s^8 + 0.2887s^7 + 0.8306s^6 + 2.3145s^5 + 4.0733s^4 + 5.9711s^3 + 5.5999s^2 + 3.5364s + 1.0126$$

The next step is to form the lossless transfer (or ABCD) matrix for the filter, which yields the result:

$$T = \frac{1}{F}\begin{bmatrix} A & B \\ C & D \end{bmatrix} \tag{7}$$

where

$$\begin{aligned}
A(s) &= 0.0563s^8 + 0.9031s^6 + 4.2052s^4 + 5.6797s^2 + 1.0000 \\
B(s) &= 0.0179s^9 + 0.4111s^7 + 2.8992s^5 + 6.8792s^3 + 3.8816s \\
C(s) &= 0.1123s^7 + 1.4436s^5 + 4.5485s^3 + 2.9583s \\
D(s) &= 0.0356s^8 + 0.7066s^6 + 3.7722s^4 + 5.3252s^2 + 1.0000 \\
F &= (-0.0275s^2 + 1)(0.0739s^4 - 0.2115s^2 + 1)
\end{aligned}$$

and where the source and load resistances are 0.7284 Ω, and 1 Ω. Next, T is factorized, to give

$$T = T_1T_2T_3T_4T_5 \tag{8}$$

where

$$
\begin{aligned}
T_1 &= \begin{bmatrix} 1 & 0.5008s \\ 0 & 1 \end{bmatrix} \\
T_2 &= \frac{1}{-0.0275s^2+1}\begin{bmatrix} 1.427\times10^{-3}s^2+1 & 0.8479s \\ 0.6923s & 0.5305s^2+1 \end{bmatrix} \\
T_3 &= \begin{bmatrix} 1 & 0 \\ 0.5016s & 1 \end{bmatrix} \\
T_4 &= \frac{1}{F}\begin{bmatrix} 0.4220s^4+2.2752s^2+1 & 0.9077s^3+2.2155s \\ 0.6299s^3+1.7644s & 0.0130s^4+1.2107s^2+1 \end{bmatrix}
\end{aligned}
\qquad (9)
$$

where $F = 0.0739s^4 - 0.2115s^2 + 1$, and

$$
T_5 = \begin{bmatrix} 1 & 0.3174s \\ 0 & 1 \end{bmatrix}
$$

If these factors are multiplied together, they give the original T matrix for the example. Each section has a fairly simple realization in terms of element values. When the sections are cascaded, and parallel capacitors added together, the resulting circuit is shown in Figure 4. The section with the 0.4240 coils produces the transmission zeros on the frequency axis. The section after that produces the complex transmission zeros. Even though there is not a common earth between input and output, the realization is efficient and avoids coupled coils.

6. FILTER PERFORMANCE

How does the data transmission filter of the above example perform in the time and frequency domains? The amplitude response is shown in Figure 5. The filter has a good (but not equiripple) passband response and offers 50 dB attenuation at twice the cutoff frequency. Since there is some attenuation at DC, the source resistance is not equal to the load resistance. In the far stopband the attenuation will increase by 18 dB/octave, because of the three transmission zeros at infinity.

The group delay response is even more interesting, and displayed in Figure 6. It is seen that the passband group delay has definite passband ripple and a peak near the cutoff frequency. It throws into doubt the belief that a good data transmission filter needs to have a very flat passband group delay.

The evidence for this statement is given by the impulse response of the filter which is shown in Figure 7. The response of the example filter has been shifted left by 3.4242 s, so that it can be directly compared with the $\sin(\pi t)/(\pi t)$ response of an ideal filter. The fit, especially in the main lobe and at the zero crossings, is remarkable. For comparison purposes, the error function between these two responses is shown in Figure 8. This graph shows in greater detail the main lobe match and the accuracy of the zero crossings in the example filter.

7. CONCLUDING REMARKS

This chapter has been concerned with the design of optimally realizable data transmission filters. An overall description of the approximation method has been given, as well as a table of results.

A nine-pole example filter has been examined in detail. A passive realization has been shown, and the filter performance displayed in both frequency and time domains. The frequency domain

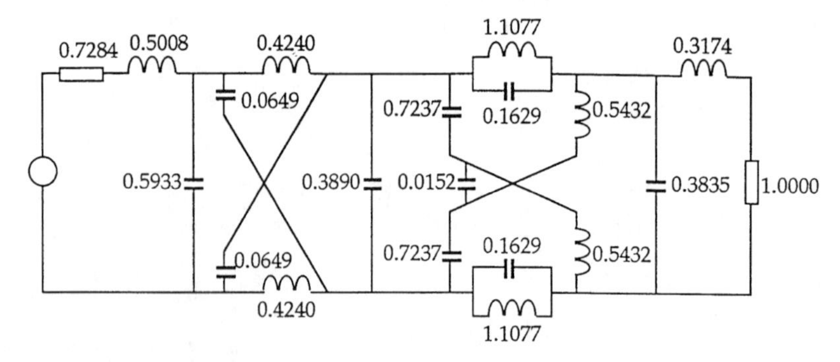

Figure 4 Final passive realization of a data transmission filter, with N = 9, E = 10^{-4}. The inductance values are in henrys, the capacitance values in farads, and the resistances in ohms. The data rate is 1 bit/s.

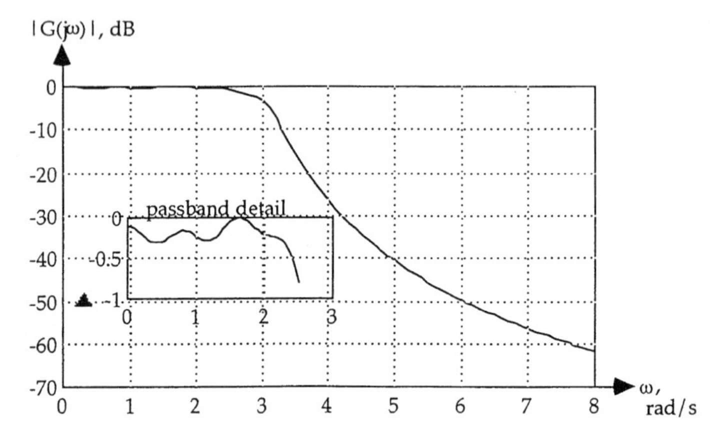

Figure 5 Amplitude response of the example filter. The filter has nine poles and a total squared isi error of 10^{-4}. Its cutoff (Nyquist) frequency is at 3.14 rad/s.

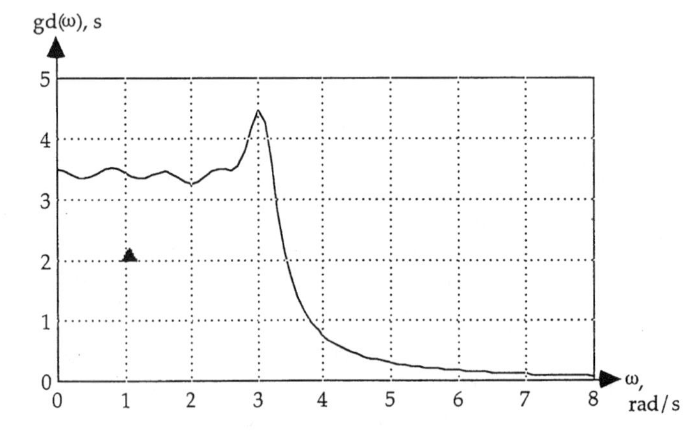

Figure 6 Group delay response of the nine-pole example filter. Its cutoff (Nyquist) frequency is at 3.14 rad/s.

responses (especially the group delay) are somewhat at variance with accepted design practice for data transmission filters. The impulse response, however, is remarkably close to that of an ideal filter and, therefore, justifies the goodness of the approximation method, and of its inclusion in this handbook!

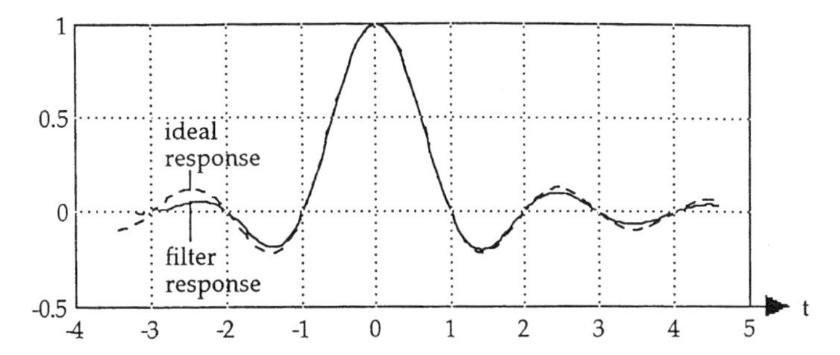

Figure 7 Impulse response of the nine-pole example filter (solid line) and an ideal filter (dotted line). The input data rate is one sample per second.

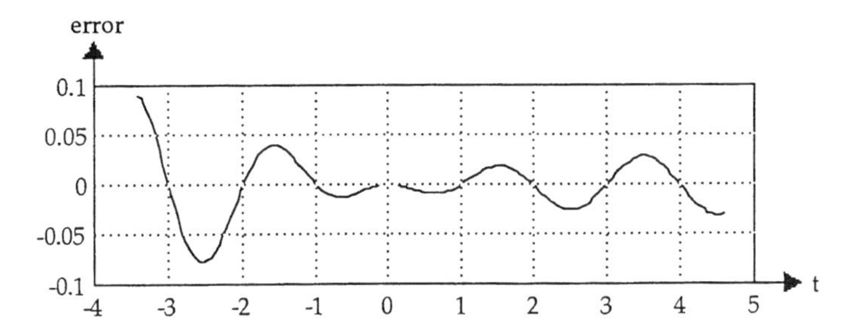

Figure 8 Error (approximate–ideal) between the two impulse responses of Figure 7.

REFERENCES

1. **Nyquist, H.** Certain topics in telegraph transmission theory, *Trans. AIEE,* 47, 617–644, February, 1928.
2. **Nader, S. E. and Lind, L. F.** Optimal data transmission filters, *IEEE Trans. Circuits Syst.,* CAS-26, 36-45, January, 1979.

Chapter 5

Continuous-Time Active RC Filters

Optimal Design of Active RC Networks

Peter Horn

CONTENTS

0-8493-8951-8/97/$0.00+$.50
© 1997 by CRC Press LLC

Abstract — This contribution describes optimization of active RC networks with respect to the finite amplifier gain. In the first section we discuss the reasons active RC networks are useful and indicate their practical application. The next section describes transfer functions while sensitivities of this function are explained in Section 3. The gain-sensitivity product is introduced in Section 4. It shows that an active RC network can be optimized for a limited open loop gain, depending on the frequency compensation of the amplifier. Finally, the optimization procedure is shown on a practical example with a second-order bandpass filter.

1. WHY AND WHERE TO USE ACTIVE RC FILTERS

RC active filters are limited by the frequency range of the amplifiers and in most cases operate below 1 MHz. They are used mostly in the acoustic frequency range, from 10 Hz to 20 kHz, or in telephone lines where the voice band is limited to 3.4 kHz. In telephone applications, where a limited dynamic range is acceptable, switched-capacitor filters have been more successful because of their low production cost and small physical dimensions. If high dynamic range and low noise are of importance, then active RC filters are preferred. In addition, fast design and low cost of components, like 1% resistors and 5% capacitors, make the use of RC filters very practical. We first list some cases where active RC filters are practical.

Active RC filters in feedback control systems. For stability reasons, first- and second-order active RC filters are often used to correct amplitude and phase responses of regulated systems. As an example, the unit step response of a heater may need individual adjustments of the regulating network when the control loop is closed. The frequency response and the gain of an integrator can be easily adjusted by an active RC network. In some cases, we may even use an allpass in order to correct the phase at a critical frequency.

General sensor signal filtering. Second-order lowpass or second-order bandpass filters are often used to filter noisy signals coming from various sensors. To save on hardware costs, these signals are often multiplexed and processed with only one A/D converter. Examples are low or noisy strain gauge signals which must be amplified with high gain to obtain measurable levels. Another example is the filtering output signals of PIR (passive infrared) sensors. Requirements on high gain (60 to 80 dB), high dynamic range, and low noise in the frequency band from 0.1 to 10 Hz make the use of active RC filters necessary.

Antialiasing lowpass filters. Input signals for digital filters must be below the sampling frequency in order to avoid aliasing of frequency spectra. Lowpass filters of higher orders are generally used here.

Equalizing allpass filters. If the frequency spectrum of signals needs processing without distortion, then only filters with linear phase and constant amplitude satisfy these requirements. Allpass filters do not change the gain, but influence the phase of incoming signals and are used for signal equalization.

Sinusoidal oscillators with low distortion and fixed frequency. Such oscillators can be built with second-order bandpass filters and with amplitude regulation achieved by a Mosfet as a voltage-controlled resistor. Fixed-frequency oscillators are used for impedance measurements of sensors.

Fixed-frequency lock-in amplifier. Such amplifiers can be built with a fixed-frequency sinusoidal oscillator, a second-order bandpass filter, a low pass filter, and a precision rectifier which is synchronized with the oscillator frequency. These amplifiers can have a gain of more than 100 dB and can process very noisy sensor signals.

Easy and low-cost design. The above examples show areas where active RC filters are widely used because of their low cost. The filters can be easily designed, once the learning priod has been mastered. Figure 1 shows a second-order bandpass active RC filter which uses only two resistors, two capacitors, and one operational amplifier.

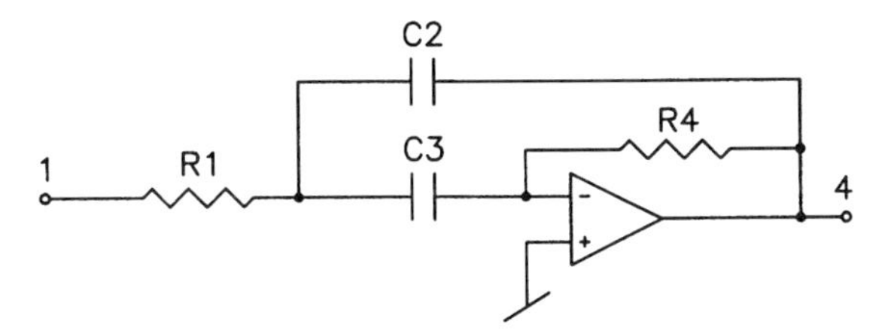

Figure 1 Second-order bandpass filter.

This filter is suitable for bandpass operation when the quality factor q_p is less than or equal to 2. If the unity-gain frequency of the amplifier is approximately 100 times higher than the frequency of the passband, finite gain of the amplifier does not influency operation. A quality factor of q_p = 1 is obtained with the values R1 = 10 k, R4 = 40 k, Cl = C2 = 39 nF and the pole frequency of 204 Hz. Because this frequency is low compared with the 1 MHz gain bandwidth of an ordinary operational amplifier, such a design will not be ciritcal. If the pole frequency and/or the quality factor q_p have to be higher, a more careful design will be needed.

2. NETWORK FUNCTIONS

Network functions are defined for linear networks only and can describe the transfer of voltage, transfer of current, or evaluate driving point and transfer impedances or admittances. In active networks, the voltage transfer function is used most often.

2.1 RATIONAL FUNCTION

All network functions of linear networks are ratios of two polynomials in terms of the frequency-domain variable $s(s = \sigma + j\omega)$. They can be written in the form:

$$T(s) = \frac{N(s)}{D(s)} = \frac{\sum_{i=0}^{m} a_i s^i}{\sum_{j=0}^{n} b_j s^j} \tag{1}$$

If we consider the voltage or the current transfer function, then the degree of the numerator, *m,* cannot be higher than the degree of the denominator, *n.*

As an example, we give the voltage transfer function of a second-order bandpass, shown in Figure 1:

$$T_{14}(s) = \frac{V_4}{V_1} = \frac{N(s)}{D(s)} = \frac{sC_3R_4}{s^2C_2C_3R_1R_4 + sC_2R_1 + sC_3R_1 + 1} \tag{2}$$

The output is taken at node 4 and the input is applied at node 1. If the operational amplifier is ideal, with infinite gain and zero phase shift, then the analysis can use a very simple method described, for instance, in Reference 1. Another posibility is to use the indefinite admittance matrix.[2] In such case, the coefficients of the transfer function are combinations of circuit elements.

If we consider separately the numerator or the denominator, we can define a general polynomial function, $F(j\omega)$, with real coefficients a_i:

$$F(j\omega) = \sum_{i=0}^{n} a_i s^i \tag{3}$$

This complex function can be decomposed into a real and an imaginary part. The real part is an even function; the imaginary part is an odd function of ω:

$$F(j\omega) = \mathrm{Re}\left[F(j\omega)\right] + \mathrm{Im}\left[F(j\omega)\right] = A(\omega) + jB(\omega) \tag{4}$$

The absolute value of the complex polynomial function $F(j\omega)$ is then:

$$\left|F(j\omega)\right| = \sqrt{A^2(\omega) + B^2(\omega)} \tag{5}$$

and its phase $\varphi(\omega)$ is

$$\varphi(\omega) = \arctan\left(\frac{B(\omega)}{A(\omega)}\right) \tag{6}$$

The polynomial function $F(j\omega)$ can also be written in the form:

$$F(j\omega) = \left|F(j\omega)\right| e^{j\varphi(\omega)} \tag{7}$$

Applying the above equations to the transfer function $T(s)$ and using frequency ω as the independent variable, we obtain the amplitude response of $T(s)$:

$$\left|T(j\omega)\right| = \frac{\left|N(j\omega)\right|}{\left|D(j\omega)\right|} \tag{8}$$

and the phase response of $T(s)$:

$$\varphi(\omega) = \varphi\left[T(\omega)\right] = \varphi\left[N(\omega)\right] - \varphi\left[D(\omega)\right] \tag{9}$$

Both responses can be easily computed from the rational function without needing the roots of the polynomials.

2.2 REAL AND COMPLEX POLES AND ZEROS

To analyze the transfer function in the time domain, we need the roots of the denominator polynomial, the residues, and the inverse Laplace transform.[3] The numerator polynomial, $N(s)$, has m roots called the *zeros* z_i. The n roots of the denominator polynomial, $D(s)$, are called the *poles* p_j.

The transfer function with known poles and zeros can be written in the form:

$$T(s) = \frac{N(s)}{D(s)} = K\frac{\prod_{i=1}^{m}\left(s - z_i\right)}{\prod_{j=1}^{n}\left(s - p_j\right)} \tag{10}$$

If all poles are simple, then residue calculus allows us to decompose the transfer function into a sum of simple rational functions of the form:

$$T(s)=\sum_{i=1}^{n}\frac{K_i}{s-\sigma_i}+\sum_{k=n+1}^{m}\left[\frac{A_k+jB_k}{s-\sigma_k-j\hat{\omega}_k}+\frac{A_k-jB_k}{s-\sigma_k+j\hat{\omega}_k}\right] \tag{11}$$

The time-domain response is found from a table of inverse Laplace transforms. The residue of the real pole, K_i, is always real and the Laplace transform inverse is

$$L^{-1}\left[\frac{K_i}{s-\sigma_i}\right]=K_i e^{\sigma_i t} \tag{12}$$

The complex conjugate pair, p_k, $\bar{p}_k$, has complex conjugate residues K_k, $\bar{K}_k$, and

$$p_k=\sigma_k+j\hat{\omega}_k,\quad \bar{p}_k=\sigma_k-j\hat{\omega}_k \tag{13}$$

$$K_k=A_k+jB_k,\quad \bar{K}_k=A_k-jB_k \tag{14}$$

The time-domain response of such a pole pair is

$$L^{-1}\left[\frac{A_k+jB_k}{s-\sigma_k-j\hat{\omega}_k}+\frac{A_k-jB_k}{s-\sigma_k+j\hat{\omega}_k}\right]=2e^{\sigma_k t}\left(A_k\cos\hat{\omega}_k t-B_k\sin\hat{\omega}_k t\right) \tag{15}$$

If the active RC network is stable, then the single pole, or the complex conjugate pair, must be in the left half of the complex s-plane and the coefficients σ_i and σ_k must be negative.

Figure 2 shows one single pole, σ_l, and one complex conjugate pair p_2; $\bar{p}_2$ of a stable active RC network in the complex s-plane.

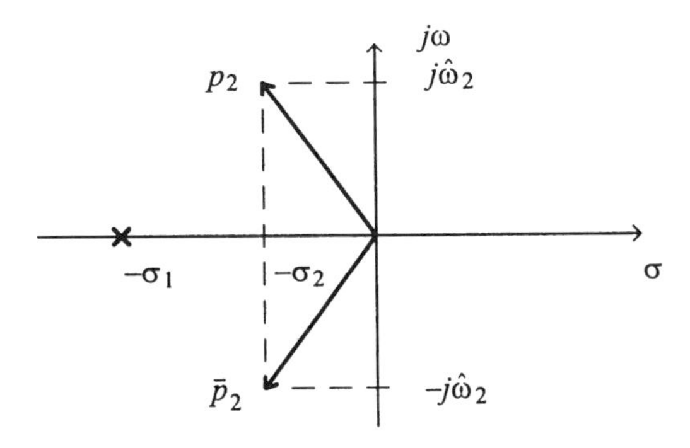

Figure 2 One real and one complex conjugate pole pair in the s plane.

2.3 TRANSFER FUNCTION AS A PRODUCT OF FIRST- AND SECOND-ORDER TRANSFER FUNCTIONS

The transfer function $T(s)$ can be also written as

$$T(s)=\prod_{i=1}^{n1}K_i\cdot\frac{s+\omega_{zi}}{s+\omega_{pi}}\cdot\prod_{i=n1+1}^{n1+n2}K_i\cdot\frac{s^2+\dfrac{\omega_{zi}}{q_{zi}}\cdot s+\omega_{zi}^2}{s^2+\dfrac{\omega_{pi}}{q_{pi}}\cdot s+\omega_{pi}^2} \tag{16}$$

where $n1$ = number of the first-order functions
$n2$ = number of the second-order functions
ω_{zi} = zero frequency
ω_{pi} = pole frequency
q_{zi} = zero quality factor
q_{pi} = pole quality factor

The transfer function in Equation 16 can be realized by cascading first- and second-order building blocks.

If we take a pair of complex conjugate poles, then we can write

$$\left(s - \sigma_k - j\hat{\omega}_k\right) \cdot \left(s - \sigma_k + j\hat{\omega}_k\right) = s^2 + \frac{\omega_{pk}}{q_{pk}} \cdot s + \omega_{pk}^2 \tag{17}$$

where

$$\sigma_k = -\frac{\omega_{pk}}{2 \cdot q_{pk}} \tag{18}$$

and

$$\hat{\omega}_k = \sqrt{\omega_{pk}^2 - \sigma_k^2} = \omega_{pk} \cdot \sqrt{1 - \frac{1}{4 \cdot q_{pk}^2}} \tag{19}$$

Equation 13 can now be written in the form

$$p_k = -\frac{\omega_{pk}}{2 \cdot q_{pk}} + j\omega_{pk} \cdot \sqrt{1 - \frac{1}{4 \cdot q_{pk}^2}} \tag{20a}$$

$$\bar{p}_k = -\frac{\omega_{pk}}{2 \cdot q_{pk}} - j\omega_{pk} \cdot \sqrt{1 - \frac{1}{4 \cdot q_{pk}^2}} \tag{20b}$$

These last expressions are important in order to derive the influence of changes of the frequency and of the pole quality factor on the pole sensitivities. Note that positive pole frequency and positive quality factor describe the negative real part of the complex pole p_k.

For one complex pole pair we can use Equation 13 and write

$$\sigma_p = \mathrm{Re}(p) \tag{21a}$$

$$\hat{\omega}_p = \mathrm{Im}(p) \tag{21b}$$

The relative pole variation,

$$\frac{dp}{p} = \mathrm{Re}\frac{dp}{p} + j \cdot \mathrm{Im}\frac{dp}{p} \tag{22}$$

is complex for a complex pole. Using Equation 22, we obtain

$$\frac{dp}{p} = \frac{d\sigma_p + j \cdot d\hat{\omega}_p}{\sigma_p + j \cdot \hat{\omega}_p} = \frac{\sigma_p \cdot d\sigma_p + \hat{\omega}_p \cdot d\hat{\omega}_p + j \cdot \left(\sigma_p \cdot d\hat{\omega}_p - \hat{\omega}_p \cdot d\sigma_p\right)}{\sigma_p^2 + \hat{\omega}_p^2} \tag{23}$$

with

$$\omega_p^2 = \sigma_p^2 + \hat{\omega}_p^2 \tag{24}$$

and

$$\omega_p \cdot d\omega_p = \sigma_p \cdot d\sigma_p + \hat{\omega}_p \cdot d\hat{\omega}_p \tag{25}$$

Comparing with Equation 23 we obtain

$$\mathrm{Re}\,\frac{dp}{p} = \frac{d\omega_p}{\omega_p} \tag{26}$$

Similarly, using

$$\frac{\hat{\omega}_p}{\sigma_p} = -\sqrt{4q_p^2 - 1} \tag{27}$$

and

$$\sigma_p \cdot d\hat{\omega}_p - \hat{\omega}_p \cdot d\sigma_p = -\frac{dq_p}{q_p} \cdot \frac{\omega_p^2}{\sqrt{4q_p^2 - 1}} \tag{28}$$

we obtain

$$\mathrm{Im}\,\frac{dp}{p} = -\frac{1}{\sqrt{4q_p^2 - 1}} \cdot \frac{dq_p}{q_p} \tag{29}$$

Taking Equations 26 and 29 together,

$$\frac{dp}{p} = \frac{d\omega_p}{\omega_p} - j \cdot \frac{1}{\sqrt{4q_p^2 - 1}} \cdot \frac{dq_p}{q_p} \tag{30}$$

and rewriting Equation 20 for the pole itself, we get

$$p_k = -\frac{\omega_p}{2 \cdot q_p} + j \cdot \frac{\omega_p}{2 \cdot q_p} \cdot \sqrt{4 \cdot q_p^2 - 1} \tag{31}$$

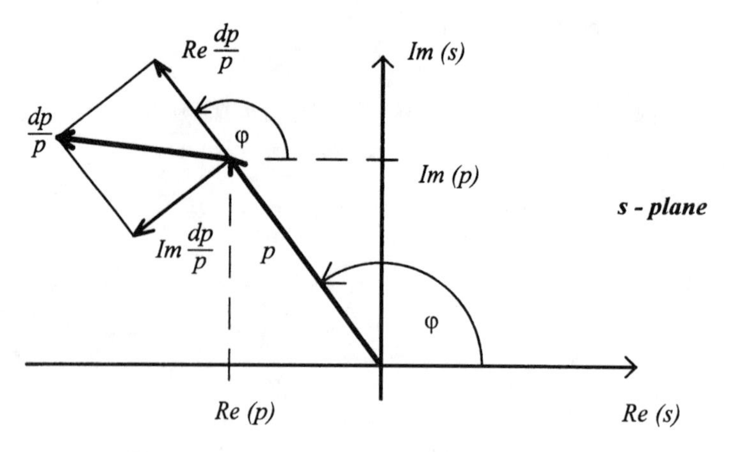

Figure 3 Pole p variation in the complex s-plane.

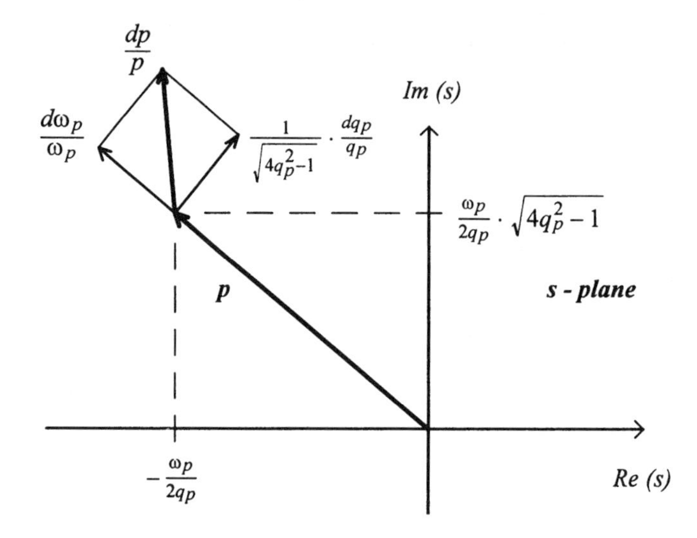

Figure 4 Pole p coordinates and variation as function of pole frequency and quality factor.

These last two equations are explained by Figures 3 and 4 for the upper-half plane complex pole. Figure 3 shows the imaginary and the real parts of the p and its variation. Please note that the real and imaginary axes of the relative change dp/p are rotated in the same angle φ as the pole p.

In Figure 4, the pole p, and its variation, are shown with respect to the pole frequency and pole quality variations, given by Equations 30 and 31. It is important to realize that the change in the pole frequency has the same direction as the *relative change dp/p* of the pole p and that the positive change of the pole quality factor increases the imaginary part of the pole p, even if the imaginary part of the *relative change dp/p* is negative.

3. SENSITIVITIES OF TRANSFER FUNCTIONS TO ELEMENTS

3.1 SENSITIVITY DEFINITIONS AND RULES

We start with a function of several variables

$$y = y(x_1, x_2, \dots, x_n) \tag{32}$$

and with its complete differential

$$dy = \frac{\delta y}{\delta x_1} dx_1 + \frac{\delta y}{\delta x_2} dx_2 + \ldots + \frac{\delta y}{\delta x_n} dx_n \tag{33}$$

The relative variation is obtained as a ratio of Equations 32 and 33. If we simultaneously multiply and divide each term on the right side by the variable x_i, we get the expression

$$\frac{dy}{y} = \frac{\delta y}{\delta x_1} \frac{x_1}{y} \frac{dx_1}{x_1} + \frac{\delta y}{\delta x_2} \frac{x_2}{y} \frac{dx_2}{x_2} + \ldots + \frac{\delta y}{\delta x_n} \frac{x_n}{y} \frac{dx_n}{x_n} \tag{34}$$

The sensitivity of the function y with respect to the variable x is defined as the multiplication factor by which the relative variation of the variable x has to be multipied to obtain the relative variation of the function y. If only one variable is considered, it is usually written in the form

$$S_x^y = \frac{dy/y}{dx/x} \tag{35}$$

Another equivalent definition has been given by Bode:[4]

$$S_x^y = \frac{d \ln y}{d \ln x} = \frac{dy/y}{dx/x} \tag{36}$$

It allows us to write Equation 34 in the form

$$\frac{dy}{y} = S_{x_1}^y \cdot \frac{dx_1}{x_1} + S_{x_2}^y \cdot \frac{dx_2}{x_2} + \ldots + S_{x_n}^y \cdot \frac{dx_n}{x_n} \tag{37}$$

This mathematical expression is general, but we are usually interested in measurable changes of the function, caused by measurable changes of the elements. In such case, it is advantageous to convert Equation 37 into an expression in increments:

$$\frac{\Delta y}{y} \approx S_{x_1}^y \cdot \frac{\Delta x_1}{x_1} + S_{x_2}^y \cdot \frac{\Delta x_2}{x_2} + \ldots + S_{x_n}^y \cdot \frac{\Delta x_n}{x_n} \tag{38}$$

Mathematical operations involving sensitivities are considerably simplified by a set of rules collected in Table 1. The formulas can be derived by applying the definitions in Equations 35 or 36. Additional sensitivity rules can be found in Reference 2.

3.2 SENSITIVITY OF AMPLITUDE AND PHASE RESPONSES

Sensitivity responses of the amplitude and phase are important for any active RC network. If we calculate the amplitude and the phase responses, the frequency-domain variable s is restricted to the imaginary axis only, by setting $s = jw$ for all following equations. The amplitude response sensitivity with respect to an element x is obtained by using the sensitivity rules 12 and 16:

$$\frac{\Delta |T(s)|}{|T(s)|} = \mathrm{Re}\left(S_x^{T(s)}\right) \cdot \frac{\Delta x}{x} \tag{39}$$

$$\alpha(\omega) = 20 \cdot \log |T(s)| \quad [\mathrm{dB}] \tag{40}$$

Table 1 Sensitivity Rules

1.	$S_x^x = S_x^{cx} = 1$	2.	$S_x^{cy} = S_x^y$	3.	$S_x^{cx^n} = n$ c and n are constants
4.	$S_x^{y^n} = n \cdot S_x^y$	5.	$S_{x^n}^y = \frac{1}{n} \cdot S_x^y$	6.	$S_x^{uv} = S_x^u + S_x^v$ u and v are functions
7.	$S_x^{u/v} = S_x^u - S_x^v$	8.	$S_x^y = S_{u1}^y \cdot S_x^{u1} + S_{u2}^y \cdot S_x^{u2} + \cdots$ where $y = y(u1, u2, \ldots)$		
9.	$S_x^{y+c} = \frac{y}{y+c} \cdot S_x^y$	10.	$S_x^{e^y} = y \cdot S_x^y$	11.	$S_x^{\ln\lvert y\rvert} = \frac{1}{\ln\lvert y\rvert} \cdot S_x^{\lvert y\rvert}$
12.	$S_x^{20 \cdot \log\lvert y\rvert} = \frac{20 \cdot \log e}{20 \cdot \log\lvert y\rvert} \cdot S_x^{\lvert y\rvert}$ where $20 \cdot \log e = 8.68588964$				
13.	$S_x^{u+v+\cdots} = \frac{1}{u+v+\cdots} \cdot (u \cdot S_x^u + v \cdot S_x^v + \ldots)$				
14.	$S_x^y = S_{u_1}^y \cdot S_{u_2}^{u_1} \cdot S_{u_3}^{u_2} \cdot \ldots \cdot S_x^{u_n}$ where $y = y(u_1(u_2(u_3(\ldots u_n(x)\ldots))))$				
Writing $y = \lvert y\rvert \cdot e^{j \cdot \varphi y}$ and $y^* = \lvert y\rvert \cdot e^{-j \cdot \varphi y}$, we obtain					
15.	$S_x^y = S_x^{\lvert y\rvert} + j \cdot \varphi y \cdot S_x^{\varphi y}$	16.	$S_x^{\lvert y\rvert} = \mathrm{Re}(S_x^y)$	17.	$S_x^{\varphi y} = \frac{1}{\varphi y} \cdot Im(S_x^y)$
18.	$S_x^{y^*} = (S_x^y)^*$				

Note: Additional sensitivity rules can be found in Reference 2.

Amplitude increment in decibels is

$$\Delta\alpha(\omega) = 8.68589 \cdot \mathrm{Re}\left(S_x^{T(s)}\right) \cdot \frac{\Delta x}{x} \quad [\mathrm{dB}] \tag{41}$$

Increment of the phase response is obtained by applying rule 17. If we convert radians into degrees, then

$$\Delta\varphi(\omega) = 57.29578 \cdot \mathrm{Im}\left(S_x^{T(s)}\right) \cdot \frac{\Delta x}{x} \quad [\mathrm{Degrees}] \tag{42}$$

The sensitivity of the transfer function $x\ S_x^{T(s)}$ with respect to the element is a complex and frequency-dependent function. The amplitude and phase responses, as well as their sensitivities, are frequency dependent, but are real functions. Equations 41 and 42 are therefore valid only for real relative changes of $\Delta x/x$. This is the case for passive elements, like resistors and capacitors. It is not generally true for amplifiers.

Using Equation 38, we obtain the overall increment of the amplitude response with respect to all passive elements:

$$\Delta\alpha(\omega) = 8.68589 \cdot \sum_{i=1}^{n} \mathrm{Re}\left(S_{x_i}^{T(s)}\right) \cdot \frac{\Delta x_i}{x_i} \quad [\mathrm{dB}] \tag{43}$$

For the overall increment of the phase response we get

$$\Delta\varphi(\omega) = 57.29578 \cdot \sum_{i=1}^{n} \mathrm{Im}\left(S_{x_i}^{T(s)}\right) \cdot \frac{\Delta x_i}{x_i} \quad [\mathrm{Degrees}] \tag{44}$$

If the transfer function is written as a rational function of two polynomials Equation 1, then the sensitivity of the transfer function with respect to a passive element, x, becomes

$$S_x^{T(s)} = S_x^{N(s)} - S_x^{D(s)} = \sum_{i=0}^{m} \frac{a_i s^i}{N(s)} \cdot S_x^{a_i} - \sum_{j=0}^{n} \frac{b_j s^j}{D(s)} \cdot S_x^{b_j} \tag{45}$$

Writing the tranfer function in the form of Equation 10 and applying the sensitivity rules, we obtain

$$S_x^{T(s)} = S_x^K - \sum_{i=1}^{m} \frac{z_i}{s - z_i} \cdot S_x^{z_i} + \sum_{j=1}^{n} \frac{p_j}{s - p_j} \cdot S_x^{p_j} \tag{46}$$

Equation 45 gives the sensitivity of the transfer function in terms of its coefficients; Equation 46 gives it in terms of its poles and zeros sensitivities. To compare various realizations of the same transfer function, the coefficient sensitivities and the sensitivities of poles and zeros have to be evaluated. In practice, sensitivities of the dominant, highest-quality factor pole, are considered first.

3.3 SENSITIVITY OF POLE FREQUENCY AND POLE QUALITY FACTOR

The relative increment of the pole frequency ω_p is obtained from Equation 26:

$$\frac{\Delta\omega_{pi}}{\omega_{pi}} = S_x^{\omega_{pi}} \cdot \frac{\Delta x}{x} = \mathrm{Re}\left(S_x^{p_i}\right) \cdot \frac{\Delta x}{x} \tag{47}$$

For the pole quality factor, q_p, we obtain

$$\frac{\Delta q_{pi}}{q_{pi}} = S_x^{q_{pi}} \cdot \frac{\Delta x}{x} = -\sqrt{4 \cdot q_{pi}^2 - 1} \cdot \mathrm{Im}\left(S_x^{p_i}\right) \cdot \frac{\Delta x}{x} \tag{48}$$

The relative increment of the pole p_i is, according to Equation 30,

$$\frac{\Delta p_i}{p_i} = S_x^{\omega_{pi}} \cdot \frac{\Delta x}{x} - j \cdot \frac{1}{\sqrt{4q_{pi}^2 - 1}} \cdot S_x^{q_{pi}} \cdot \frac{\Delta x}{x} \tag{49}$$

Consequently, the pole sensitivity is

$$S_x^{pi} = S_x^{\omega_{pi}} - j \cdot \frac{1}{\sqrt{4q_{pi}^2 - 1}} \cdot S_x^{q_{pi}} \tag{50}$$

In practice, the pole frequency and the pole quality factor sensitivities are calculated first. The resulting pole shift is given by Equation 49, sensitivities of the transfer function by Equation 46, and increments of the amplitude or phase by Equations 43 and 44. All equations are valid only for real changes of the passive elements. Changes of the operational amplifier, and their influence on performance, depend on the internal compensation of the amplifier and are, in general, frequency dependent and complex.

4. GAIN-SENSITIVITY PRODUCT

4.1 REASON FOR GAIN-SENSITIVITY PRODUCT

Network functions are usually calculated for an ideal operational amplifier with infinite open loop gain, $A \to \infty$. In reality, this gain is finite, complex, and frequency dependent. Each network function,

F, which is stable and is differentiable in the vinicity of $A \to \infty$, can be expanded into a Taylor series. The first two terms of this expansion have the following form:

$$F = F_{id} + \frac{F_a}{A(s)} \tag{51}$$

Both F_{id} and F_a are functions of passive elements only. Calculation of the sensitivity with respect to the open loop gain gives

$$S_A^F = -\frac{F_a}{F \cdot A(s)} / _{A(s)\to\infty} = 0 \tag{52}$$

Because this expression tends to zero for the gain approaching infinity, a new definition, the gain-sensitivity product, was introduced. It is finite even if the operational amplifer gain approaches infinity:

$$\Gamma_A^F = A \cdot S_A^F = -\frac{F_a}{F} / _{A(s)\to\infty} = -\frac{F_a}{F_{id}} \tag{53}$$

The gain-sensitivity product is a function of passive network elements only.

4.2 GAIN-SENSITIVITY PRODUCT OF POLE FREQUENCY AND POLE QUALITY FACTOR

Relative changes of the pole frequency and of the pole quality factor with respect to the finite open loop gain of the amplifier are[5]

$$\frac{\Delta\omega_{pi}}{\omega_{pi}} = \mathrm{Re}\left[\left(\Gamma_A^{\omega_{pi}} - j\frac{\Gamma_A^{q_{pi}}}{\sqrt{4q_{pi}^2 - 1}}\right)\cdot\left(\mathrm{Re}\left[\frac{\Delta A}{A^2}\right] + j\,\mathrm{Im}\left[\frac{\Delta A}{A^2}\right]\right)\right] \tag{54}$$

$$\frac{\Delta q_{pi}}{q_{pi}} = -\sqrt{4q_{pi}^2 - 1}\cdot \mathrm{Im}\left[\left(\Gamma_A^{\omega_{pi}} - j\frac{\Gamma_A^{q_{pi}}}{\sqrt{4q_{pi}^2 - 1}}\right)\cdot\left(\mathrm{Re}\left[\frac{\Delta A}{A^2}\right] + j\,\mathrm{Im}\left[\frac{\Delta A}{A^2}\right]\right)\right] \tag{55}$$

These equations can be used to optimize the individual internal frequency compensation, like in the 748 operational amplifier. If the operational amplifier is internally compensated, we assume a 20-dB roll-off and constant phase shift –90° of the open loop gain. For example, the TL071 operational amplifier has the phase shift of 90° in the frequency range from 100 Hz to 1 MHz and a gain-bandwidth product, GB, of about 3 MHz. In this practical frequency range we can write the open loop gain as

$$A(s)/_{s=j\omega} \approx -j\,\frac{GB}{\omega} \tag{56}$$

and obtain for the amplifier gain changes:

$$\mathrm{Re}\left[\frac{\Delta A}{A^2}\right] = 0 \tag{57}$$

and

$$\text{Im}\left[\frac{\Delta A}{A^2}\right]_{s=j\omega} = \frac{\omega}{GB} \cdot \frac{\Delta A}{A} \tag{58}$$

For the single-pole roll-off of a frequency-compensated amplifier, we obtain

$$\frac{\Delta\omega_{pi}}{\omega_{pi}} = \frac{\Gamma_A^{q_{pi}}}{\sqrt{4q_{pi}^2 - 1}} \cdot \text{Im}\left[\frac{\Delta A}{A^2}\right] \tag{59}$$

$$\frac{\Delta q_{pi}}{q_{pi}} = -\sqrt{4q_{pi}^2 - 1} \cdot \Gamma_A^{\omega_{pi}}\, \text{Im}\left[\frac{\Delta A}{A^2}\right] \tag{60}$$

In most practical cases the gain-sensitivity product of the pole frequency is zero, and only Equation 59 is considered for optimization of an active RC network. The amplifier type determines the values given in Equation 58. To get the best results, the gain-sensitivity product of the pole quality factor must be optimized. Such optimization will minimize the pole frequency shift caused by the nonideal amplifier and by the changes of its open loop gain. In some cases, the shift of the pole frequency must be avoided. This will be the case in a second-order allpass filter where the operational amplifier must be compensated externally in order to achieve a phase shift of the open loop amplifier gain approximately –180°. Under these conditions a nonideal amplifier will cause a quality factor shift instead of a pole frequency shift [reference 5, pages 174–176].

5. OPTIMIZATION STRATEGY WITH A PRACTICAL EXAMPLE

As a practical example we consider the second-order bandpass filter in Figure 5. It is used for medium-quality factors q_p between 2 and 20.

5.1 CALCULATION OF THE TRANSFER FUNCTION

If we consider a nonideal operational amplifier with an open loop gain A, then the transfer function is

$$T_{15}(s) = -\frac{1}{\frac{1}{A} + \frac{R_6}{R_5 + R_6}} \cdot \frac{s \cdot \frac{1}{R_1 C_2}}{s^2 + s\left[\frac{1}{R_4 C_2} + \frac{1}{R_4 C_3} + \frac{1}{R_1 C_2} \cdot \frac{\frac{1}{A} - \frac{R_5}{R_5 + R_6}}{\frac{1}{A} + \frac{R_6}{R_5 + R_6}}\right] + \frac{1}{R_1 C_2 C_3 R_4}} \tag{61}$$

This form is necessary to calculate sensitivities to network elements and the gain-sensitivity product of the amplifier. Standard nodal analysis or signal flow graph theory can be used to obtain the function. It is recommended that the function be normalized to the form:

$$T(s) = K \cdot \frac{\frac{\omega_p}{q_p} \cdot s}{s^2 + \frac{\omega_p}{q_p} \cdot s + \omega_p^2} \tag{62}$$

In this form, the function will be equal to the constant K at the pole frequency. We will not consider variations of K because it influences only the gain of the filter. Important for us is the pole stability, given by the variations of the pole frequency and of the pole quality factor.

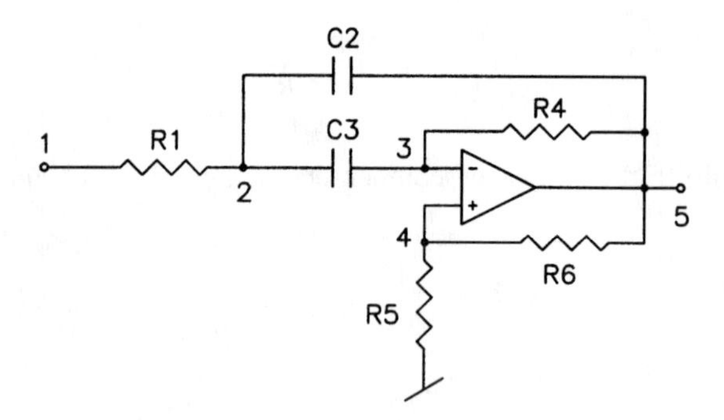

Figure 5 Medium q_p second-order bandpass filter.

Comparison of Equations 61 and 62 provides

$$\omega_p = \frac{1}{\sqrt{R_1 C_2 C_3 R_4}} \tag{63}$$

$$q_p = \frac{1}{\sqrt{R_1 C_2 C_3 R_4}} \cdot \frac{1}{\dfrac{1}{R_4 C_2} + \dfrac{1}{R_4 C_3} + \dfrac{1}{R_1 C_2} \cdot \dfrac{\dfrac{1}{A} - \dfrac{R_5}{R_5 + R_6}}{\dfrac{1}{A} + \dfrac{R_6}{R_5 + R_6}}} \tag{64}$$

$$K = -\frac{\sqrt{\dfrac{C_3 R_4}{R_1 C_2}}}{\dfrac{1}{A} + \dfrac{R_6}{R_5 + R_6}} \cdot q_p \tag{65}$$

For an ideal operational amplifier with an infinite gain, the last two equations reduce to

$$q_p = \frac{\sqrt{\dfrac{R_4 C_2}{R_1 C_3}}}{1 + \dfrac{C_2}{C_3} - \dfrac{R_4 R_5}{R_1 R_6}} \tag{66}$$

$$K = q_p \cdot \left(1 + \frac{R_5}{R_6}\right) \cdot \sqrt{\frac{R_4 C_3}{R_1 C_2}} \tag{67}$$

5.2 SENSITIVITY CALCULATION

If we apply Equations 63 and 66 and the sensitivity rules from Table 1, the pole frequency and the pole quality factor sensitivities will be

$$S_{R_1}^{\omega_p} = S_{C_2}^{\omega_p} = S_{C_3}^{\omega_p} = S_{R_4}^{\omega_p} = -\frac{1}{2} \tag{68}$$

$$S_{R_1}^{q_p} = -\frac{1}{2} - \frac{\frac{R_4 R_5}{R_1 R_6}}{1 + \frac{C_2}{C_3} - \frac{R_4 R_5}{R_1 R_6}} \tag{69}$$

$$S_{R_4}^{q_p} = \frac{1}{2} + \frac{\frac{R_4 R_5}{R_1 R_6}}{1 + \frac{C_2}{C_3} - \frac{R_4 R_5}{R_1 R_6}} \tag{70}$$

$$S_{R_5}^{q_p} = \frac{\frac{R_4 R_5}{R_1 R_6}}{1 + \frac{C_2}{C_3} - \frac{R_4 R_5}{R_1 R_6}} \tag{71}$$

$$S_{R_6}^{q_p} = -\frac{\frac{R_4 R_5}{R_1 R_6}}{1 + \frac{C_2}{C_3} - \frac{R_4 R_5}{R_1 R_6}} \tag{72}$$

$$S_{C_2}^{q_p} = \frac{1}{2} - \frac{\frac{C_2}{C_3}}{1 + \frac{C_2}{C_3} - \frac{R_4 R_5}{R_1 R_6}} \tag{73}$$

$$S_{C_3}^{q_p} = -\frac{1}{2} + \frac{\frac{C_2}{C_3}}{1 + \frac{C_2}{C_3} - \frac{R_4 R_5}{R_1 R_6}} \tag{74}$$

5.3 GAIN-SENSITIVITY PRODUCT CALCULATION

By using Equation 64, the gain-sensitivity product of the quality factor becomes

$$\Gamma_A^{q_p} = A \cdot \frac{-\frac{1}{R_1 C_2} \cdot \frac{\frac{1}{A} - \frac{R_5}{R_5 + R_6}}{\frac{1}{A} + \frac{R_6}{R_5 + R_6}}}{\frac{1}{R_4 C_2} + \frac{1}{R_4 C_3} + \frac{1}{R_1 C_2} \cdot \frac{\frac{1}{A} - \frac{R_5}{R_5 + R_6}}{\frac{1}{A} + \frac{R_6}{R_5 + R_6}}} \cdot \left[\frac{-\frac{1}{A}}{\frac{1}{A} - \frac{R_5}{R_5 + R_6}} + \frac{\frac{1}{A}}{\frac{1}{A} + \frac{R_6}{R_5 + R_6}} \right] \tag{75}$$

Should the gain be infinitely large, then we get the following simplifications:

$$\Gamma_A^{q_p} = q_p \cdot \left(1 + \frac{R_5}{R_6}\right)^2 \cdot \sqrt{\frac{R_4 C_3}{R_1 C_2}} \tag{76}$$

and

$$\Gamma_A^{\omega_p} = 0 \tag{77}$$

5.4 OPTIMIZATION STRATEGY

Suppose that we are asked to match a given pole frequency and a given pole quality factor for the bandpass filter specified in Equation 62. We do not care about variations of the constant K. We must select six passive elements, but we have only two Equations (63) and (66). From a practical point of view, it is best to postpone the selection of capacitor values. Looking at Equation 63, we see that we must specify the resistor product $R_1 * R_4$. In addition, an objective function must be defined for optimization.

It is advantageous to select practical capacitor and resistor ratios. We define the following steps in our calculations:

1. In: f_p, q_p, C_2, R_6
2. The ratio of C_3/C_2 can be in the range from 0.1 to 10
3. The ratio of R_4/R_1 can be in the range from 1 to 100
4. Calculate:

$$R_1 = \frac{1}{2 \cdot \pi \cdot f_p \cdot \sqrt{\frac{R_4}{R_1} \cdot C_2 C_3}} \tag{78}$$

$$R_5 = R_6 \cdot \left[\frac{R_1}{R_4} \cdot \left(1 + \frac{C_2}{C_3}\right) - \frac{1}{q_p} \cdot \sqrt{\frac{R_1 C_2}{R_4 C_3}}\right] \tag{79}$$

5. Calculate the gain-sensitivity product and the sensitivities to passive elements
6. Calculate objective functions as discussed in the following

It can be shown [reference 5] that the minimum of the gain-sensitivity product is obtained for the following ratios of R_4/R_1 and C_2/C_3:

$$\frac{R_4}{R_1} = \frac{\frac{C_2}{C_3}}{4q_p^2} \cdot \left[\sqrt{1 + 12q_p^2\left(1 + \frac{C_3}{C_2}\right)} - 1\right]^2 \tag{80}$$

However, it may turn out that the minimal gain-sensitivity product is not a good choice for the best dimensioning of the network. To get a better insight, we define an objective function which considers the sensitivities to passive elements and the influence of the nonideal amplifier on the gain-sensitivity product. One possibility is to minimize the shift of the pole vector shown in Figures 3 and 4.

In order to minimize the pole shift, the real and imaginary parts of the following equations must be minimized. The pole shift with respect to the relative changes of passive elements

$$\frac{\Delta p}{p} = \sum S_{x_i}^{\omega_p} \cdot \frac{\Delta x_i}{x_i} \mp j \cdot \frac{1}{\sqrt{4q_p^2 - 1}} \cdot \sum S_{x_i}^{q_p} \cdot \frac{\Delta x_i}{x_i} \tag{81}$$

and the pole shift with respect to the relative change of the open loop gain, A,

$$\frac{\Delta p}{p} = \frac{\Delta\omega_p}{\omega_p}[\text{eqn. } 54] - j\cdot\frac{1}{\sqrt{4q_p^2-1}}\cdot\frac{\Delta q_p}{q_p}[\text{eqn. } 55] \tag{82}$$

We now define three objective functions:

$$O_1 = \sum_{x_i} \mathrm{Re}\left(\left|\frac{\Delta p}{p}\right|\right)_{x_i} = \sum_{x_i}\left|S_{x_i}^{\omega_p}\right|\cdot\frac{\Delta x_i}{x_i} \tag{83}$$

$$O_2 = \sum_{x_i} \mathrm{Im}\left(\left|\frac{\Delta p}{p}\right|\right)_{x_i} = \frac{1}{\sqrt{4q_p^2-1}}\sum_{x_i}\left|S_{x_i}^{q_p}\right|\cdot\frac{\Delta x_i}{x_i} \tag{84}$$

$$O_3 = \mathrm{Re}\left(\left|\frac{\Delta p}{p}\right|\right)_A = \frac{\Gamma_A^{q_p}}{\sqrt{4q_p^2-1}}\cdot\frac{\omega}{GB}\cdot\frac{\Delta A}{A} \tag{85}$$

and an overall objective function:

$$O_4 = O_1 + O_2 + O_3 \tag{86}$$

In the evaluation of the objective functions we take into account the temperature coefficients of the resistors, of the capacitors, and of the amplifier gain. We also consider the gain bandwidth product of the amplifier.

5.5 RESULTS IN EXCEL

The spread sheet EXCEL was used to optimize the bandpass network in Figure 5. Any other general-purpose program with three-dimensional plotting can be used. The following data were used:

$$f_p = 65 \text{ kHz};\ q_p = 10;\ C_2 = 1 \text{ nF};\ R_6 = 10 \text{ k};\ GB = 3 \text{ MHz};$$

$$dC/C = 100 \text{ ppm/K};\ dR/R = 50 \text{ ppm/K};\ dA/A = 10\ 000 \text{ ppm/K}$$

We consider the temperature range from 25 to 75°C. The open loop gain and the gain bandwidth product were taken from the data sheet of the TI amplifier TL 071.

Calculation of the first objective function, Equation 83, used results from Equation 68

$$O_1 = \frac{\Delta R}{R} + \frac{\Delta C}{C} = 1.5E-04 \tag{87}$$

This objective function turns out to be a constant and gives the pole frequency shift with respect to the temperature coefficients of the resistors and capacitors. Since it is a constant, there is no need for plotting.

Equations 69 through 74 were used in the objective function O_2, Equation 84. After some steps we get

$$O_2 = \frac{1}{\sqrt{4q_p^2-1}}\cdot\left[\left(2+4q_p\frac{R_5}{R_6}\sqrt{\frac{R_4C_3}{R_1C_2}}\right)\frac{\Delta R}{R} + \left(2q_p\sqrt{\frac{R_1C_2}{R_4C_3}}-1\right)\frac{\Delta C}{C}\right] \tag{88}$$

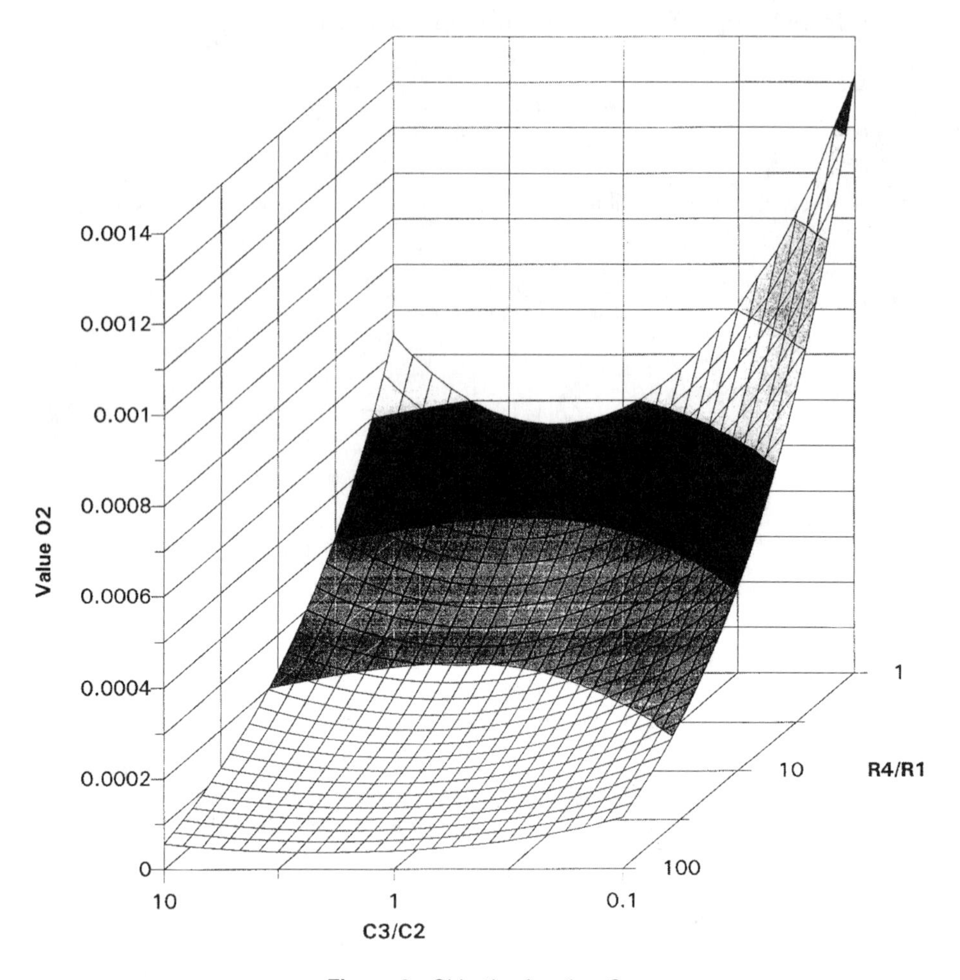

Figure 6 Objective function O_2.

The plot of this objective function is in Figure 6, and a zoomed part around its minimum is in Figure 7. For our range of element values (R_4/R_1 between 1 to 100; C_3/C_2 between 0.1 to 10), the minimum is found for $R_4/R_1 = 100$ and $C_3/C_2 = 2.2$. This minimum represents the best choice of the passive element values to minimize their influence on the pole quality factor.

The objective function O_3 is given by Equation 85. It is proportional to the gain-sensitivity product, and Figures 8 and 9 show its plots. Comparing the objective functions O_2 and O_3 we see that their minima are at different places. This means that minimizing the influence of passive element variations will not minimize the influence of the operational amplifier gain variations, and vice versa. We clearly need the sum of all influences, defined in the objective function O_4. Its plot is in Figures 10 and 11. It is similar to the plot of O_3 because of 65 kHz the variations of the open loop gain have more influence than the variations of passive elements. To show this effect, the objective function O_4 was also evaluated for the same bandpass at the frequency 1 kHz and the results are shown in Figure 12. We see that the minimum is shifting in the direction of the minimum of the objective function O_2 (see Figures 6 and 7).

The resulting optimal designs are

Bandpass filter at 65 kHz; $R_4/R_1 = 38.3$; $C_3/C_2 = 0.1$; Objective function $O_4 = \text{le}^{-03}$ — One possible solution is R_1 = lk251; R_4 = 47k91; C_2 = 1 nF; C_3 = 100 pF; R_5 = 2k36; R_6 = 10 k.

Bandpass filter at 1 kHz; $R_4/R_1 = 100$; $C_3/C_2 = 1$; Objective function $O_4 = 2.2\text{e}^{-04}$ — One possible solution is R_1 = 1k592; R_4 = 159k2; C_2 = 10 nF; C_3 = 10 nF; R_5 = 100E; R_6 = 10 k.

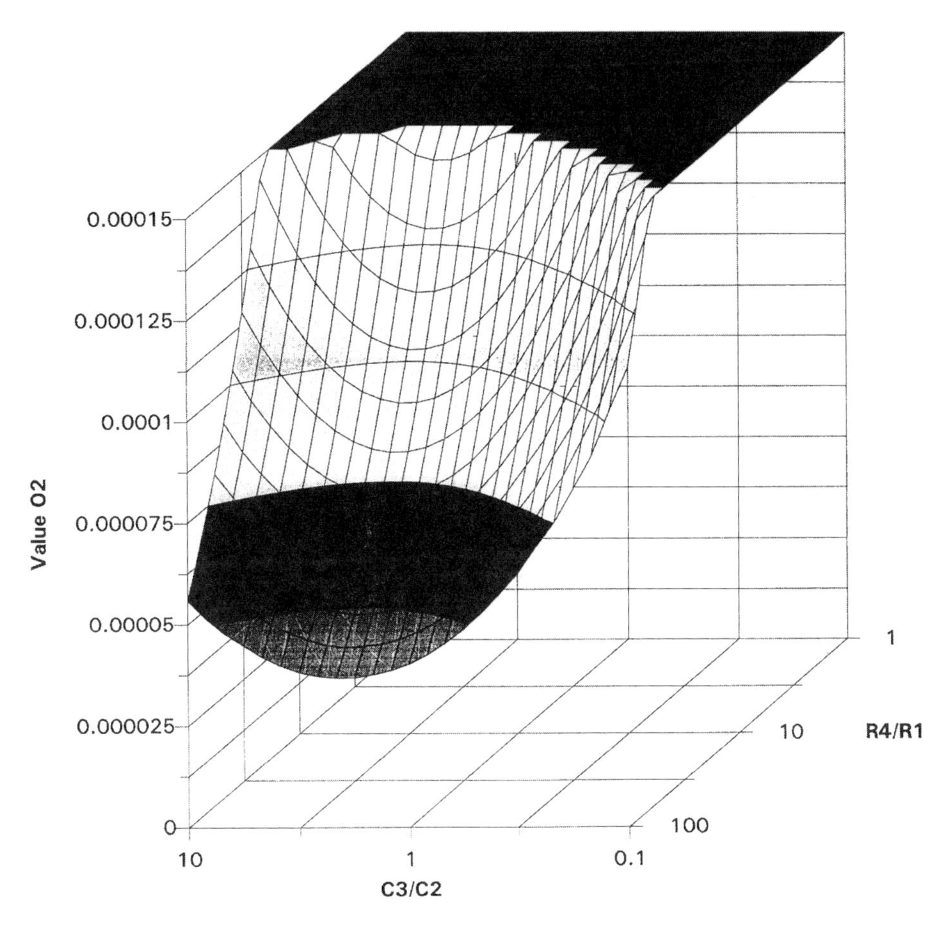

Figure 7 Objective function O_2 zoomed.

5.6 TUNING STRATEGY

The tuning strategy is based on the sensitivity calculations. We see from Equations 68 to 74 that the pole frequency tuning has to be done first, with either R_1 or R_4. Afterward, the pole quality will be tuned using the resistors R_5 or R_6; this should not influence the pole frequency. In practice, as can be seen from Equation 59, the pole frequency will also be shifted. As a result, the tuning steps will have to be repeated until the match of the desired values is reached.

5.7 SUMMARY

We have shown how to optimize a second-order active RC network using sensitivities of passive elements and the gain-sensitivity product of ideal amplifiers. Optimization is performed visually, and the user needs only a three-dimensional plotting package, widely available in many computer programs. The objective functions take into account temperature coefficients of the elements. To proceed, we must

1. Specify ranges of acceptable values for the resistors and capacitors,
2. Know the gain, A, and the gain bandwidth product, GB, of the operational amplifier, and
3. Know the temperature coefficients of passive (dR/R, dC/C) active (dA/A) elements.

Once these data are available, we evaluate the objective functions for ratios of capacitors and resistors, as shown on the plots. Minima of the plots are determined visually, eventually by zooming to the area of the minimum.

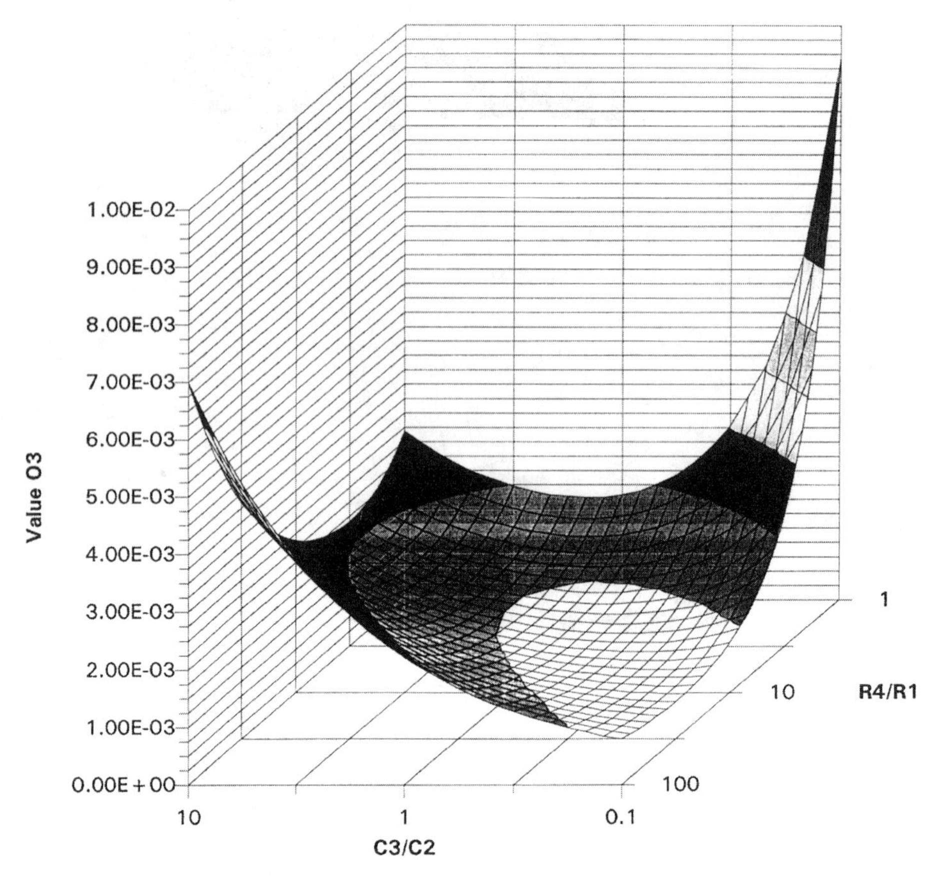

Figure 8 Objective function O_3.

5.8 SPECIALIZED PROGRAMS

The above strategy for active RC network optimization gives very good results, but is time-consuming because the equations have to be derived by hand. It is applicable only to second-order network functions. Some programs can help by providing the symbolic function in the form of Equation 61, but a considerable amount of work still remains. We wish to mention here a program, NETOPT, which was specially written for the design of analog filters. It has the usual features, like frequency- and time-domain analysis, sensitivity analysis, poles and zeros, and their sensitivities. In addition, it is coupled interactively with an optimizer and can be used to design filters in the presence of parasitics, nonideal operational amplifiers, etc. To get the desired result, the user provides the filter topology, specifies which elements can be changed, and defines requirements on the passband, the stopband, and the phase response (if the filter has the capability to control independently the amplitude and phase). Up to 75 element values can be simultaneously modified to get the desired result. The program is available for personal computers; for information contact the author.

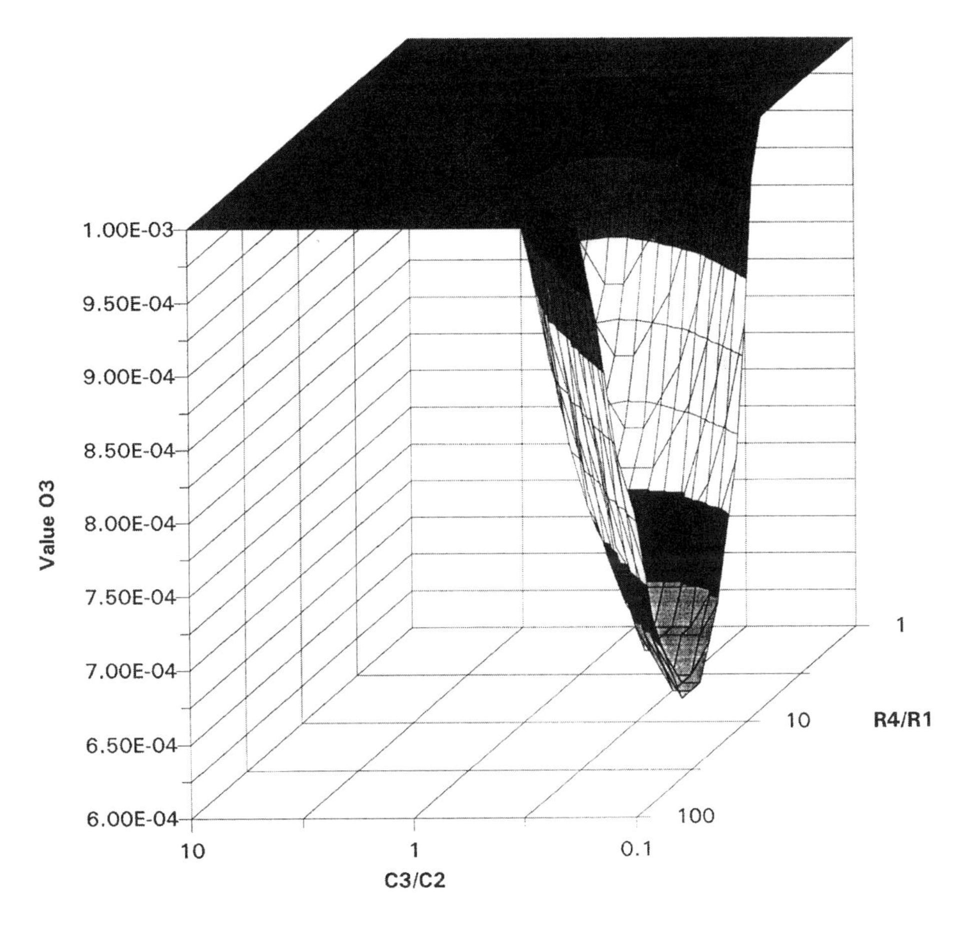

Figure 9 Objective function O_3 zoomed.

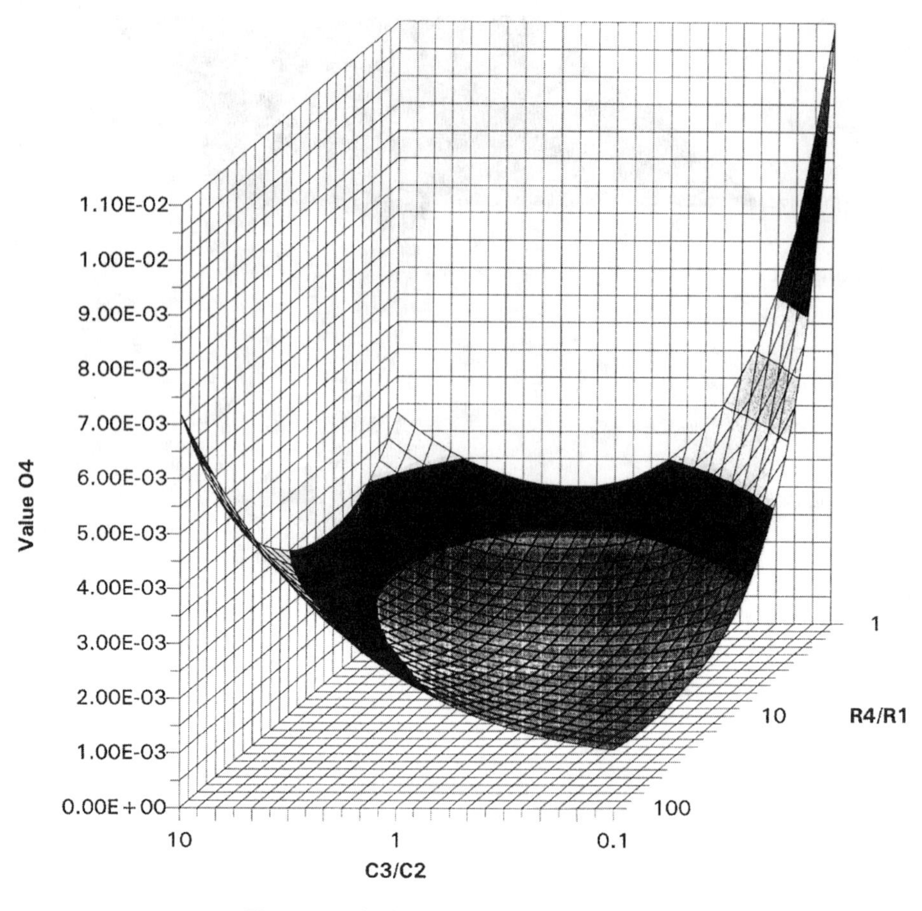

Figure 10 Objective function O_4, f_p = 65 kHz.

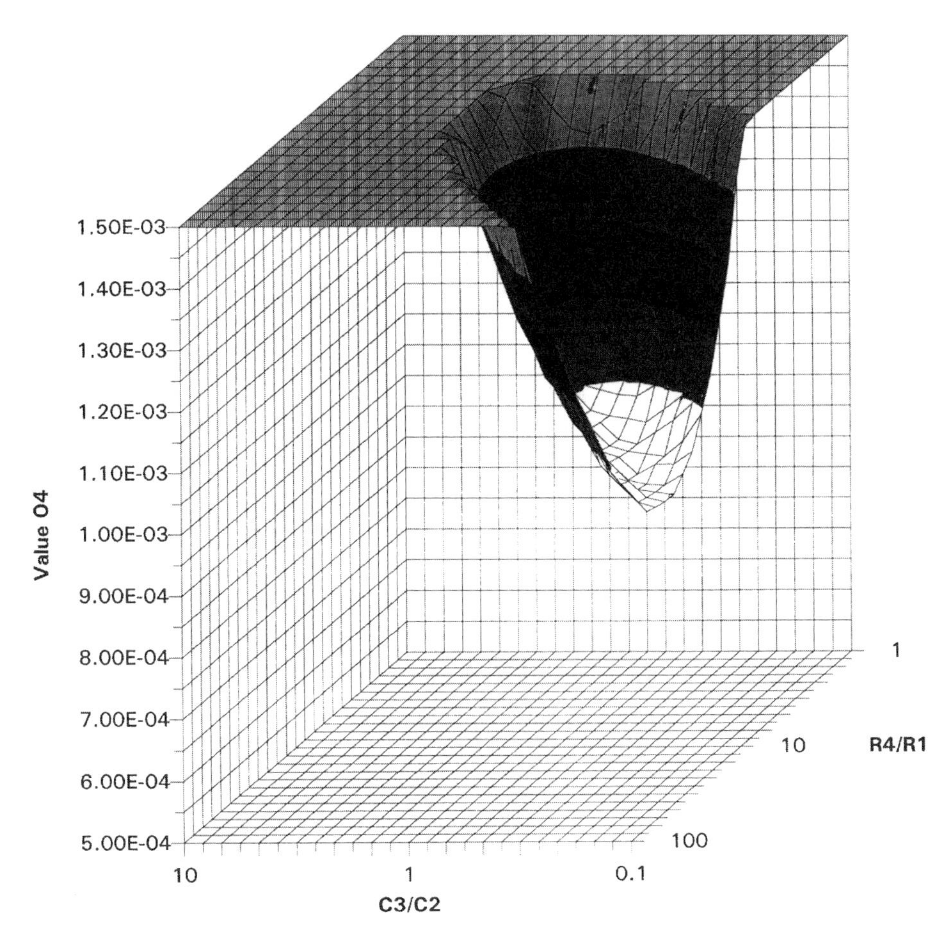

Figure 11 Objective function O_4 zoomed, f_p = 65 kHz.

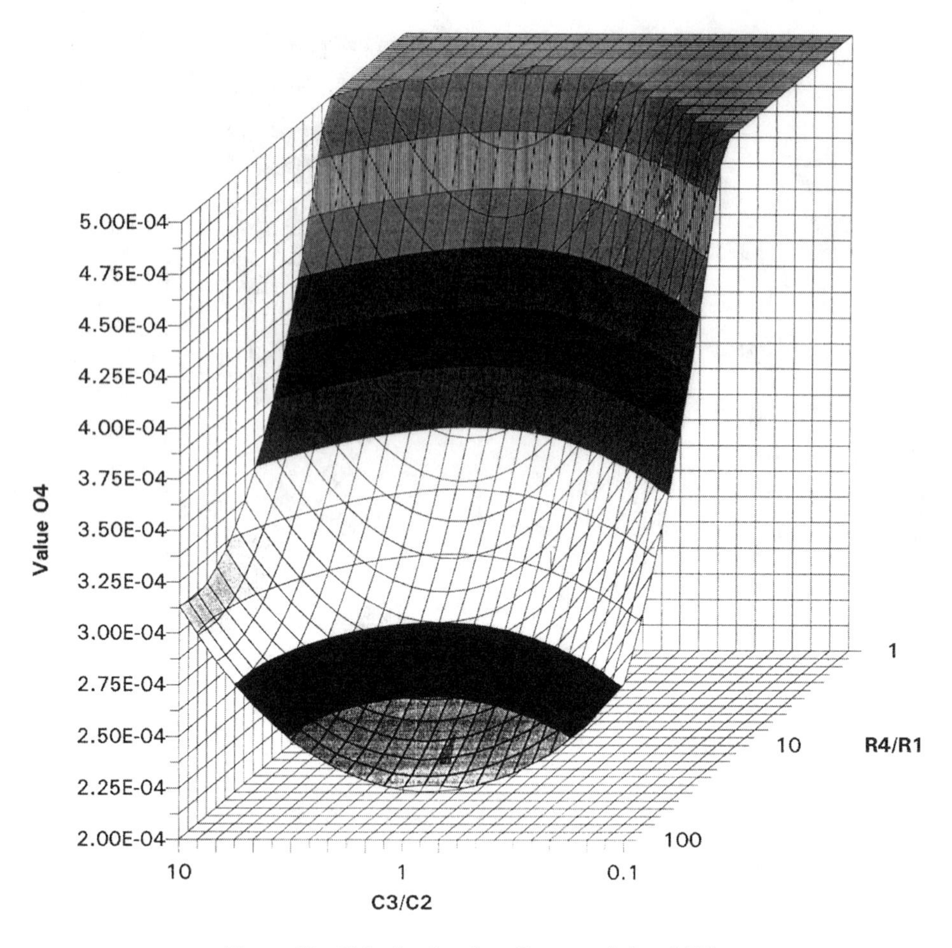

Figure 12 Objective function O_4 zoomed, $f_p = 1$ kHz.

REFERENCES

1. **J. Vlach,** Network theory and CAD, *IEE Trans.,* Educations, 36(1), 23–27, February, 1993.
2. **G. S. Moschytz,** *Linear Integrated Networks, Fundamentals,* Van Nostrand Reinhold, New York, 1974.
3. **J. Vlach, K. Singhal,** *Computer Methods for Circuit Analysis and Design,* Van Nostrand Reinhold, New York, 1983.
4. **H. W. Bode,** *Network Analysis and Feedback Amplifier Design,* Van Nostrand, New York, 1945, 52.
5. **G.S. Moschytz and P. Horn,** *Active Filter Design Handbook,* John Wiley & Sons, New York, 1981, 25.

Higher-Order Filters

F. William Stephenson and William B. Kuhn

CONTENTS

1. INTRODUCTION

There are two broad approaches to the design of higher-order active filters. Cascade design entails the interconnection of first- and second-order sections to yield the desired characteristic. These noninteracting stages allow the designer to isolate pole and zero pairs so that the tuning of each section is routine and precise. However, the resulting filter is relatively more sensitive than designs based on LC passive structures. This latter approach makes use of the fact that LC filters are minimally sensitive in the passband. Therefore, direct inductor replacement methods or multi-loop/active analogs yield less-sensitive realizations than the cascade of biquads. However, tuning is necessarily more difficult because of the strong interdependence of components within the structure.

This chapter reviews the major features of cascade, direct replacement, and multiloop synthesis of high-order active filters. Several examples are presented to illustrate the techniques. A summary of implementation technologies and available components concludes the discussion.

2. HIGHER-ORDER DESIGN METHODS

2.1 CASCADE DESIGN

This very simple approach is extremely convenient for nonexacting filter designs. Any of the second-order sections described earlier may be used to yield a higher-order filter by the method shown in

0-8493-8951-8/97/$0.00+$.50
© 1997 by CRC Press LLC

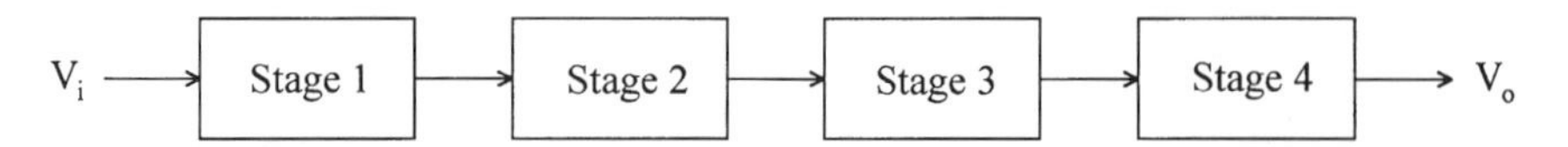

Figure 1 High-order filter realized as the cascade of second-order stages.

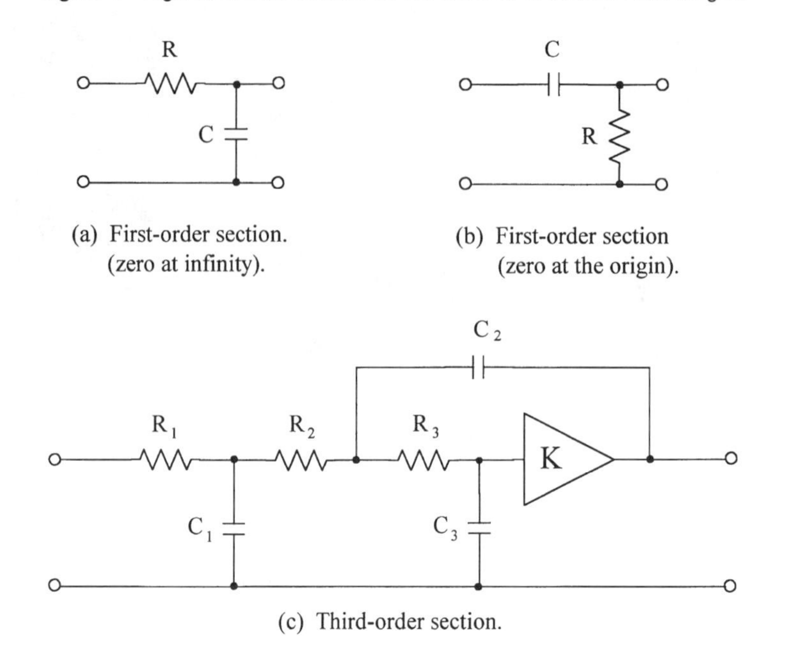

Figure 2 Odd-order sections for use in cascade designs. (a) First-order section (zero at infinity), (b) first-order section (zero at the origin); and (c) third-order section.

Figure 1. The additional pole required for an odd-order realization may be incorporated as a first-order RC section or by a third-order structure as shown in Figure 2. In the latter case, the real pole should be associated with the lowest-Q pole pair.

When designing a cascade structure, both the pole–zero pairing and the order of sections in the cascade are important. In the case of maximizing dynamic range, the rules are conveniently summarized[1] as follows:

1. Place the stages in the order of increasing pole Q, counting a first-order section as zero Q.
2. Begin with the highest pole Q and pair it with the zero pair closest in frequency.
3. Equalize the maximum gain in each section.

A more complete discussion of section ordering is presented elsewhere.[2,3] For convenience, the simple rules listed above will be used in the design examples presented in this section.

The basic element of the cascade is the biquadratic section, described as:

$$T(s) = \frac{K\left\{s^2 + \frac{\omega_z}{Q_z}s + \omega_z^2\right\}}{\left\{s^2 + \frac{\omega_p}{Q_p}s + \omega_p^2\right\}} \tag{1}$$

For the lowpass, bandpass, and highpass derivatives of Equation 1 it is possible to use any of the basic second-order structures, (e.g., Sallen and Key (S&K), multiple feedback (MFB), for low-Q ($Qp \leq 10$) realizations. Higher-Q requirements are more readily affected by means of the generalized impedance converter (GIC) biquad or state variable structures. Neither S&K nor MFB should be used to yield notch or full biquads since they require the use of twin-T or parallel ladder networks which are quite sensitive and difficult to tune.

While the Friend biquad is quite versatile for such applications, its reliance on a single amplifier renders it less effective than the state variable or GIC biquad in more exacting situations. Nevertheless, especially when coupled with laser trimming and thin film realization, the Friend circuit is widely used in practice.[4]

The cascade technique is illustrated by several examples. Normalized solutions will be presented in all cases.

Example 1

Design an eighth-order lowpass Butterworth filter having a passband gain of 0 dB.

Solution

The transfer function is given by:

$$T(s) = \frac{1}{\left\{s^8 + 5.125831s^7 + 13.137071s^6 + 21.846151s^5 + 25.688356s^4 + 21.846151s^3 + 13.137071s^2 + 5.125831s + 1\right\}} \quad (2)$$

The pole–zero locations and Q of each second-order section (placed in order of ascending Q are as follows:[5]

	Pole Location	Pole Q
Stage 1	–0.98079 ± j0.19509	0.50980
Stage 2	–0.83147 ± j0.55557	0.60134
Stage 3	–0.55557 ± j0.83147	0.89998
Stage 4	–0.19509 ± j0.98079	2.56292

Since the Q factors are so low, it is convenient to use the S&K structure for the individual sections. For an overall gain of 0 dB, each stage has unity passband gain which allows the use of a voltage follower in each stage.

A single stage of the cascade filter realization is shown in Figure 3. The transfer function for this stage with gain K set to unity is

$$T(s) = \frac{\dfrac{1}{R_1R_2C_1C_2}}{s^2 + \left(\dfrac{1}{R_1C_1} + \dfrac{1}{R_2C_1}\right)s + \dfrac{1}{R_1R_2C_1C_2}} \quad (3)$$

Component values can be found by matching coefficients with Equation 1, provided two of the values are selected in advance (e.g., the two resistor values or the two capacitor values). Since the range of available capacitor values is usually more constrained than that of available resistors, we will specify the capacitor values and solve for the resistors. The resulting design equations are:

$$R_{1,2} = \frac{1}{2\omega_0 C_2 Q}\left[1 \pm \left(1 - 4Q^2\frac{C_2}{C_1}\right)^{1/2}\right] \quad (4)$$

This can be solved provided $C_1/C_2 \geq 4Q^2$.

For a low sensitivity design, we select a ratio of C_1/C_2 close to $4Q^2$, consistent with available values. The factor $4Q^2$ together with the normalized capacitor values selected for this design are

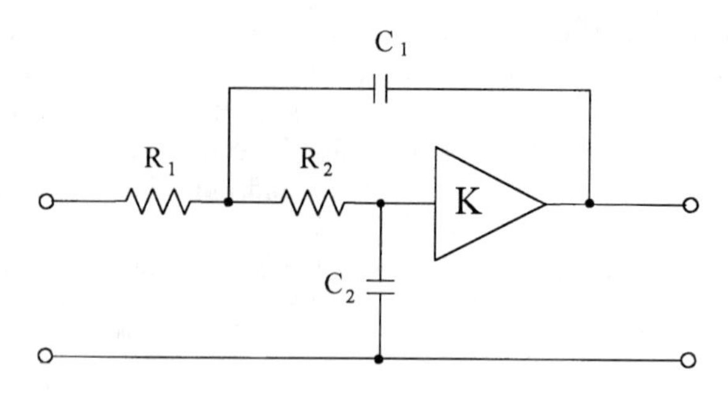

Figure 3 Lowpass S&K second-order section.

shown below for each of the four stages. Note that for convenience, we have set $C_2 = 1.0$ in all cases, so only C_1 is given in the table.

Stage	Pole Q	$4Q^2$	C_1
1	0.50980	1.03958	1.2
2	0.60134	1.44644	1.5
3	0.89998	3.23986	3.3
4	2.56292	26.2742	27.0

Using the values selected for the capacitors and the known Q values, Equation 4 can be solved for the required resistances (the pole frequency for all sections is 1.0 for the Butterworth transfer function). The normalized component values computed are:

Stage	C_1	C_2	R_1	R_2
1	1.2	1.0	1.33937	0.62218
2	1.5	1.0	0.98860	0.67436
3	3.3	1.0	0.63057	0.38057
4	27.0	1.0	0.22708	0.16311

The resistor spreads in this example are reasonable. However, if the spread between R_1 and R_2 were found to be too large (due to C_1/C_2 too far from $4Q^2$), then C_1 or C_2 could be changed and the resistor values recomputed.

Final capacitor and resistor values are chosen in conjunction with frequency denormalization and impedance scaling. Performing frequency denormalization first allows selection of the impedance-scaling constant to preserve the selected capacitance ratio. Resistors may then be selected from available values and possibly laser trimmed if required to meet demanding specifications. Expected deviations in the filter response resulting from capacitor and resistor tolerances may be determined from a sensitivity analysis or with the aid of an analog circuit simulator capable of performing a Monte Carlo or worst-case small signal analysis.

Example 2

Design a fifth-order lowpass elliptic characteristic filter having a passband gain of 10 dB, a stopband frequency (ω_s) of 1.1, and a passband ripple of 0.1 dB.

Solution

The required transfer function is

$$T(s) = \frac{1.17572\left(s^2 + 1.25932\right)\left(s^2 + 2.193093\right)}{\left(s + 0.932112\right)\left(s^2 + 0.099067s + 1.082694\right)\left(s^2 + 0.659383s + 1.017475\right)} \tag{5}$$

Using the rules outlined earlier, the section ordering and pole–zero pairing is as follows:

Section 1

$$T_1(s) = \frac{K_1}{(s + 0.932112)} \tag{6a}$$

Section 2 — $Q_p = 1.52976$

$$T_2(s) = \frac{K_2(s^2 + 2.193093)}{(s^2 + 0.659383s + 1.017475)} \tag{6b}$$

Section 3 — $Q_p = 10.5033$

$$T_3(s) = \frac{K_3(s^2 + 1.25932)}{(s^2 + 0.099067s + 1.082694)} \tag{6c}$$

Section 1 is realized by a simple RC buffer stage and the remaining sections are most conveniently designed as a combination of the Friend circuit (section 2) and state variable (section 3). The Friend circuit can readily handle the lower-Q section, whereas the easier tuning afforded by the state variable structure is desirable for the more exacting response of section 3.

With the pole–zero pairing and the ordering of the filter sections completed, the remaining task in optimizing the dynamic range is selection of the gains K_1, K_2, and K_3. Selection of these gains is complicated by the fact that different sections have gain peaks at different frequencies. Hence, more than one selection criterion exists.[6]

One commonly used criterion is to select the gains so that the onset of saturation in the different filter sections occurs at a common input amplitude. Each section will saturate at this amplitude at only one frequency which is found according to:

$$\max_{\omega} |H_i(\omega)| \equiv \max_{\omega} \left| \prod_{j=1}^{i} T_j(\omega) \right| \tag{7}$$

where the $T_j(\omega)$ are the individual section transfer functions (including the gain constants K_j), and where $H_i(\omega)$ is the transfer function from the input to the output of the ith stage.

Following Reference 6, we may define $\hat{T}_i(\omega)$ as the ith section transfer function without the gain K_i included:

$$\hat{T}_i(\omega) = \frac{T_i(\omega)}{K_i} \tag{8a}$$

and $\hat{H}_i(\omega)$ as

$$\hat{H}_i(\omega) = \prod_{j=1}^{i} \hat{T}_j(\omega) \tag{8b}$$

Then, setting the maximum of $H_i(\omega)$ over frequency equal for all i, the following iterative formula may be derived for the gains K_i of an N-stage cascade of filter sections:

$$K_i = \frac{\max_\omega |H_N(\omega)|}{\left(\prod_{j=1}^{i-1} K_j\right)\left(\max_\omega |\hat{H}_i(\omega)|\right)} \qquad i = 1,\ldots,N \tag{9}$$

To compute the K_i for this example, we need to know the maxima over frequency of the transfer function magnitudes $|\hat{H}_i(\omega)|$ and of the overall transfer function $|H_3(\omega)| = |T(\omega)|$. The latter is known to be 3.1623 in this example from the filter gain specification of 10 dB. The remaining maxima can be found with a computer or programmable calculator and then used to solve Equation 9 above. The results are shown below:

Section	$\max_\omega |\hat{H}_i(\omega)|$	K_i
1	1.0728	2.9478
2	2.3124	0.4639
3	2.6896	0.8597

With the filter section gains known, design of the individual sections can proceed.

Section 1: RC Buffer — The frequency-normalized design for section 1 consists of an RC lowpass filter buffered by a noninverting amplifier. The result is shown in Figure 4a, where the element values are in ohms and farads.

Section 2: Friend Biquad — The frequency-normalized design for section 2 consists of an opamp-based Friend biquad. By using established design procedures,[6] together with coefficient matching, the circuit of Figure 4b was obtained. Again, element values are in ohms and farads.

Section 3: State Variable — The frequency-normalized design for section 3 consists of the state variable filter shown in Figure 4c. The design of this section is more involved than that of sections 1 and 2 and requires some discussion.

The core of the state variable filter is built around opamps A_1, A_2, and A_3 whose outputs provide highpass, bandpass, and lowpass responses, respectively. For the present example, the lowpass and highpass outputs are summed with appropriate weighting by opamp A_4 to produce the overall section 3 transfer function $T_3(s)$. Since the circuit contains four opamps, there are four possible points of saturation and the gains *within* the state variable filter must be carefully selected.

We shall assume that the two integrators are identical and each has gain ω_p/s to optimize the state variable core. Then the task is to split the overall gain K_3 into a gain K_{3a} associated with the core and a gain K_{3b} associated with the output summer. The transfer function $T_{3(s)}$ then becomes

$$T_3(s) = \frac{K_{3b}K_{3a}\left(s^2 + \omega_z^2\right)}{s^2 + \frac{\omega_p}{Q}s + \omega_p^2} \tag{10}$$

For the circuit shown, K_{3a} and K_{3b} may be independently set using R_2 and R_{10}, respectively. The values of K_{3a} and K_{3b} may be found using a selection criterion such as that used previously to find K_1, K_2, and K_3 . If this is done after selecting K_3, then the following (suboptimal) solution results:

$$K_{3a} = \frac{\max_\omega |T_3(\omega)|}{\max_\omega |\hat{T}_{3HP}(\omega)|} \tag{11a}$$

$$K_{3b} = \frac{K_3}{K_{3a}} \tag{11b}$$

where

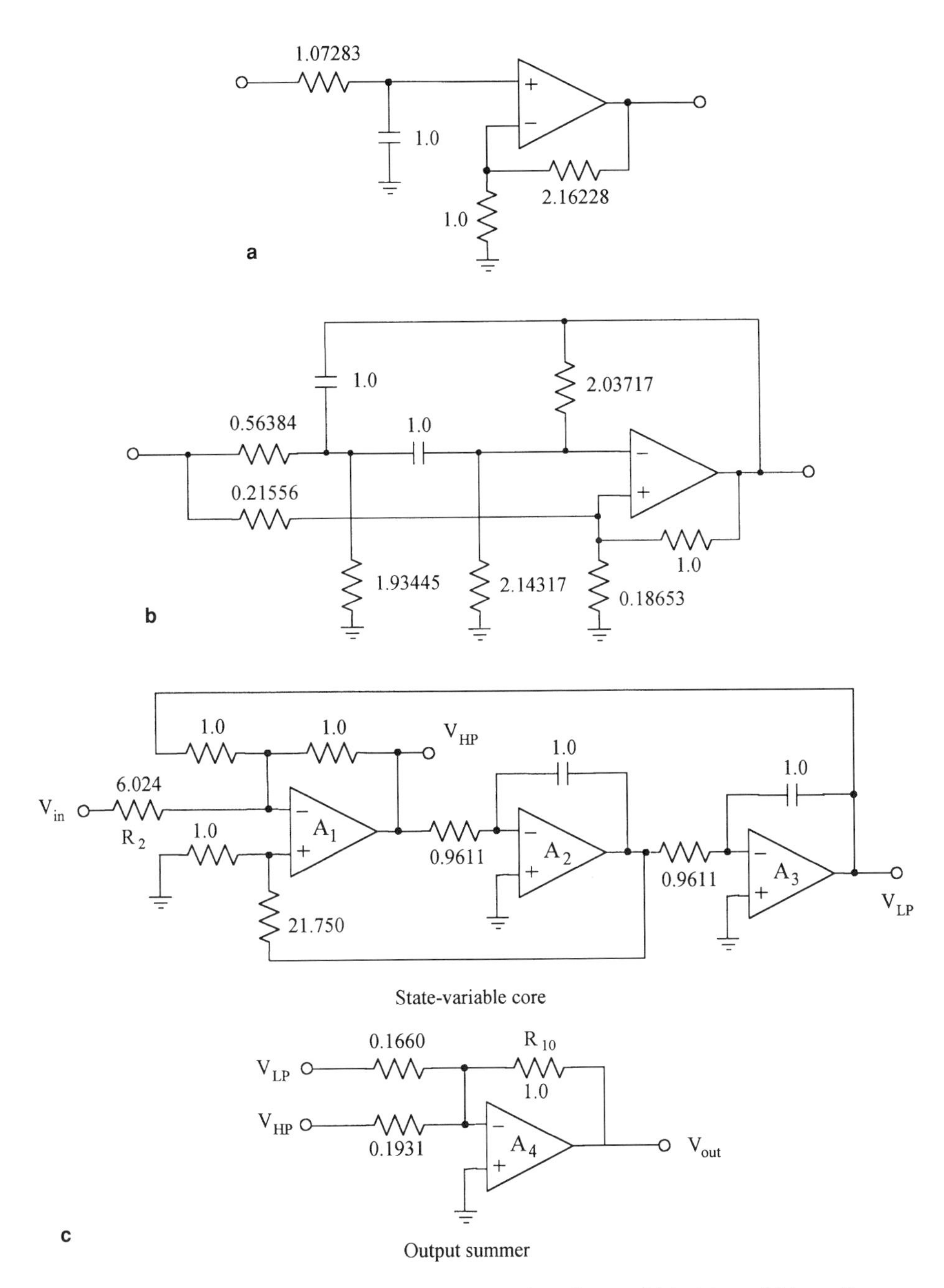

Figure 4 (a) RC-buffer section; (b) Friend biquad section; and (c) state variable section.

$$\hat{T}_{3HP}(s) = \frac{s^2}{s^2 + \frac{\omega_p}{Q}s + \omega_p^2}, \qquad s = j\omega \tag{11c}$$

With this approach, we find the following gains which were used in the design of Figure 4c:

$$K_{3a} = 0.166$$
$$K_{3b} = 5.179$$

Alternatively, the state variable filter could be treated as two individual sections in cascade, and the procedure used previously to find K_1, K_2, and K_3 could be applied to find K_1, K_2, K_{3a}, and K_{3b}.

Example 3

Design a Butterworth bandpass filter having a Q of 5 and a midband gain of 20 dB.

Solution

The frequency normalized transfer function for this filter can be found from a second-order Butterworth lowpass prototype by replacing s with $5(s + 1/s)$. The result may be written in factored form as

$$T(s) = 0.4 \frac{s^2}{\left(s^2 + 0.15142s + 1.15218\right)\left(s^2 + 0.13142s + 0.86792\right)} \tag{12}$$

This transfer function can be implemented with two cascaded bandpass sections. From $T(s)$ we find the Q values and resonant frequencies of the individual sections to be

Section	Q	ω_p
1	7.089	1.0734
2	7.089	0.9316

Since the Qs of the individual sections are equal and relatively low, each section may be implemented with the multiple feedback circuit of Figure 5 and the ordering of the sections is not critical. The dynamic range of the cascade filter may be optimized as in the previous example by appropriately allocating the overall filter gain between the two sections. The filter transfer function then becomes

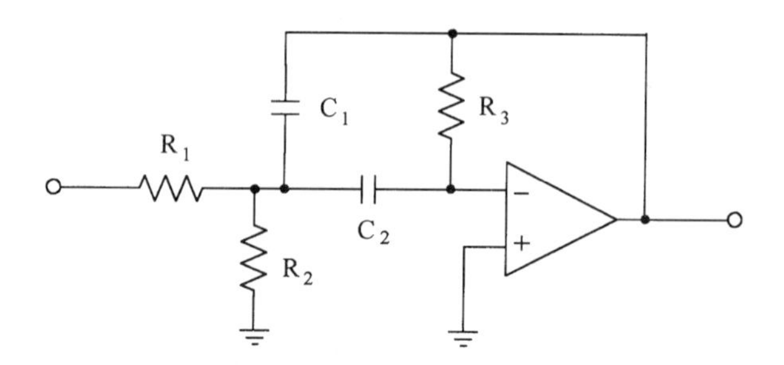

Figure 5 Two-pole multiple feedback bandpass section.

$$T(s) = \frac{-K_1 s}{\left(s^2 + 0.15142s + 1.15218\right)} \cdot \frac{-K_2 s}{\left(s^2 + 0.13142s + 0.86792\right)} \tag{13}$$

where the product $K_1 K_2 = 0.4$. The peak gain of the first section, excluding the gain constant K_1 is found to be:

$$\max_{\omega} \left|\hat{T}_1(\omega)\right| = \frac{Q_1}{\omega_{p1}} = 6.6043$$

Since the overall filter gain is specified as 20 dB, the gain K_1 is found from

$$K_1 = \frac{10}{6.6043} = 1.5142$$

and the gain K_2 is found from

$$K_2 = \frac{0.4}{K_1} = 0.2642$$

Section 1 thus provides the majority of the filter gain, guaranteeing that the output of each stage saturates at the same filter input level (although at different frequencies). Additionally, the filter output noise is kept low since section 2 attenuates out-of-band noise generated in section 1 while introducing a relatively low noise level itself due to its lower gain.

With the gain constants selected, the component values for the individual sections can be determined using established design procedures. The frequency normalized values are found to be

Section	R_1	R_2	R_3	C_1	C_2
1	0.6604	0.0730	13.209	1.0	1.0
2	3.7850	0.0773	15.219	1.0	1.0

Note the relatively large spread in resistor values for the multiple feedback configuration ($R_3/R_2 \approx 4Q^2$). In addition, this design requires that the gain of the operational amplifier used be significantly greater than $4Q^2$ at the frequency of operation. For high-frequency operation, or for higher Qs, an alternative filter topology such as a state variable design may be more appropriate.

2.2 DIRECT REALIZATION METHODS

Direct realization methods are based on passive LCR prototypes from which we can be assured of low passband sensitivity. Furthermore, we can utilize the extensive tabulation[7] of passive structures without the need for formal synthesis. Realization depends upon the *direct* or *indirect* replacement of inductors in the passive LCR structure.

The earliest inductor simulations were based on the gyrator and were most readily applied to "all-pole" highpass designs because of the exclusive presence of grounded inductors. Extension to general sections became possible as a result of the floating inductor simulation.[8] However, the performance and versatility of the GIC[9,10] have resulted in its widespread use for direct realization. The GIC shown in Figure 6 has a conversion factor, $k(s)$, given by

$$k(s) = \frac{Z_1 Z_3}{Z_2 Z_4} \tag{14}$$

By appropriate choice of $Z_1 \rightarrow Z_4$, $k(s)$ may take the form $k's^n$ where $n = \pm 1$ or ± 2. Of particular interest are the cases $n = 1$, $n = -2$.

If $n = 1$, $k(s) = sk'$ and the input impedance of the GIC is made inductive by means of a resistive termination, $Z_2' = R_2'$. In this way, it is possible to realize a grounded inductor of high quality and high value.

Example 4

Realize a simulated (grounded) inductor of value 450 mH.

Solution

The input impedance of the resistively terminated GIC of Figure 6 is:

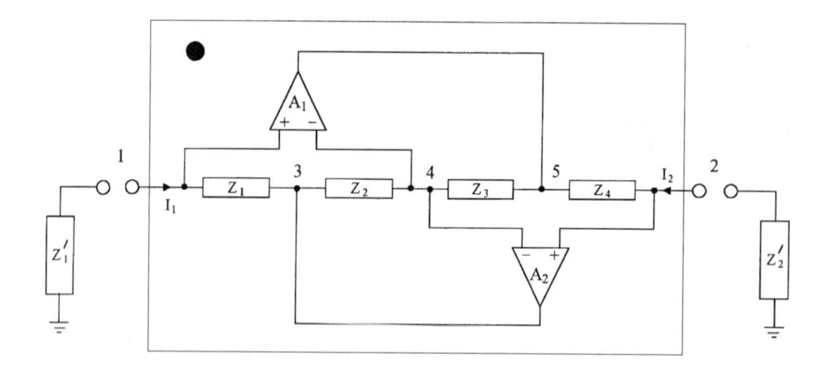

Figure 6 GIC.

$$Z_{in} = \frac{Z_1 Z_3 R_2'}{Z_2 Z_4} \tag{15a}$$

Replacing Z_4 by a capacitor and the remaining elements by resistors,

$$Z_{in} = \frac{sR_1 R_3 R_2' C_4}{R_2} \tag{15b}$$

Hence, the equivalent inductance is:

$$L_{eq} = \frac{R_1 R_3 R_2' C_4}{R_2} \tag{15c}$$

By setting the internal elements to standard (and equal) values, the terminating resistance can be adjusted to achieve the desired inductance.

Hence, if $R_1 = R_3 = R_2 = 10$ kΩ and $C_4 = 1$ nF,

$$R_2' = \frac{R_2 L_{eq}}{R_1 R_3 C_4} = (0.45) \cdot 10^5 = 45 \text{ k}\Omega$$

For discrete component designs, the nearest preferred value is 45.3 kΩ.

The grounded inductor realization may be used as a direct replacement element in an LCR filter. This is illustrated in the following example.

Example 5

Design a doubly terminated highpass filter having a seventh-order Butterworth characteristic.

Solution

From design tables,[7] we obtain the lowpass prototype shown in Figure 7a. For the normalized case, the lowpass-to-highpass transformation on the elements reduces to $L_i = 1/C_i$ and $C_j = 1/L_j$. The resulting highpass realization is shown in Figure 7b. The filter can be denormalized, and each inductor is then replaced by a simulated element designed according to the procedure described in Example 4.

If the GIC of Figure 6 is terminated by Z_1' at port 1, the input impedance viewed from port 2 is given by

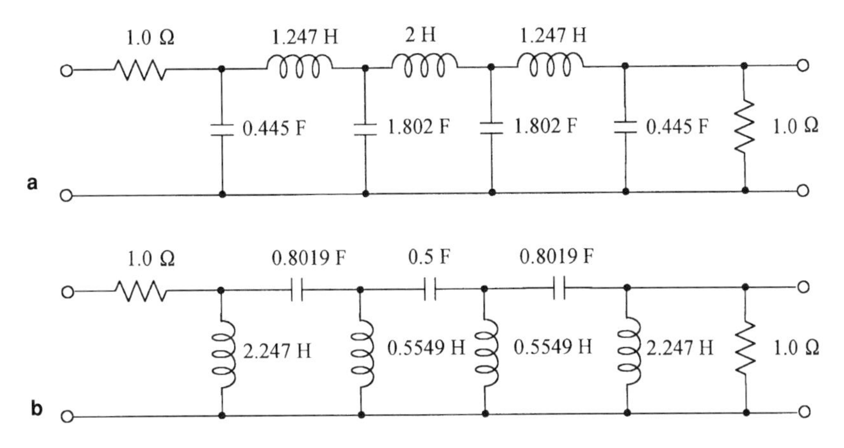

Figure 7 (a) Lowpass prototype; and (b) highpass LC ladder transformed from (a).

$$Z_{22} = \frac{Z_2 Z_4 Z_1'}{Z_1 Z_3} \tag{16}$$

Retaining the internal element forms used in Example 4, and selecting Z_1' to be capacitive, we obtain

$$Z_{22} = \frac{R_2}{s^2 R_1 R_3 C_4 C_1'} = \frac{1}{s^2 D} \tag{17a}$$

At real frequencies,

$$Z_{22}\Big|_{s=j\omega} = \frac{-1}{\omega^2 D} \tag{17b}$$

This is negative, frequency dependent, and (although strictly having units of farad-seconds) has been called resistive. Hence, the term FDNR (frequency-dependent negative resistance). Two FDNRs (with conversion factor ks^2) may be used in the configuration of Figure 8 to simulate a floating (series arm) inductor of value k/R. However, since two active (and balanced) blocks are required to replace each inductor, this approach is rarely used in practice. Instead, the indirect replacement technique has proved to be widely applicable. Its power is illustrated in the following examples.

Example 6

Realize a normalized sixth-order double-terminated lowpass Butterworth filter.

Solution

From design tables, we obtain the filter shown in Figure 9a. The network is transformed[10] by scaling admittances by a factor s. This yields the DCR network of Figure 9b in which floating inductors have become floating resistors and the FDNRs are all grounded. Each FDNR may be replaced as described earlier, and the circuit may be denormalized as appropriate.

Practical biasing problems may be encountered because of the presence of the input series capacitor. If so, resistive shunts can be employed as shown in Figure 9c. In order to evaluate the shunt elements, the DC level is set to be equivalent in each circuit. Hence, the DC gain of 0.5 in the prototype must also be obtained from the DCR network. Thus,

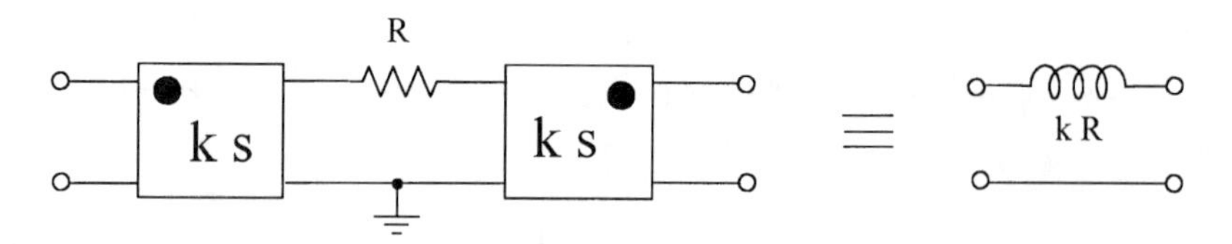

Figure 8 Simulation of a "floating" inductor using FDNR elements.

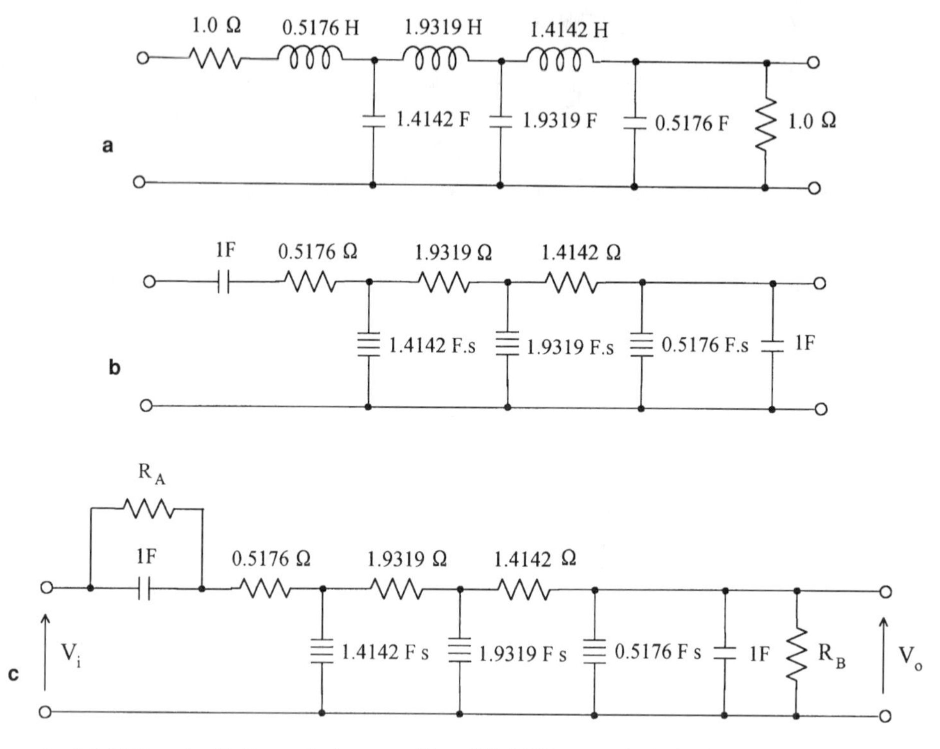

Figure 9 (a) Sixth-order Butterworth lowpass filter; (b) DCR network derived from (a); and (c) addition of resistive shunts to overcome biasing problems of (b).

$$\frac{R_B}{R_A + R_B + 3.864} = 0.5$$

A second relationship may be established by ensuring that R_A, R_B do not load the capacitors. For example, setting $R_A = 100\ \Omega$ yields $R_B = 103.864\ \Omega$.

The FDNR approach is extremely versatile, and can clearly be applied to more complex examples than the one illustrated above. Some modification to the approach is necessary if floating capacitors are present in the LCR prototype since these lead to floating FDNRs in the DCR network. One possibility is to use the direct replacement of floating elements as shown in Figure 10. This is not convenient if lots of floating FDNRs are present. An alternative is to transform only part of the original filter and then embed this transformed subsection between two GICs, as shown in Figure 11.[11,12]

Example 7

Use the embedded FDNR approach to realize the bandpass section of Figure 12a.

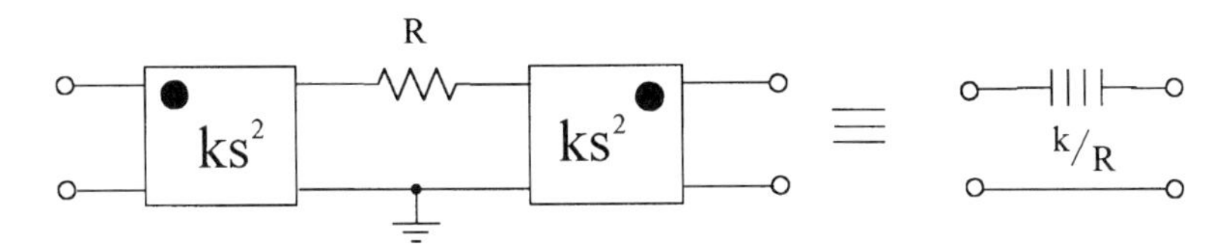

Figure 10 Simulation of a floating FDNR.

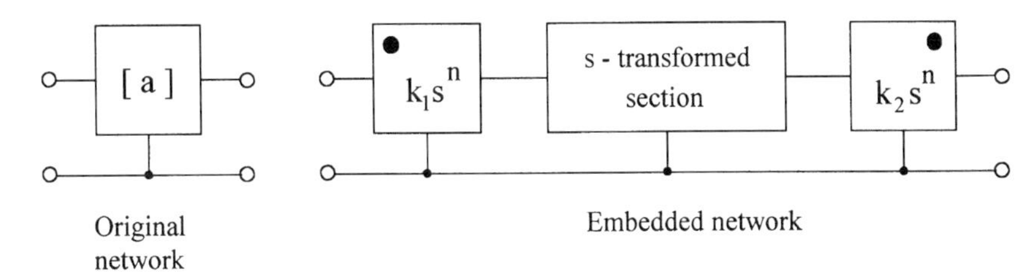

Figure 11 Partial transformation technique.

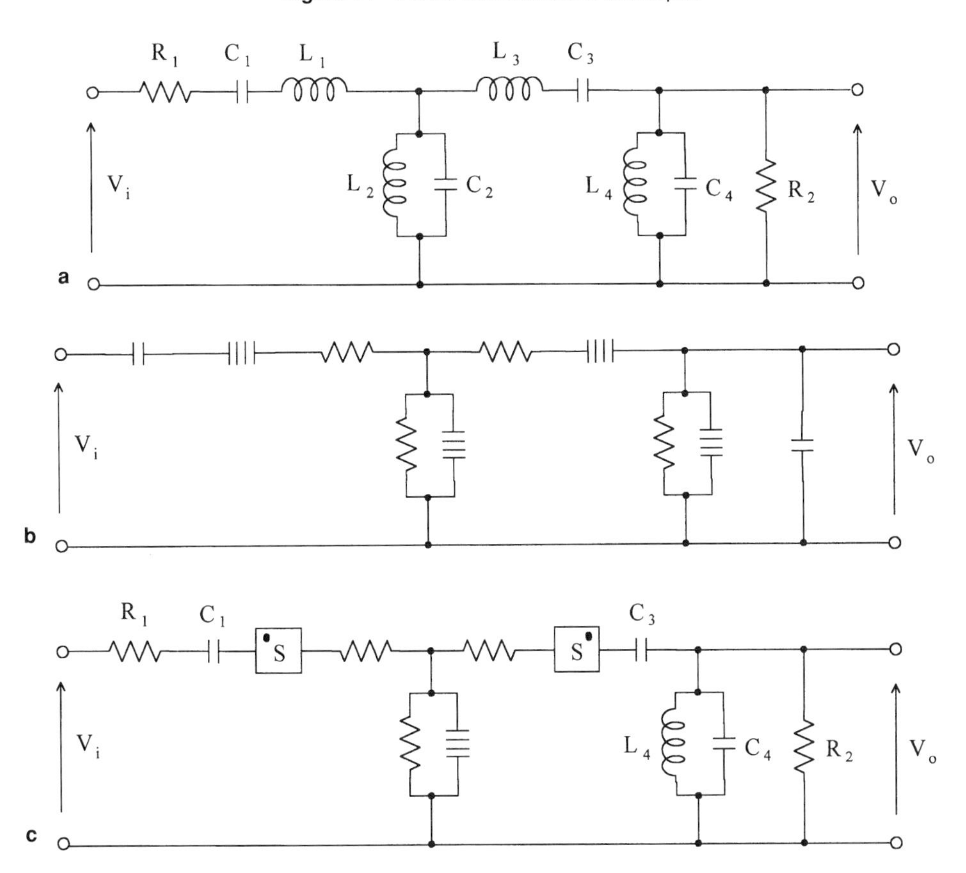

Figure 12 (a) Bandpass prototype; (b) DCR network derived by complete transformation of (a); and (c) network resulting from use of the embedded FDNR transformation.

Solution

Application of the direct FDNR transformation yields the DCR network of Figure 12b. This circuit would require the use of six GICs to replace the two floating FDNRs and the two grounded FDNRs.

By contrast, the embedded approach allows for the selective transformation of Figure 12a to yield Figure 12c which requires a total of four GICs, one of which directly simulates L_4.

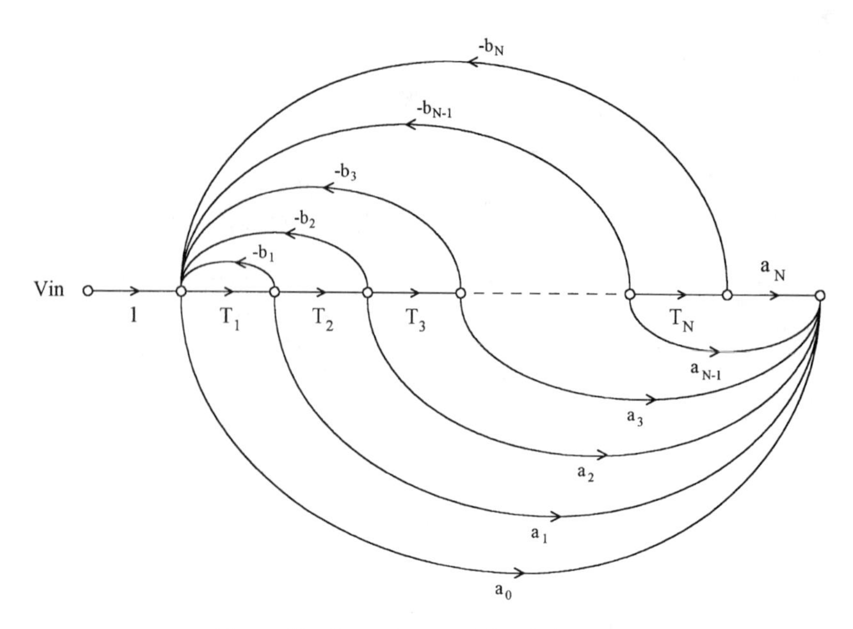

Figure 13 Signal flowgraph for FLF system.

2.3 MULTILOOP DESIGN METHODS

A variety of higher-order filter design methods have been derived from the generalized flowgraph of a state variable system. For example, the signal flowgraph of Figure 13 has a voltage transfer ratio given by:

$$T = \frac{V_{out}}{V_{in}} = \frac{a_o + \sum_{i=1}^{N} a_i \prod_{j=1}^{i} T_j}{1 + \sum_{k=1}^{N} b_k \prod_{j=1}^{k} T_j} \tag{18}$$

In general, the T_j can represent transfer functions rather than the simple integrators of the state variable solution. Most commonly, second-order sections are used since they are readily tuned. Examples of multiloop structures[13-15] derived from Figure 13 are shown in Figure 14.

Multiloop synthesis techniques yield low-sensitivity structures while retaining a certain degree of the modularity afforded by the cascade technique. The primary resonator block (PRB) method is particularly convenient since it incorporates identical second-order sections. However, all multiloop designs are less convenient to tune than their cascade-design counterparts.

The leapfrog technique[16] will be used to illustrate multiloop design since it is not only the earliest technique, but arguably the most widely used. The advent of switched-capacitor technology in particular led to widespread use of this method. The approach is based on the simulation of an LC ladder by an active analog in which the passive reactances are replaced on a one-to-one basis by integrators. The technique is particularly suited to the realization of lowpass responses which are the most difficult to realize using the direct replacement methods of the previous section.

The technique may be understood by noting the analogy between the passive ladder of Figure 15a and the multiloop feedback system of Figure 15b. Each may be described as follows:

$$V_2 = I_2 Z_2 \tag{19a}$$

$$I_2 = (V_1 - V_2) Y_2 \tag{19b}$$

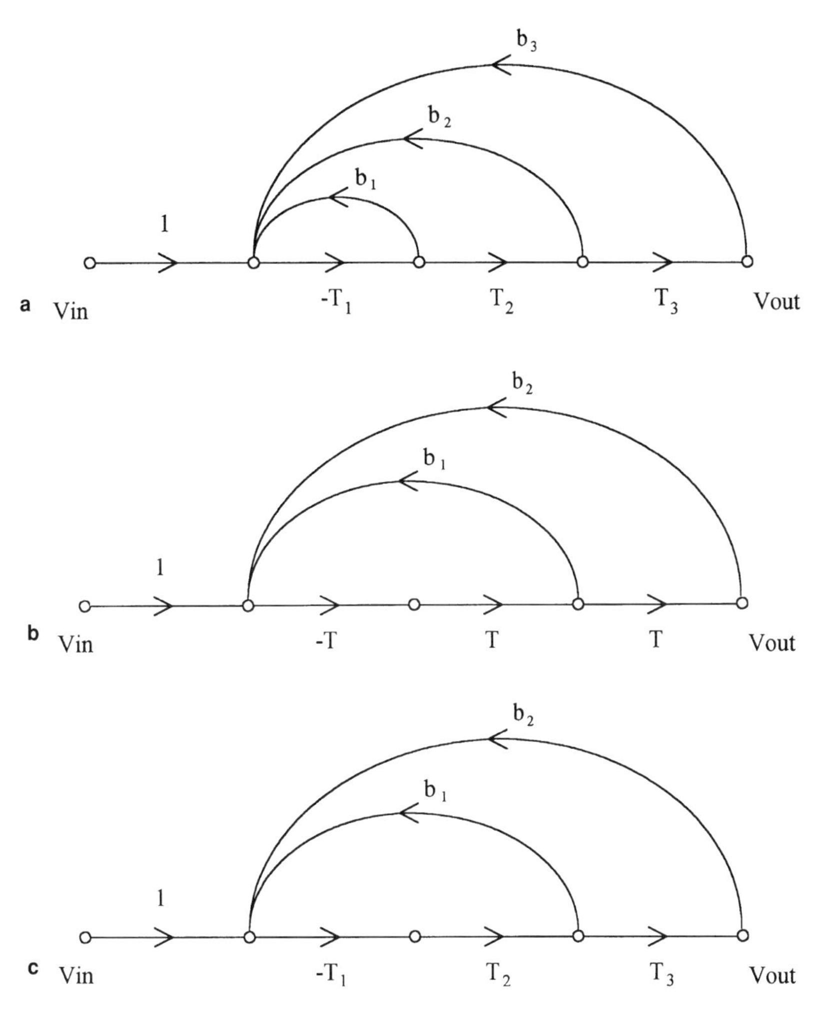

Figure 14 Signal flowgraphs of multiloop filter structures. (a) Follow-the-leader feedback (FLF); (b) primary resonator block; and (c) shifted companion-form (SCF).

$$V_1 = (I_1 - I_2)Z_1 \tag{19c}$$

$$I_1 = (V_{in} - V_1)Y_1 \tag{19d}$$

Analog computer principles may thus be used to simulate the passive ladder. The analog variables will all be voltages, representing both voltages and currents in the original ladder. Furthermore, immittances in the ladder are replaced by transfer function blocks in the analog. In practice, these blocks are realized by simple integrator stages.

Example 8

Realize a fifth-order lowpass filter using the leapfrog technique.

Solution

The general filter structure is shown in Figure 16a, from which

$$Y_1 = \frac{1}{R_s + sL_1}; \quad Z_1 = \frac{1}{sC_2}; \quad Y_2 = \frac{1}{sL_3}; \quad Z_2 = \frac{1}{sC_4}; \quad \text{and} \quad Y_3 = \frac{1}{R_L + sL_5}$$

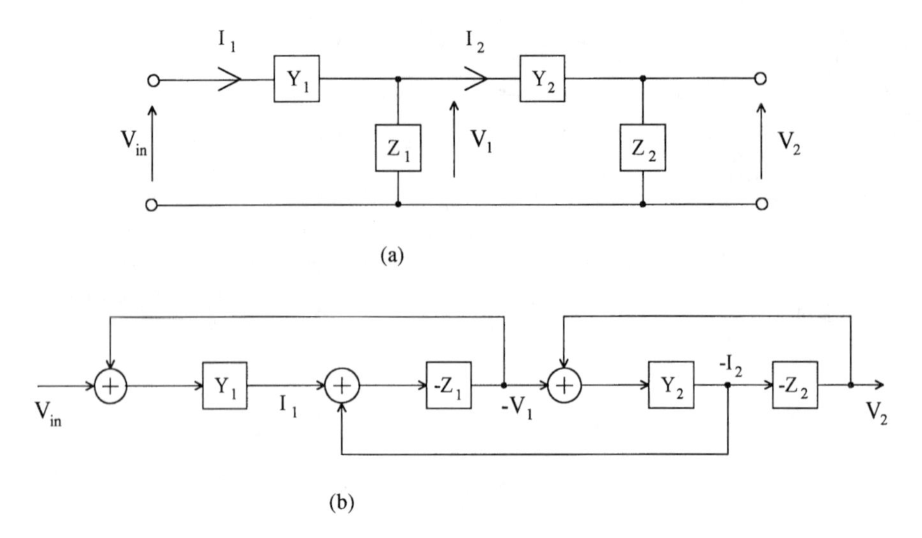

Figure 15 The leapfrom technique. (a) Passive ladder and (b) multiloop feedback system.

If the ladder is normalized with respect to the terminating resistance ($R = R_L = R_S$), the immittances became dimensionless and may be represented by simple integrating/summer blocks. Then,

$$Y_1 \rightarrow T_1(s) = \frac{\frac{R}{L_1}}{s + \frac{R}{L_1}}$$

$$-Z_1 \rightarrow T_2(s) = \frac{-1}{sC_2R}$$

$$Y_2 \rightarrow T_3(s) = \frac{R}{sL_3}$$

$$-Z_2 \rightarrow T_4(s) = \frac{-1}{sC_4R}$$

$$Y_3 \rightarrow T_5(s) = \frac{\frac{R}{L_5}}{s + \frac{R}{L_5}}$$

The final circuit is shown in Figure 17.

The leapfrog technique can readily be extended to bandpass realizations[5] derived from all-pole lowpass prototypes. Realization of more-complex structures is also possible, but not widely practiced.

3. IMPLEMENTATION TECHNOLOGIES AND COMPONENT SELECTION

A variety of technologies exist through which high-order active filters can be realized. Some of the major alternatives are outlined in this section. Which technology is the most appropriate for a particular situation will usually depend on a number of factors, including system design, manufacturing, and economic considerations.

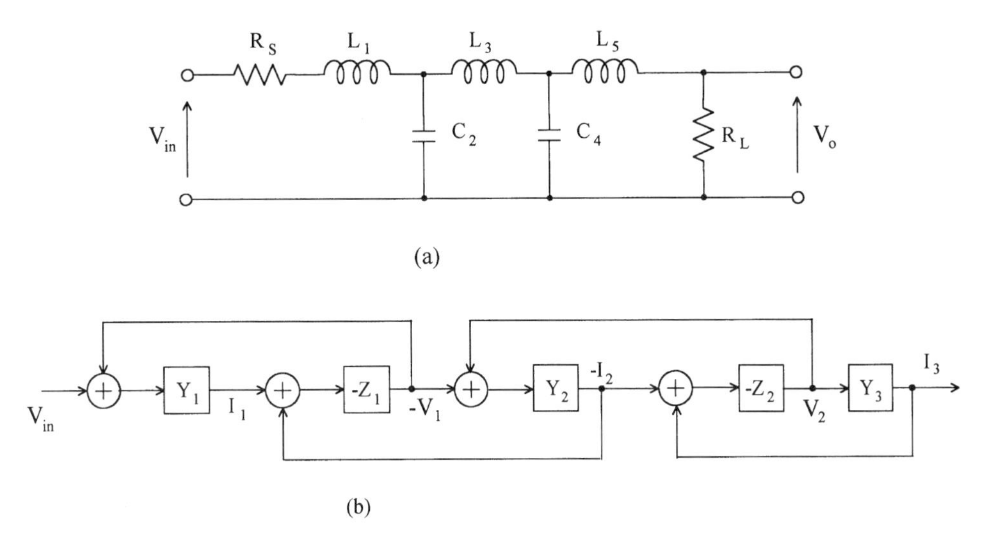

Figure 16 Leapfrog realization of a fifth-order lowpass filter. (a) LC prototype and (b) leapfrog model of (a).

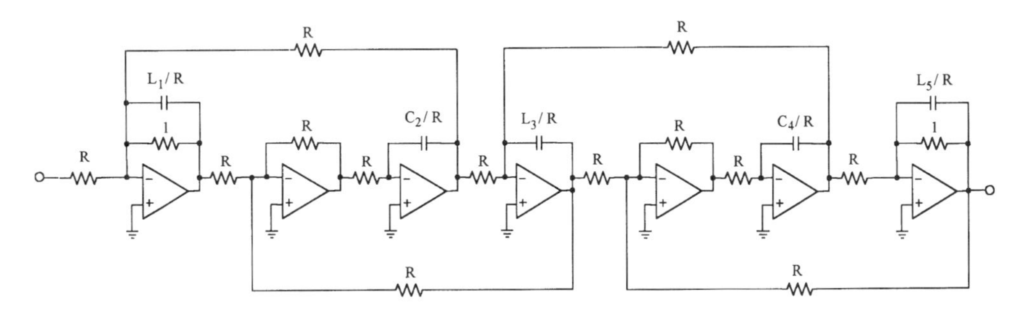

Figure 17 Fifth-order leapfrog realization.

3.1 OFF-THE-SHELF SOLUTIONS

Several companies offer ready-to-use filters ranging from rack-mountable chassis and associated plug-in modules to hybrid and integrated circuit (IC) components that can be incorporated into printed circuit board designs. Most of these filters are designed for use within the audio frequency range or up to a few hundred kilohertz, although a few operate into the megahertz region. A sampling of offerings available is shown in Table 1.

The offerings shown can be grouped into two main categories — application-specific and general-purpose designs. Where there are high-volume demands in particular applications, precision, fixed-tuned filters are often available. Such applications include digital audio, computer disk drives, and communications equipment. For general-purpose use, including data acquisition, filters may be found that are either fixed tuned (selected by part number), or programmable by resistor selection, digital words, control voltages/currents, or clock frequency. In addition, some filter vendors will provide quotes on request for developing custom designs to fit customer special needs.

3.2 DISCRETE COMPONENT DESIGNS

For situations where off-the-shelf solutions are not appropriate, custom filters may be designed using the techniques discussed in this chapter. This section discusses components applicable to board-level implementations including active devices, resistors, and capacitors. Implementation in hybrid or IC form is discussed in Sections 3.3 and 3.4, respectively.

The active components used in designing the filter may consist of discrete transistors or IC devices. Discrete transistors may provide an economic solution in situations where high precision and high Q are not required. Operational amplifier designs generally enable more accurate gains

Table 1 Sample Filter Offerings Available Off-the-Shelf

Description	Applications	Packaging	Example Vendors
Continuous-time audio lowpass with cutoff frequencies of 15–20 kHz and orders up to 11	Digital audio	Hybrid	TDK, TOKO
Lowpass and bandpass speech continuous-time and switched-capacitor designs	Telephone, modems, facsimile	Hybrid/IC	TOKO, TOYOCOM, MX-COM, EXAR
High-order Bessel continuous-time lowpass with programmable boost and cutoff frequency operating into the megahertz range	Disk drives	IC	Analog Devices
Switched-capacitor universal active filters; cascadable second-order sections; digitally, resistor, and clock programmable	General	IC	EG&G Reticon, EXAR, Maxim, National Semiconductor, Texas Instruments
Continuous-time universal active filters; a collection of opamps and thin-film resistors and capacitors for constructing state variable realizations; resistor programmable	General	Hybrid/IC	Burr Brown, National Semiconductor
Switched-capacitor lowpass, bandpass, highpass, and notch filters; digitally and clock programmable	General	IC	EG&G Reticon, EXAR, Maxim, National Semiconductor, Texas Instruments
Fixed and programmable frequency, continuous-time, lowpass, highpass, bandpass, and notch filters in Butterworth, Chebyshev, Bessel, and Cauer types with orders from 2nd through 17th	General	Rackmount, hybrid, and IC	Avens, Datel, Frequency Devices, Maxim

Table 2 Typical Operational Amplifier Data

Part Number	uA741	OP-27A	LF353	TL074A
Type	Single	Single	Dual	Quad
GBWP	1 MHz	8 MHz	4 MHz	3 MHz
Slew rate	0.5 V/μS	2.8 V/μS	13 V/μS	13 V/μS
Supply current	1.7 mA	2.8 mA	3.6 mA	5.6 mA
Offset voltage	1 mV	0.025 mV	5 mV	3 mV
Bias current	80 nA	40 nA	50 pA	65 pA
Input resistance	2 MΩ	6 MΩ	10^{12} Ω	10^{12} Ω
Voltage gain	200 K	1000 K	100 K	200 K
Noise	20 nV/Hz$^{1/2}$	3 nV/Hz$^{1/2}$	16 nV/Hz$^{1/2}$	18 nV/Hz$^{1/2}$
	2 pA/Hz$^{1/2}$	0.4 pA/Hz$^{1/2}$	0.01 pA/Hz$^{1/2}$	0.01 pA/Hz$^{1/2}$

to be implemented and result in improved filter performance. When selecting an operational amplifier, important characteristics to consider include gain–bandwidth product (GBWP) (which determines the maximum frequency at which the filter can be accurately designed), the slew rate (which limits large signal response), and noise (which limits dynamic range). Operational amplifiers are available from many different manufacturers and are available in a variety of packages, including dual in-line, surface mount, and chip forms. Table 2 provides a sampling of some commonly available, internally compensated, operational amplifiers and the performance characteristics that can be achieved.

The performance of passive components used in the filter is critical in achieving the desired transfer function. The most obvious performance parameters to consider include resistor and capacitor tolerances, but other parameters, such as temperature coefficient, effects of aging, humidity, and soldering heat, should be taken into account. For demanding applications laser trimming

Table 3 Typical Resistor Performance Data

Performance Parameters	Resistor Type		
	Carbon Film	**Thick Film (RuO_2)**	**Metal Film**
Resistance range	1 Ω to 2.2 MΩ	10 Ω to 1 MΩ	10 Ω to 10 MΩ
Best initial tolerance	±2%	±1%	±0.25%
Temperature coefficient	–700 ppm/°C	±100 ppm/°C	±50 ppm/°C
Resistance to soldering	±1%	±1%	±0.1%
Humidity load life	±3%	±3%	±0.5%
Load life (aging)	±2%	±2%	±0.5%

Table 4 Typical Capacitor Performance Data

Performance Parameter	Dielectric Type					
	NPO/COG Ceramic	**Mica**	**Polystyrene**	**Metal Film**	**Polyester**	**X7R Ceramic**
Range of values	0.5 pF to 0.047 uF	1–1000 pF	20 pF to 1 uF	0.01–1 uF	1000 pF to 1 uF	330 pF to 0.22 uF
Best initial tolerance	±1%	±1%	±1%	±5%	±5%	±10%
Temperature coefficient	<50 ppm/°C	50 ppm/°C	–150 ppm/°C	±200 ppm/°C	500 ppm/°C	±1000 ppm/°C
Dissipation factor	0.1%	0.05%	0.05%	1%	0.3%	2.5%

of resistors can also be used. Trimming procedures may be either deterministic (or passive) in which the value of individual resistors is adjusted before applying power to the circuit, or functional in which the resistors are trimmed in operation to yield the desired response characteristic. A combination of deterministic trimming followed by functional trimming may be used to reduce the problem complexity and number of iterations required. Table 3 lists representative performance parameters for a variety of common resistor types.

Carbon film and thick film (RuO_2) are commonly available in tolerances of 2% and 5%, and are relatively inexpensive. However, soldering heat, moisture, and aging can result in further deviations from nominal values up to 2% or more under worst-case conditions. Metal film resistors offer substantially better performance and are available with initial tolerances of 1%, and better, at moderate increases in cost.

Capacitor performance is more variable than that of resistors, especially in the area of temperature coefficient which is a strong function of the dielectric material used. Table 4 lists representative performance for a variety of common capacitor types.

The best stability is generally provided by NPO or COG ceramic and mica dielectrics, although other types may perform well over restricted temperature ranges. As with resistors, effects such as aging, humidity, and resistance to soldering heat should be taken into account.

3.3 HYBRID DESIGNS

For exacting performance, hybrid circuit construction offers significant advantages over PC board construction. Hybrid techniques involve fabricating circuits on high-quality ceramic substrates such as alumina. Conductors, resistors, and (sometimes) capacitors are printed or deposited using thick-film or thin-film techniques. Chip capacitors and active components are most commonly attached by reflow soldering or epoxy, and the finished circuit is then typically packaged in a form that can be incorporated into printed circuit board designs.

The advantages offered by hybrid construction include miniaturization, stable resistors and capacitors, laser trimming, good heat transfer and temperature tracking, and potential cost savings in high quantity. Various companies offer design and manufacturing services from prototype development through production manufacturing runs.

Table 5 lists some performance characteristics typical of hybrid processes. More-detailed information may be obtained from a variety of processing handbooks.[17,18]

Table 5 Typical Characteristics of Hybrid Processes

Thick-film resistors	Range of values	10 Ω to >10 MΩ
	Resistivity	10–100 K Ω/square
	Tolerance	±10% to 0.1% (trimmed)
	Temperature coefficient	±100 ppm/°C
	Tracking	±5 ppm/°C
Thin-film resistors	Range of values	10 Ω to 1 MΩ
	Resistivity	20–300 Ω/square
	Tolerance	±10% to 0.1% (trimmed)
	Temperature coefficient	±50 ppm/°C
	Tracking	±5 ppm/°C

Table 6 High-Order, Integrated Filters

Type	Frequency	Order	Application
Bandpass	200 kHz	18th	Narrowband FM receiver
Elliptic lowpass	10 kHz	3rd + 7th	Radio paging receiver
Bessel lowpass	10 to 30 MHz	7th	Disk drive

3.4 MONOLITHIC DESIGNS

For very high volume applications, it may be cost-effective to realize as much circuitry as possible in integrated circuit form. In low frequency applications, switched-capacitor filter designs [see Chapter 7] have provided an effective solution to this problem for many years. However, for reasons ranging from frequency of operation to power consumption, considerable attention has been directed in recent years at realizing on-chip continuous-time filters. Applications for which such filters have been designed include television and VCR signal processing, computer disk drive electronics, telephone systems, and radio pagers. Table 6 provides a sample listing of high-order, on-chip, continuous-time filters that have been reported in the literature.[19-21]

Integrated continuous-time filters differ from board-level and hybrid designs in two important respects. First, on-chip filters usually employ architectures based on operational transconductance amplifiers (OTAs) and capacitors rather than the more traditional operational amplifiers, resistors, and capacitors. This difference results from a number of factors including lack of suitable on-chip resistors in typical IC processes, a desire for high-frequency operation, the need for tunability, and the desire to reduce chip area.

The second important difference is the need for real-time, on-chip tuning of the filter transfer function. This requirement results from difficulties in realizing accurate time constants. While precise absolute component values are difficult to achieve, tight ratios and tracking are readily achieved. Ratio accuracies and tracking of 1% or better can be realized depending on the process and size/geometries of the components involved. This feature is exploited by the most commonly used tuning technique — the master–slave approach. In this technique, an on-chip controlled oscillator (the master) is constructed using components similar or identical to those used in the filter. The frequency of this oscillator is then phase-locked with appropriate on-chip circuitry to a known reference (usually provided by an off-chip frequency reference). The filter (slave) receives the same tuning voltages or currents as the master and is therefore tuned to the desired frequency of operation. Tuning of filter Q, although less commonly used, can also be achieved. Details of integrated circuit filter design and related tuning techniques can be found in the references.[22]

REFERENCES

1. **B.D. Nelin,** Design of high-order active filters, in *RC Active Filter Design*, F.W. Stephenson (Ed.), Prentice Hall, Englewood Cliffs, NJ, 1988, chapter 11.
2. **E. Leuder,** A decomposition of a transfer function minimizing distortion and inband losses, *Bell Syst. Tech. J.*, 49(3), 455–469, 1970.

3. **G.S. Moschytz,** A second-order pole-zero pair selection for nth-order minimum sensitivity networks, *IEEE Trans.*, CT-17(4), 527–534, 1970.
4. **R.A. Friedenson, R.W. Daniels, R.J. Dow, and P.H. McDonald,** RC active filters for the D3 channel bank, *Bell Syst. Tech. J.*, 54(3), 507–529, 1975.
5. **L.P. Huelsman,** *Active and Passive Analog Filter Design*, McGraw-Hill, New York, 1993.
6. **M.S. Ghausi and K.R. Laker,** *Modern Filter Design*, Prentice-Hall, Englewood Cliffs, NJ, 1981, chapter 5.
7. **A.L. Zverev,** *Handbook of Filter Synthesis*, John Wiley & Sons, New York, 1967.
8. **A.G.J. Holt and J.R. Taylor,** Method of replacing ungrounded inductors by grounded gyrators, *Electron. Lett.*, 1(4), 105, 1965.
9. **A. Antoniou,** Realization of gyrators using operational amplifiers and their use in RC-active-network synthesis, *Proc. IEE,* 116(11), 1838–1850, 1969.
10. **L.T. Bruton,** Network transfer functions using the concept of frequency-dependent negative resistance, *IEEE Trans.*, CT-16, 406–408, 1969.
11. **A.W. Keen and J.L. Glover,** Active RC equivalents of RCL networks by similarity transformation, *Electron. Lett.*, 7(11), 288–290, 1971.
12. **L.T. Bruton and A.B. Haase,** Sensitivity of generalized immittance converter-embedded ladder structures, *IEEE Trans.*, CAS-21(2), 245–250, 1974.
13. **K.R. Laker and M.S. Ghausi,** Synthesis of low-sensitivity multiloop feedback active RC filter, *IEEE Trans.*, CAS-21(2), 252–259, 1974.
14. **J. Tow,** Design and evaluation of shifted-companion form of active filters, *Bell Syst. Tech. J.*, 54(3), 545–568, 1975.
15. **G. Hurtig,** The primary resonator block technique of filter synthesis, *Proc. IEEE International Filter Symposium*, Santa Monica, CA, p 84, April 1972 [US Patent 3,720,881, March, 1973].
16. **F.E. Girling and E.F. Good,** Active Filters, *Wireless World*, 76, 341–345 and 445–450, 1970. [The Leapfrog Method was first described by the authors in *RRE Memo*. No 1177, September 1955.]
17. **J.J. Licari and L.R. Enlow,** *Hybrid Microcircuit Technology Handbook*, Noyles Publications, Pork Ridge, NJ, 1988.
18. **C.A. Harper (Ed.),** *Electronic Packaging and Interconnection Handbook*, McGraw-Hill, New York, 1991.
19. **F. Krummenacher and G. Van Ruymbeke,** Integrated selectivity for narrow-band FM IF systems, *IEEE J. Solid-State Circuits*, SC-25(3), 757–760, June 1990.
20. **J.F. Wilson, R. Youell, J.H. Richards, G. Luff, and R. Pilaski,** A single-chip VHF and UHF receiver for radio paging, *IEEE J. Solid-State Circuits*, SC-26(12), 1944–1950, December 1991.
21. **G.A. De Veirman and R.G. Yamasaki,** Monolithic 10-30 MHz tunable bipolar Bessel lowpass filter, *IEEE Proc. ISCAS*, 1444–1447, 1991.
22. **Y.P. Tsividis and J.O. Voorman (Eds.),** *Integrated Continuous-Time Filters*, IEEE Press, Piscataway, NJ, 1993.

Chapter 6

Digital Filters

6.1

Introduction to Digital Filters

Andreas Antoniou

CONTENTS

1. INTRODUCTION

The digital filter is a digital system that can be used to process discrete-time signals. It can be represented by a recursive or nonrecursive difference equation. Its time-domain analysis can be carried out by solving the difference equation or by performing a convolution summation. Frequency-domain analysis can be carried out by finding the steady state sinusoidal response as in analog filters and systems.

Typically, in practice, certain frequency-domain specifications are prescribed which need to be satisfied, and a network or structure is obtained by realizing an appropriate transfer function. The specifications describe the passband(s) and stopband(s) of the filter in terms of the required passband and stopband edges, the maximum passband ripple, and the minimum stopband attenuation.

This chapter provides an overview of the basic principles associated with the analysis and design of digital filters. The characterization of digital filters in terms of difference equations and their representation in terms of networks are described. Digital filters are analyzed and designed through the use of the z transform which plays the same role as the Laplace transform in analog filters. The highlights of the z transform are summarized in Section 5. The z transform is then applied in the derivation of the transfer function and in the time-domain and frequency-domain analysis of digital

0-8493-8951-8/97/$0.00+$.50
© 1997 by CRC Press, Inc.

Figure 1 The digital filter as a system.

filters in Sections 6 to 8. Some key issues in filter design, namely, amplitude and phase distortion and the importance of a linear phase response, are examined in Section 9. The necessary steps to design a digital filter are discussed in Section 10. The last section, Section 11, provides a brief comparison between analog and digital filters.

2. CHARACTERIZATION

A digital filter is a system that can be used to manipulate the frequency spectrum of a discrete-time signal, e.g., to select or reject a specific frequency component, enhance or attenuate a range of frequency components, etc. A digital filter can be represented by the block diagram of Figure 1 where $x(nT)$ is the input or excitation and $y(nT)$ is the output or response. The excitation $x(nT)$ may be generated by sampling a continuous-time signal $x(t)$ at instants $t = 0, T, 2T, \ldots$, but sometimes it may exist in its own right without a continuous-time counterpart, e.g., the closing price of a stock at the stock exchange as a function of time. The response is related to the excitation by some rule of correspondence which must of necessity entail some frequency-domain specifications.

Analog filters are characterized in terms of differential equations. Digital filters, on the other hand, are characterized in terms of *difference* equations. An arbitrary linear, time-invariant, and causal digital filter can be represented by the Nth-order difference equation

$$y(nT) = \sum_{i=0}^{N} a_i x(nT - iT) - \sum_{i=1}^{N} b_i y(nT - iT) \tag{1}$$

where a_i and b_i are constant coefficients that determine the performance and characteristics of the filter. Integer N is said to be the filter order. As can be seen, the response at instant nT depends on previous values of the response and, in effect, Equation 1 is a recursive equation. For this reason, the filter represented is often referred to as a *recursive* one.

If $b_i = 0$ for $i = 1, 2, \ldots, N$, Equation 1 simplifies to

$$y(nT) = \sum_{i=0}^{N} a_i x(nT - iT) \tag{2}$$

In this case, the response at instant nT depends only on the values of the excitation at instants nT, $nT - T, \ldots, nT - NT$. Evidently, the equation is no longer recursive and the filter obtained is often referred to as a *nonrecursive* one.

3. NETWORK REPRESENTATION

The basic elements of digital filters are the *unit delay*, the *adder*, and the *multiplier*. Their characterizations and symbols are given in Table 1. Ideally, the adder will produce the sum of its inputs and the multiplier will multiply its input by a constant instantaneously. The unit delay, on the other hand, will record its input at instant nT and deliver its previous input to the output. In effect, the unit delay is a memory element. The implementation of the digital filter elements can assume

Table 1 Digital Filter Elements

Element	Symbol	Equation
Unit delay	$x(nT)$ → [T] → $y(nT)$	$y(nT) = x(nT - T)$
Adder	$x_1(nT)$, $x_2(nT)$, …, $x_K(nT)$ → (+) → $y(nT)$	$y(nT) = \sum_{i=1}^{K} x_i(nT)$
Multiplier	$x(nT)$ → (⊗, m) → $y(nT)$	$y(nT) = mx(nT)$

From A. Antoniou, *Digital Filters: Analysis, Design, and Applications,* 2nd Edition, McGraw-Hill, New York, 1993. With permission of McGraw-Hill, Inc.

various forms, depending on the representation of the signals to be processed. Normally, the signals are sequences of binary numbers, the adder and multiplier are combinational or sequential digital circuits, and the unit delay assumes the form of some type of register. Collections of unit delays, adders, and multipliers can be interconnected to form *digital filter networks.*

The analysis of digital filter networks, namely, the process of generating the filter difference equation from the network is almost always much simpler than that in analog filters. It can be carried out by using signal flow graphs,[1,2] Mason's gain formula,[1-3] state-space methods,[1] and quite often by simply writing down the network equations and then solving them.

4. CONVOLUTION SUMMATION

The impulse response of a digital filter, designated as $h(nT)$, is the response produced by the unit impulse defined as

$$\delta(nT) = \begin{cases} 1 & n = 0 \\ 0 & n \neq 0 \end{cases}$$

If the impulse response of a linear and time-invariant digital filter is known, then the response, $y(nT)$, produced by an arbitrary excitation, $x(nT)$, can be deduced as

$$y(nT) = \sum_{k=-\infty}^{\infty} x(kT)h(nT - kT) = \sum_{k=-\infty}^{\infty} h(kT)x(nT - kT) \tag{3}$$

as can be readily shown.[1] This most important relationship is said to be the *convolution summation,* and it plays the same key role as the convolution integral in analog filters and systems in general.

Some special cases of the convolution summation are of particular interest. If the filter is causal, $h(nT) = 0$, for $n < 0$, and so

$$y(nT) = \sum_{k=-\infty}^{\infty} x(kT)h(nT - kT) = \sum_{k=0}^{\infty} h(kT)x(nT - kT)$$

If, in addition, $x(nT) = 0$ for $n < 0$, we have

$$y(nT) = \sum_{k=0}^{n} x(kT)h(nT - kT) = \sum_{k=0}^{n} h(kT)x(nT - kT) \tag{4}$$

If the impulse response of a digital filter is of finite duration such that $h(nT) = 0$ for $n > N$, then Equation 4 assumes the form

$$y(nT) = \sum_{k=0}^{N} h(kT)x(nT - kT)$$

This equation is of the same form as Equation 2 with $h(0) = a_0$, $h(T) = a_1, \ldots, h(NT) = a_N$ and, in effect, the filter is nonrecursive. Conversely, it readily follows from Equation 2 that if the filter is nonrecursive, the impulse response is of finite duration. In recursive filters, the impulse response is usually, but not always, of infinite duration, but if the impulse response is of infinite duration, then the filter is always recursive. Evidently, one may classify digital filters as *finite-duration impulse response* (FIR) filters and *infinite-duration impulse response* (IIR) filters. Here, we shall be referring to digital filters as nonrecursive and recursive although the alternative classification is equally valid and widely used.

The convolution summation has several applications. It can be used to find the constraints on the impulse response that will guarantee the stability of a digital filter, it can be used along with the z transform to generate a complex-domain characterization for digital filters, namely, the transfer function, and so on.

A digital filter is said to be *stable* if and only if any bounded excitation results in a bounded response, i.e., if

$$|x(nT)| < \infty \qquad \text{for all } n$$

then

$$|y(nT)| < \infty \qquad \text{for all } n$$

Through the use of the convolution summation in Equation 3, a necessary and sufficient condition for stability can be obtained as

$$\sum_{k=-\infty}^{\infty} |h(kT)| < \infty$$

that is, the impulse response of a stable digital filter is absolutely summable; conversely, if the impulse response is absolutely summable, the filter is stable (see Reference 1 for proof).

5. THE *z* TRANSFORM

The analysis of linear, time-invariant digital filters is almost invariably carried out by using the *z transform.* The principal reason for this is that upon the application of the *z* transform, the difference equations characterizing digital filters are transformed into algebraic equations which are usually much easier to mainpulate and solve.

The *z* transform, like the Laplace and Fourier transforms, is useful because it has an inverse, namely, the inverse *z* transform. The application of the *z* transform to a discrete-time signal $x(nT)$ yields a representation of the signal in terms of a rational function $X(z)$, where z is a complex variable. If the signal is to be processed by a digital filter, then the required processing can be carried out in the *z* domain through algebraic mainpulations. In this way, a transformed version of $X(z)$, say $Y(z)$, is obtained, and by applying the inverse *z* transform to $Y(z)$, the processed signal is obtained.

The *z* transform of a discrete-time signal $x(nT)$ is defined as

$$X(z) = \sum_{n=-\infty}^{\infty} x(nT)z^{-n}$$

where $z = re^{j\varphi}$ is a complex variable. The *z* transform is defined for all z for which $X(z)$ converges and is often represented by the simplified notation

$$X(z) = \mathcal{Z}x(nT)$$

Signal $x(nT)$ can be uniquely recovered from $X(z)$ by using the inverse *z* transform given by

$$x(nT) = \frac{1}{2\pi j}\oint_{\Gamma} X(z)z^{n-1}dz \tag{5}$$

where Γ is a contour in the counterclockwise sense enclosing all the singularities of $X(z)z^{n-1}$. Various types of singularities exist in complex analysis but those encountered in linear digital filters are always poles. Function $x(nT)$ is often represented by the simplified notation

$$x(nT) = \mathcal{Z}^{-1}X(z)$$

If

$$X(z)z^{n-1} = X_0(z) = \frac{N(z)}{\Pi_{i=1}^{M}(z-p_i)^{m_i}}$$

where M and m_i are positive integers, then by using the residue theorem of complex analysis,[4] we have

$$x(nT) = \sum_{i=1}^{M} \operatorname{res}_{z=p_i} X_0(z)$$

where

Table 2 Standard z Transform Relations

Property	Relation
Linearity	$z[ax(nT) + by(nT)] = aX(z) + bY(z)$
Translation	$zx(nT + mT) = z^m X(z)$
Complex scale change	$z[w^{-n}x(nT)] = X(wz)$
Complex differentiation	$z[nTx(nT)] = -Tz \dfrac{dX(z)}{dz}$
Real convolution	$z \sum_{k=-\infty}^{\infty} h(kT)x(nT - kT) = H(z)X(z)$
	$z \sum_{k=-\infty}^{\infty} h(nT - kT)x(kT) = H(z)X(z)$
Complex convolution	$Y(z) = z[h(nT)x(nT)] = \dfrac{1}{2\pi j} \oint_{\Gamma_1} H(v)X\left(\dfrac{z}{v}\right) v^{-1}\, dv$
	$Y(z) = z[h(nT)x(nT)] = \dfrac{1}{2\pi j} \oint_{\Gamma_2} H\left(\dfrac{z}{v}\right) X(v)v^{-1}\, dv$

$$\text{res}_{z=p_i} X_0(z) = \frac{1}{(m_i - 1)!} \lim_{z \to p_i} \frac{d^{m_i - 1}}{dz^{m_i - 1}} \left[(z - p_i)^{m_i} X_0(z)\right]$$

for a pole of order m_i and

$$\text{res}_{z=p_i} X_0(z) = \lim_{z \to p_i} \left[(z - p_i) X_0(z)\right]$$

for a simple pole.

Note that $X_0(z)$ may have a simple pole at the origin when $n = 0$ and possibly higher-order poles for $n < 0$. This fact must be taken into account in the determination of $x(0)$, $x(-T)$, $x(-2T)$, … .

The z transform satisfies certain relations which are often very useful in deriving the z transforms of signals. These are usually presented as theorems, but are listed here in tabular form in Table 2. In this table,

$$\mathcal{Z}x(nT) = X(z), \qquad \mathcal{Z}h(nT) = H(z), \qquad \text{and} \qquad \mathcal{Z}y(nT) = Y($$

a, b, and T are constants, and Γ_1 (or Γ_2) is a contour in the common region of convergence of $H(v)$ and $X(z/v)$ [or $H(z/v)$ and $X(v)$]. Proofs can be found in Reference 1.

Using the relations in Table 2, the z transforms of some frequently encountered elementary signals can be easily obtained, as shown in Table 3.

The z transform of $x(nT)$ is actually a Laurent series of $X(z)$ about the origin of the z plane,[4] and, in effect, finding the inverse z transform amounts to finding the coefficients of a Laurent series. In consequence, the methods available for constructing Laurent series can also be used to obtain the inverse z transform of a signal, for example, by equating coefficients, performing long division, expressing $X(z)$ in terms of binomial series or a partial-fraction expansion, or applying the convolution theorem.

6. THE TRANSFER FUNCTION

Through the use of the z transform, a digital filter can be characterized by a *discrete-time transfer function* which plays the same key role as the continuous-time transfer function in an analog filter. In this section, the discrete-time transfer function is defined and some of its properties are examined.

Table 3 Standard z Transforms

$x(nT)$	$X(z)$
$\delta(nT)$	1
$u(nT)$	$\dfrac{z}{z-1}$
$u(nT-T)K$	$\dfrac{K}{z-1}$
$u(nT)Kw^n$	$\dfrac{Kz}{z-w}$
$u(nT-T)Kw^{n-1}$	$\dfrac{K}{z-w}$
$u(nT)e^{-\alpha nT}$	$\dfrac{z}{z-e^{-\alpha T}}$
$r(nT)$	$\dfrac{Tz}{(z-1)^2}$
$r(nT)e^{-\alpha nT}$	$\dfrac{Te^{-\alpha T}z}{\left(z-e^{-\alpha T}\right)^2}$
$u(nT)\sin\omega nT$	$\dfrac{z\sin\omega T}{z^2-2z\cos\omega T+1}$
$u(nT)\cos\omega nT$	$\dfrac{z(z-\cos\omega T)}{z^2-2z\cos\omega T+1}$
$u(nT)e^{-\alpha nT}\sin\omega nT$	$\dfrac{ze^{-\alpha T}\sin\omega T}{z^2-2ze^{-\alpha T}\cos\omega T+e^{-2\alpha T}}$
$u(nT)e^{-\alpha nT}\cos\omega nT$	$\dfrac{z\left(z-e^{-\alpha T}\cos\omega T\right)}{z^2-2ze^{-\alpha T}\cos\omega T+e^{-2\alpha T}}$

The transfer function of a digital filter is defined as the ratio of the z transform of the response to the z transform of the excitation. Consider a linear, time-invariant digital filter, and let $x(nT)$, $y(nT)$, and $h(nT)$ be the excitation, response, and impulse response, respectively. From the convolution summation in Equation 3, we have

$$y(nT)=\sum_{k=-\infty}^{\infty}x(kT)h(nT-kT)$$

and, therefore, by applying the z transform, we obtain

$$Y(z)=,\mathcal{Z}(nT)=\mathcal{Z}h(nT)\mathcal{Z}x(nT)=H(z)X(\tag{6}$$

or

$$\frac{Y(z)}{X(z)}=H(z)$$

(see real convolution in Table 2). In effect, the transfer function of a digital filter is the z transform of the impulse response.

Other authors define the transfer function as the z transform of the impulse response and then show that it happens to be the ratio of the z transform of the response to the z transform of the excitation, but the two definitions are entirely equivalent.

The transfer function can be derived from the difference equation characterizing the filter or the filter network. For causal, recursive filters

$$y(nT) = \sum_{i=0}^{N} a_i x(nT - iT) - \sum_{i=1}^{N} b_i y(nT - iT)$$

and by using the linearity and translation relations in Table 2, we obtain

$$Zy(nT) = \sum_{i=0}^{N} a_i z^{-i} Zx(nT) - \sum_{i=1}^{N} b_i z^{-i} Zy(nT$$

or

$$\frac{Y(z)}{X(z)} = H(z) = \frac{\sum_{i=0}^{N} a_i z^{N-i}}{z^N + \sum_{i=1}^{N} b_i z^{N-i}} \tag{7}$$

Thus the transfer function is a ratio of two polynomials in z with real coefficients. Note that for a causal filter the degree of the numerator polynomial is equal to or less than that of the denominator polynomial.

By factorizing the numerator and denominator polynomials in Equation 7, $M(z)$ can be put in the form

$$H(z) = \frac{N(z)}{D(z)} = H_0 \frac{\prod_{i=1}^{N} (z - z_i)}{\prod_{i=1}^{N} (z - p_i)^{m_i}} \tag{8}$$

where $z_1, z_2, \ldots, z_N$ are the zeros and $p_1, p_2, \ldots, p_N$ are the poles of $H(z)$, m_i is the order of pole p_i, and H_0 is a multiplier constant. Thus, digital filters, like analog filters, can be represented by zero-pole plots like the one in Figure 2.

In nonrecursive filters, $b_i = 0$ for $i = 1, 2, \ldots, N$, and so the poles in these filters are all located at the origin of the z plane.

The z domain characterizations of the unit delay, the adder, and the multiplier are obtained from Table 1 as

$$Y(z) = z^{-1} X(z), \qquad Y(z) = \sum_{i=1}^{K} X_i(z), \qquad \text{and} \qquad Y(z) = mX(z)$$

respectively. By using these relations, $H(z)$ can often be derived directly from the filter network.

The transfer function can also be obtained from a state-space characterization, if one is available.[1]

7. TIME-DOMAIN ANALYSIS

The *time-domain* response of a digital filter to any excitation $x(nT)$ can be readily obtained from Equation 6 as

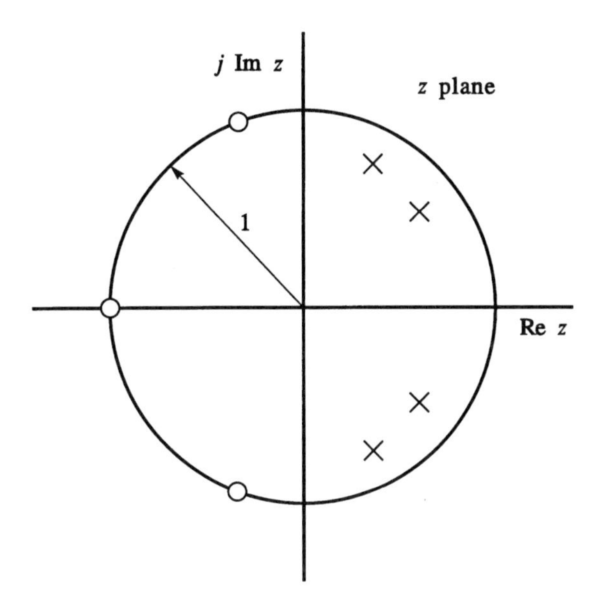

Figure 2 Typical zero-pole plot for a digital filter.

$$y(nT) = \mathcal{Z}^{-1}\left[H(z)X(z)\right]$$

It can be evaluated by using the inverse z transform given by Equation 5 or, more frequently, by expressing the product $H(z)X(z)$ or $H(z)X(z)/z$ as a partial-fractions expansion and then using Table 3 to invert each of the individual partial fractions.

The impulse response can be obtained as the inverse z transform of $H(z)$ as

$$y_i(nT) = h(nT) = \mathcal{Z}^{-1}H(z) = \frac{1}{2\pi j}\oint_{\Gamma} H(z)z^{n-1}dz$$

For $n = 0$

$$h(0) = R_0 + \sum_{i=1}^{N} \text{res}_{z=p_i}\left[\frac{H(z)}{z}\right]$$

where

$$R_0 = \text{res}_{z=0}\left[\frac{H(z)}{z}\right]$$

if $H(z)$ has no zeros or poles at $z = 0$, and $R_0 = 0$ in all other cases. On the other hand, for $n > 0$

$$h(nT) = \sum_{i=1}^{N} \text{res}_{z=p_i}\left[H(z)z^{n-1}\right] = \sum_{i=1}^{N} p_i^{n-1}\text{res}_{z=p_i} H(z)$$

It can be shown that if the poles of the filter are inside the unit circle $|z| = 1$ of the z plane, then the impulse response is absolutely summable and, if at least one pole is on or outside the unit circle, the impulse response cannot be absolutely summable.[1] Therefore, a necessary and sufficient condition for the stability of a digital filter is that *the poles be located inside the unit circle* $|z| = 1$. Thus, if $p_i = r_i e^{j\phi_i}$ where $r_i = |p_i| < 1$ and $\phi_i = \arg p_i$ for $i = 1, 2, \ldots, N$, then the filter is stable. Since the poles in nonrecursive filters are all located at the origin of the z plane, these filters are always stable.

Another time-domain response of significant usefulness in filter design is the unit-step response. This is obtained as

$$y_u(nT) = \mathcal{Z}^{-1}\left[H(z)\frac{z}{z-1}\right] = \frac{1}{2\pi j}\oint_\Gamma \frac{H(z)z^n}{z-1}dz$$

The unit-step response can be used to determine the overshoot, the delay time, and the rise time, as in other types of filters.

The *overshoot* is the difference between the peak value and the asymptotic value of the step response in percent as $n \to \infty$. The *delay time* is the time required for the step response to reach 50% of the asymptotic value. The *rise time* is the time required for the step response to increase from 10 to 90% of the asymptotic value.

Yet another important time-domain response of a digital filter is its sinusoidal response which is the response of the filter to a sinusoid. This is given by

$$y(nT) = \mathcal{Z}^{-1}\left[H(z)X(z)\right]$$

where

$$X(z) = \mathcal{Z}\left[u(nT)\sin\omega nT\right] = \frac{z\sin\omega T}{\left(z - e^{j\omega T}\right)\left(z - e^{-j\omega T}\right)} \tag{9}$$

Hence

$$y(nT) = \frac{1}{2\pi j}\oint_\Gamma H(z)X(z)z^{n-1}d\mathcal{Z} = \sum \text{res}\left[H(z)X(z)z^{n-1}\right] \tag{10}$$

For $n > 0$, Equations 9 and 10 yield

$$y(nT) = \frac{1}{2j}\left[H\left(e^{j\omega T}\right)e^{j\omega nT} - H\left(e^{-j\omega T}\right)e^{-j\omega nT}\right] + \sum_{i=1}^{N} X(p_i)p_i^{n-1}\text{res}_{z=p_i}H(z)$$

Assuming that the filter is stable, we have $|p_i| = |r_i e^{j\phi_i}| < 1$ or $r_i < 1$ for $i = 1, 2, \ldots, N$ and hence as $n \to \infty$, the summation part in the above equation tends to zero since

$$\lim_{n\to\infty} p_i^{n-1} = \lim_{n\to\infty}\left[r_i^{n-1}e^{j(n-1)\phi_i}\right] \to 0$$

Therefore, the steady state sinusoidal response, designated as $\tilde{y}(nT)$, can be expressed as

$$\tilde{y}(nT) = \lim_{n\to\infty} y(nT) = \frac{1}{2j}\left[H\left(e^{j\omega T}\right)e^{j\omega nT} - H\left(e^{-j\omega T}\right)e^{-j\omega nT}\right]$$

After some manipulation, one can deduce

$$\tilde{y}(nT) = M(\omega)\sin[\omega nT + \theta(\omega)]$$

where

$$M(\omega) = \left|H\left(e^{j\omega T}\right)\right| \tag{11}$$

and

$$\theta(\omega) = \arg H\left(e^{j\omega T}\right) \tag{12}$$

Clearly, the effect of a digital filter on a sinusoidal excitation, like that of an analog filter, is to introduce a *gain* $M(\omega)$ and a *phase shift* $\theta(\omega)$.

8. FREQUENCY-DOMAIN ANALYSIS

A digital filter, like an analog filter, can be represented in the frequency domain by its *amplitude* and *phase responses* which are the gain and phase shift of the filter as functions of frequency. The amplitude and phase responses together consitute the *frequency response.* The main difference between analog and digital filters is that in the first case the frequency response is obtained by evaluating the continuous-time transfer function $H(s)$ on the imaginary axis of the s plane, whereas in the second case this is obtained by evaluating the discrete-time transfer function $H(z)$ on the unit circle $|z| = 1$ of the z plane, as can be seen in Equations 11 and 12.

For a transfer function expressed in terms of its zeros and poles as in Equation 8

$$H\left(e^{j\omega T}\right) = M(\omega)e^{j\theta(\omega)} = H_0 \frac{\Pi_{i=1}^{N}\left(e^{j\omega T} - z_i\right)}{\Pi_{i=1}^{N}\left(e^{j\omega T} - p_i\right)^{m_i}} \tag{13}$$

and by letting

$$e^{j\omega T} - z_i = M_{z_i} e^{j\psi_{z_i}} \tag{14}$$

$$e^{j\omega T} - p_i = M_{p_i} e^{j\psi_{p_i}} \tag{15}$$

we obtain

$$M(\omega) = \left|H_0\right| \frac{\Pi_{i=1}^{N} M_{z_i}}{\Pi_{i=1}^{N} M_{p_i}^{m_i}} \tag{16}$$

$$\theta(\omega) = \arg H_0 + \sum_{i=1}^{N} \psi_{z_i} - \sum_{i=1}^{N} m_i \psi_{p_i} \tag{17}$$

where $\arg H_0 = \pi$ if H_0 is negative. Thus $M(\omega)$ and $\theta(\omega)$ can be determined graphically by using the following procedure:

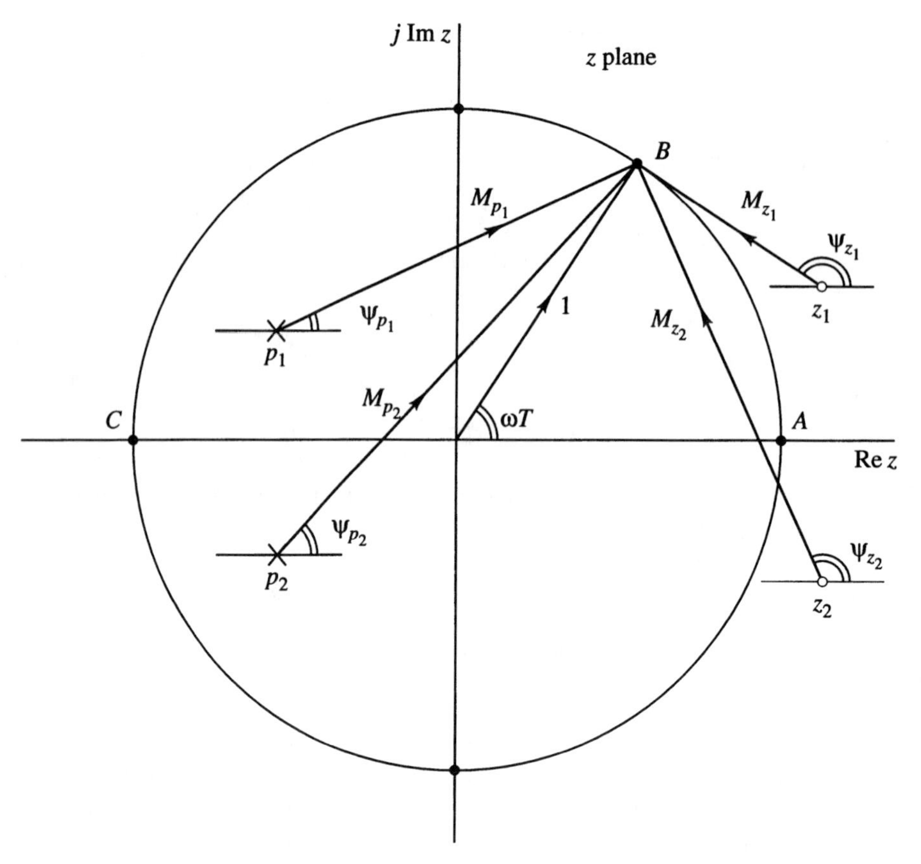

Figure 3 Phasor diagram for a second-order filter. (From A. Antoniou, *Digital Filters: Analysis, Design, and Applications,* 2nd Edition, McGraw-Hill, New York, 1993. With permission of McGraw-Hill, Inc.)

1. Mark the zeros and poles of the filter in the z plane.
2. Draw the unit circle.
3. Draw the phasor $e^{j\omega T}$ where ω is the frequency of interest.
4. Draw a phasor of the type in Equation 14 for each simple zero of $H(z)$.
5. Draw m_i phasors of the type in Equation 15 for each pole of order m_i.
6. Measure the magnitudes and angles of the phasors in steps 4 and 5 and use Equations 16 and 17 to calculate the gain $M(\omega)$ and phase shift $\theta(\omega)$, respectively.

The amplitude and phase responses of a filter can be determined by reapeating the above procedure for frequencies $\omega = \omega_1, \omega_2, \ldots$, in the range 0 to π/T. This method of analysis is illustrated in Figure 3 for a second-order filter with simple zeros and poles.

The modern approach for the analysis of digital filters is, of course, through the use of digital computers. Nevertheless, the above graphical method is of interest and merits consideration for two reasons. First, it illustrates some of the fundamental properties of digital filters. Second, it provides a certain degree of intuition about the expected amplitude or phase response of a filter. For example, if a filter has a pole close to the unit circle, then as ωT approaches the angle of the pole, the magnitude of the phasor from the pole to the unit circle decreases rapidly to a very small value and then increases as ωT increases above the angle of the pole. As a result, the amplitude response will exhibit a large peak in that frequency range. On the other hand, a zero close to the unit circle will lead to a notch in the amplitude response when ωT is equal to the angle of the zero.

Points A and C in Figure 3 correspond to the frequencies 0 and π/T, and one complete revolution of $e^{j\omega T}$ about the origin corresponds to a frequency increment of $\omega_s = 2\pi/T$ rad/s. Since T is the period between samples, ω_s is said to be the *sampling frequency.*

If phasor $e^{j\omega T}$ in Figure 3 is rotated k complete revolutions, the values of $M(\omega)$ and $\theta(\omega)$ will obviously remain unchanged, and as a result

$$H\left(e^{j(\omega+k\omega_s)T}\right) = H\left(e^{j\omega T}\right)$$

In effect, the *frequency response,* namely, $H(e^{j\omega T})$, is a *periodic function* of frequency with a period ω_s. This is to be expected, of course, since discrete-time signals sin ωnT and sin $[(\omega + k\omega_s)nT]$ are numerically identical, i.e.,

$$\begin{aligned}\sin\left[(\omega + k\omega_s)nT\right] &= \sin\omega nT\cos kn\omega_s T + \cos\omega nT\sin kn\omega_s T \\ &= \sin\omega nT\cos 2kn\pi + \cos\omega nT\sin 2kn\pi \\ &= \sin\omega nT\end{aligned}$$

and thus the responses produced by the two signals must be identical.

In practice, one is almost exclusively concerned with the fundamental period of $H(e^{j\omega T})$, which extends from $-\omega_s/2$ to $\omega_s/2$. This frequency range is called the *baseband.* The frequency $\omega_s/2$, which corresponds to point C in Figure 3, is often called the *Nyquist frequency.* Different types of filter responses, e.g., lowpass, highpass, bandpass, and bandstop, are defined in relation to the baseband, that is, a digital filter that will reject frequency components in the range $0 \le \omega \le \omega_{c1}$ and pass components with frequencies in the range $\omega_{c2} \le \omega \le \omega_s/2$ and where $\omega_{c1} < \omega_{c2} < \omega_s/2$ is essentially a highpass filter.

9. AMPLITUDE AND DELAY DISTORTION

In practice, a digital filter can distort the information content of the signal to be processed as will now be shown. Consider a digital filter charactrized by a transfer function $H(z)$ and assume that its input and output signals are $x(nT)$ and $y(nT)$, respectively. The frequency response of the filter is given by Equations 11 and 12, where $M(\omega)$ and $\theta(\omega)$ are the gain and phase shift at frequency ω. The group delay of the filter at frequency ω is defined as

$$\tau(\omega) = -\frac{d\theta(\omega)}{d\omega}$$

and $\tau(\omega)$ as a function of ω is usually referred to as the *delay characteristic.*

The *frequency spectrum* of a discrete-time signal is its z transform evaluated on the unit circle of the z plane. Like the frequency response of a digital filter, it is a periodic function of ω with period ω_s. Assume that the information content of $x(nT)$ is concentrated in frequency band B given by the set

$$B = \left\{\omega : \omega_L \le \omega \le \omega_H\right\}$$

and that its frequency spectrum is zero elsewhere. If

$$M(\omega) = G_0 \qquad \text{for } \omega \in B \tag{18}$$

and

$$\theta(\omega) = -\tau_g\omega + \theta_0 \qquad \text{for } \omega \in B \tag{19}$$

or

$$\tau(\omega) = \tau_g \qquad \text{for } \omega \in \mathrm{B}$$

where τ_g is a constant, then the frequency spectrum of the output signal $y(nT)$ can be obtained from Equations 6, 13, 18, and 19 as

$$Y(e^{j\omega T}) = H(e^{j\omega T})X(e^{j\omega T})$$
$$= G_0 e^{-j\omega\tau_g + j\theta_0} X(e^{j\omega T})$$

Now, if $\tau_g = mT$ where m is a constant, we have

$$Y(z) = G_0 e^{j\theta_0} z^{-m} X(z)$$

and from the translation relation in Table 2, we deduce

$$y(nT) = G_0 e^{j\theta_0} x(nT - mT)$$

In effect, if the amplitude response of the filter is flat and its phase response is a linear function of ω (i.e., the delay characteristic is flat) in frequency band B, then the output signal is a delayed replica of the input signal except that a gain G_0 and a phase shift θ_0 are introduced.

If the amplitude response of the filter is not flat in band B, then *amplitude distortion* will be introduced since different frequency components of the signal will be amplified by different amounts. On the other hand, if the delay characteristic is not flat, different frequency components will be delayed by different amounts, and *phase (or delay) distortion* will be introduced. Amplitude distortion can be quite objectionable in practice. Consequently, in each frequency band that carries information, the amplitude response is required to be constant to within a prescribed tolerance. If the ultimate receiver of the signal is the human ear, e.g., when a speech or music signal is to be processed, phase distortion is quite tolerable. However, in other applications it can be as objectionable as amplitude distortion, and the delay characteristic is required to be fairly flat. Applications of this type include data transmission, where the signal is to be interpreted by ditital hardware, and image processing, where the signal is used to reconstruct an image which is to be interpreted by the human eye.

10. INTRODUCTION TO THE DESIGN PROCESS

The design of digital filters comprises four general steps, as follows:

1. Approximation
2. Realization
3. Study of arithmetic errors
4. Implementation

The *approximation step* is the process of generating a transfer function that satisfies the desired specifications which may concern the amplitude, phase, and possibly the time-domain response of the filter.

The available methods for the solution of the approximation problem can be classified as *direct* or *indirect.* In direct methods, the problem is solved directly in the z domain. In indirect methods, a continuous-time transfer function is first obtained and is then converted into a corresponding discrete-time transfer function. Nonrecursive filters are always designed through direct methods, whereas recursive filters can be designed either through direct or indirect methods. Approximation methods can also be classified as *closed form* or *iterative.* In closed-form methods, the problem is solved through a small number of design steps using a set of closed-form formulas. In iterative methods, an initial solution is assumed, and through the application of optimization methods a series of progressively improved solutions is obtained until some design cirterion is satisfied. In general, the designer is interested in simple and reliable approximation methods that yield precise designs with the minimum amount of computation.

The *synthesis* of a digital filter is the process of converting the transfer function or some other characterization of the filter into a network. This process is also referred to as the *realization step.* The network obtained is said to be the realization of the transfer function. As for approximation methods, realization methods can be classified as *direct* or *indirect.* In direct methods, the transfer function is put in some form that allows the identification of an interconnection of elemental digital filter subnetworks. In indirect methods, an analog filter network is converted into a topologically related digital filter network. Many realization methods have been proposed in the past which lead to structures of varying complexity and properties. The designer is usually interested in realizations which require the minimum number of unit delays, adders, and multipliers and which are not seriously affected by the use of finite-precision arithmetic in the implementation.

During the approximation step the coefficients of the transfer function are determined to a high degree of precision. In practice, however, digital hardware has a finite precision which depends on the length of registers used to store numbers; the type of number system used (e.g., signed-magnitude, two's complement); the type of arithmetic used (e.g., fixed point or floating point), etc. Consequently, filter coefficients must be quantized (e.g., rounded or truncated) before they can be stored in registers. When the transfer function coefficients are quantized, errors are introduced in the amplitude and phase responses of the filter. In extreme cases, the required specifications can actually be violated. Similarly, signals to be processed, as well as the internal signals of a digital filter (e.g., the products generated by multipliers), must be quantized. Since errors introduced by the quantization of signals are actually sources of noise, they can have a dramatic effect on the performance of the filter. Under these circumstances, the design process cannot be deemed to be complete until the *effects of arithmetic errors* on the performance of the filter are investigated.

The *implementation* of a digital filter can assume two forms: *software* or *hardware.* In the first case, implementation involves the simulation of the filter network on a general-purpose digital computer, workstation, personal computer, or digital signal processing (DSP) chip. In the second case, it involves the conversion of the filter network into a dedicated piece of hardware.

The choice of implementation is usually cirtically dependent on the application at hand. In *non-real-time* applications where a record of the data to be processed is available, a software implementation may be entirely satisfactory. In *real-time* applications, however, where data must be processed at a very high rate (e.g., in communication systems), a hardware implementation is mandatory. Often the best engineering solution might be partially in terms of software and partially in terms of hardware, since software and hardware are highly exchangeable nowadays.

The design of digital filters may often involve steps that do not appear explicitly in the above list. For example, if a digital filter is required to process continuous-time signals, the *effect of interfacing devices* (e.g., analog-to-digital and digital-to-analog converters) on the accuracy of processing must be investigated.

11. DIGITAL VS. ANALOG FILTERS

A band-limited continuous-time signal can be transformed into a discrete-time signal by means of sampling. Conversely, the discrete-time signal so generated can be used to regenerate the original continuous-time signal by means of interpolation, by virtue of Shannon's sampling theorem. As a consequence, hardware digital filters can be used to perform real-time filtering tasks which in the not too distant past were performed almost exclusively by analog filters. The advantages to be gained are the traditional advantages associated with digital systems in general:

1. Component tolerances are uncritical.
2. Component drift and spurious environmental signals have no influence on the system performance.
3. Accuracy is high.
4. Physical size is small.
5. Reliability is high.
6. The cost is low.

A very important additional advantage of digital filters is the ease with which their characteristics can be changed or adapted by simply changing the contents of a finite number of registers. Owing to this feature, digital filters are naturally suited for the design of programmable filters that can be used to perform a multiplicity of filtering tasks, and for the design of adaptive filters that can be used in a diverse range of applications, such as system identification, channel equalization, signal enhancement, and signal prediction.

The basic limitation of hardware digital filters is associated with the speed of operation of the processing elements used. Basically, a digital filter receives its input and produces its output as series of numbers. When a new input value is received, the digital filter computes the next output by performing a number of multiplications and additions. The total number of arithmetic operations is directly related to the filter order. As the sampling frequency is reduced, the sampling period is increased, and, therefore, it is possible to perform more and more arithmetic operations per sampling period. This allows the processing elements to be time-shared, and thus a very economical digital filter can be achieved.

In RC active filters, the quality of the active elements improves while the cost and quality of capacitors tend to remain constant down to about 1.0 kHz. In passive filters, on the other hand, the quality of inductors deteriorates and their cost increases below 5.0 kHz. Thus, between RC active and LC passive filters, active filters are preferred for low-frequency applications even though they require power supplies to operate.

The choice between active and digital filters is less obvious and depends critically on the application. At frequencies below 1.0 kHz, the cost and size of capacitors tend to increase, whereas in digital filters it is possible to apply more and more time-sharing of one or more processing elements without increasing the size or number of very large scale integrated (VLSI) chips. Hence, fairly economical filters can be constructed that have the many advantages associated with digital systems in general, as stated above. On the negative side, when used to process continuous-time signals, digital filters require interfacing devices such as analog-to-digital and digital-to-analog converters and sometimes band-limiting and reconstruction *analog* lowpass filters.

In certain filtering applications the obvious or only choice is the digital filter. Some of these applications are as follows:

1. When the signal to be processed is already in digital format, e.g., multirate signal processing;[5]
2. In non-real-time applications where the signal is recorded on a mass-storage device;
3. In quasi-real-time applications where a certain amount of processing time is allowed;
4. In two-dimensional and multidimensional filtering[6,7] and image processing;
5. In adaptive filters;[8,9]
6. To realize filters that have constant group delay, e.g., linear-phase Hilbert transformers;[1]
7. To realize certain filter types that cannot be realized as analog filters, e.g., fan filters for seismic signal prospecting.[6,7]

REFERENCES

1. **A. Antoniou,** *Digital Filters: Analysis, Design, and Applications,* 2nd ed., McGraw-Hill, New York, 1993.
2. **N. Balabanian and T. A. Bickart,** *Electrical Network Theory,* Wiley, New York, 1969.
3. **S. J. Mason,** Feedback theory — further properties of signal-flow graphs, *Proc. IRE,* 44, 920–926, July, 1956.
4. **E. Kreyszig,** *Advanced Engineering Mathematics,* Wiley, New York, 1972.
5. **R. E. Crochiere and L. R. Rabiner,** *Multirate Digital Signal Processing,* Prentice-Hall, Englewood Cliffs, NJ, 1983.
6. **D. E. Dudgeon and R. M. Mersereau,** *Multidimensional Digital Signal Processing,* Prentice-Hall, Englewood Cliffs, NJ, 1984.
7. **W.-S. Lu and A. Antoniou,** *Two-Dimensional Digital Filters,* Marcel Dekker, New York, 1992.
8. **B. Widrow and S. D. Stearns,** *Adaptive Signal Processing,* Prentice-Hall, Englewood Cliffs, NJ, 1985.
9. **S. Haykin,** *Adaptive Filter Theory,* 2nd ed., Prentice-Hall, Englewood Cliffs, NJ, 1991.

Basic Design of One-Dimensional Digital Filters

Majid Ahmadi

CONTENTS

1. INTRODUCTION

Digital filters are computational algorithms that transform input sequences of numbers into output sequences of numbers according to prespecified rules, hence yielding some desired modification to the characteristics of the input sequences. These filters are generally used for signal enhancement, noise reduction, etc.

Digital filters are divided into two classes, namely, finite impulse response (FIR) and infinite impulse response (IIR). They have also been called nonrecursive and recursive filters, respectively.

An arbitrary linear, time-invariant causal FIR filter can be represented by its difference equation:

$$y(nT) = \sum_{i=0}^{N} a_i x(nT - iT) \tag{1}$$

0-8493-8951-8/97/$0.00+$.50
© 1997 by CRC Press LLC

where x(·) and y(·) are input and output sequences, respectively, T is the sampling period, N is the order of the filter, and a_i are the coefficients of the filter to be determined during the design process. FIR filters can alternatively be characterized by its transfer function:

$$H(z) = y(z)/x(z) = \sum_{i=0}^{N} a_i z^{-i} \tag{2}$$

or

$$H(z) = \frac{\sum_{i=0}^{N} a_i z^{N-i}}{z^N} \tag{3}$$

where $z = e^{j\omega T}$.

As can be seen from Equation 3, all poles of a one-dimensional (1-D) FIR filter are located at the origin of the unit circle in the z-plane and, therefore, they are always stable. Stability of 1-D FIR filters can also be proved because of the summability of their impulse responses.

On the other hand, an arbitrary, linear, time-invariant causal 1-D IIR filter can be represented by a recursive difference equation of the form:

$$y(nT) = \sum_{i=0}^{N} a_i x(nT - iT) - \sum_{i=1}^{N} b_i y(nT - iT) \tag{4}$$

where x(·) and y(·) are the input and the output sequences, T is the sampling period, N is the order of the filter, while a_i and b_i are coefficients of the filters to be determined through the design process. The transfer function of an IIR digital filter is obtained by taking the z transform from both sides of Equation 4 and deriving y(z)/x(z), which is

$$H(z) = \frac{y(z)}{x(z)} = \frac{A(z)}{B(z)} = \frac{\sum_{i=0}^{N} a_i z^{-i}}{\sum_{i=0}^{N} b_i z^{-i}} \tag{5}$$

As can be seen from Equation 5, the poles of the transfer function can lie anywhere in the z-plane; however, to have a stable IIR filter we should have[1,2]

$$B(z) \neq 0 \qquad \text{for} \qquad |z| \geq 1 \tag{6}$$

Evaluating H(z) at equally spaced frequency points on the unit circle in the z-plane yields the frequency response of the digital filter, which is

$$H(z)\big|_{z=e^{j\omega T}} = H\left(e^{j\omega T}\right) = M(\omega)e^{j\phi(\omega)} \tag{7}$$

where

$$M(\omega) = \left|H\left(e^{j\omega T}\right)\right| \tag{8}$$

is the magnitude response, and the phase and group delay of the filter are defined, respectively, as

$$T_p = -\frac{\phi(\omega)}{\omega} \tag{9}$$

$$T_g = -\frac{d\phi(\omega)}{d\omega} \tag{10}$$

where

$$\phi(\omega) = \arg H\left(e^{j\omega T}\right) \tag{11}$$

It can easily be shown[1] that for a FIR filter to have linear phase characteristics we must have

$$a_i = a_{N-i} \quad \text{for} \quad 0 \le i \le N \tag{12}$$

where a_i is the coefficient of the FIR filter (Equation 1).

It can also be shown that linear phase characteristics cannot be achieved over the entire band of frequency for linear, causal, stable IIR filters. In the following sections, various design techniques for 1-D FIR and IIR digital filters are presented.

2. DESIGN OF 1-D FIR FILTERS USING WINDOW FUNCTIONS

The frequency response of a 1-D FIR filter is a periodic function of ω with the period of ω_s rad/s. ω_s is the sampling frequency of the filter. One can now express a Fourier series expansion of $H(e^{j\omega T})$ as

$$H\left(e^{j\omega T}\right) = \sum_{i=-\infty}^{\infty} h(iT) e^{-j\omega iT} \tag{13}$$

where

$$h(iT) = \frac{1}{\omega_s} \int_{-\omega_s/2}^{\omega_s/2} H\left(e^{j\omega T}\right) e^{j\omega iT} d\omega \tag{14}$$

The transfer function of the filter is obtained by inserting $e^{j\omega T} = z$ in Equation 13, which yields

$$H(z) = \sum_{i=-\infty}^{\infty} h(iT)\, z^{-i} \tag{15}$$

Equations 14 and 15 describe a method for derivation of the transfer function of a 1-D FIR filter with the frequency response $H(e^{j\omega T})$. Assuming that an analytical representation for $H(e^{j\omega T})$ exists and the integral of Equation 14 is feasible, then the transfer function of the filter is obtained by Equation 15. It should be noted that this transfer function is of infinite order and noncausal. To obtain a FIR filter with finite order, the summation of Equation 15 should be truncated by

$$h(iT) = 0 \qquad \text{for} \qquad |i| > \frac{N-1}{2}$$

in which case

$$H(z) = h(0) + \sum_{i=1}^{(N-1)/2} \left[h(-iT)z^{i} + h(iT)z^{-i} \right] \tag{16}$$

Translation of Equation 16 by $z^{-(N-1)/2}$ can make the filter causal. That is

$$H'(z) = \bar{z}^{(N-1)/2} H(z) \tag{17}$$

Unfortunately, the truncation of the impulse response of the filter will cause undesirable oscillations in the passband and stopband of the filter which are known as Gibbs oscillations. To reduce the effect of Gibbs oscillations, the impulse response of the FIR filter h(iT), Equation 14 can be preconditioned by utilization of Window functions. The procedure then requires

1. Given $H(e^{j\omega T})$, obtain h(iT) using Equation 14.
2. Truncate h(iT) to an appropriate size.
3. Multiply h(iT) by W(iT), which is one of the following Window functions:
   a. Rectangular Window

$$W_R(iT) = \begin{cases} 1 & \text{for } |i| \le \dfrac{N-1}{2} \\ 0 & \text{otherwise} \end{cases} \tag{18}$$

   b. Triangular Window

$$W_T(iT) = \begin{cases} \dfrac{2i}{N-1} & \text{for} \quad 0 \le i \le \dfrac{N-1}{2} \\ 2 - \dfrac{2i}{N-1} & \text{for} \quad \dfrac{N-1}{2} < i \le N-1 \end{cases} \tag{19}$$

   c. Hann and Hamming Windows

$$W_H(iT) = \begin{cases} \alpha + (1-\alpha)\cos\dfrac{2\pi i}{N-1} & \text{for} \quad |i| \le \dfrac{N-1}{2} \\ 0 & \text{otherwise} \end{cases} \tag{20}$$

   $\alpha = 0.5$ and $\alpha = 0.54$ represent Hann and Hamming Windows, respectively.

   d. Blackman Window

$$W_B(iT) = \begin{cases} 0.42 + 0.5\cos\dfrac{2\pi i}{N-1} + 0.08\cos\dfrac{4\pi i}{N-1} & \text{for} \quad |n| \le \dfrac{N-1}{2} \\ 0 & \text{otherwise} \end{cases} \tag{21}$$

e. Kaiser Window

$$W_K(iT) = \begin{cases} I_0(\beta)/I_0(\alpha) & \text{for} \quad |i| \le \frac{N-1}{2} \\ 0 & \text{otherwise} \end{cases} \tag{22}$$

where α is an independent parameter and

$$\beta = \alpha\left[1 - \left(\frac{2i}{N-1}\right)^2\right]^{1/2} \tag{23}$$

$I_o(x)$ is the zeroth-order Bessel function of the first kind and represented by the following rapidly converging series

$$I_0(x) = 1 + \sum_{n=1}^{\infty}\left[\frac{1}{n!}\left(\frac{x}{2}\right)^2\right]^2 \tag{24}$$

To evaluate the above series with a desired accuracy, a finite number of terms needs to be included in the calculation since this represents a fast converging series.

4. Form the noncausal transfer function of the designed filter by

$$H_D(z) = \sum_{i=-(N-1)/2}^{(N-1)/2} [h(iT)W(iT)]z^{-i} \tag{25}$$

5. To obtain the causal transfer function of the designed 1-D FIR filter, Equation 25 should be translated by $z^{-(N-1)/2}$, which yields

$$H_D'(z) = z^{(N-1)/2} H_D(z) \tag{26}$$

It should be noted that in this design approach only Kaiser Window has an independent parameter α which needs to be calculated in addition to N, which is the order of the filter.

As an example, we design a lowpass digital filter with the following specification

$$H\left(e^{j\omega T}\right) = \begin{cases} 1 & \text{for} \quad |\omega| \le 2.5 \text{ rad/s} \\ 0 & \text{for} \quad 2.5 < |\omega| \le 5 \text{ rad/s} \end{cases} \tag{27}$$

using (a) Hann, (b) Hamming, (c) Blackman, and (d) Rectangular Windows. Assume $\omega_s = 10$ rad/s and N = 15.

From Equation 4 one obtains the impulse response of the filter, which is

$$h(iT) = \frac{1}{i\pi}\int_{-\omega_c}^{\omega_c} e^{j\omega iT}\, d\omega = \frac{i}{i\pi} \sin \omega_c\, iT$$

with $\omega_c = 2.5$ rad/s we get

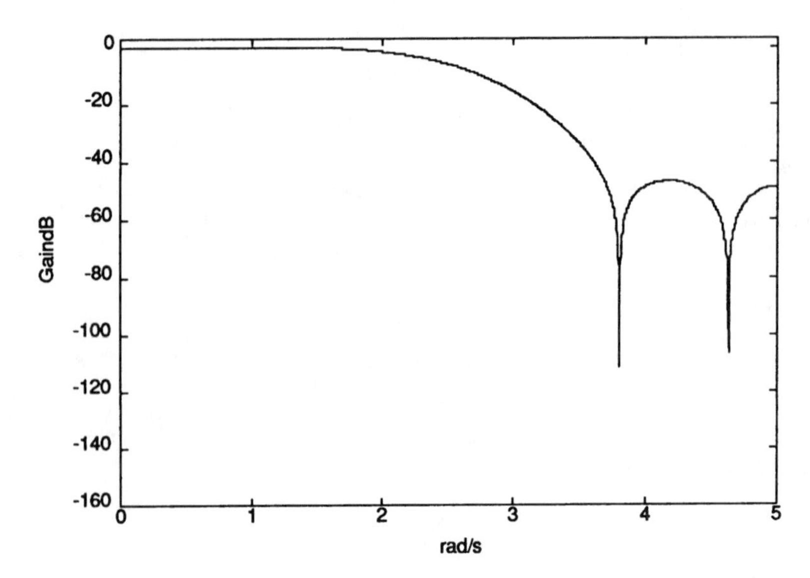

Figure 1 Magnitude response of the 1-D lowpass filter using Hamming Window.

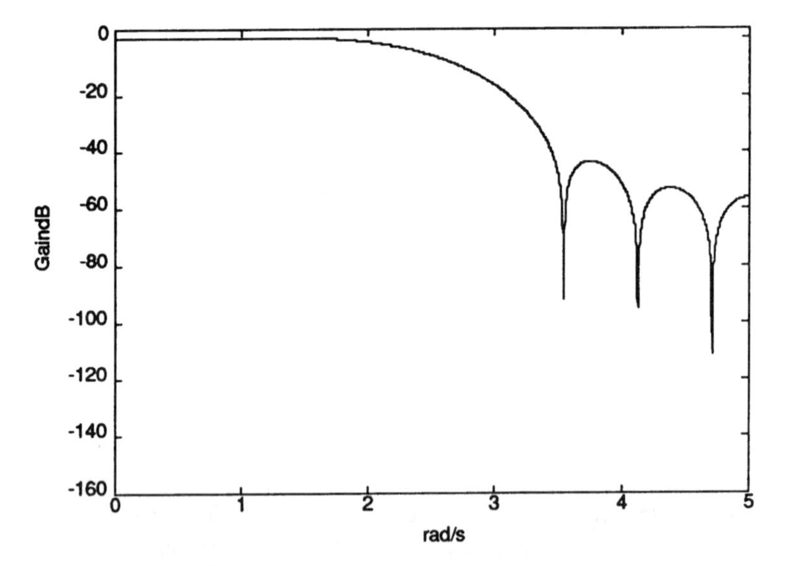

Figure 2 Magnitude response of the 1-D lowpass filter using Hann Window.

$$h(iT) = \frac{1}{i\pi} \sin 2.5\ iT$$

Now by following steps 2 to 5 of this section the designed filter is obtained. Figures 1 through 4 show the magnitude responses of the designed filter, respectively.

3. DESIGN OF 1-D FIR FILTERS USING KAISER WINDOW

1-D FIR filters satisfying prescribed magnitude specifications can be readily designed using the procedure outlined in Reference 1, which is as follows.

Let us assume the required passband gain and stopband attenuation is given in decibels as

$$A_p = 20 \log \frac{1 + \delta_1}{1 - \delta_1} \tag{28}$$

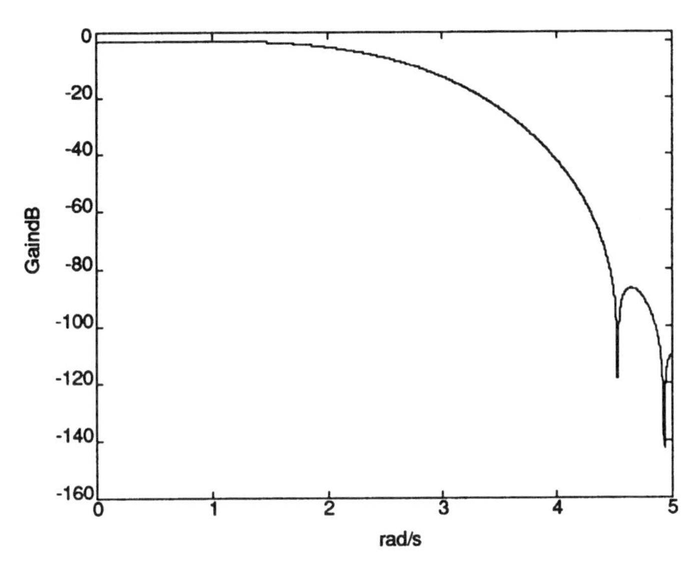

Figure 3 Magnitude response of the 1-D lowpass filter using Blackman Window.

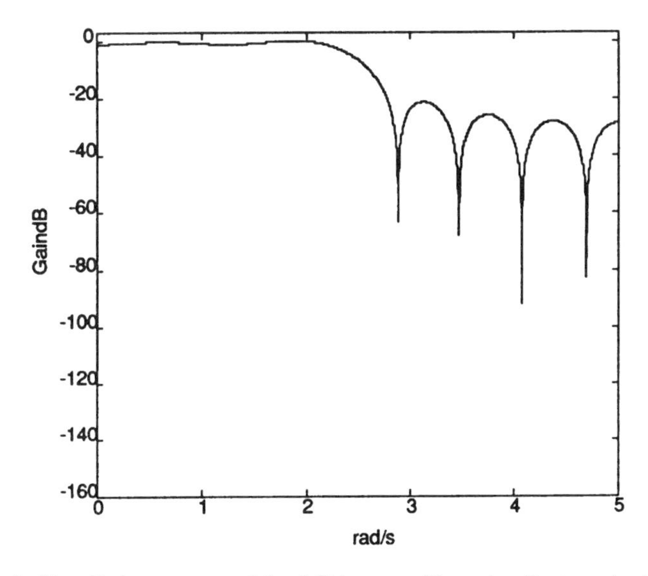

Figure 4 Magnitude response of the 1-D lowpass filter using Rectangular Window.

and

$$A_a = -20 \log \delta_2 \tag{29}$$

where δ_1 and δ_2 are as shown in Figure 5.

To design the filter

1. Determine h(iT) using Equation 14 by assuming that

$$\omega_c = 0.5(\omega_p + \omega_a) \tag{30}$$

where ω_p and ω_a are passband and stopband edges of the filter, respectively.

2. Choose Δ using Equations 28 and 29 as

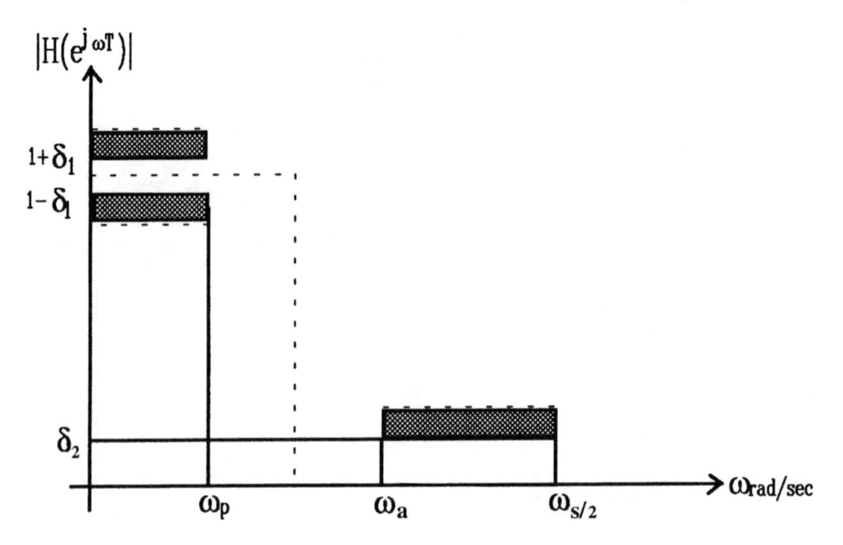

Figure 5 Specification for a 1-D lowpass filter.

$$\Delta = \min\left(\Delta_1, \Delta_2\right) \tag{31}$$

where

$$\Delta_1 = 10^{-0.05A} a \tag{32}$$

$$\Delta_2 = \frac{10^{0.05A} p - 1}{10^{0.05A} p + 1} \tag{33}$$

3. Calculate A using

$$A = 20 \log \Delta \tag{34}$$

4. Choose parameter α using the following empirical formula

$$\alpha = \begin{cases} 0 & \text{for} \quad A \le 21 \\ 0.5842(A-21)^{0.4} + 0.07886(A-21) & \text{for} \quad 21 < A \le 50 \\ 0.1102(A-8.7) & \text{or} \quad A > 50 \end{cases} \tag{35}$$

5. Choose the lowest odd value of N (order of the filter) satisfying the inequality

$$N \ge \frac{\omega_s D}{B_t} + 1 \tag{36}$$

where

$$D = \begin{cases} 0.9222 & \text{for} \quad A \le 21 \\ \dfrac{A - 7.95}{14.36} & \text{for} \quad A > 21 \end{cases} \tag{37}$$

and using Figure 5

$$B_t = \omega_a - \omega_p \tag{38}$$

6. Calculate $W_k(iT)$
7. Form $H'_D(z) = z^{-(N-1)/2} H_D(z)$, where

$$H_D(z) = \sum_{i=0}^{N-1} [h(iT)W_k(iT)]z^{-i} \tag{40}$$

As an example, we design a FIR filter satisfying the following specifications:

$$A_p \leq 0.1 \text{ dB}$$

$$A_a \geq 44.0 \text{ dB}$$

$$\omega_p = 20 \text{ rad/s}$$

$$\omega_a = 30 \text{ rad/s}$$

$$\omega_s = 100 \text{ rad/s}$$

Solution:

$$\omega_c = 0.5(\omega_p + \omega_a) = 0.5(20 + 30) = 25 \text{ rad/s}$$

$$h(iT) = \frac{1}{i\pi} \sin 25iT$$

$$\Delta_1 = 0.006310$$

$$\Delta_2 = 0.005756$$

$$\Delta = \min(\Delta_1, \Delta_2) = 0.005756$$

$$A = -20 \log \Delta = 44.795780$$

Using Equation 35 gives

$$\alpha = 3.92425$$

Using Equation 37 gives

$$D = 2.565862$$

$$B_t = 30 - 20 = 10 \text{ rad/s}$$

Equation 36 yields the order of the filter, which is

$$N \geq \frac{100 \times 2.565862}{10} + 1$$

$$N = 27$$

Table 1 Value of the Coefficients of the Designed FIR Filter

+i or –i	W(iT)h(iT)
0	5.000000×10^{-1}
1	3.151149×10^{-1}
2	$-2.370441 \times 10^{-17}$
3	-9.680709×10^{-2}
4	2.094073×10^{-17}
5	4.908526×10^{-2}
6	$-1.687857 \times 10^{-17}$
7	-2.684133×10^{-2}
8	1.220080×10^{-17}
9	1.412885×10^{-2}
10	$-7.634676 \times 10^{-18}$
11	-6.555915×10^{-3}
12	3.792885×10^{-18}
13	2.257268×10^{-3}

Table 1 gives the value of W(iT)h(iT) which are the coefficients of the FIR filter. Small modification is needed to the above procedure in order to include design of bandpass or bandstop filters using Kaiser Window.

For calculation of the order of the filter using Equation 36, B_t should be chosen as the narrower of the two transition bands, i.e.,

$$B_t = \min\left\{\left(\omega_{p1} - \omega_{a1}\right), \left(\omega_{a2} - \omega_{p2}\right)\right\} \tag{41}$$

Also the two cutoff frequencies for the passband edges are obtained using the following relationship:

$$\omega_{c1} = \omega_{p1} - \frac{B_t}{2} \tag{42}$$

$$\omega_{c2} = \omega_{p2} + \frac{B_t}{2} \tag{43}$$

The rest of the steps are identical to those of the design for the lowpass filter. Readers are encouraged to refer to Reference 1 for further details.

4. EQUIRIPPLE DESIGN OF A FIR FILTER

In this section a method for the design of a FIR filter with equiripple magnitude response using the Remez exchange algorithm is presented.[3] This algorithm is based on the following important theorem.

4.1 THEOREM 1

If $P(\omega)$ is a linear combination of n cosine functions of the form

$$P(\omega) = \sum_{i=0}^{n-1} \alpha(i) \cos i\omega \tag{44}$$

then a necessary and sufficient condition that P(ω) be the unique, best-weighted Chebyshev approximation to a continuous function D(ω) on A, a compact subset of [0,π], is that the weighted error function E(ω) exhibits at least n + 1 extremal frequencies in A; i.e., there must exist n + 1 points ω_i in A such that

$$\omega_0 < \omega_1 < \ldots < \omega_n \tag{45}$$

$$E(\omega_i) = -E(\omega_{i+1}) \tag{46}$$

and

$$|E(\omega_i)| = \max_{\omega \varepsilon A} |E(\omega)| \; i = 0, 1, \ldots, n \tag{47}$$

4.2 PROBLEM FORMULATION

Let us focus our attention to the design of linear lowpass FIR filter with symmetrical impulse response and odd N. It can easily be shown that the extension to include bandpass, highpass, and bandstop filters is trivial. The transfer function of this filter is of the form:

$$H(z) = \sum_{i=0}^{N+1} a_i z^{-i} \tag{48}$$

Assuming that $\omega_s = 2\pi$ rad/s and

$$a_i = a_{(N+1)-1} \tag{49}$$

Equation 48 can be written as

$$H(e^{j\omega}) = e^{j\omega(n-1)} P(\omega) \tag{50}$$

where function P(ω) has identical form as Equation 44

$$P(\omega) = \sum_{i=0}^{n-1} a(i) \cos i\omega \qquad n = \frac{N+1}{2} \tag{51}$$

The magnitude response of an equiripple lowpass filter is shown in Figure 6, where δ_1 and δ_2 are the magnitude of the passband and stopband ripples, respectively. In the same figure ω_p and ω_a represent passband edge and stopband edge, respectively.

The desired value of P(ω) for a lowpass filter, designated D(ω), is given by

$$D(\omega) = \begin{cases} 1 & \text{for} \quad 0 \le \omega \le \omega_p \\ 0 & \text{for} \quad \omega_a \le \omega \le \pi \end{cases} \tag{52}$$

One can hence define an error function in the form of

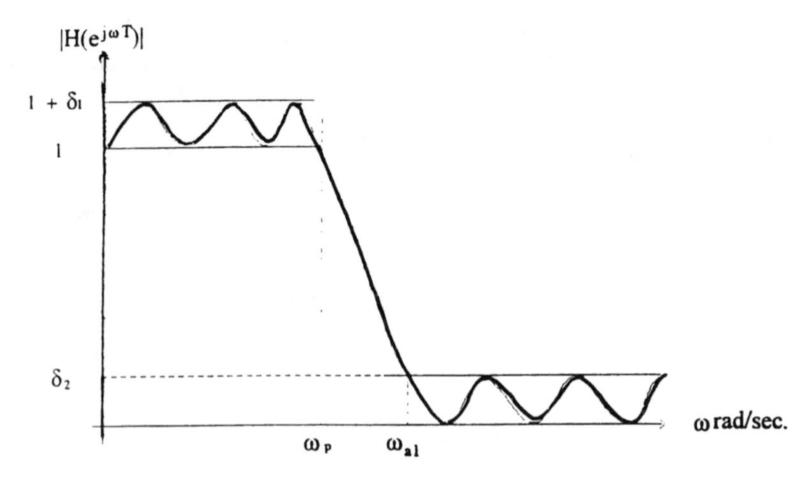

Figure 6 Magnitude response of an equiripple 1-D filter.

$$E_0(\omega) = D(\omega) - P(\omega) \tag{53}$$

where

$$|E_0(\omega)| = \begin{cases} \delta_1 & \text{for} \quad 0 \le \omega \le \omega_p \\ \delta_2 & \text{for} \quad \omega_a \le \omega \le p \end{cases} \tag{54}$$

Alternatively, one can write

$$|E(\omega)| \le \delta_1 \tag{55}$$

where

$$E(\omega) = W(\omega)E_0(\omega) \tag{56}$$

$$W(\omega) = \begin{cases} 1 & \text{for} \quad 0 \le \omega \le \omega_p \\ \delta_1/\delta_2 & \text{for} \quad \omega_a \le \omega \le \pi \end{cases} \tag{57}$$

Combining Equations 53 and 56 gives

$$E(\omega) = W(\omega)[D(\omega) - P(\omega)] \tag{58}$$

From the alternation property cited in Theorem 1, a set of extremal frequencies $\{\omega_i\}$ and a cosine polynomial $P(\omega)$ exist, such that

$$E(\omega_i) = W(\omega_i)[D(\omega_i) - P(\omega_i)] = (-1)^i \delta \tag{59}$$

for i = 0, 1, 2, …, n, where δ is a constant. This relation can be expressed in the following matrix form

$$\begin{bmatrix} 1 & \cos\omega_0 & \cos 2\omega_0 & \cdots & \cos[(n-1)\omega_0] \; 1/W(\omega_0) \\ 1 & \cos\omega_1 & \cos 2\omega_1 & & \vdots \\ \vdots & \vdots & \vdots & & \vdots \\ \vdots & \vdots & \vdots & & \vdots \\ \vdots & \vdots & \vdots & & \dfrac{(-1)^n}{W(\omega_n)} \end{bmatrix} \begin{bmatrix} \alpha(0) \\ \alpha(1) \\ \vdots \\ \alpha(n-1) \\ \delta \end{bmatrix} = \begin{bmatrix} D(\omega_0) \\ D(\omega_1) \\ \vdots \\ \vdots \\ D(\omega_n) \end{bmatrix} \tag{60}$$

Therefore, if the extremal frequencies are known, the FIR filter coefficients α(n) and, hence, the frequence response of the filter can be computed by inverting the above matrix. Inevitability of this matrix is guaranteed by the Harr condition on the basic functions.

4.3 REMEZ EXCHANGE ALGORITHM

A powerful algorithm which can be used to deduce a unique lowpass filter for given values of ω_p, ω_a and N is the so-called Remez exchange algorithm which is given in Reference 4 and can be described in the following steps.

Step 1 Select n + 1 extremal frequencies $\omega_0, \omega_1, \ldots, \omega_n$ such that m + 1 of them, namely, ω_0, ω_1, …., ω_m, are uniformly spaced in the frequency range 0 to ω_p and n – m of them, namely, $\omega_{m+1}, \omega_{m+2}, \ldots, \omega_n$, are uniformly spaced in the frequency range ω_a to π. In addition, it is required that $\omega_m = \omega_p$ and $\omega_{m+1} = \omega_a$. Further, the frequency interval $\omega_i - \omega_{i-1}$ is required to be approximately the same in the two bands.

Step 2 Compute function E(ω) over a dense set of frequencies $\omega = 0, i_p, i_{2p}, \ldots, \omega_p, \omega_a, \omega_a + i_a, \omega_a + 2_{ia}, \ldots, \pi$ where i_p and i_a are the passband and stopband frequency increments, respectively.

Step 3 Let $\hat{\omega}_0, \hat{\omega}_1, \ldots, \hat{\omega}_{kp}, \hat{\omega}_\mu, \hat{\omega}_{\mu+1}, \ldots, \hat{\omega}_{ka}, \ldots, \omega_\xi$ be potential extremal frequencies for the next iteration, and assume that $\hat{\omega}_0, \hat{\omega}_\mu$, to $\hat{\omega}_\xi$ as follows:

a. If $[\{E(0) >$ and $E(0) > E(i_p)\}$
   or $\{E(0) < 0$ and $E(0) < E(i_p)\}]$
   and $|E(0)| \ge |\delta|$, set $0 \to \hat{\omega}_0$.

b. If $[\{E(\omega) > 0$ and $E(\omega - i_p) < E(\omega) > E(\omega + i_p)\}$
   or $\{E(\omega) < 0$ and $E(\omega - i_p) > E(\omega) < E(\omega + i_p)\}]$,
   set $\omega \to \omega_{kp}$.

c. If $[\{E(\omega_p) > 0$ and $E(\omega_p) > E(\omega_p - i_p)\}$
   or $\{E(\omega_p) < 0$ and $E(\omega_p) < E(\omega_p + i_p)\}]$
   and $|(E\omega_p)| \ge |\delta|$, set $\omega_p \to \hat{\omega}_\mu$.

d. If $[\{E(\omega_a) > 0$ and $E(\omega_a) > E(\omega_a - i_a)\}$
   or $\{E(\omega_a) < 0$ and $E(\omega_a) < E(\omega_a + i_a)\}]$
   and $|E(\omega_a)| \ge |\delta|$, set $\omega_a \to \hat{\omega}_{\mu+1}$.

e. If $[\{E(\omega) > 0$ and $E(\omega - i_a) < E(\omega) > E(\omega + i)\}$
   or $\{E(\omega) < 0$ and $E(\omega - i_a) > E(\omega) < E(\omega + i_a)\}]$,
   set $\omega \to \hat{\omega}_{ka}$.

f. If $[\{E(\pi) > 0$ and $E(\pi) > E(\pi - i_a)\}$
   or $\{E(\pi) < 0$ and $E(\pi) < E(\pi - i_a)\}]$
   and $|E(\pi)| \ge |\delta|$, set $\pi \to \hat{\omega}_\xi$.

Step 4 Compute

$$Q = \frac{\max\left|E(\hat{\omega}_k)\right| - \min\left|E(\hat{\omega}_k)\right|}{\max\left|E(\hat{\omega}_k)\right|}$$

where $k = 0, 1, \ldots, \xi$

Step 5 Reject the $\xi - n$ frequencies $\hat{\omega}_k$ for which $|E(\hat{\omega}_k)|$ is lowest, and renumber the remaining passband frequencies from 0 to m and the remaining stopband frequencies from m + 1 to n. Then, update extremal frequencies by letting $\omega_k = \hat{\omega}_k$, for $k = 0, 1, \ldots, n$.

Step 6 If $Q > 0.01$, repeat from Step 2. Otherwise, continue to Step 7.

Step 7 Compute $P(\omega)$ using the last set of extremal frequencies.

Step 8 Deduce the impulse response $h(n)$ of the filter.

The preceding algorithm converges very rapidly, usually within four to eight iterations.

4.4 IMPLEMENTATION OF ALGORITHM

In Step 1 of the preceding algorithm, it is convenient to set $\omega_0 = 0$ and $\omega_n = \pi$.

Suitable passband and stopband intervals I_p and I_a to achieve uniform distribution of extremal frequencies in the passband as well as the stopband can be computed as follows:

Let

$$I = \frac{\omega_p + \pi - \omega_a}{n} \tag{61}$$

$$m = \text{int}\left[\frac{\omega_p}{1} + 0.5\right] \tag{62}$$

$$m_1 = \text{int}\left[\frac{\pi - \omega_a}{I} + 0.5\right] \tag{63}$$

where int [ ] represents the closest integer by rounding up the value inside the bracket. Then compute

$$I_p = \frac{\omega_p}{m} \qquad I_a = \frac{\pi - \omega_a}{m_1}$$

In Step 2, a dense set of frequencies is required in order to achieve sufficient accuracy in the determination of frequencies $\hat{\omega}_k$ in Step 3. A reasonable number of frequency points are $8(N - 1)$.[4] This corresponds to about 16 frequency points per ripple of $|E(\omega)|$. The frequency set should be chosen carefully to ensure that extremal frequencies belong to the set in order to avoid numerical ill conditioning in the evaluation of $E(\omega)$. Suitable passband and stopband frequency intervals are $i_p = I_p/16$ and $i_a = I_a/16$, respectively.

The solution of Equation 60 can be obtained if precisely n + 1 extremal frequencies are available. By differentiating $E(\omega)$ with respect to ω, one can show that the number of maxima in $|E(\omega)|$ (potential extremal frequencies in Steps 3a, b, e, and f) can be as high as n + 1.

Since two additional potential extremal frequencies can occur at $\omega = \omega_p$ and ω_a (see Steps 3c and d), a maximum of two superfluous potential extremal frequencies can in principle occur in Step 3. This problem is overcome by rejecting superfluous values of $\hat{\omega}_k$, if any, as described in Step 5 of the algorithm.

In Steps 2 and 7, $E(\omega)$ and $P(\omega)$ can be evaluated by determining coefficients $\alpha(n)$ through matrix inversion. However, this approach can slow down the computation and possibly cause numerical ill conditioning, in particular if δ is small and N is large. An alternative and more efficient approach is to deduce δ analytically and then to interpolate P(ω) on the n frequency points using the barycentric form of the Lagrange interpolating formula. The necessary formulation is as follows.

Parameter δ can be deduced as

$$\delta = \frac{\sum_{k=0}^{n} a_k D(\omega_k)}{\sum_{k=0}^{n} \frac{(-1)^k a_k}{W(\omega_k)}} \tag{64}$$

and P(ω) is given by

$$P(\omega) = \begin{bmatrix} C_k & \text{for} \quad \omega = \omega_0, \omega_1, \ldots, \omega_{n-1} \\ \dfrac{\sum_{k=0}^{n-1} \frac{\beta_k}{x - x_k} C_k}{\sum_{k=0}^{n-1} \frac{\beta_k}{x - x_k}} & \text{otherwise} \end{bmatrix} \tag{65}$$

where

$$a_k = \prod_{\substack{i=0 \\ i\neq k}}^{n} \frac{1}{x - x_i} \tag{66}$$

$$C_k = D(\omega_k) - (-1)^k \frac{\delta}{W(\omega_k)} \tag{67}$$

$$B_k = \prod_{\substack{i=0 \\ i\neq k}}^{n-1} \frac{1}{x_k - x_i} \tag{68}$$

$$x = \cos\omega \qquad x_k = \cos(\omega_k) \qquad k = 0, 1, 2, \ldots, n-1 \tag{69}$$

The impulse response h(i) in Step 8 can be deduced by computing P(kΩ) for k = 0, 1, …, n, where $\Omega = 2\pi/N$, and then using the inverse Fourier transform. For a symmetrical impulse response, it can be shown that

$$h(i) = h(-i) = \frac{1}{N}\left\{P(0) + \sum_{k=1}^{n} 2P(k\Omega)\cos\left(\frac{2\pi ki}{N}\right) \quad \text{for} \quad i = 0, 1, 2, \ldots, n\right\}$$

Example: Design a 21st-order highpass FIR filter with $\omega_p = 2$ rad/s, $\omega_a = 1$ rad/s, $\omega_s = 2\pi$ rad/s, and $\delta_1/\delta_2 = 18$ using the Remez exchange algorithm. Table 2 shows the values of the coefficients of the designed filter while Figure 7 shows the magnitude plot of the designed filter.

Table 2 Values of the Coefficients of the Designed Filter

h(−n) = h(n)	
h(0) = 4.976192	E-01
h(1) = −3.120628	E-01
h(2) = 2.46299	E-03
h(3) = 8.853907	E-02
h(4) = −2.605336	E-03
h(5) = −3.790087	E-02
h(6) = 2.553469	E-03
h(7) = 1.553835	E-02
h(8) = −2.126568	E-03
h(9) = −5.222708	E-03
h(10) = 1.898114	E-03

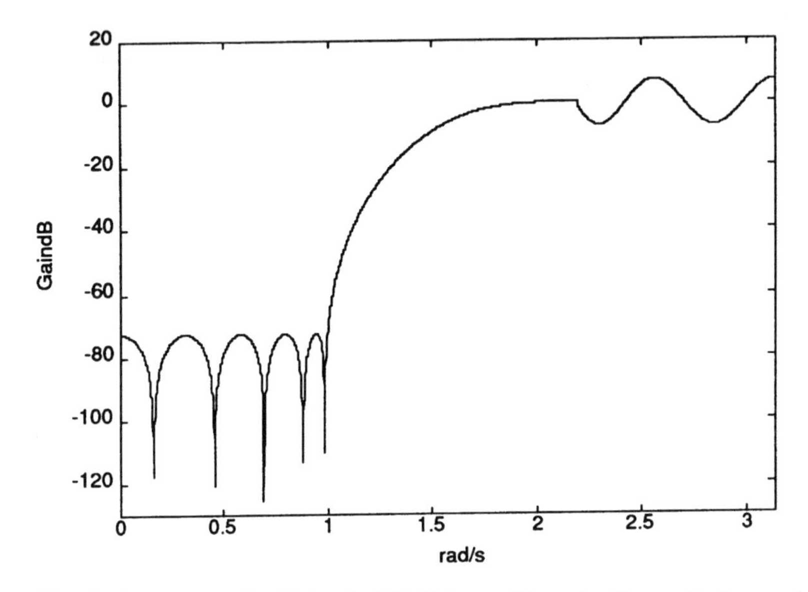

Figure 7 Magnitude response of a 21st-order FIR highpass filter using Remez Exchange algorithm.

5. DESIGN OF IIR DIGITAL FILTERS

There are two approaches to the design of IIR digital filters, the indirect and the direct methods. In indirect methods, one of the classical analog filter approximations (such as Butterworth, Chebyshev, elliptic) is used to generate an analog transfer function, which is subsequently discretized by the application of bilinear transformation. Although other discretization methods such as mapping of differentials, impulse invariant transform, and matched-z transform, have also been mentioned in the literature, due to their limited practicality they have never been widely used and therefore are omitted for further discussion in this book. In the direct method, on the other hand, the desired discrete-time transfer function is obtained directly from the given magnitude, with or without a specified phase specification, generally through the use of an iterative method based on linear or nonlinear programming.[1,2,5] In the next sections, several methods for the design of stable IIR filters will be presented.

5.1 BILINEAR TRANSFORMATION

A simple conformal mapping from the s plane to the z plane, and vice versa, which preserves the desired algebraic form of the analog filter is the bilinear transformation defined by

$$s \rightarrow \frac{2}{T}\frac{1-z^{-1}}{1+z^{-1}} \tag{70}$$

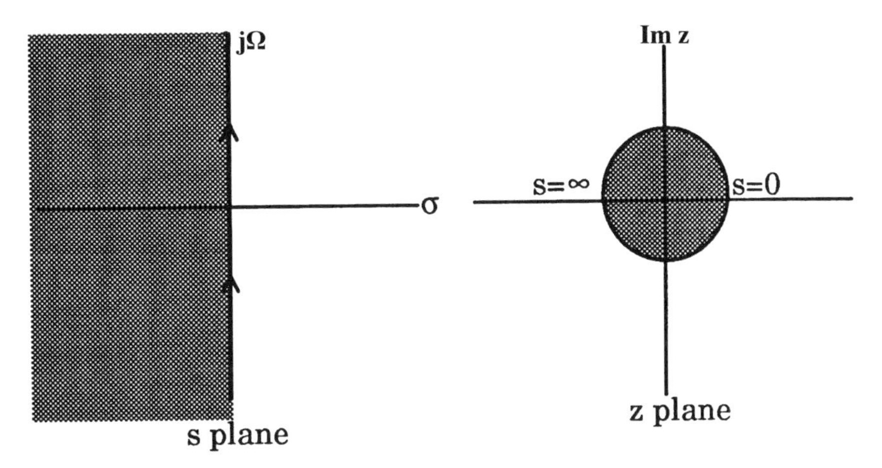

Figure 8 Mapping from s-plane to the z-plane using bilinear transformation.

and

$$z \to \frac{2/T + s}{2/T - s} \tag{71}$$

Using this transformation, the entire jΩ axis in the s plane is mapped onto the circumference of the unit circle in the z-plane; the left-half s plane is mapped inside the unit circle and the right-half s-plane is mapped outside the unit circle in the z plane, as shown in Figure 8.

The preservation of stability through this transformation can be verified by replacing $s = j\Omega$ in Equation 71, which yields

$$z = \frac{2/T + j\Omega}{2/T - j\Omega} \tag{72}$$

It can be verified from Equation 72 that

1. $z = 1$ when $\Omega = 0$,
2. $z = -1$ when $\Omega = \infty$, and
3. Between the above limits the angle of z varies monotonically from 0 to Ω.

Letting $s = \sigma + j\Omega$ gives

$$z = \frac{2/T + \sigma + j\Omega}{2/T + \sigma - j\Omega} \tag{73}$$

One can deduce from the above that for $\sigma < 0$ (left-half plane), $|z| < 1$, and for $\sigma > 0$ (right-half plane), $|z| > 1$. This means the poles on the left-half s plane are mapped inside the unit circle while the poles on the right half are mapped outside the unit circle in the z plane. Hence, a stable filter in the analog domain will be transformed to a stable filter in the digital domain, and vice versa. While the frequency response of the analog filter and digital filter have the same amplitudes, there is a nonlinear relationship between corresponding digital and analog frequencies which can be derived by letting $z = e^{j\omega T}$ and $s = j\Omega$ in Equation 71:

$$j\Omega = \frac{2}{T}\frac{1 - e^{j\omega T}}{1 + e^{j\omega T}} \tag{74}$$

or

$$\Omega = \frac{2}{T}\tan\frac{\omega T}{2} \tag{75}$$

Using the above, the inverse relationship can be derived as

$$\omega = \frac{2}{T}\tan^{-1}\frac{\Omega T}{2} \tag{76}$$

The above relationship indicates that if the analog filter has a set of critical frequencies, Ω_i, $i = 1, 2, \ldots, k$, the bilinearly transformed filter will have

$$\omega_i = \frac{2}{T}\tan^{-1}\frac{\Omega_i T}{2} \qquad i = 1, 2, \ldots, k \tag{77}$$

The above equation indicates that unless $\Omega_i T/2$ is so small that $\tan^{-1} \Omega_i T/2$ can be approximated by $\Omega_i T/2$ a nonlinear distortion is applied to these critical frequencies known as the warping effect.

This warping effect can be compensated if the critical frequencies of the analog prototype filter are prewarped prior to bilinear transformation. This requires that Equation 75 be used to generate the new set of critical frequencies for the analog prototype filter.

5.2 DESIGN OF IIR DIGITAL FILTERS DERIVED FROM THEIR ANALOG COUNTERPARTS

The procedure for the design of a digital filter derived by discretization of an analog counterpart using bilinear transformation can be summarized in the following steps:

1. Prewarp the digital critical frequencies.
2. Design an analog filter to meet the prewarped specification.
3. Apply the bilinear transformation to obtain H(z), the digital filter transfer function.

This technique, though simple to use, has a major drawback; that is, if the phase response of the analog filter is linear, the bilinearly transformed IIR filter will not have a linear phase characteristic. In this case, the iterative methods which will be discussed in the next section, should be used to design an IIR digital filter satisfying a desired magnitude and phase (or group-delay) response.

6. AN ITERATIVE APPROACH TO THE DESIGN OF IIR FILTERS SATISFYING A GIVEN MAGNITUDE AND CONSTANT GROUP-DELAY RESPONSE

There are two approaches to the design of constant group-delay 1-D IIR digital filters. In the first approach, a 1-D analog filter is designed to meet the desired magnitude specification using any of the classical analog filters, namely, Butterworth, Chebyshev, and elliptic filters. Then this analog filter is discretized by the application of the bilinear transformation and cascaded with a stable digital allpass transfer function. In the second step, the parameters of the allpass function are determined through minimization of an objective function to achieve desired group-delay characteristics.[7] In the second approach, on the other hand, the desired magnitude and group-delay responses are obtained through a one-step optimization procedure using linear programming,[8] or nonlinear programming.[6,9]

In this section the method of Reference 6 is presented because of its generality and simplicity of implementation. In this method a Hurwitz polynomial (HP) is generated using properties of positive definite matrices.

HP has all its zeros on the left-hand side of the $j\omega$ axis in the s plane, which indicates stability. If this polynomial is discretized using bilinear transformation, one will obtain a rational function in z with all its zeros inside the unit circle and all its poles at –1, in the z plane. The numerator of this rational function therefore represents a stable polynomial in z. This polynomial can be assigned to the denominator of Equation 5, while the numerator is left unchanged to obtain a stable transfer function in the z variable. Parameters of the desired stable transfer function can now be calculated by minimizing an error function which gives a measure of closeness to the desired magnitude and group-delay responses of the filter.

6.1 GENERATION OF 1-D HP

It is well known that a network described by a symmetric positive definite (or positive semidefinite) immittance matrix is always physically realizable. It is further known that any positive definite matrix P can always be decomposed as a product of two matrices Q and Q^T, where Q is either an upper-triangular or a lower-triangular matrix. Let us define the matrix

$$D = C\Gamma C^T s + G + R\Delta R^T \tag{78}$$

where C and R are upper-triangular matrices with unity elements in their diagonal, Γ and Δ are diagonal matrices with non-negative elements, and G is a skew symmetric matrix shown below.

$$C = \begin{bmatrix} 1 & C_{12} & C_{13} & \dots & C_{1n} \\ 0 & 1 & C_{23} & \dots & C_{2n} \\ 0 & 0 & 1 & \dots & C_{3n} \\ . & & & & \\ . & & & & \\ . & & & & \\ . & & & & \\ 0 & 0 & . & \dots & 1 \end{bmatrix} \tag{79}$$

$$\Gamma = \begin{bmatrix} \gamma_1^2 & 0 & 0 & \dots & 0 \\ 0 & \gamma_2^2 & 0 & \dots & 0 \\ 0 & 0 & \gamma_3^2 & \dots & 0 \\ . & & & & \\ . & & & & \\ . & & & & \\ . & & & & \\ 0 & 0 & . & \dots & \gamma_n^2 \end{bmatrix} \tag{80}$$

$$R = \begin{bmatrix} 1 & r_{12} & r_{13} & \cdots & r_{1n} \\ 0 & 1 & r_{23} & \cdots & r_{2n} \\ 0 & 0 & 1 & \cdots & r_{3n} \\ \cdot & & & & \\ \cdot & & & & \\ \cdot & & & & \\ \cdot & & & & \\ 0 & 0 & \cdot & \cdots & 1 \end{bmatrix} \tag{81}$$

$$\Delta = \begin{bmatrix} \delta_1^2 & 0 & 0 & \cdots & 0 \\ 0 & \delta_2^2 & 0 & \cdots & 0 \\ 0 & 0 & \delta_3^2 & \cdots & 0 \\ \cdot & & & & \\ \cdot & & & & \\ \cdot & & & & \\ \cdot & & & & \\ 0 & 0 & \cdot & \cdots & \delta_n^2 \end{bmatrix} \tag{82}$$

$$G = \begin{bmatrix} 0 & g_{12} & g_{13} & \cdots & g_{1n} \\ -g_{12} & 0 & g_{23} & \cdots & g_{2n} \\ -g_{13} & 0 & -g_{23} & \cdots & g_{3n} \\ \cdot & & & & \\ \cdot & & & & \\ \cdot & & & & \\ \cdot & & & & \\ -g_{1n} & -g_{2n} & \cdot & \cdots & 0 \end{bmatrix} \tag{83}$$

It can easily be shown that a 1-D HP can be generated by simply taking the determinant of matrix D in Equation 78.

As an example, we generate a second-order HP using the above method as follows:

$$D = \begin{bmatrix} 1 & C \\ 0 & 1 \end{bmatrix} \begin{bmatrix} \gamma_1^2 & 0 \\ 0 & \gamma_2^2 \end{bmatrix} \begin{bmatrix} 1 & 0 \\ C & 1 \end{bmatrix} s + \begin{bmatrix} 0 & g \\ -g & 0 \end{bmatrix} + \begin{bmatrix} 1 & r \\ 0 & 1 \end{bmatrix} \begin{bmatrix} \delta_1^2 & 0 \\ 0 & \delta_2^2 \end{bmatrix} \begin{bmatrix} 1 & 0 \\ r & 1 \end{bmatrix} \tag{84}$$

Taking the determinant of Equation 84 gives

$$B(s) = \det D = \gamma_1^2\gamma_2^2 s^2 + \left\{\delta_1^2\gamma_2^2 + \delta_2^2\gamma_1^2 + (c-r)^2\gamma_2^2\delta_2^2\right\}s + \left\{\delta_1^2\delta_2^2 + g^2\right\} \tag{85}$$

which is a second-order 1-D HP.

It should be noted that in the above example δ_1^2 can be set equal to zero without changing the properties of the HP. Higher-order HPs can be generated by either taking higher-order matrices in Equation 78 or by cascading lower order HPs.

6.2 FORMULATION OF THE DESIGN PROBLEM

In the method to be presented here, a 1-D HP is generated using the method described earlier and assigned to the denominator of a 1-D analog reference filter of the form:

$$H_a(s) = \frac{N(s)}{D(s)} = \frac{\sum_{i=0}^{M} n(i)s^i}{\sum_{i=0}^{M} d(i)s^i} \tag{86}$$

while the numerator is left unchanged. The discrete version of the filter is obtained by the application of the bilinear transformation.

The error between the ideal and the designed magnitude response of the 1-D filter is calculated using the following relationship:

$$E_M(j\omega_m) = \left|H_I\left(e^{j\omega_m T}\right)\right| - \left|H_D\left(e^{j\omega_m T}\right)\right| \tag{87}$$

where E_M is the error in the magnitude response at discrete points m and $|H_I|$ and $|H_D|$ are the magnitude responses of the ideal and the designed filter, respectively.

In the same fashion, the error between the ideal and the designed group-delay responses of the 1-D filter is calculated using the formula:

$$E_\tau(j\omega_m) = \tau_I T - \tau_W\left[e^{j\omega_m T}\right] \tag{88}$$

where τ_I is a constant representing the group-delay response of the ideal filter, its value normally chosen equal to the order of the filter,[9] or by taking it as one of the parameters of the optimization, and τ_ω is the group-delay response of the designed filter.

We now introduce the general mean square error E_G as an objective function to be minimized as follows:

$$E_G(j\omega_m, \psi) = \alpha \sum_{m \in I_{ps}} E_M^2(j\omega_m) + (1-\alpha)\sum_{m \in I_p} E_\tau^2(j\omega_m) \tag{89}$$

where I_{ps} is the set of all discrete frequency points along the ω axis in the passband and stopband, α is a weighting factor to emphasize the magnitude or the group-delay characteristics, I_p is the set of all discrete frequency points along ω axis in the passband of the filter, and ψ is the coefficient vector to be determined.

Table 3 Values of the Parameters of the Designed Filter

$\gamma_1 = 1.7171$	$n_0 = 2.114$
$\gamma_2 = 1.2342$	$n_1 = -0.1062$
$\delta_1 = 0.9953$	$n_2 = -1.7062$
$\delta_2 = 0.9343$	$n_3 = -0.2937$
$c_1 = 1.1940$	$n_4 = -0.1891$
$r_1 = 0.8219$	
$g_1 = 0.7688$	
$\gamma_3 = 0.9516$	
$\gamma_4 = 1.1312$	
$\delta_3 = 0.7125$	
$\delta_4 = 0.6969$	
$c_2 = 1.2250$	
$r_2 = 0.9250$	
$g_2 = 1.0811$	

In this method, the conjugate gradient optimization technique proposed by Fletcher and Powell[10] can be used to calculate the coefficients of the filter by minimizing Equation 89.

As an example, we design a fourth-order constant group-delay IIR filter with the following magnitude specification:

$$\left|H_I\left(e^{j\omega T}\right)\right| = \begin{cases} 1 & 0 \le \omega \le 0.7 \text{ rad/s} \\ 0 & 1.4 \le \omega \le \pi \text{ rad/s} \end{cases}$$

In this example, the sampling frequency ω_s is taken to be equal to 2π, while α in Equation 89 is set to $1/2$.

To generate a fourth-order HP we cascaded two sections of a second-order HP derived in Equation 85 as follows:

$$B(s) = B_1(s) \cdot B_2(s)$$

where

$$B_1(s) = \gamma_1^2\gamma_2^2 s^2 + \left\{\delta_1^2\gamma_2^2 + \delta_2^2\gamma_1^2 + (c_1 - r_1)\gamma_2^2\delta_2^2\right\}s + \left\{\delta_1^2\delta_2^2 + g_1^2\right\}$$

$$B_2(s) = \gamma_3^2\gamma_4^2 s^2 + \left\{\delta_3^2\gamma_4^2 + \delta_4^2\gamma_3^2 + (c_2 - r_2)\gamma_4^2\delta_4^2\right\}s + \left\{\delta_3^2\delta_4^2 + \gamma_2^2\right\}$$

Table 3 shows the parameters of the designed filter, while Figure 9a and b show the magnitude and group-delay response of the designed filter, respectively.

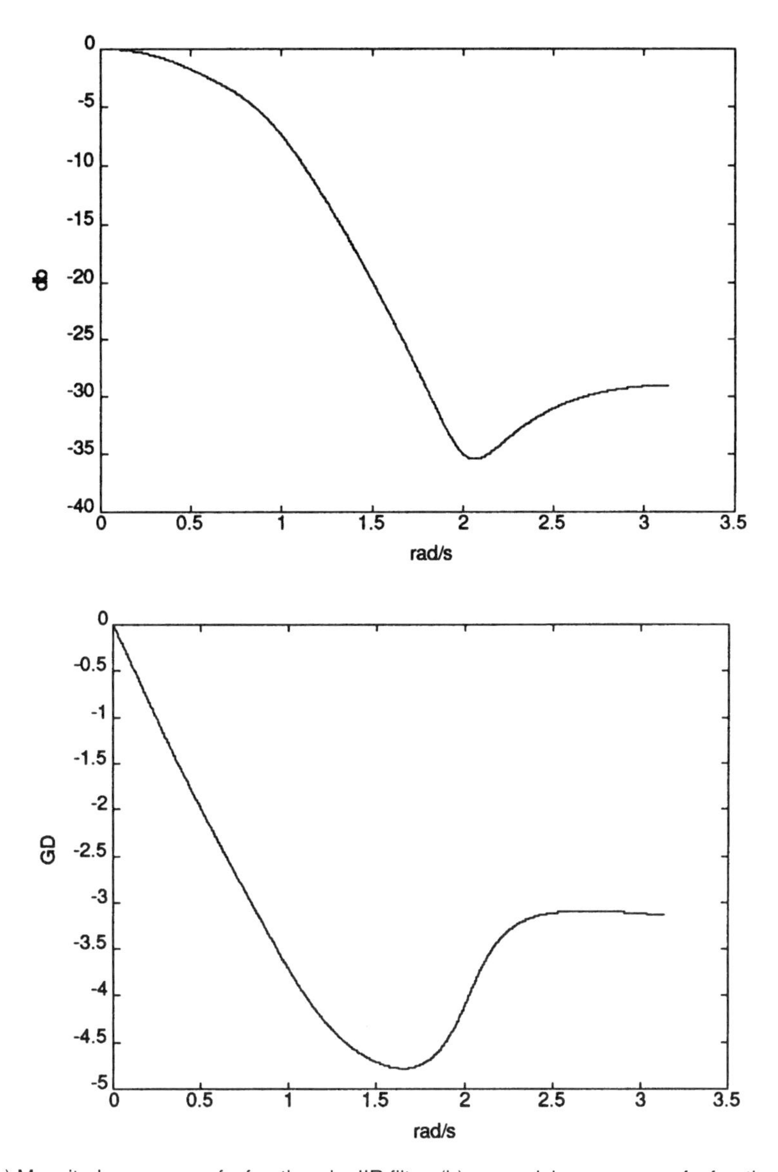

Figure 9 (a) Magnitude response of a fourth-order IIR filter; (b) group-delay response of a fourth-order IIR filter.

REFERENCES

1. **A. Antoniou,** *Digital Filters, Analysis, Design, and Applications,* 2nd ed., McGraw-Hill, New York, 1993.
2. **R. King, M. Ahmadi, R. Gorgui-Naguib, A. Kwabwe, M. Azimi-Sadjadi,** *Digital Filtering in One and Two Dimensions, Design and Applications,* Plenum Press, New York, 1989.
3. **A. Antoniou,** Accelerated procedure for the design of equiripple nonrecursive digital filters, *Proc. IEE,* Part G, 129(1), 1–10, 1982.
4. **L. R. Rabiner, J. H. McClellan and T. W. Parks,** FIR digital filter design techniques using weighted Chebyshev approximation, *Proc. IEEE,* 67, 595–610, 1975.
5. **L. R. Rabiner, B. Gold,** *Theory and Application of Digital Signal Processing,* Prentice Hall, Englewood Cliffs, NJ, 1975.
6. **M. Ahmadi, M. Shridhar, H. J. J. Lee, V. Ramachandran,** A method for the design of 1-D recursive digital filters satisfying a given magnitude and constant group-delay response, *J. Franklin Inst.,* 326(3), 381–393, 1989.
7. **C. Charalambous, A. Antoniou,** Equalization of recursive digital filters, *Proc. IEE,* Part G, 177(5), 219–225, 1980.
8. **A. Chottera, G. A. Jullien,** A linear programming approach to recursive digital filter design with linear phase, *IEEE Trans. Circuits, Syst.,* CAS-29, 139–149, March, 1982.
9. **V. Ramachandran, C. S. Gargour, M. Ahmadi, M. T. Boraie,** Direct design of recursive digital filters based on a new stability test, *J. Franklin Inst.,* 318(6), 407–413, 1984.
10. **R. Fletcher, M.J.D. Powell,** A rapid descent method for minimization, *Comput. J.,* 6, 163–168, 1963.

Design of Two-Dimensional Digital Filters

Wu-Sheng Lu and Andreas Antoniou

CONTENTS

1. TWO-DIMENSIONAL SIGNALS AND SYSTEMS

A continuous two-dimensional (2-D) signal is a physical or contrived quantity that is a function of two independent variables defined over a 2-D continuous domain. The light intensity in a photograph, the depth of the ocean in a specified area, and the temperature distribution in a metal plate are examples of continuous 2-D signals. A 2-D discrete signal, on the other hand, is a physical or contrived quantity that depends on two independent integer variables. Although there are 2-D signals that are inherently discrete, most 2-D discrete signals are obtained by sampling corresponding continuous signals. Two such examples are a digitized image and the ocean depth in a specified

0-8493-8951-8/97/$0.00+$.50
© 1997 by CRC Press LLC

Figure 1 2-D digital filter as a system.

region at discrete coordinates. Discrete signals are much preferred over continuous signals because they can be processed by powerful digital computers using some highly sophisticated analysis and design algorithms. Discrete signals are usually generated from continuous signals through the processes of sampling and discretization. Since practically useful continuous 2-D signals are band limited, their discretized versions contain all the information if the sampling frequency used is sufficiently high, by virtue of Shannon's sampling theorem.

The types of processing that can be applied may range from noise removal, feature extraction, signal restoration, to subband decomposition. For example, a digitized image may be contaminated by noise that was present in the picture before digitization and by noise introduced during transmission. In such a case one has the task of removing as much of the noise as possible without blurring the edges. If the camera was moving or out of focus when the picture was taken, the image is likely to be degraded severely. One then needs to develop techniques to restore the image from the degraded picture. In the context of image compression, an effective means is to decompose the image into several subimages over different frequency bands so that each subimage can be subsampled and efficiently coded.

In order to perform such tasks, the 2-D signal in question is treated as the input to a 2-D discrete system known as a 2-D digital filter that produces an output which is an improved version of an input signal in a certain sense. A 2-D digital filter can be represented by the block diagram of Figure 1 where $x(n_1, n_2)$ and $y(n_1, n_2)$ are the input and output, respectively.

2. LINEAR SHIFT-INVARIANT 2-D DIGITAL FILTERS

Like their one-dimensional (1-D) counterparts, 2-D digital filters can be classified as linear or nonlinear, causal or noncausal, and shift invariant or shift dependent. In Figure 1, the output is related to the input by a certain rule of correspondence that can be represented analytically by $y(n_1, n_2) = \mathcal{F}x(n_1, n_2)$.

A 2-D digital filter is said to be linear if

$$\mathcal{F}\left[\alpha x_1(n_1, n_2) + \beta x_2(n_1, n_2)\right] = \alpha \mathcal{F} x_1(n_1, n_2) + \beta \mathcal{F} x_2(n_1, n_2) \tag{1}$$

for all possible values of α and β and all possible inputs $x_1(n_1, n_2)$ and $x_2(n_1, n_2)$. The 2-D filter is called nonlinear if Equation 1 is violated for some signals or constants of proportionality.

An initially relaxed 2-D digital filter is said to be shift invariant if

$$y(n_1 - k_1, n_2 - k_2) = \mathcal{F}x(n_1 - k_1, n_2 - k_2) \tag{2}$$

for all inputs and all possible integers k_1 and k_2. A filter that does not satisfy this condition is said to be shift dependent.

A 2-D digital filter is said to be causal if its output for $n_1 \le k_1$ and $n_2 \le k_2$ is independent of the input for the values of $n_1 > k_1$ or $n_2 > k_2$ for all integers k_1 and k_2.

The properties of linear, shift-invariant, and causal 2-D digital filters are well understood, and many reliable algorithms for the design of these filters have been developed. More importantly, these filters have found numerous applications in engineering practice. It is for this reason that we shall focus our attention on this class of filters in the rest of this chapter.

For the class of linear, shift-invariant, and causal 2-D filters, two types of filters can be identified, namely, recursive and nonrecursive. These filters can be characterized in terms of difference equations.

A linear, shift-invariant 2-D filter with a quarter-plane support is said to be nonrecursive if the output $y(n_1, n_2)$ can be expressed as

$$y(n_1, n_2) = \sum_{i=0}^{N_1} \sum_{j=0}^{N_2} a_{ij} x(n_1 - i, n_2 - j) \tag{3}$$

where a_{ij} are constants and (N_1, N_2) is called the order of the filter.

A linear, shift-invariant 2-D filter with a quarter-plane support is said to be recursive if the output $y(n_1, n_2)$ can be expressed as

$$y(n_1, n_2) = \sum_{i=0}^{N_1} \sum_{j=0}^{N_2} a_{ij} x(n_1 - i, n_2 - j) - \sum_{i=0}^{N_1} \sum_{j=0}^{N_2} b_{ij} y(n_1 - i, n_2 - j) \tag{4}$$

where a_{ij} and b_{ij} are constants, $b_{00} = 0$, and (N_1, N_2) is the order of the filter. If $b_{ij} = 0$ for all i, j, Equation 4 reduces to Equation 3. It follows that the nonrecursive digital filter is a special case of the recursive one.

3. 2-D *z* TRANSFORM AND ITS APPLICATION TO 2-D FILTERS

3.1 DEFINITION AND THEOREMS

The z transform of a 2-D discrete signal $g(n_1, n_2)$ is defined as

$$G(z_1, z_2) = \sum_{n_1=-\infty}^{\infty} \sum_{n_2=-\infty}^{\infty} g(n_1, n_2) z_1^{-n_1} z_2^{-n_2} \tag{5}$$

for all (z_1, z_2) for which the double summation in Equation 5 converges. For simplicity, the z transform of $g(n_1, n_2)$ is often denoted by $G(z_1, z_2) = zg(z_1, z_2)$. The 2-D signal $g(n_1, n_2)$ can be recovered by using the inverse z transform, i.e.,

$$g(n_1, n_2) = \frac{1}{(2\pi j)^2} \oint_{\Gamma_2} \oint_{\Gamma_1} F(z_1, z_2) z_1^{n_1-1} z_2^{n_2-1} dz_1 dz_2 \tag{6}$$

where Γ_1 and Γ_2 are two simple closed contours that encircle the origin of the z_1 and z_2 planes, respectively, and the two integrals are evaluated in counterclockwise sense over Γ_1 and Γ_2.

The properties of the 2-D z transform can be described by several theorems that are straightforward extensions of their 1-D counterparts.[1,2]

Theorem 1 (Linearity)

$$Z[a_1 g_1(n_1, n_2) + a_2 g_2(n_1, n_2)] = a_1 G_1(z_1, z_2) + a_2 G_2(z_1, z_2)$$

Theorem 2 (Translation)

$$Zg(n_1 + k_1, n_2 + k_2) = z_1^{k_1} z_2^{k_2} G(z_1, z_2)$$

Theorem 3 (Complex Scale Change)

$$\mathcal{Z}\left[w_1^{-n_1} w_2^{-n_2} g(n_1, n_2)\right] = G(w_1 z_1, w_2 z_2)$$

Theorem 4 (Initial-Value theorem) If $g(n_1, n_2) = 0$ for $n_1 < 0$ or $n_2 < 0$, then

$$\lim_{z_1 \to \infty} G(z_1, z_2) = \sum_{n_2=-\infty}^{\infty} g(0, n_2) z_2^{-n_2}$$

$$\lim_{z_2 \to \infty} G(z_1, z_2) = \sum_{n_1=-\infty}^{\infty} g(n_1, 0) z_1^{-n_1}$$

Theorem 5 (Real Convolution)

$$\mathcal{Z} \sum_{k_1=-\infty}^{\infty} \sum_{k_2=-\infty}^{\infty} f(k_1, k_2) g(n_1 - k_1, n_2 - k_2) = F(z_1, z_2) G(z_1, z_2)$$

Theorem 6 (Complex Convolution)

$$\mathcal{Z}\left[f(n_1, n_2) g(n_1, n_2)\right] = \frac{1}{(2\pi j)^2} \oint_{\Gamma_2} \oint_{\Gamma_1} F(v_1, v_2) G\left(\frac{z_1}{v_1}, \frac{z_2}{v_2}\right) \frac{dv_1}{v_1} \frac{dv_2}{v_2}$$

where Γ_1 and Γ_2 are contours in the common region of convergence of $F(v_1, v_2)$ and $G(z_1/v_1, z_2/v_2)$.

Theorem 7 (Parseval's Formula) If $(z_1, z_2) = (1, 1)$ is in the region of convergence of the z transform of $f(n_1, n_2)\, g^*(n_1, n_2)$, then

$$\sum_{n_1=-\infty}^{\infty} \sum_{n_2=-\infty}^{\infty} f(n_1, n_2) g^*(n_1, n_2) = \frac{1}{(2\pi j)^2} \oint_{\Gamma_2} \oint_{\Gamma_1} F(v_1, v_2) G^*\left(\frac{1}{v_1^*}, \frac{1}{v_2^*}\right) \frac{dv_1}{v_1} \frac{dv_2}{v_2}$$

In particular, with $f(n_1, n_2) = g(n_1, n_2)$ and Γ_1, Γ_2 being unit circles,

$$\sum_{n_1=-\infty}^{\infty} \sum_{n_2=-\infty}^{\infty} \left|g(n_1, n_2)\right|^2 = \frac{1}{4\pi^2} \int_{-\pi}^{\pi} \int_{-\pi}^{\pi} \left|G\left(e^{j\omega_1}, e^{j\omega_2}\right)\right|^2 d\omega_1 d\omega_2$$

The reader is referred to Reference 2 for a detailed exposition of the 2-D z transform and inverse z transform.

3.2 TRANSFER FUNCTION AND FREQUENCY RESPONSE OF A 2-D FILTER

The transfer function of a 2-D digital filter is defined as the ratio of the z transform of the output to the z transform of the input. By applying the z transform to Equation 4, we obtain

$$\frac{Y(z_1, z_2)}{X(z_1, z_2)} = H(z_1, z_2) = \frac{N(z_1, z_2)}{D(z_1, z_2)} = \frac{\sum_{i=0}^{N_1} \sum_{j=0}^{N_2} a_{ij} z_1^{-i} z_2^{-j}}{1 + \sum_{i=0}^{N_1} \sum_{j=0}^{N_2} b_{ij} z_1^{-i} z_2^{-j}} \tag{7}$$

Hence the transfer function of a recursive 2-D filter is a rational function in two complex variables. The transfer function of a nonrecursive 2-D filter can be obtained by setting $b_{ij} = 0$ for all i and j in Equation 7, which leads to

$$H(z_1, z_2) = \sum_{i=0}^{N_1} \sum_{j=0}^{N_2} a_{ij} z_1^{-i} z_2^{-j} \tag{8}$$

While the difference equations in Equations 3 and 4 characterize nonrecursive and recursive 2-D filters in the spatial domain, the transfer functions given by Equations 8 and 7 characterize them in the frequency domain. In effect, it can be readily shown[2] that for a sinusoidal input $\sin(n_1\omega_1 T_1 + n_2\omega_2 T_2)$ the output from a 2-D filter with a transfer function $H(z_1, z_2)$ is also a sinusoidal signal with an amplitude $M(\omega_1, \omega_2)$ and a phase angle $\theta(\omega_1, \omega_2)$ given by

$$M(\omega_1, \omega_2) = \left| H\left(e^{j\omega_1 T_1}, e^{j\omega_2 T_2}\right) \right|$$

$$\theta(\omega_1, \omega_2) = \arg H\left(e^{j\omega_1 T_1}, e^{j\omega_2 T_2}\right)$$

For this reason $H(e^{j\omega_1 T_1}, e^{j\omega_2 T_2})$, $M(\omega_1, \omega_2)$, and $\theta(\omega_1, \omega_2)$ are called the frequency response, amplitude response, and phase response of the filter, respectively.

4. DESIGN SPECIFICATIONS

A 2-D discrete signal can be represented by a frequency spectrum that can be modified, reshaped, or manipulated through filtering, and this type of processing can be carried out by means of 2-D digital filters. The requirements for the design of a 2-D digital filter are, therefore, often specified in the frequency domain. Since the frequency response of a 2-D filter is periodic with periods $\omega_{s_1} = 2\pi/T_1$ and $\omega_{s_2} = 2\pi/T_2$ where T_1 and T_2 are the sampling periods in the horizontal and vertical directions, the desired frequency response is usually specified in the baseband B defined by

$$B = \left\{ (\omega_1, \omega_2) : -\frac{\omega_{s_1}}{2} \le \omega_1 \le \frac{\omega_{s_1}}{2} \quad -\frac{\omega_{s_2}}{2} \le \omega_2 \le \frac{\omega_{s_2}}{2} \right\}$$

The frequency responses that can be achieved with 2-D filters can assume a variety of forms and often possess certain types of symmetries. These include lowpass, highpass, bandpass, and bandstop frequency responses with circular, rectangular, fan-shaped, and diamond-shaped passband and stopband boundaries. 2-D filters are often described on the basis of the type of amplitude response realized as circularly symmetric, fan-, or diamond-shaped, etc. Some typical idealized amplitude responses are shown in Figure 2. The performance of a 2-D filter is usually measured in terms of the maximum ripple in the amplitude response in the passband and the minimum attenuation in the stopband.

It has been shown[3] that a recursive 2-D filter with quadrantally symmetric frequency response always possesses a separable denominator, namely, its denominator is of the form $D(z_1^{-1}, z_2^{-1}) = d_1(z_1^{-1})\, d_2(z_2^{-1})$. This property greatly helps the designer in reducing the number of design parameters when the filter being designed is quadrantally symmetric.

While there are many kinds of amplitude responses that may arise from design practice, the phase response of a filter is often required to be linear, i.e.,

$$\theta(\omega) = -k_1\omega_1 - k_2\omega_2 \tag{9}$$

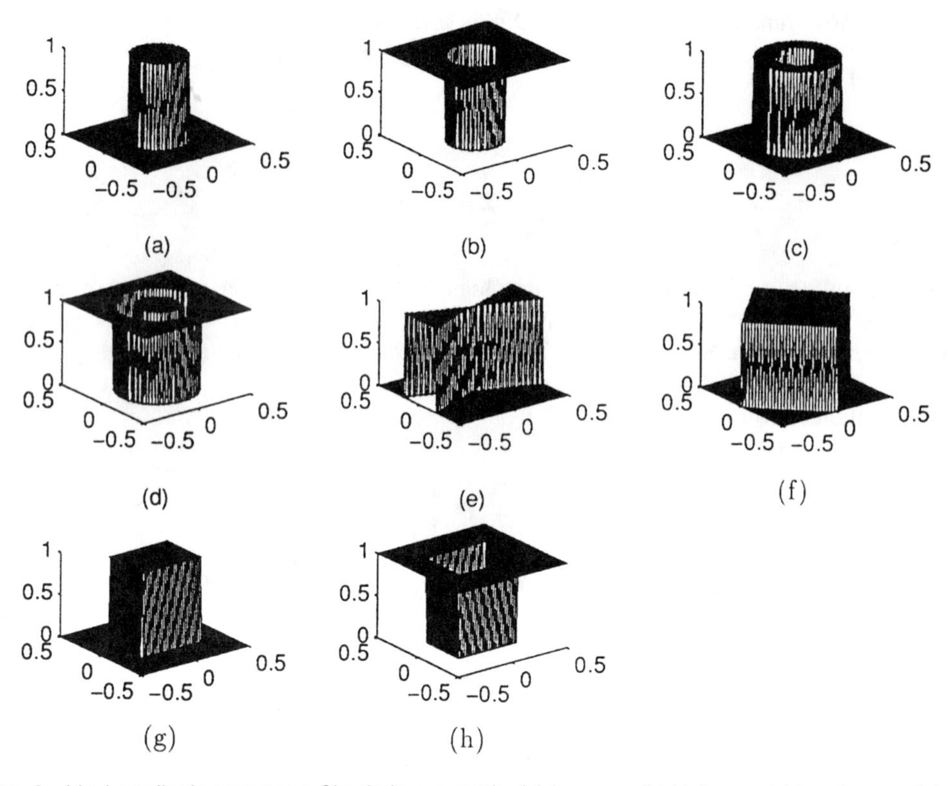

Figure 2 Ideal amplitude responses. Circularly symmetric: (a) lowpass, (b) highpass, (c) bandpass, (d) bandstop; (e) fan-shaped, (f) diamond-shaped; and rectangularly symmetric: (g) lowpass and (h) highpass.

where k_1 and k_2 are constants. This is particularly the case for many applications in image processing in order to prevent phase distortion. Associated with the phase response is a pair of parameters called the group delays. They are defined as

$$\tau_1 = -\frac{\partial\theta(\omega_1, \omega_2)}{\partial\omega_1}, \qquad \tau_2 = -\frac{\partial\theta(\omega_1, \omega_2)}{\partial\omega_2} \tag{10}$$

In effect, a filter with linear phase response has constant group delays τ_1 and τ_2.

5. METHODS FOR THE DESIGN OF 2-D NONRECURSIVE FILTERS

5.1 PHASE RESPONSE OF A NONRECURSIVE FILTER

Let the transfer function of a nonrecursive filter to be designed be

$$H(z_1, z_2) = \sum_{n_1=0}^{N_1-1} \sum_{n_2=0}^{N_2-1} h(n_1, n_2) z_1^{-n_1} z_2^{-n_2} \tag{11}$$

where N_1 and N_2 are odd integers.

An important feature of nonrecursive filters is that a linear phase response can be easily achieved if the coefficients $h(n_1, n_2)$ in Equation 11 satisfy certain symmetry conditions. If

$$h(n_1, n_2) = h(N_1 - 1 - n_1, N_2 - 1 - n_2) \quad \text{for} \quad 0 \le i \le N_1 - 1, 0 \le j \le N_2 - 1 \tag{12a}$$

then the phase response is given by Equation 9 with $k_1 = (N_1 - 1)T_1/2$ and $k_2 = (N_2 - 1)T_2/2$. If the coefficients $h(n_1, n_2)$ are antisymmetrical, i.e.,

$$h(n_1, n_2) = -h(N_1 - 1 - n_1, N_2 - 1 - n_2) \quad \text{for} \quad 0 \le i \le N_1 - 1, 0 \le j \le N_2 - 1 \tag{12b}$$

then the phase response is given by

$$\theta(\omega_1, \omega_2) = -k_1\omega_1 - k_2\omega_2 \pm \pi / 2 \tag{13}$$

with $k_1 = (N_1 - 1)T_1/2$ and $k_2 = (N_2 - 1)T_2/2$.

5.2 DESIGN BASED ON FOURIER SERIES

By using the 2-D Fourier series along with 2-D window functions, the design of 2-D nonrecursive filters can be accomplished.

A 2-D window function is a 2-D discrete function with the following properties:

1. $w_2(n_1, n_2) = 0$ for $|n_1| > (N_1 - 1)/2$ or $|n_2| > (N_2 - 1)/2$
2. It is symmetrical with respect to the n_1 and n_2 axes, i.e.,

$$w_2(-n_1, -n_2) = w_2(n_1, n_2)$$

The amplitude spectrum of a 2-D window function consists of a main lobe centered at the origin of the (ω_1, ω_2) plane and a number of side lobes such that the volume of the side lobes is small relative to the volume of the main lobe.

Proceeding as in the case of 1-D filters,[1] the impulse response obtained by applying the 2-D Fourier series is multiplied by the 2-D window function to yield a modified impulse response as

$$h(n_1, n_2) = w_2(n_1, n_2)h_d(n_1, n_2) \tag{14}$$

where $h_d(n_1, n_2)$ are the Fourier series coefficients of the desired frequency response, $H_d(e^{j\omega_1 T_1}, e^{j\omega_2 T_2})$, i.e.,

$$h_d(n_1, n_2) = \frac{1}{4\pi^2}\int_{-\pi}^{\pi}\int_{-\pi}^{\pi} H_d(e^{j\mu_1}, e^{j\mu_2})e^{j(\mu_1 n_1 + \mu_2 n_2)}\, d\mu_1 d\mu_2 \tag{15}$$

There are two ways to construct a 2-D window function. If $w_{1A}(n_1)$ and $w_{1B}(n_2)$ are 1-D window functions, then

$$w_2(n_1, n_2) = w_{1A}(n_1)w_{1B}(n_2) \tag{16}$$

has the required properties, and, therefore, it can serve as a 2-D window function, where $w_{1A}(n_1)$ and $w_{1B}(n_2)$ can be any one of the commonly used 1-D window functions, such as the Hamming, Kaiser, etc. window.[1] Another way to construct a 2-D window function is to transform a 1-D window by a rotational transformation as suggested in Reference 3, namely,

$$w_2(n_1, n_2) = w_1(n)\Big|_{n=\sqrt{n_1^2+n_2^2}} \tag{17}$$

From Equations 14 through 17, it follows that the window method offers a feasible design approach if the Fourier coefficients $h_d(n_1, n_2)$ given by Equation 15 can be evaluated efficiently. It turns out that for circularly symmetric lowpass, highpass, bandpass, bandstop filters, and fan filters, $h_d(n_1, n_2)$ can be expressed in closed form, which leads to fast evaluation.[2] For example, for the idealized frequency response of a circularly symmetric lowpass filter

$$H_d\left(e^{j\omega_1}, e^{j\omega_2}\right) = \begin{cases} 1 & \text{for } \sqrt{\omega_1^2 + \omega_2^2} \le \omega_c \\ 0 & \text{otherwise} \end{cases}$$

the impulse response is given by

$$h_d(n_1, n_2) = \begin{cases} \dfrac{\omega_c^2}{4\pi} & (n_1, n_2) = (0, 0) \\ \dfrac{\omega_c^2}{2\pi} \dfrac{J_1\left(\omega_c \sqrt{n_1^2 + n_2^2}\right)}{\sqrt{n_1^2 + n_2^2}} & \text{otherwise} \end{cases}$$

where J_1 is the first-order Bessel function of the first kind.

For the frequency response of an idealized fan filter characterized by

$$H_d\left(e^{j\omega_1}, e^{j\omega_2}\right) = \begin{cases} 1 & \text{for } -k\omega_1 \le \omega_2 \le k\omega_1,\ \omega_1 \ge 0 \\ & \text{or } k\omega_1 \le \omega_2 \le -k\omega_1,\ \omega_1 < 0 \\ 0 & \text{otherwise} \end{cases}$$

we have

$$h_d(n_1, n_2) = \begin{cases} k/2 & \text{for } (n_1, n_2) = (0, 0) \\ C_1 & n_1 \ne 0,\ n_2 = 0 \\ 0 & n_2 \ne 0,\ n_1^2 - n_2^2 k^2 = 0 \\ C_2 & \text{otherwise} \end{cases}$$

where

$$C_1 = \frac{k\left[(-1)^{n_1} - 1\right]}{n_1^2 \pi^2}$$

$$C_2 = \frac{k}{\left(n_2^2 k^2 - n_1^2\right)\pi^2} - \frac{1}{2 n_2 \pi^2}\left\{\frac{\cos\left[(n_2 k - n_1)\pi\right]}{n_2 k - n_1} + \frac{\cos\left[(n_2 k + n_1)\pi\right]}{n_2 k + n_1}\right\}$$

In summary, given the desired frequency response $H_d(e^{j\omega_1}, e^{j\omega_2})$ and the order of the filter, the Fourier series coefficients $h_d(n_1, n_2)$ are computed using Equation 15 and are then multiplied by a 2-D window function evaluated using either Equation 16 or 17 to obtain the modified impulse response $h(n_1, n_2)$ given by Equation 14. The transfer function of the filter designed is given by

$$H(z_1, z_2) = z_1^{-(N_1-1)/2} z_2^{-(N_2-1)/2} \sum_{n_1=-(N_1-1)/2}^{(N_1-1)/2} \sum_{n_2=-(N_2-1)/2}^{(N_2-1)/2} h(n_1, n_2) z_1^{-n_1} z_2^{-n_2} \tag{18}$$

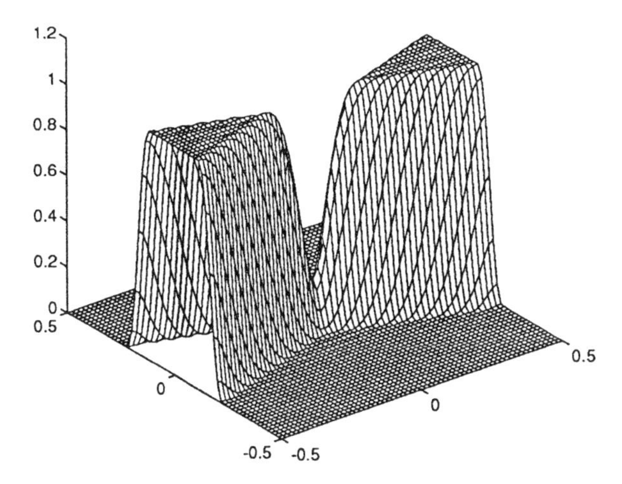

Figure 3 Amplitude response of a nonrecursive fan filter of order (29, 29).

Because of the symmetry property of $h_d(n_1, n_2)$ and $w_2(n_1, n_2)$, the nonrecursive filter designed has linear phase response. Figure 3 shows the amplitude response of a fan filter of order (29, 29) designed using this method.

The window method as described in this section can be used to design nonrecursive 2-D filters satisfying prescribed specifications. A detailed exposition of the design procedure is presented in Reference 2.

5.3 DESIGN BASED ON THE McCLELLAN TRANSFORMATION

Another method for the design of nonrecursive filters is based on a transformation proposed by McClellan.[4] This is given by

$$\begin{aligned}\cos\omega T &= A\cos\omega_1 T_1 + B\cos\omega_2 T_2 + C\cos\omega_1 T_1 \cos\omega_2 T_2 + D \\ &= F(\omega_1 T_1, \omega_2 T_2)\end{aligned} \tag{19}$$

where A, B, C, and D are constants. We begin the design by applying the transformation to the frequency response of a 1-D nonrecursive filter, namely,

$$H_1(e^{j\omega T}) = \sum_{n=0}^{(N-1)/2} a_0(nT)\cos n\omega T = \sum_{n=0}^{(N-1)/2} a_1(nT)(\cos\omega T)^n \tag{20}$$

which yields

$$\begin{aligned}H_2(e^{j\omega T}) &= \sum_{n_1=0}^{(N-1)/2}\sum_{n_2=0}^{(N-1)/2} a_2(nT)(\cos\omega_1 T_1)^{n_1}(\cos\omega_2 T_2)^{n_2} \\ &= \sum_{n_1=0}^{(N-1)/2}\sum_{n_2=0}^{(N-1)/2} a_3(nT)\cos n_1\omega_1 T_1 \cos n_2\omega_2 T_2\end{aligned} \tag{21}$$

In the above equation, coefficients $a_1(nT)$ are obtained from coefficients $a_0(nT)$ and coefficients $a_3(nT)$ are obtained from coefficients $a_2(nT)$. Expression 21 may be considered as the frequency

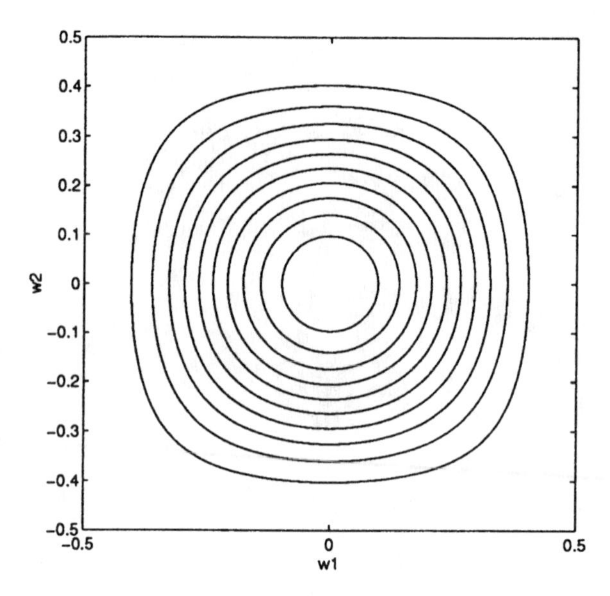

Figure 4 Contours of the McClellan transformation with $A = B = C = -D = 0.5$.

response of a zero-phase nonrecursive filter, and the corresponding 2-D transfer function can be obtained by substituting $(z_i^{n_i} + z_i^{-n_i})/2$ for $\cos n_i\omega_i T_i$.

With suitably chosen parameters A, B, C, D, and a fixed frequency ω, the transformation in Equation 19 defines a curve in the (ω_1, ω_2) plane. For example, if $A = B = C = -D = 0.5$, Equation 19 maps the frequency interval $0 \le \omega \le \omega_s/2$ onto a set of contours as shown in Figure 4, which shows good circularity at low frequencies. It follows that if the 1-D prototype filter described by Equation 20 is a lowpass or highpass or bandpass filter, then Equation 21 gives a circularly symmetric lowpass or highpass or bandpass frequency response.

Obviously, the frequency response of the 2-D filter obtained depends critically on the choice of the transformation parameters. Techniques for the optimal choice of these parameters can be found in References 2 and 5.

5.4 MINIMAX DESIGN

In the design of nonrecursive and recursive filters, minimax optimization methods are often preferred because their application minimizes the maximum of the approximation error. The design in this case is transformed into an unconstrained optimization problem by defining the objective function as

$$F(x) = \max_{1 \le i \le m} f_i(\mathbf{x})$$

$$f_i(x) = \left| M(\mathbf{x}, \omega_1, \omega_{2i}) - M_I(\omega_{1i}, \omega_{2j}) \right| \tag{22}$$

where $M(\mathbf{x}, \omega_1, \omega_2)$ and $M_I(\omega_1, \omega_2)$ denote the actual and desired amplitude responses of the filter, $\{(\omega_{1i}, \omega_{2i}) : i = 1, \ldots, m\}$ is the set of grid points at which $f_i(\mathbf{x})$ is evaluated, and $\mathbf{x}$ is the column vector whose elements are the coefficients of the filter to be designed.

The problem described above can be solved efficiently by minimizing an alternative objective function defined by

$$\Phi(\mathbf{x}, \xi, \lambda) = \frac{1}{2} \sum_{i \in I_1} \lambda_i [\phi_i(\mathbf{x}, \xi)]^2 + \frac{1}{2} \sum_{i \in I_2} [\phi_i(\mathbf{x}, \xi)]^2 \tag{23}$$

where

$$\phi_i(\mathbf{x}, \xi) = f_i(\mathbf{x}) - \xi$$

$$I_1 = \{i : \phi_i(\mathbf{x}, \xi) > 0 \text{ and } \lambda_i > 0\}$$

$$I_2 = \{i : \phi_i(\mathbf{x}, \xi) > 0 \text{ and } \lambda_i = 0\}$$

and parameters ξ and λ_i ($i = 1, \ldots, m$) are constants. It can be shown[6] that a local minimum point $\mathbf{x}^*$ can be found by forcing λ_i to approach λ_i^*, which are the minimax multipliers corresponding to $\mathbf{x}^*$, and making $F(\mathbf{x}^*) - \xi$ sufficiently small. These can be achieved by applying the following algorithm.

Algorithm

Step 1: Set $\xi = 0$ and $\lambda_i = 1$ for $i = 1, \ldots, m$, and initialize $\mathbf{x}$.
Step 2: Minimize $\Phi(\mathbf{x}, \xi, \lambda)$ to obtain $\tilde{\mathbf{x}}$.
Step 3: Set

$$S = \sum_{i \in I_1} \lambda_i \phi_i(\tilde{\mathbf{x}}, \xi) + \sum_{i \in I_2} \phi_i(\tilde{\mathbf{x}}, \xi) \tag{24}$$

and update λ_i and ξ as

$$\lambda_i = \begin{cases} \lambda_i \phi_i(\tilde{\mathbf{x}}, \xi)/S & \text{if } \phi_i(\tilde{\mathbf{x}}, \xi) > 0, \lambda_i > 0 \\ \phi_i(\tilde{\mathbf{x}}, \xi)/S & \text{if } \phi_i(\tilde{\mathbf{x}}, \xi) > 0, \lambda_i = 0 \\ 0 & \text{if } \phi_i(\tilde{\mathbf{x}}, \xi) < 0 \end{cases}$$

$$\xi = \sum_{i=1}^{m} \lambda_i f_i(\tilde{\mathbf{x}})$$

Step 4: If $[F(\tilde{\mathbf{x}}) - \xi]/F(\tilde{\mathbf{x}}) < \varepsilon$ where ε is a prescribed tolerance, stop; otherwise, go to Step 2.

The unconstrained minimization required in Step 2 of the algorithm can be carried out by using, for example, one of the quasi-Newton methods[7] which require the evaluation of the gradient of $\Phi(\mathbf{x}, \xi, \lambda)$ with respect to $\mathbf{x}$. It can be readily verified that the gradient of $\Phi(\mathbf{x}, \xi, \lambda)$ is given by

$$\nabla\Phi(\mathbf{x}, \xi, \lambda) = \sum_{i \in I_1} \lambda_i \phi_i(\mathbf{x}, \xi) \nabla\phi_i(\mathbf{x}, \xi) + \sum_{i \in I_2} \phi_i(\mathbf{x}, \xi) \nabla\phi_i(\mathbf{x}, \xi)$$

$$\nabla\phi_i(\mathbf{x}, \xi) = \operatorname{sgn}\left[E(\mathbf{x}, \omega_{1i}, \omega_{2i})\right] \nabla M(x, \omega_{1i}, \omega_{2i})$$

$$\operatorname{sgn}(E) = \begin{cases} 1 & \text{if } E \geq 0 \\ -1 & \text{if } E < 0 \end{cases} \tag{25}$$

where ∇M is the gradient of the amplitude response with respect to $\mathbf{x}$. A more detailed exposition of the method and several design examples can be found in Reference 2.

5.5 DESIGN BASED ON THE SINGULAR-VALUE DECOMPOSITION

The singular-value decomposition (SVD) is a matrix decomposition in which a given complex matrix $\mathbf{F}$ is represented by its singular values and singular vectors in the form of

$$\mathbf{F} = \mathbf{U}\Sigma\mathbf{V}^H = \sum_{i=1}^{r} \tilde{\mathbf{u}}_i \tilde{\mathbf{v}}_i^H \tag{26}$$

where $\mathbf{U} = [\mathbf{u}_1 \cdots \mathbf{u}_r \cdots]$ and $\mathbf{V} = [\mathbf{v}_1 \cdots \mathbf{v}_r \cdots]$ are unitary matrices, $\Sigma = \text{diag}\{\sigma_1, \ldots \sigma_r, 0 \ldots 0\}$ with singular values $\sigma_1 \geq \sigma_2 \geq \cdots \geq \sigma_r > 0$, $\tilde{\mathbf{u}}_i = \sigma^{1/2}\mathbf{u}_i$, $\tilde{\mathbf{v}}_i = \sigma^{1/2}\mathbf{v}_i$, and r is the rank of $\mathbf{F}$. The vectors $\mathbf{u}_i$ and $\mathbf{v}_i$ in Equation 26 are called the ith left and right singular vectors, respectively. In a digital filter context, the importance of the SVD can be appreciated by noticing the following properties:

1. If a matrix $\mathbf{F}_K$ of rank K is defined by

$$\mathbf{F}_K = \sum_{i=1}^{K} \tilde{\mathbf{u}}_i \tilde{\mathbf{v}}_i^H \qquad \text{for } 1 \leq K \leq r \tag{27}$$

then $\mathbf{F}_K$ is the best approximation of $\mathbf{F}$ of rank K in the two-norm or Frobenius-norm sense,[8] namely,

$$\left\|\mathbf{F} - \mathbf{F}_K\right\|_{2,F} = \min_{\substack{\mathbf{F}_K \text{with} \\ \text{rank}(\hat{F}_K)=K}} \left\|F - \hat{F}_K\right\|_{2,F}$$

where the two-norm and F-norm are defined by

$$\|\mathbf{A}\|_2 = \left[\max \lambda\left(\mathbf{A}^H\mathbf{A}\right)\right]^{1/2}$$

$$\|\mathbf{A}\|_F = \left[\text{trace}\left(\mathbf{A}^H\mathbf{A}\right)\right]^{1/2}$$

2. If $\mathbf{F}$ in Equation 26 is obtained by uniformly sampling the desired frequency response, then $\tilde{\mathbf{u}}_i$ and $\tilde{\mathbf{v}}_i$ $(i = 1, \ldots, r)$ are mirror-image complex-conjugate symmetric. Therefore, each $\tilde{\mathbf{u}}_i$ and $\tilde{\mathbf{v}}_i$ can be interpreted as the sampled frequency response of a 1-D digital filter.[9,10]

From the above properties, it follows that the design of a 2-D filter can be accomplished by obtaining a set of 1-D polynomials $f_i(z_1)$ and $g_i(z_2)$ that approximate the weighted singular vectors $\tilde{\mathbf{u}}_i$ and $\tilde{\mathbf{v}}_i$ $(i = 1, \ldots, K)$, respectively. The transfer function of the nonrecursive 2-D filter is then obtained as

$$H(z_1, z_2) = \sum_{i=1}^{K} f_i(z_1) g_i(z_2) \tag{28}$$

The performance of the 2-D filter designed depends on two factors: the number of sections, K, used in the design, and the approximation errors introduced in the design of the 1-D filters represented by $f_i(z_1)$ and $g_i(z_2)$. Evidently, with the order of the filter fixed and the 1-D design method chosen, improved designs can be obtained by increasing the number of sections at the cost of increasing the complexity of the software or hardware implementation. On the other hand, with a fixed K, improved designs can be obtained by increasing the order of the filter or by using better techniques for the design of the 1-D filters. A quantitative error analysis for this design methodology can be found in Reference 9.

There is an additional step that one can take to obtain a more efficient SVD-based design. In this refined approach one keeps all r pairs of singular vectors ($\tilde{\mathbf{u}}_i$ and $\tilde{\mathbf{v}}_i$) that are associated with a nonzero singular values. Let N_1 and N_2 be the orders of transfer functions $f_i(z_1)$ and $g_i(z_2)$, respectively, and write

$$f_i(z_1) = \mathbf{z}_1^T \mathbf{c}_{1i} \tag{29a}$$

$$g_i(z_2) = \mathbf{c}_{2i}^T \mathbf{z}_2 \tag{29b}$$

where $\mathbf{z}_1 = [1\ z_1^{-1} \ldots z_1^{-N_1}]^T$, $z_2 = [1\ z_2^{-1} \ldots z_2^{-N_1}]^T$, and $\mathbf{c}_{1i}$, $\mathbf{c}_{2i}$ are the coefficient vectors of $f_i(z_1)$ and $g_i(z_2)$, respectively. The transfer function of the 2-D filter obtained can now be expressed as

$$H(z_1, z_2) = \mathbf{z}_1^T \mathbf{C} \mathbf{z}_2$$

where

$$\mathbf{C} = \sum_{i=1}^{r} \mathbf{c}_{1i} \mathbf{c}_{2i}^T \tag{30}$$

is an $N_1 \times N_2$ real-valued coefficient matrix. The SVD is now applied to $\mathbf{C}$ to yield

$$\mathbf{C} = \mathbf{U}_c \Sigma_c \mathbf{V}_v^T = \sum_{i=1}^{r_c} \hat{\mathbf{u}}_i \hat{\mathbf{v}}_i^T \tag{31}$$

It turns out that for a variety of 2-D filters the number of significant singular values of matrix $\mathbf{C}$, K_C, is often considerably less than that of matrix F. By keeping the first K_C pairs of weighted singular vectors $\{\hat{\mathbf{u}}_i, \hat{\mathbf{v}}_i, i = 1, \ldots, K_C\}$ and neglecting the rest of the $\hat{\mathbf{u}}_i$'s and $\hat{\mathbf{v}}_i$'s, we obtain a 2-D transfer function

$$H_{K_C}(z_1, z_2) = \mathbf{z}_1^T \left(\sum_{i=1}^{K_C} \hat{\mathbf{u}}_i \hat{\mathbf{v}}_i^T \right) \mathbf{z}_2 = \sum_{i=1}^{K_C} (\mathbf{z}_1^T \hat{\mathbf{u}}_i)(\hat{\mathbf{v}}_i^T \mathbf{z}_2) = \sum_{i=1}^{K_C} \hat{f}_i(z_1) \hat{g}_i(z_2) \tag{32}$$

which can be realized with K_C filter sections. It can be shown[2] that the L_2-norm approximation error $\|H(z_1, z_2) - H_{K_C}(z_1, z_2)\|_2$ is equal to $\{\Sigma_{i=K_C+1}^{r} \mu_i^2\}^{1/2}$, where μ_i ($i = K_c + 1, \ldots, r_c$) are the $(r_c - K_c)$ least-significant singular values of $\mathbf{C}$. A step-by-step description of this design method is given in terms of the following algorithm.

Algorithm

Step 1: Obtain the sampled frequency response matrix $\mathbf{F}$ from the given design specifications.
Step 2: Perform the SVD of $\mathbf{F}$.
Step 3: Find nonrecursive 1-D transfer functions $f_i(z_1)$ and $g_i(z_2)$ whose frequency responses approximate the weighted singular vectors $\tilde{\mathbf{u}}_i$ and $\tilde{\mathbf{v}}_i$, respectively, for $i = 1, \ldots, r$.
Step 4: Compute matrix $\mathbf{C}$ using Equations 29 and 30 and perform the SVD of $\mathbf{C}$.
Step 5: Obtain the nonrecursive 2-D transfer function $H_{K_C}(z_1, z_2)$ using Equation 32, where K_c is the number of significant singular values of $\mathbf{C}$ that the designer wants to keep.

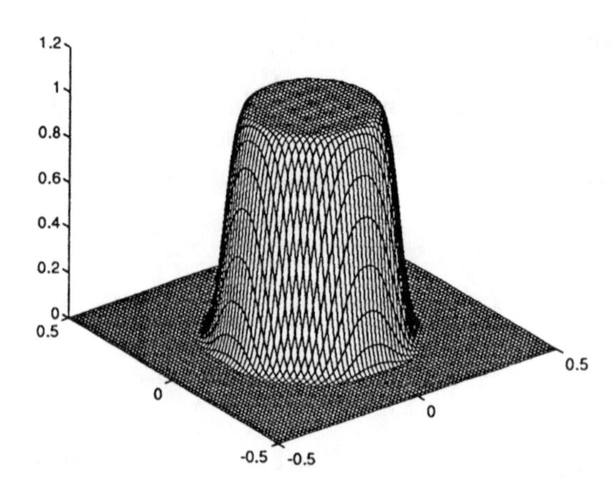

Figure 5 Amplitude response of lowpass filter designed using the SVD method.

A MATLAB program that enables the user to design a nonrecursive 2-D filter using the above algorithm in an interactive manner can be found in the appendix of this chapter. Since linear phase response is often desirable, the MATLAB program is written for the design of linear phase, nonrecursive 2-D filters. It accepts a sampled amplitude response matrix and the order of the filter as the input data. At a certain point the program constructs a figure showing the distribution of the singular values of the coefficient matrix **C** from which the user can select the right value for K_C. Upon the receipt of the user's input for K_C, the program continues, and eventually generates the coefficient matrix for $H_{K_C}(z_1, z_2)$ and a 3-D plot of the amplitude response of the filter designed. Figure 5 shows a 3-D plot of the amplitude response of a circular symmetric lowpass filter designed using the above algorithm. The order of the filter is (29, 29) and the number of section K_C used in the design is six.

6. METHODS FOR THE DESIGN OF 2-D RECURSIVE FILTERS

6.1 DESIGN BY LEAST-*p*TH OPTIMIZATION

As in the nonrecursive case, optimization techniques can be applied to the design of recursive filters. In particular, the least-*p*th optimization method has been used extensively in the past in the design of both 1-D and 2-D filters. The design algorithm addressed in this section is based on the work in Reference 11.

The transfer function of the filter to be designed is assumed to be of the form

$$H(z_1, z_2) = H_0 \prod_{k=1}^{K} \frac{N_k(z_1, z_2)}{D_K(z_1, z_2)} = H_0 \prod_{k=1}^{K} \frac{\sum_{l=0}^{2} \sum_{m=0}^{2} a_{lm}^{(k)} z_1^{-l} z_2^{-m}}{\sum_{l=0}^{2} \sum_{m=0}^{2} b_{lm}^{(k)} z_1^{-l} z_2^{-m}} \tag{33}$$

where $a_{00}^{(k)} = b_{00}^{(k)} = 1$ for $k = 1, \ldots, K$. The parameter vector **x** is defined by

$$\mathbf{x} = \left[\mathbf{a}^T \ \mathbf{b}^T \ H_0\right] \tag{34}$$

with

$$\mathbf{a} = \left[a_{10}^{(1)} \ldots a_{22}^{(1)} \ a_{10}^{(2)} \ldots a_{22}^{(2)} \ldots a_{22}^{(K)}\right]^T$$

$$\mathbf{b} = \left[b_{10}^{(1)} \ldots b_{22}^{(1)} \ b_{10}^{(2)} \ldots b_{22}^{(2)} \ldots b_{22}^{(K)}\right]^T$$

If amplitude response is the only design specification, the least-*p*th objective function can be defined as

$$J(\mathbf{x}) = \sum_{n_1=1}^{N_1} \sum_{n_2=1}^{N_2} \left[M(\mathbf{x}, n_1, n_2) - M_I(n_1, n_2)\right]^p \tag{35}$$

where $M(\mathbf{x}, n_1, n_2)$ and $M_I(n_1, n_2)$ are the actual and desired amplitude responses at the grid point $(\omega_{1n_1}, \omega_{2n_2})$, and p is an even positive integer. If the design specifications also include the desired phase response, the terms $M(\mathbf{x}, n_1, n_2)$ and $M_I(n_1, n_2)$ in Equation 35 are replaced by the actual and desired frequency responses, respectively, and the expression $[\cdot]^p$ in Equation 35 needs to be replaced by $|\cdot|^p$.

The design problem can now be solved by minimizing $J(\mathbf{x})$ in Equation 35 over the parameter space, such that each parameter vector corresponds to a stable transfer function. The well-established quasi-Newton methods[7] can be used to carry out the optimization. With $J(\mathbf{x})$ given by Equation 35, we compute the gradient of $J(\mathbf{x})$ as

$$\nabla J = \left[\frac{\partial J}{\partial x_1} \cdots \frac{\partial J}{\partial x_N}\right]^T$$

where $N = 16K + 1$, and

$$\frac{\partial J}{\partial x_k} = p\sum_{n_1=1}^{N_1} \sum_{n_2=1}^{N_2} \left[M(n_1, n_2) - M_I(n_1, n_2)\right]^{p-1} M^{-1}(n_1, n_2)\,\mathrm{Re}\left(\overline{H}\frac{\partial H}{\partial x_k}\right)$$

where $H \cong H(e^{j\omega_1 T_1}, e^{j\omega_2 T_2})$ and $\overline{H}$ is the complex conjugate of H.

In order to ensure that the filter designed be stable, one can start the optimization with an initial point $\mathbf{x}_0$ that corresponds to a stable 2-D transfer function. For example, an N-dimensional vector of the form $\mathbf{x}_0 = [\mathbf{a}_0^T\ \mathbf{0}^T\ H_0]^T$ with arbitrary $\mathbf{a}_0$ and H_0 leads to a nonrecursive transfer function which is always stable. In a numerical optimization algorithm, an improved parameter vector at the kth iteration is obtained by adding an appropriate vector increment $\delta\mathbf{x}_k$ to the preceding point $\mathbf{x}_{k-1}$, i.e.,

$$\mathbf{x}_k = \mathbf{x}_{k-1} + \delta\mathbf{x}_k$$

If $\mathbf{x}_{k-1}$ is a point associated with a stable transfer function but the transfer function obtained using $\mathbf{x}_k$ is unstable, then the increment $\delta\mathbf{x}_k$ is reduced by half and a new point $\mathbf{x}_k$ is obtained as

$$\mathbf{x}_k = \mathbf{x}_{k-1} + \frac{1}{2}\delta\mathbf{x}_k$$

This process continues until a stable $\mathbf{x}_k$ is found. However, the designer should be cautious with this approach since its outcome is likely to be a transfer function that has a very small stability margin. Alternatively, the designer may try different initial points and choose the best stable design achieved. Design methods that eliminate the problem of stability altogether are described in Section 6.3 and Reference 2.

6.2 MINIMAX OPTIMIZATION ALGORITHMS

We consider again the transfer function given by Equation 33 with the parameter vector defined by Equation 34. The minimax design seeks a vector $\mathbf{x}$ that minimizes the objective function $F(\mathbf{x})$ defined by

$$F(\mathbf{x}) = \max_{1 \le i \le m} f_i(\mathbf{x})$$

$$f_i(\mathbf{x}) = \left| E(\mathbf{x}, \omega_i, \omega_{2i}) \right|$$

$$E(\mathbf{x}, \omega_1, \omega_2) = M(\mathbf{x}, \omega_1, \omega_2) - M_I(\omega_1, \omega_2) \tag{36}$$

where $\{(\omega_{1i}, \omega_{2i}) : i = 1, \ldots, m\}$ is a set of grid points over the baseband at which the objective function is evaluated. Evidently, the algorithm described in Section 5.4 is also applicable to the design problem here. An alternative approach to the design problem at hand is to formulate an objective function $\Phi(\mathbf{x})$ for increasing values of p.[12] The objective function $\Phi(\mathbf{x})$ is defined by

$$\Phi(\mathbf{x}) = F(\mathbf{x}) \left\{ \sum_{i=1}^{m} \left[\frac{E(\mathbf{x}, \omega_{1i}, \omega_{2i})}{F(\mathbf{x})} \right]^p \right\}^{1/p}$$

where p is an even positive integer. The algorithm given below provides a step-by-step description of the minimax design method.

Algorithm

Step 1: Input $\mathbf{x}_0$. Set $k = 1$, $p = 2$, $\mu = 2$, $F_0 = 10^{99}$.
Step 2: Initialize frequencies ω_{1i}, ω_{2i}, for $i = 1, \ldots, m$.
Step 3: Using $\mathbf{x}_{k-1}$ as the initial point, minimize $\Phi(\mathbf{x})$ with respect to $\mathbf{x}$ to obtain $\mathbf{x}_k$. Set $F_k = F(\mathbf{x}_k)$.
Step 4: If $|F_k - F_{k-1}| < \varepsilon$, then output $\mathbf{x}^* = \mathbf{x}_k$ and F_k and stop; otherwise, set $p = \mu p$, $k = k + 1$, and go to Step 3.

The minimization in Step 3 can be carried out using a quasi-Newton method.[7] The gradient of the objective function required by the quasi-Newton method is given by

$$\nabla\Phi(\mathbf{x}) = \left\{ \sum_{i=1}^{m} \left[\frac{\left|E(\mathbf{x}, \omega_{1i}, \omega_{2i})\right|}{F(\mathbf{x})} \right]^p \right\}^{(1/p)-1} \times \sum_{i=1}^{m} \left[\frac{\left|E(\mathbf{x}, \omega_{1i}, \omega_{2i})\right|}{F(\mathbf{x})} \right]^{p-1} \nabla\left|E(\mathbf{x}, \omega_{1i}, \omega_{2i})\right|$$

Two remarks on the above algorithm are now in order. First, note that the power of the objective function, p, increases by a factor of 2 after each iteration loop is completed, and $\Phi(\mathbf{x})$ gets closer and closer to $\max_{1 \le i \le m} |E(\mathbf{x}, \omega_{1i}, \omega_{2i})|$ as p increases. Second, at the kth iteration, the algorithm makes good use of the vector x_{k-1} obtained from the preceding minimization as a reasonable initial point. Finally, concerning the stability of the filter designed, the strategy described in Section 6.1 can also be used in Step 3 of this algorithm, and as long as each $\mathbf{x}_k$ obtained corresponds to a stable transfer function, the stability of the recursive filter designed is guaranteed. Design examples using this algorithm are presented in Reference 2.

6.3 DESIGN BASED ON THE SINGULAR-VALUE DECOMPOSITION

The SVD-based method addressed in Section 5.5 is also applicable to the design of recursive 2-D filtrs. In what follows, we describe two SVD-based approaches that can be used to obtain a recursive 2-D filter that approximates a given frequency response with guaranteed stability.

Following the first two steps of the algorithm in Section 5.5, stable, recursive 1-D digital filters with transfer functions $f_i(z_1)$ and $g_i(z_2)$ are obtained, whose frequency responses approximate $\tilde{\mathbf{u}}_i$ and $\tilde{\mathbf{v}}_i$, respectively, for $i = 1, \ldots, K$, where K is a positive integer such that the last

$r - K$ singular values of $\mathbf{F}$ are insignificant and can be neglected. Well-established techniques can be used to obtain the designs of the 1-D filters involved,[1] and the stability of the 2-D filter obtained is assured as long as all the 1-D filters used are stable. Design examples based on this approach can be found in Reference 13.

In the second approach, we begin by designing a higher-order nonrecursive filter using the algorithm of Section 5.5. Let the transfer function of the nonrecursive filter obtained be

$$H(z_1, z_2) = \sum_{n_1=0}^{N_1} \sum_{n_2=0}^{N_2} h(n_1, n_2) z_1^{-n_1} z_2^{-n_2} \tag{37}$$

In Roesser's local state-space charactrization, $H(z_1, z_2)$ can be represented by

$$\begin{bmatrix} \mathbf{x}^h(i+1, j) \\ \mathbf{x}^v(i, j+1) \end{bmatrix} = \begin{bmatrix} \mathbf{A}_1 & \mathbf{A}_2 \\ \mathbf{A}_3 & \mathbf{A}_4 \end{bmatrix} \begin{bmatrix} \mathbf{x}^h(i, j) \\ \mathbf{x}^v(i, j) \end{bmatrix} + \begin{bmatrix} \mathbf{b}_1 \\ \mathbf{b}_2 \end{bmatrix} u(i, j) \equiv \mathbf{A}x + \mathbf{b}u \tag{38a}$$

$$y(i, j) = \begin{bmatrix} \mathbf{c}_1 & \mathbf{c}_2 \end{bmatrix} \begin{bmatrix} \mathbf{x}^h(i, j) \\ \mathbf{x}^v(i, j) \end{bmatrix} + du(i, j) \equiv \mathbf{c}\mathbf{x} + du \tag{38b}$$

where

$$\mathbf{A}_1 = \begin{bmatrix} \mathbf{0} & \mathbf{I}_{N_1-1} \\ 0 & 0 \end{bmatrix}, \ \mathbf{A}_2 = \begin{bmatrix} h_{1N_2} & \cdots & h_{11} \\ \vdots & \cdots & \vdots \\ k_{N_1N_2} & \cdots & h_{N_11} \end{bmatrix}, \mathbf{A}_3 = \mathbf{0}, \ \mathbf{A}_4 = \begin{bmatrix} \mathbf{0} & \mathbf{I}_{N_2-1} \\ 0 & \mathbf{0} \end{bmatrix}$$

$$\mathbf{b} = \begin{bmatrix} h_{10} \ldots h_{N_10} \,|\, 0 \ldots 0 \ 1 \end{bmatrix}^T$$

$$\mathbf{c} = \begin{bmatrix} 1 \ 0 \ldots 0 \,|\, h_{0N_2} \ldots h_{01} \end{bmatrix}$$

$$d = h_{00}$$

The reachability and observability gramians of the filter are defined by

$$\mathbf{K} = \frac{1}{(2\pi j)^2} \oint_{|z_1|=1} \oint_{|z_2|=1} \mathbf{F}(z_1, z_2) \mathbf{F}^*(z_1, z_2) \frac{dz_1}{z_1} \frac{dz_2}{z_2}$$

$$\mathbf{W} = \frac{1}{(2\pi j)^2} \oint_{|z_1|=1} \oint_{|z_2|=1} \mathbf{G}^*(z_1, z_2) \mathbf{G}(z_1, z_2) \frac{dz_1}{z_1} \frac{dz_2}{z_2}$$

where

$$\mathbf{F}(z_1, z_2) = \left[\mathbf{I}(z_1, z_2) - \mathbf{A}\right]^{-1} \mathbf{b}$$

$$\mathbf{G}(z_1, z_2) = \mathbf{c}\left[I(z_1, z_2) - \mathbf{A}\right]^{-1}$$

$$\mathbf{I}(z_1, z_2) = z_1 \mathbf{I}_{N_1} \oplus z_2 \mathbf{I}_{N_2}$$

Let us denote the $N_1 \times N_1$ upper left blocks and the $N_2 \times N_2$ lower right blocks of $\mathbf{K}$ and $\mathbf{W}$ by $\mathbf{K}_{11}$, $\mathbf{W}_{11}$, and $\mathbf{K}_{22}$, $\mathbf{W}_{22}$, respectively. The system represented by Equation 38 is said to be balanced if

$$\mathbf{K}_{11} = \mathbf{W}_{11} = \text{diag}\left\{\sigma_{11}, \sigma_{12}, \ldots, \sigma_{1N_1}\right\} \tag{39a}$$

$$\mathbf{K}_{22} = \mathbf{W}_{22} = \text{diag}\left\{\sigma_{21}, \sigma_{22}, \ldots, \sigma_{2N_2}\right\} \tag{39b}$$

It is known that the order of a stable balanced system can be reduced with a relatively small reduction error and guaranteed stability.[14] In order to obtain a balanced state-space realization, $\mathbf{K}_{ii}$ and $\mathbf{W}_{ii}$ ($i = 1, 2$) need to be evaluated. Because of the fact that $\mathbf{A}_3 = 0$, these matrices can be easily computed as[9]

$$\mathbf{K}_{11} = \sum_{i=0}^{N_1-1} \mathbf{A}_1^i \mathbf{P}\left(\mathbf{A}_1^T\right)^i \tag{40}$$

where

$$\mathbf{P} = \mathbf{H}_b \mathbf{H}_b^T$$

$$\mathbf{H}_b = \begin{bmatrix} h_{10} & h_{11} & \cdots & h_{1N_2} \\ h_{20} & h_{21} & \cdots & h_{2N_2} \\ \vdots & \vdots & & \vdots \\ h_{N_1 0} & h_{N_1 1} & \cdots & h_{N_1 N_2} \end{bmatrix}$$

and

$$\mathbf{K}_{22} = \mathbf{I}_{N_2} \tag{41}$$

$$\mathbf{W}_{11} = \mathbf{I}_{N_1} \tag{42}$$

$$\mathbf{W}_{22} = \sum_{i=0}^{N_1-1} \left(\mathbf{A}_4^T\right)^i Q \mathbf{A}_4^i \tag{43}$$

where

$$\mathbf{Q} = \mathbf{H}_c^T \mathbf{H}_c$$

$$\mathbf{H}_c = \begin{bmatrix} h_{0N_2} & \cdots & h_{02} & h_{01} \\ h_{1N_2} & \cdots & h_{12} & h_{11} \\ \vdots & & \vdots & \vdots \\ h_{N_1 N_2} & \cdots & h_{N_1 2} & h_{N_1 1} \end{bmatrix}$$

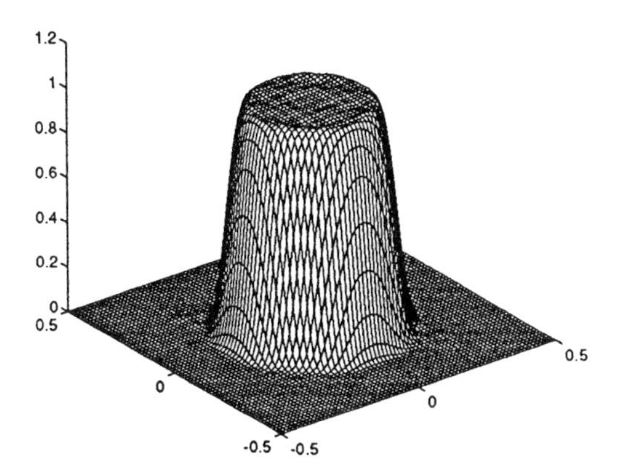

Figure 6 Amplitude response of recursive lowpass filter of order (14, 14).

Once $\mathbf{K}_{ii}$ and $\mathbf{W}_{ii}$ (i = 1, 2) are evaluated, a balancing transformation $\mathbf{T}$ can be computed[14] to obtain a balanced realization ($\hat{\mathbf{A}}$, $\hat{\mathbf{b}}$, $\hat{\mathbf{c}}$, $\hat{d}$) = ($\mathbf{T}^{-1}$ $\mathbf{AT}$, $\mathbf{T}^{-1}\mathbf{b}$, $\mathbf{cT}$, d). A reduced state-space model ($\mathbf{A}_r$, $\mathbf{b}_r$, $\mathbf{c}_r$, d) of order (r_1, r_2) can be obtained by partitioning $\hat{\mathbf{A}}$, $\hat{\mathbf{b}}$, $\hat{\mathbf{c}}$, as

$$\hat{\mathbf{A}} = \begin{bmatrix} \hat{\mathbf{A}}_1 & \hat{\mathbf{A}}_2 \\ \hat{\mathbf{A}}_3 & \hat{\mathbf{A}}_4 \end{bmatrix} = \begin{bmatrix} \mathbf{A}_{1r} & \mathbf{A}_{12} & \mathbf{A}_{2r} & \mathbf{A}_{22} \\ \mathbf{A}_{13} & \mathbf{A}_{14} & \mathbf{A}_{23} & \mathbf{A}_{24} \\ \mathbf{0} & \mathbf{0} & \mathbf{A}_{4r} & \mathbf{A}_{42} \\ \mathbf{0} & \mathbf{0} & \mathbf{A}_{43} & \mathbf{A}_{44} \end{bmatrix}$$

$$\hat{\mathbf{b}} = \begin{bmatrix} \hat{\mathbf{b}}_1 \\ \hat{\mathbf{b}}_2 \end{bmatrix} = \begin{bmatrix} \mathbf{b}_{1r} \\ * \\ \mathbf{b}_{2r} \\ * \end{bmatrix}, \quad \hat{\mathbf{c}} = \left[\hat{\mathbf{c}}_1 \mid \hat{\mathbf{c}}_2\right] = \left[\mathbf{c}_{1r} \; * \mid \mathbf{c}_{2r} \; *\right]$$

A transfer function can now be constructed as

$$H_r(z_1, z_2) = \mathbf{c}_r\left[\mathbf{I}_r(z_1, z_2) - \mathbf{A}_r\right]^{-1}\mathbf{b}_r + d \tag{44}$$

$H_r(z_1, z_2)$ describes a stable, recursive 2-D filter of order (r_1, r_2), and the error introduced in this balanced approximation is controlled by the sum of the Hankel singular values that are neglected, i.e., $\sigma_{1,r_1+1}, \ldots, \sigma_{1N_1}$ and $\sigma_{2,r_1+1}, \ldots, \sigma_{2N_2}$. Therefore, if the nonrecursive filtr designed is of good quality and if the Hankel singular values neglected are all small, then the recursive filter represented by Equation 44 will also be of good quality. As an example, if the nonrecursive lowpass filter of order (29, 29) whose amplitude response is depicted in Figure 5 is balanced and then approximated by a stable recursive filter, the order is reduced to (14, 14). The amplitude response obtained is shown in Figure 6. Since the original nonrecursive filter has linear phase response, the recursive filter obtained has a nearly linear phase response in the passband as illustrated in Figure 7.

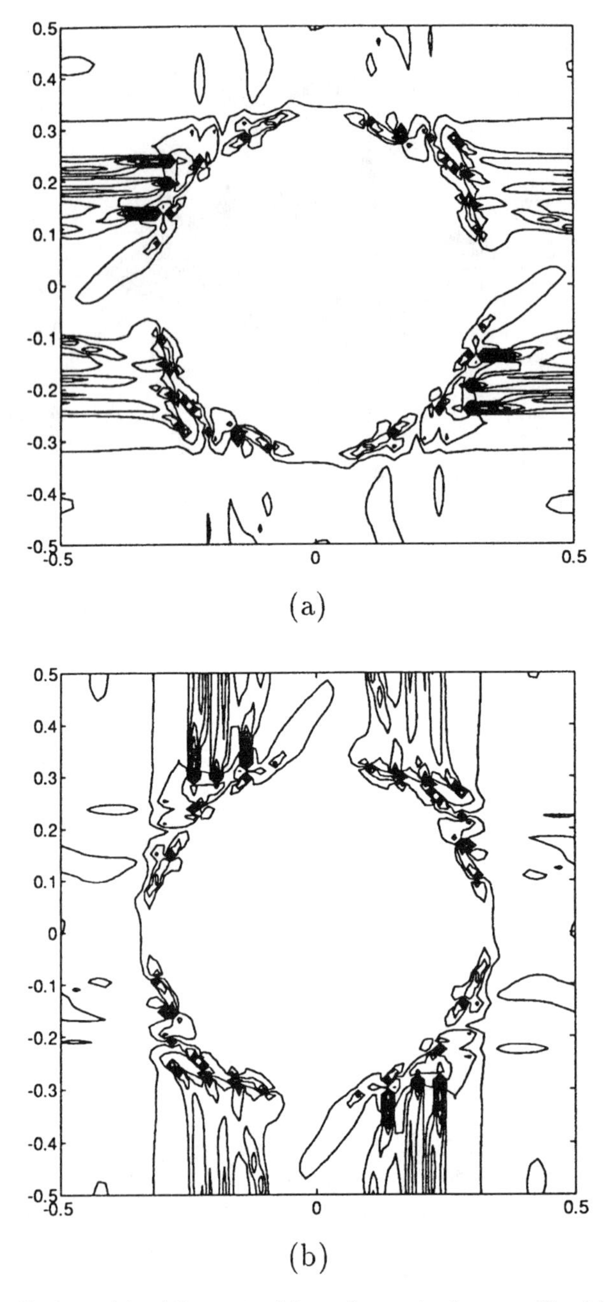

Figure 7 Contour plots of the group delays of recursive lowpass filter (a) τ_1; (b) τ_2.

REFERENCES

1. **A. Antoniou,** *Digital Filters: Analysis, Design, and Applications,* 2nd ed., New York: McGraw-Hill, 1993.
2. **W.-S. Lu and A. Antoniou,** *Two-Dimensional Digital Filters,* New York: Marcel Dekker, 1992.
3. **T. S. Huang,** Two-dimensional windows, *IEEE Trans. Audio Electroacoust.,* 20, 88–89, March 1972.
4. **J. H. McClellan,** The design of two-dimensional filters by transforms, in *Proc. 7th Annual Princeton Conf. Information Sciences and Systems,* 247–251, 1973.
5. **R. M. Mersereau, W. F. G. Mecklenbräuker, and T. F. Quatieri, Jr.,** McClellan transforms for two-dimensional digital filtering: I — Design, *IEEE Trans. Circuits Syst.,* 23, 405–413, July 1976.
6. **C. Charalambous,** Acceleration of the least *p*th algorithm for minimax optimization with engineering applications, *Math Program,* 17, 270–297, 1979.
7. **R. Fletcher,** *Practical Methods of Optimization,* 2nd ed., Chichester: John Wiley, 1987.

8. **G. W. Stewart,** *Introduction to Matrix Computations,* New York: Academic Press, 1974.
9. **W.-S. Lu, H.-P. Wang, and A. Antoniou,** Design of 2-D digital filters using the singular value decomposition and balanced approximation, *IEEE Trans. Signal Process.,* 39, 2253–2262, Oct. 1991.
10. **W.-S. Lu and A. Antoniou,** Design of 2-D digital filters with arbitrary amplitude and phase responses by using the singular value decomposition, in *Proc. Int. Symp. Circuits Syst.,* 618–621, Singapore, May 1991.
11. **G. A. Maria and M. M. Fahmy,** An l_p design technique for two-dimensional digital recursive filters, *IEEE Trans. Acoust., Speech, Signal Process.,* 22, 15–21, Feb. 1974.
12. **C. Charalambous,** A unified review of optimization, *IEEE Trans. Microwave Theor. Tech.,* 22, 289–300, March 1974.
13. **A. Antoniou and W.-S. Lu,** Design of 2-D digital filters by using the singular value decomposition, *IEEE Trans. Circuits Syst.,* 34, 1191–1198, Oct. 1987.
14. **W.-S. Lu, E. B. Lee, and Q.-T. Zhang,** Balanced approximation of two-dimensional and delay-differential systems, in *Proc. IEEE Conf. Decision and Control,* 917–922, Athens, Greece, Dec. 1986.

APPENDIX A: A MATLAB PROGRAM FOR THE DESIGN OF NONRECURSIVE 2-D FILTERS USING THE SVD METHOD

Listed below is an interactive MATLAB program named fir2d.m that can be used to generate a linear phase, nonrecursive 2-D transfer function whose amplitude response approximates a desired amplitude response. The input data to the program is a sampled amplitude response matrix over the baseband, which is assumed to be square, and the order of the filter to be designed. The output of the program is the coefficient matrix $\mathbf{C}_k$ of the nonrecursive filter designed.

```
% fir2d.m
% Input the desired sampled amplitude response matrix.
%
am_d=input('desired amplitude response (square, odd size) = ');
%
% Input the order of the filter.
%
order=input ('order of the filter (odd-valued-vector) = ');
%
M=max(size(am_d));
N1=order(1);
N2=order(2);
%
% Perform the SVD of the desired amplitude response matrix.
%
[u,s,v]=svd(am_d);
U1=u*sqrt(s);
V1=v*sqrt(s);
r=rank(am_d);
U=U1((M+1)/2:M,1:r);
V=V1((M+1)/2:M,1:r);
f=(0:2/(M-1):1)';
%
% Find 1-D FIR transfer functions fi(z1) and gi(z2) to
% approximate ui~ and vi~, and construct the coefficient
% matrix C.
%
C=zeros(N1,N2);
for i=1:r,
 fi=fir2(N1-1,f,U(:,i))';
 gi=fir2(N2-1,f,V(:,i));
 C=C+fi*gi;
end
```

```
clear order u s v U1 V1 U V fi gi f r i
%
% Perform the SVD of the coefficient matrix and display
% its singular values.
%
[uc,sc,vc]=svd(C);
Sc=diag(sc);
plot(Sc,'x')
%
% Select a value for Kc -- the number of sections
% to be used.
%
Kc=input(number of sections to be used = );
Ck=zeros(N1,N2);
for i=1:Kc,
 Ck=Ck+Sc(i)*uc(:,i)*vc(:,i)';
end
clear M N1 N2 C uc sc vc Sc i
%
% Evaluate and display the amplitude response of the
% filter designed.
%
am=abs(fftshift(fft2(Ck,71,71)));
f=-0.5:1/70:0.5;
mesh(f,f,am)
axis('square')
clear f
```

Adaptive Filters

Marcello L. R. de Campos and Andreas Antoniou

CONTENTS

1. INTRODUCTION

An adaptive filter is a time-dependent filter that has its coefficients, and in some cases its order, automatically adjusted in real time in order to improve its performance. Although adaptive filtering is a relatively young area in signal processing, recursively adjusting the parameters of a function based on observed data goes back centuries. The idea is to compare the solution obtained from the parameterized function with some reference data and to produce a new set of parameters that is closer in some sense to the optimal solution. Although intuitively sound, the formalization of the method can become very involved. For instance, carefully chosen criteria must guide the actions of the algorithm, and interference by the designer must be absent or be kept to a minimum. Furthermore, adaptive filters must be reliable because in some applications failure can be disastrous.

A good example of an adaptation method was proposed by Gauss[1] at the beginning of the 19th century for the determination of orbits of heavenly bodies satisfying as nearly as possible any number of observations. His approach was to correct the available parameters representing the orbit using new data in order to satisfy all the observations in the most accurate manner possible. The more data gathered, the more accurate the parameters describing the orbit should be.

In the area of signal processing, the need for better predictors, equalizers, interference cancelers, etc. requires systems that can adapt themselves to the characteristics of a particular environment and, therefore, can yield more satisfactory results. Moreover, if the environment changes, the adaptive filter must adjust its parameters automatically to the new situation. In these applications,

0-8493-8951-8/97/$0.00+$.50
© 1997 by CRC Press LLC

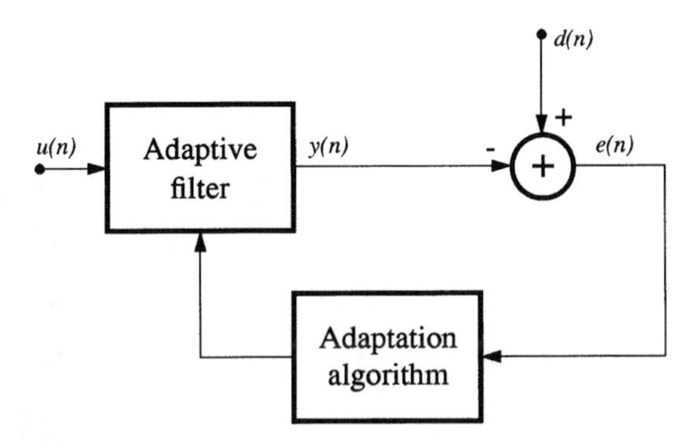

Figure 1 General adaptive filter structure.

as in the trajectory predictions by Gauss, the adaptive filter must rely on new observations to achieve the best solution possible.

Although adaptive filters can be analog or digital, finite-duration impulse response (FIR) digital filter structures have become the most widely used because of their simplicity and reliability. Adaptive infinite-duration impulse response (IIR) digital filters are still being investigated as possible alternatives to FIR adaptive filters in applications where a high filter order is required. The recursive structure of IIR adaptive filters, however, poses extra problems with regard to robustness and stability.[2,3] For these reasons, the use of IIR adaptive filters is still the subject of considerable research.

A typical adaptive digital filter configuration is illustrated in Figure 1. The external data provided to the system at every iteration n are the input signal, $u(n)$, and the reference (or desired) signal, $d(n)$. The adaptation algorithm adjusts the coefficients of the filter such that some norm of the error signal defined as

$$e(n) = d(n) - y(n)$$

is minimized. Signal $y(n)$ is the output of the filter for the given input $u(n)$. The updating is, in general, carried out at every iteration according to the equation

$$\mathbf{w}(n) = \mathbf{w}(n-1) + \mu(n)\mathbf{c}(n) \tag{1}$$

where $\mathbf{w}(n)$ or simply $\mathbf{w}$ is the coefficient vector of the filter, $\mu(n)$ is the convergence factor, and $\mathbf{c}(n)$ is the correction to be applied to the coefficients. Sometimes the amount of computation is excessive and the coefficients are updated only every N iterations; this is referred to as block adaptation.

For readers familiar with optimization procedures, the format of Equation 1 should come as no surprise. The differences among the several available adaptive-filtering schemes are in the way $\mathbf{w}$ is determined as well as how μ and $\mathbf{c}$ are calculated. The most common structure, and the most simple, is the transversal filter (TF), sometimes called the tapped delay line or linear combiner, depicted in Figure 2. In this case, the output of the filter is the result of the inner product operation between the coefficient vector

$$\mathbf{w} = \left[w_1 \; w_2 \cdots w_M\right]^T$$

and an input-signal vector

$$\mathbf{u}(n) = \left[u(n) \; u(n-1) \cdots u(n-M+1)\right]^T$$

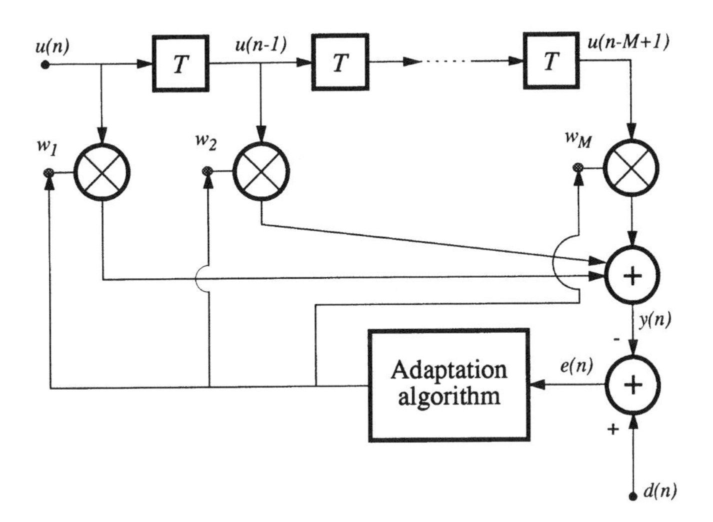

Figure 2 Transversal filter structure.

For the TF structure, the derivation of algorithms and their analysis are relatively simple as compared with those of more-complex structures. These reasons are strong enough to make the TF structure popular and a perfect example for introducing the subject.

Once the structure is chosen, the next step is to choose among the available adaptation algorithms the one that best suits the application at hand. Before we go on with the study of the different algorithms and their properties, we briefly discuss some applications where adaptive filters have proved particularly useful. This will help the reader to better understand how signals $u(n)$ and $d(n)$ are obtained and will also emphasize the importance, flexibility, and usefulness of adaptive filters.

Echo calcellation — either in loudspeaker-enclosure-microphone systems (LEMS) or in satellite communication systems — is one application where the nonstationarity of the signals involved and changes in the medium make adaptive filters especially useful. In the telephone network, although local calls are established through two-wire circuits which serve for transmission in both directions, long-distance calls are made through four-wire circuits where separate paths are provided for each direction of transmission. The connection between the two-wire and the four-wire circuits is made by a hybrid circuit. Impedance mismatch at the hybrid circuit results in coupling of the incoming and outgoing branches of the four-wire circuit, and, consequently, part of the signal is reflected and returned to the original talker. Traditional methods for echo control can degrade significantly the quality of the conversation, especially when both parties are talking simultaneously. High-quality communication systems call for alternative solutions which suppress the echo without compromising the quality of the transmission. The adaptive filter employed as echo canceler simulates the transfer characteristics of the echo path and produces a copy of the echo to be subtracted from the return signal.[4] In the case of acoustic cancellation in an LEMS, the adaptive filter compensates for any echo generated when the microphone captures the signal generated at the loudspeaker.[5]

Another application where adaptive filters are very useful is channel equalization. When the frequency response in a telephone or radio channel is nonideal, e.g., if the amplitude response is not constant and/or the phase response is not linear with respect to the channel bandwidth, time dispersion can arise which causes symbols to overlap thereby resulting in intersymbol interference (ISI). The use of an adaptive filter for equalization is illustrated in Figure 4 where the equalizer corrects ISI caused by channels with different and often time-dependent characteristics. In the past, equalization was done largely with fixed linear equalizers which work well when tuned to a specific channel whose characteristics do not change with time. The underlying adaptation process essentially adjusts the transfer function of the equalizer such that the equalized channel has a constant amplitude response and linear phase response with respect to the channel bandwidth. In this application, the reference signal is, in general, a training sequence known at the receiver. After the

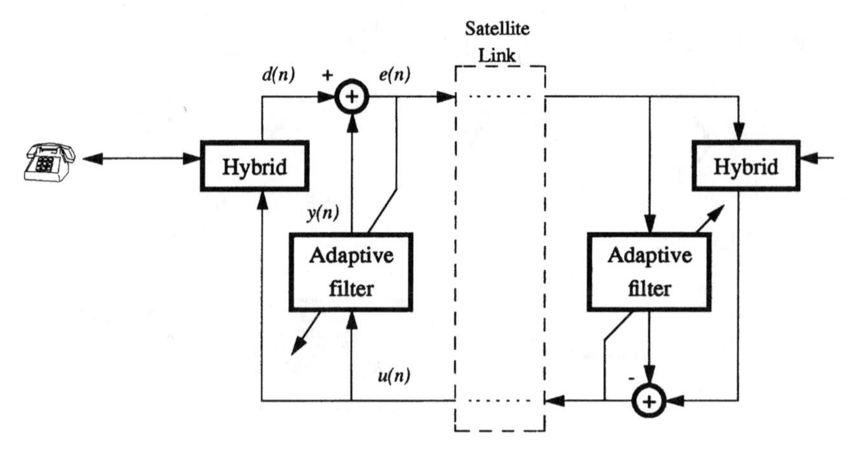

Figure 3 Adaptive echo canceler.

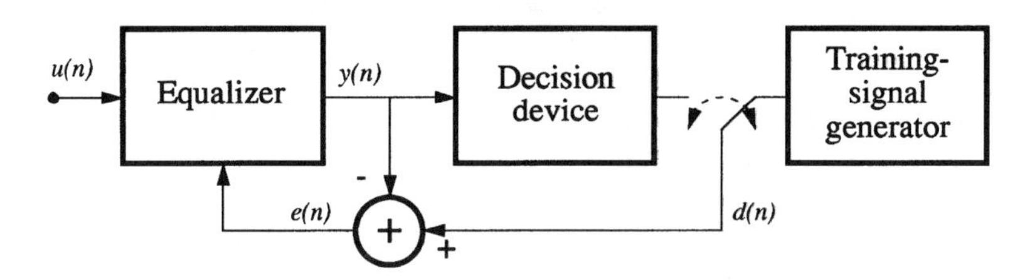

Figure 4 Simplified adaptive channel equalizer.

training period, decision-directed techniques are used to continuously provide a reference signal and, therefore, to allow for tracking slow variations of the channel.[6] Better results in mitigating the effects of time dispersion can be achieved with more-sophisticated adaptive equalizers than with the simplified model shown in Figure 4. The basics of the adaptation of the coefficients, however, are the same.

Adaptive filters can be employed in numerous other applications, such as system identification, automatic regulation, linear prediction, etc. The need for more-efficient and robust algorithms is, therefore, continuously increasing. Next, we shall present an overview of some of the known algorithms and their basic properties.

2. WIENER FILTERING

Linear least-squares (LS) estimation is a vast field that goes beyond adaptive filters. Some knowledge of the Wiener–Hopf equation,[7] however, is essential for the development and understanding of adaptation algorithms. Therefore, a brief introduction of Wiener filtering is appropriate at this point.*

Let the *objective function* $J(n)$ denote the mean-squared error (MSE) between the reference signal and the output of a transversal filter with coefficient vector **w**, i.e.,

$$J(n) = E\left[\left|d(n) - \mathbf{u}(n)^T \mathbf{w}\right|^2\right] \tag{2}$$

The goal of optimal filtering in this case is to find the optimal coefficient vector, $\mathbf{w} = \mathbf{w}_o$, that minimizes $J(n)$, i.e., $J(n) = J_{\min}$. By differentiating Equation 2 with respect to the elements of **w** and equating the first derivatives to zero, we obtain

* For a more complete discussion on linear estimation, the reader is referred to Reference 8.

$$E\{\mathbf{u}(n)[d(n) - \mathbf{u}^T(n)\mathbf{w}_o]\} = 0 \tag{3}$$

Since Equation 2 is quadratic in $\mathbf{w}$, a unique solution is guaranteed. Furthermore, it is easy to verify that it corresponds to a minimum as long as $E[\mathbf{u}(n)\mathbf{u}^T(n)]$ is positive definite. Equation 3 is often referred to as the *orthogonality principle*,[7] which means that for the optimal set of coefficients, the estimation error is orthogonal to the previous M input-signal samples. The solution, $\mathbf{w}_o$, satisfies the Wiener–Hopf equation which can be expressed in terms of the input-signal autocorrelation matrix

$$\mathbf{R} = E[\mathbf{u}(n)\mathbf{u}^T(n)]$$

and the crosscorrelation between the input signal and the reference signal

$$\mathbf{p} = E[d(n)\mathbf{u}(n)]$$

as

$$\mathbf{R}\mathbf{w}_o = \mathbf{p}$$

The *minimum MSE,* obtained when $\mathbf{w} = \mathbf{w}_o$, is given by

$$J_{\min} = E[|d(n)|^2] - \mathbf{w}_o^T\mathbf{R}\mathbf{w}_o \tag{4}$$

Since $\mathbf{R}$ and $\mathbf{p}$ are often unknown, the task of the adaptation algorithm is to recursively produce estimates of $\mathbf{w}_o$, i.e., $\mathbf{w}(n)$. For signals that are stationary, at least in the wide sense,[9] convergence of $\mathbf{w}(n)$ to $\mathbf{w}_o$ in the mean can be expected in most cases. The covariance, however, depends on several different factors and is, in general, one of the chief figures of merit for comparing the performance of different algorithms. A study of the first two moments of $\mathbf{w}(n)$ are key to assuring convergence and to establishing a performance criterion. In general, performance is measured in terms of the *excess MSE* $J_{\text{ex}}(n)$, which is defined as

$$J_{\text{ex}}(n) = J(n) - J_{\min} \tag{5}$$

From Equations 4 and 5 it can be easily shown that

$$J_{ex}(n) = E\{[\mathbf{w}(n) - \mathbf{w}_o]^T\mathbf{R}[\mathbf{w}(n) - \mathbf{w}_o]\}$$

In effect, $J_{\text{ex}}(n)$ is closely related to the covariance of $\mathbf{w}(n)$. For convenience, performance is sometimes measured in terms of a dimensionless parameter called *misadjustment,* defined as the excess MSE divided by the minimum MSE, i.e.,

$$M(n) = \frac{J_{\text{ex}}(n)}{J_{\min}}$$

Function $J(n)$ can be represented by a *performance surface* which describes the MSE as a function of the M-dimensional coefficient vector. The shape of the performance surface in the (M

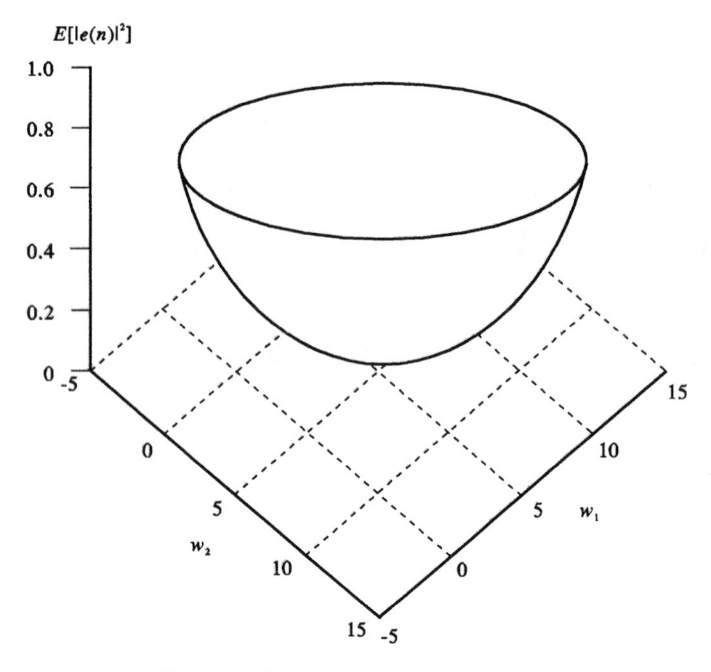

Figure 5 Performance surface for $M = 2$.

+ 1)-dimensional hyperspace depends on the input signal; the more spread in the eigenvalues of **R**, the more asymmetric is the performance surface. Figure 5 shows an example of a performance surface for $M = 2$.

When nonstationary signals are present, either **R** or **p** or both are time dependent, and the algorithm is required to track the solution. In practice, nonstationarities are identified in various forms (e.g., fading channels, changes in the position of furniture and people in LEMS, etc.). Modeling the optimal coefficients as first- or second-order Markov processes is a valuable tool to verify the performance of an algorithm under these conditions. Reconvergence testing can be performed by introducing abrupt changes in the coefficients. Proper operation can be observed only when the time constant of the algorithm is smaller than that of the nonstationarity.

3. LEAST MEAN SQUARES ALGORITHM

An adaptation algorithm known as the least-mean-squares (LMS) algorithm has become very popular since it was proposed by Widrow and Hoff,[10,11] and its simplicity and reliability have a great deal to do with its popularity. In this algorithm, the coefficients are updated as

$$\mathbf{w}(n) = \mathbf{w}(n-1) + 2\mu e(n)\mathbf{u}(n) \tag{6}$$

where μ is the convergence factor and $e(n)$ is referred to as the *a priori* output error and is defined as

$$e(n) = d(n) - \mathbf{u}^T(n)\mathbf{w}(n-1) \tag{7}$$

If the computational complexity of the algorithm is measured in terms of the number of multiplications required per iteration, the LMS algorithm has a complexity which is proportional to the order of the filter, or $O(M)$ in mathematical notation. The algorithm can be further simplified by replacing the output error by its sign. The resulting *sign-error* algorithm can be described as

$$\mathbf{w}(n) = \mathbf{w}(n-1) + 2\mu \ \text{sign}[e(n)]\mathbf{u}(n) \tag{8}$$

where

$$\text{sign}[e(n)] = \begin{cases} -1 & \text{for } e(n) < 0 \\ 0 & \text{for } e(n) = 0 \\ +1 & \text{for } e(n) > 0 \end{cases}$$

In Equations 6 and 8 the coefficients are corrected in the direction of the negative of the gradient of the instantaneous squared error,[12] $e^2(n)$, and the magnitude of $e(n)$, $|e(n)|$, respectively. The LMS algorithm and its variations are often referred to as gradient-based algorithms.

The LMS algorithm is not consistent from the statistical point of view, for measurement noise and nonstationarity imply nonzero covariance of the coefficient vector.[13] For instance, μ is proportional to the portion of the misadjustment due to measurement noise, and it is inversely proportional to that due to *lag* in tracking nonstationarities. In stationary environments, zero misadjustment can be attained if the convergence factor is gradually reduced to zero after convergence. In most cases, though, this procedure should be avoided, for the algorithm will have no power to track most of the changes in the environment. The convergence factor is intrinsically related to stability, and a sufficient condition for convergence of the coefficient vector in the mean, assuming stationary signals, is

$$0 < \mu < \frac{1}{tr(\mathbf{R})} \tag{9}$$

where tr(·) denotes trace. This condition is both sufficient and necessary for the convergence of the variance.[13]

The optimal choice for the convergence factor depends on prior knowledge of the statistics of the signals involved, which can be difficult or even impossible to obtain. For this reason, many methods for estimating μ online have been proposed in the past (see, for example, Reference 14). Although most of these methods introduce other parameters to which the algorithm is less sensitive, there is strong indication that a convergence factor that is inversely proportional to the instantaneous power of the input-signal vector would yield a potentially faster algorithm when limited knowledge is available.[15] Such an algorithm is called normalized LMS (NLMS) algorithm, and the coefficients are updated as

$$\mathbf{w}(n) = \mathbf{w}(n-1) + 2\frac{\bar{\mu}}{\|\mathbf{u}(n)\|_2^2} e(n)\mathbf{u}(n) \tag{10}$$

Note that a convergence factor is kept in Equation 10, but as a rule of thumb $\bar{\mu} = {}^1/_2$ can be used with good results in most cases.

The major problem affecting both the LMS and NLMS algorithms is their slow convergence when the input signal is highly correlated. For the LMS algorithm, for instance, the coefficient-error vector converges in the mean to zero according to the relation.[13]

$$E[\mathbf{v}(n)] = (\mathbf{I} - 2\mu\mathbf{R})E[\mathbf{v}(n-1)] \tag{11}$$

where $\mathbf{I}$ denotes the identity matrix and $\mathbf{v}(n)$ is defined as

$$\mathbf{v}(n) = \mathbf{w}(n) - \mathbf{w}_o$$

Applying the spectral decomposition to $\mathbf{R}$, i.e.,

$$\mathbf{R} = \mathbf{Q}\Lambda\mathbf{Q}^T$$

with $\mathbf{Q}$ formed by the orthonormal eigenvectors of $\mathbf{R}$ and Λ is a diagonal matrix formed by its eigenvalues, Equation 11 can be rewritten in terms of the modified coordinates $\bar{\mathbf{v}}$ as

$$E[\bar{\mathbf{v}}(n)] = (\mathbf{I} - 2\mu\Lambda)E[\bar{\mathbf{v}}(n-1)] \tag{12}$$

where $\bar{\mathbf{v}}(n) = \mathbf{Q}^T\mathbf{v}(n)$. The relaxation process in Equation 12 is a geometric progression with a ratio of $(1 - 2\mu\lambda_i)$ for each mode. An exponential envelope of the type e^{-n/τ_i} can then be fitted to each mode with a time constant approximately equal to[12]

$$\tau_i \approx \frac{1}{2\mu\lambda_i}$$

where $\mu\lambda_i \ll 1$. Therefore, the overall speed of convergence is dominated by the smallest eigenvalue, while from Equation 9 the upper limit for the convergence factor is dominated by the largest eigenvalue. In the case where the eigenvalues of $\mathbf{R}$ are widely spread, unacceptably slow convergence can result. Several attempts to increase the speed of convergence of the LMS algorithm have been reported, invariably at the cost of increasing its computational complexity. Since simple real-time methods for optimizing the convergence factor still suffer from slow convergence when $\mathbf{R}$ is ill-conditioned, more elaborate estimation schemes have to be employed in such situations. Transform-domain adaptive filters constitute the next step toward the study of algorithms which are less sensitive to input-signal correlation.

4. TRANSFORM-DOMAIN ADAPTIVE FILTERS

Transform-domain FIR adaptive filters (TDAF) are filters using gradient-based algorithms operating on a transformed input signal, as depicted in Figure 6. Equation 12 suggests that maximum convergence speed can be achieved if the input signal is transformed to have a diagonal autocorrelation matrix, and a different convergence factor is used for each mode. In this way the speed of convergence of each mode can be independently maximized. In other words, the transformation, combined with the power nomalization introduced by different convergence factors, provides the means for changing the aspect of the performance surface associated to the transformed input signal. Usually, it can be said that the more symmetrical the performance surface, the faster is the convergence of a gradient-based algorithm. It can also be shown that the use of an orthogonal transformation followed by normalization never degrades the performance of the LMS algorithm.[16]

A transformation that achieves a perfect decoupling of the modes of the process described by Equation 11 is the Karhunen–Loéve transformation (KLT).[7] Unfortunately, its implementation requires knowledge of the eigenvalues and eigenvectors of $\mathbf{R}$ which are not available in most applications. It is possible to find other orthogonal transformations that render the approach feasible, but the performance may be suboptimal. In general, transformations that lead to a fast implementation are preferred because of the real-time constraint. For the transform-domain LMS algorithm, the coefficient vector is updated according to the equation

$$\bar{\mathbf{w}} = \bar{\mathbf{w}}(n-1) + 2\mu e(n)\Omega^{-1}\mathbf{T}^T\mathbf{u}(n) \tag{13}$$

where $\bar{\mathbf{w}}(n) = \mathbf{T}^T\mathbf{w}(n)$ and Ω^{-1} is a diagonal matrix with an individual power normalization for each coefficient. Although the coefficient vector $\bar{\mathbf{w}}(n)$ converges in the mean to a transformed version of $\mathbf{w}_o$, it is worthwhile to note that the output of the TF with coefficient vector $\bar{\mathbf{w}}(n)$ and input-signal vector $\mathbf{T}^T\mathbf{u}(n)$ is not affected by the orthogonal transformation $\mathbf{T}$.

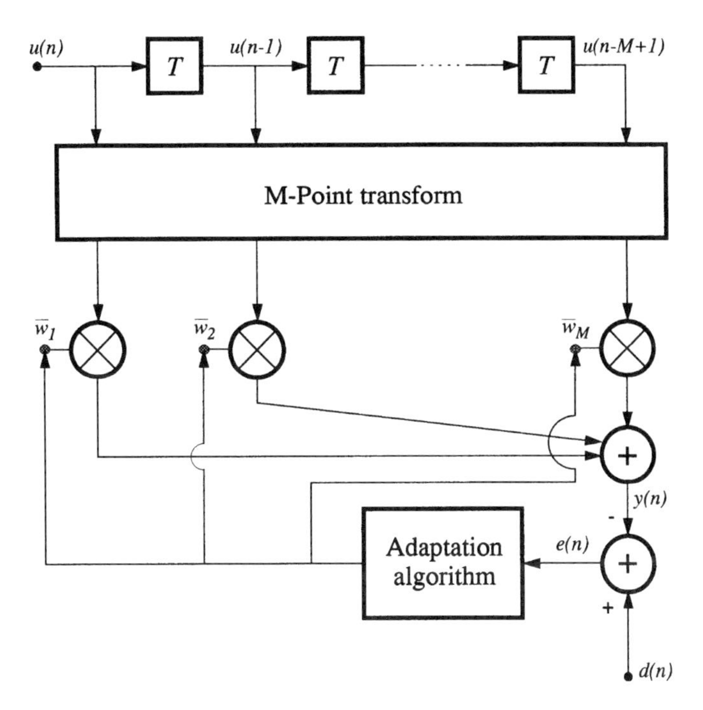

Figure 6 Transform-domain adaptive filter.

The choice of a good transformation for a particular application is by no means an easy task. Computer simulations comparing different transformations show that knowledge of the input-signal characteristics is key to assuring a substantial improvement.[17] For instance, since **R** is approximately circulant for large values of M, the discrete Fourier transform (DFT) should be a good choice in this case, because the eigenvectors of an $M \times M$ circulant matrix can be obtained from an M-point DFT. For small values of M, however, the DFT might be a bad choice as compared whith other transforms. Furthermore, its complex coefficients impose restrictions to its applicability. A potentially more efficient solution is to provide means to the adaptation algorithm to estimate the KLT in real time. An algorithm that employs such a technique is called LMS-Newton (LMSN) algorithm.

5. LMS-NEWTON ALGORITHM

The LMSN algorithm inherits its name from the analogy to Newton's method for finding the minimum of a function. The method has also been successfully used for many decades as a powerful optimization technique. The improvement in performance when compared with gradient-based techniques is due to a more accurate description of the objective function based on second-order information. For quadratic convex functions the method converges to the optimal solution in a single step. Since Equation 2 is quadratic in **w**, the Newton method would, in theory, give the solution to the Wiener–Hopf equation instantaneously. In practice, however, the objective function is not known and an iterative solution must be sought.

The LMSN algorithm updates the filter coefficients according to the equation

$$\mathbf{w}(n) = \mathbf{w}(n-1) + 2\mu e(n)\hat{\mathbf{R}}^{-1}(n)\mathbf{u}(n)$$

where $\hat{\mathbf{R}}(n)$ is a positive definite estimate of the input-signal autocorrelation matrix. This estimate must also allow an efficient inversion, for it must be performed in real time. Good results can be achieved with a stochastic approximation method like the one proposed by Robbins and Monro[18] in which

$$\hat{\mathbf{R}}(n) = \hat{\mathbf{R}}(n-1) + \alpha(n)\left[\mathbf{u}(n)\mathbf{u}^T(n) - \hat{\mathbf{R}}(n-1)\right] \tag{14}$$

where $\alpha(n)$ is a sequence of positive scalars. For practical purposes, some of the conditions for consistency are, in general, relaxed so that variance is traded for the ability to track nonstationarities in the input signal. The most common scheme uses a small positive constant $\alpha(n) = \alpha$, for which Equation 14 can be viewed as an exponentially weighted average of the outer product of vectors $\mathbf{u}(n)$ and $\mathbf{u}^T(n)$, i.e.,

$$\hat{\mathbf{R}}(n) = \alpha \sum_{i=1}^{n} (1-\alpha)^{n-i} \mathbf{u}(n)\mathbf{u}^T(n) + (1-\alpha)^n \hat{\mathbf{R}}(0) \tag{15}$$

In order to facilitate the inversion of $\hat{\mathbf{R}}(n)$ when $n < M$, a positive definite $\hat{\mathbf{R}}(0)$ must be provided since it is assumed that $u(n) = 0$ for $n < 0$. The influence of $\hat{\mathbf{R}}(0)$ decays exponentially for large values of n as can be seen from Equation 15. This scheme combines a very good approximation for sufficiently small values of α with the advantage that an inverse can be more easily obtained. The matrix inversion lemma[7] applied to Equation 14 yields

$$\hat{\mathbf{R}}^{-1}(n) = \frac{1}{1-\alpha}\left\{\hat{\mathbf{R}}^{-1}(n-1) - \frac{\hat{\mathbf{R}}^{-1}(n-1)\mathbf{u}(n)\mathbf{u}^T(n)\hat{\mathbf{R}}^{-1}(n-1)}{\frac{1-\alpha}{\alpha} + \mathbf{u}^T(n)\hat{\mathbf{R}}^{-1}(n-1)\mathbf{u}(n)}\right\}$$

Evidently, this technique is recursive in time and of complexity $O(M^2)$ which is considerably better than that for more-general inversion techniques. Matrix $\hat{\mathbf{R}}(n)$ is positive definite provided that the input signal is at least persistently exciting of order M, i.e., its frequency spectrum is nonzero at M or more points in the frequency range $-\omega_s/2$ to $\omega_s/2$, where ω_s is the sampling rate.[19]

A slightly more efficient implementation of the LMSN algorithm can be obtained if the relation

$$\hat{\mathbf{R}}^{-1}(n)\mathbf{u}(n) = \frac{1}{\alpha} \cdot \frac{\hat{\mathbf{R}}^{-1}(n-1)\mathbf{u}(n)}{\frac{1-\alpha}{\alpha} + \mathbf{u}^T(n)\hat{\mathbf{R}}^{-1}(n-1)\mathbf{u}(n)}$$

is used. The algorithm can be rewritten as

$$\mathbf{t}(n) = \hat{\mathbf{R}}^{-1}(n-1)\mathbf{u}(n)$$

$$z(n) = \frac{1-\alpha}{\alpha} + \mathbf{u}^T(n)\mathbf{t}(n)$$

$$\mathbf{w}(n) = \mathbf{w}(n-1) + \frac{2\mu}{\alpha} \cdot \frac{e(n)\mathbf{t}(n)}{z(n)}$$

$$\hat{\mathbf{R}}^{-1}(n) = \frac{1}{1-\alpha}\left\{\hat{\mathbf{R}}^{-1}(n-1) - \frac{\mathbf{t}(n)\mathbf{t}^T(n)}{z(n)}\right\}$$

where $\hat{\mathbf{R}}^{-1}(n)$, being symmetric, does not need to be fully calculated.

It can be easily verified that the LMSN algorithm is a transform-domain LMS algorithm that uses an estimate of the KLT.[20] Let $\mathbf{T} = \hat{\mathbf{Q}}(n)$ be the matrix of the eigenvectors of $\hat{\mathbf{R}}(n)$ and $\Omega = \hat{\Lambda}(n)$ be the diagonal matrix with the eigenvalues of $\hat{\mathbf{R}}(n)$. From Equation 13 we have

$$
\begin{aligned}
\mathbf{T}\overline{\mathbf{w}}(n) = \mathbf{w}(n) &= \mathbf{w}(n-1) + 2\mu e(n)\mathbf{T}\Omega^{-1}\mathbf{T}^T\mathbf{u}(n) \\
&= \mathbf{w}(n-1) + 2\mu e(n)\hat{\mathbf{Q}}(n)\hat{\Lambda}^{-1}(n)\hat{\mathbf{Q}}^T(n)\mathbf{u}(n) \\
&= \mathbf{w}(n-1) + 2\mu e(n)\hat{\mathbf{R}}^{-1}(n)\mathbf{u}(n)
\end{aligned}
$$

Although it can be argued that the gradient direction is noisy, for it is based on the instantaneous squared error rather than on the MSE, the correction applied with a good estimate of $\mathbf{R}^{-1}$ greatly improves the convergence speed when the performance surface is asymmetric. Furthermore, uncorrelated input signals yield symmetric performance surfaces and $\hat{\mathbf{R}}^{-1}(n)$, being diagonal, should not modify the gradient direction.

Parameter α is solely responsible for the memory in estimating $\mathbf{R}$, i.e., it controls how previous input vectors are weighted to form $\hat{\mathbf{R}}(n)$. Its influence on the coefficient adaptation is, therefore, indirect. Control over misadjustment and speed of convergence can be more effectively done through μ. It is interesting to note that for the specific case where $2\mu = \alpha$, the LMSN algorithm is equivalent to the recursive least-squares (RLS) algorithm, which will be discussed in the next section.

6. RECURSIVE LEAST-SQUARES ALGORITHM

The LS method was probably first used by Gauss in the late 18th century,[21] and since then it has been applied in a vast number of areas. In every case the motivation is the same: to estimate the set of parameters that best fits a model to an observed phenomenon. It comes as no surprise that the method should be tried in adaptive filtering as well, because, except from the real-time constraint, the problem is basically parameter fitting. What is really a surprise is that the method suits the application so well and that it can be exactly and efficiently implemented in real time. In addition, schemes for dealing with time-dependent statistics can be easily incorporated through the use of different weights, as was done by Gauss[1] in his work.

Let $\varepsilon(i)$ be defined as the *a posteriori* output error calculated when the output signal of a transversal filter with coefficient vector $\mathbf{w}(n)$ and input-signal vector $\mathbf{u}(i)$ is compared with a reference signal $d(i)$, $i \le n$, i.e.,

$$\varepsilon(i) = d(i) - \mathbf{w}^T(n)\mathbf{u}(i) \qquad \text{(ue18)}$$

A weighted sum of the *a posteriori* errors for $i = 1, \ldots, n$ can be constructed as

$$\mathcal{E}(n) = \sum_{i=1}^{n} \lambda^{n-i}|\varepsilon(i)|^2 \qquad (16)$$

which represents the objective function to be minimized. A *forgetting factor* $\lambda \in (0,1]$ is introduced in order to exponentially reduce the effect of older samples in the function. Usually a constant λ close to unity gives excellent results, although time-dependent values can be used. If we differentiate $E(n)$ with respect to the elements of $\mathbf{w}(n)$ and equate the result to zero, the solution

$$\hat{\mathbf{R}}(n)\mathbf{w}(n) = \hat{\mathbf{p}}(n) \qquad (17)$$

is obtained, where

$$\hat{\mathbf{R}}(n) = \sum_{i=1}^{n} \lambda^{n-i}\mathbf{u}(i)\mathbf{u}^T(i) \tag{18}$$

$$\hat{\mathbf{p}}(n) = \sum_{i=1}^{n} \lambda^{n-i} d(i)\mathbf{u}(i) \tag{19}$$

are the sample averages of the autocorrelation matrix and the crosscorrelation vector between the input signal and the reference signal, respectively. Equation 17 is sometimes referred to as the deterministic counterpart of the Wiener–Hopf equation.[7]

As λ controls the amount of memory of the algorithm, $\lambda = 1$ corresponds to the case where all the samples are equally weighted. Although in this case of infinite memory the algorithm cannot track variations in the environment, it can be shown that the excess MSE reduces asymptotically to zero as n goes to infinity.

A recursive solution of Equation 17, which is based on previously obtained values of $\mathbf{w}(n)$ and $\hat{\mathbf{R}}(n)$, can be derived by simple algebraic manipulation of Equations 18 and 19. From Equation 18 it can be verified that

$$\hat{\mathbf{R}}(n) = \lambda\hat{\mathbf{R}}(n-1) + \mathbf{u}(n)\mathbf{u}^T(n) \tag{20}$$

Similarly, from Equation 19

$$\hat{\mathbf{p}}(n) = \lambda\hat{\mathbf{p}}(n-1) + d(n)\mathbf{u}(n) \tag{21}$$

Combining Equations 17, 20, and 21, we have

$$\begin{aligned}
\hat{\mathbf{R}}(n)\mathbf{w}(n) &= \lambda\hat{\mathbf{p}}(n-1) + d(n)\mathbf{u}(n) \\
&= \lambda\hat{\mathbf{R}}(n-1)\mathbf{w}(n-1) + d(n)\mathbf{u}(n) \\
&= \left[\hat{\mathbf{R}}(n) - \mathbf{u}(n)\mathbf{u}^T(n)\right]\mathbf{w}(n-1) + d(n)\mathbf{u}(n) \\
&= \hat{\mathbf{R}}(n)\mathbf{w}(n-1) + \mathbf{u}(n)\left[d(n) - \mathbf{u}^T(n)\mathbf{w}(n-1)\right] \\
&= \hat{\mathbf{R}}(n)\mathbf{w}(n-1) + e(n)\mathbf{u}(n)
\end{aligned} \tag{22}$$

where the *a priori* output error, $e(n)$, is defined as in Equation 7. Assuming that $\hat{\mathbf{R}}(n)$ is invertible, then from Equation 22, we have

$$\mathbf{w}(n) = \mathbf{w}(n-1) + e(n)\hat{\mathbf{R}}^{-1}(n)\mathbf{u}(n)$$

The matrix inversion lemma[7] can now be applied to $\hat{\mathbf{R}}(n)$ to yield

$$\hat{\mathbf{R}}^{-1}(n) = \frac{1}{\lambda}\left\{\hat{\mathbf{R}}^{-1}(n-1) - \frac{\hat{\mathbf{R}}^{-1}(n-1)\mathbf{u}(n)\mathbf{u}^T(n)\hat{\mathbf{R}}^{-1}(n-1)}{\lambda + \mathbf{u}^T(n)\hat{\mathbf{R}}^{-1}(n-1)\mathbf{u}(n)}\right\} \tag{23}$$

where $\hat{\mathbf{R}}^{-1}(0)$ must be a positive definite matrix. As for the LMSN algorithm, for $\lambda < 1$ the introduction of $\hat{\mathbf{R}}^{-1}(0)$ has a decreasing effect on Equation 16, which becomes

$$\mathcal{E}(n) = \sum_{i=1}^{n} \lambda^{n-i} |\varepsilon(i)|^2 + \lambda^n \mathbf{w}^T(n)\hat{\mathbf{R}}(0)\mathbf{w}(n) \tag{24}$$

The advantages of a recursive algorithm as described by the above equations are well worth the negligible effects introduced by $\hat{\mathbf{R}}^{-1}(0)$.

Using the recursive relation

$$\hat{\mathbf{R}}^{-1}(n)\mathbf{u}(n) = \frac{\hat{\mathbf{R}}^{-1}(n-1)\mathbf{u}(n)}{\lambda + \mathbf{u}^T(n)\hat{\mathbf{R}}^{-1}(n-1)\mathbf{u}(n)}$$

the RLS algorithm can be implemented as

$$\mathbf{t}(n) = \hat{\mathbf{R}}^{-1}(n-1)\mathbf{u}(n)$$

$$z(n) = \lambda + \mathbf{u}^T(n)\mathbf{t}(n)$$

$$\mathbf{w}(n) = \mathbf{w}(n-1) + \frac{e(n)\mathbf{t}(n)}{z(n)}$$

$$\hat{\mathbf{R}}^{-1}(n) = \frac{1}{\lambda}\left\{\hat{\mathbf{R}}^{-1}(n-1) - \frac{\mathbf{t}(n)\mathbf{t}^T(n)}{z(n)}\right\}$$

The influence of the forgetting factor λ on the misadjustment of the RLS algorithm is similar to that of μ and α for the LMSN algorithm. As $\lambda \to 1$, the algorithm is given more memory, which translates into a better prformance in stationary environments. On the other hand, better tracking of nonstationarities is achieved for smaller values of λ. Similarities between the LMSN and RLS algorithms go even further, since it can be easily shown that for $2\mu = \alpha = 1 - \lambda$ the LMSN algorithm minimizes a scaled version of Equation 24. Under these conditions both algorithms should have similar performance. Another interesting parallel can be established between the RLS algorithm and the Kalman filter. The RLS algorithm can be viewed as a special case of the Kalman filter, if the correspondences given in Table 1 are assumed.[7] As compared with the LMS algorithm, the RLS algorithm presents a much better convergence rate when the input signal is correlated. Therefore, improvement in convergence speed is seldom necessary for the RLS algorithm. It is much more important to reduce the number of operations per iteration and assure robustness.

6.1 FINITE-PRECISION EFFECTS

One aspect that has not been considered in the previous sections is the effect of finite precision in the implementation of the various algorithms on digital signal processors.

For the LMS algorithm, quantization effects compromise the algorithm performance through a higher misadjustment. As machine precision decreases, the influence of the excess MSE due to quantization becomes more noticeable as a third source of errors in the solution. In addition, very small values of μ can cause the adaptation of the coefficients to stop prior to convergence. This phenomenon is especially common when fixed-precision representation is used.[22] For all these reasons the choice of convergence factor becomes even more difficult.

For the LMSN and RLS algorithms, quantization errors introduced by fixed- or floating-point arithmetic can, in addition to the effects described above, lead to divergence of the coefficients,

Table 1 Correspondence between the RLS Algorithm and the Kalman Filter

Kalman Filter	RLS Algorithm
Observation	$d(n)$
State transition matrix	I
Measurement matrix	$\mathbf{u}^T(n)$
Estimated state vector	$\mathbf{w}(n)$
Innovation	$e(n)$
Filtered state-error correlation matrix	$\lambda\, \hat{\mathbf{R}}^{-1}(n)$
Predicted state-error correlation matrix	$\hat{\mathbf{R}}^{-1}(n)$
Kalman gain	$\mathbf{t}(n)/z(n)$
Estimated process-noise correlation matrix	$(\lambda^{-1} - 1)\lambda\hat{\mathbf{R}}^{-1}(n)$
Estimated measurement-noise correlation matrix	1

which is unacceptable in many applications whether or not the algorithm can recover. Several researchers have attempted to identify the causes of algorithm divergence and to propose solutions. Ljung and Ljung[23] have shown that the RLS algorithm with $\lambda < 1$ is exponentially stable, since a single error introduced at time instant n_o will decay exponentially as time progresses. Notwithstanding this important contribution, it was later verified[24] that interaction and accumulation of errors can destroy the positive definiteness of matrix $\hat{\mathbf{R}}^{-1}(n)$, which can result in divergence. Maintaining the symmetry of $\hat{\mathbf{R}}^{-1}(n)$ is essential to improve robustness; however, in many situations this may not be sufficient. In these cases, rescue procedures, such as adding white noise to the input signal or reinitializing $\hat{\mathbf{R}}^{-1}(n)$, must be used.[25] One possibility is to let $\hat{\mathbf{R}}^{-1}(n) = \gamma\,\mathbf{I}$, where $\mathbf{I}$ is the identity matrix and γ is a positive constant.

Since divergence of the RLS algorithm is usually related to the lack of positive definiteness in matrix $\hat{\mathbf{R}}^{-1}(n)$, more-robust methods for solving the system of linear equations described in Equation 17 can be employed to achieve better results. Given the similarities among the equations that describe the LMSN and RLS algorithms, it is not difficult to find the set of linear equations which is minimized by the LMSN as well. The methods that are applicable to the RLS algorithm can usually be extended to the LMSN algorithm with some effort, if desired.

6.2 QR-RLS ALGORITHM

An RLS algorithm using the QR decomposition, known as the QR-RLS algorithm, is one of the most popular examples of a technique which has better numerical properties than the conventional RLS algorithm and yet solves exactly the same set of equations. The orthogonal decomposition is also a fundamental tool in the development and understanding of systolic array structures,[7] which are attractive for VLSI implementation. Combined, these properties make the study of the QR-RLS algorithm important and enlightening.

The structure of $\hat{\mathbf{R}}(n)$ given in Equation 20 leads to an elegant and relatively simple method for solving Equation 17 using the inverse of $\hat{\mathbf{R}}(n)$. In consequence, it is tempting to adopt the RLS algorithm in its conventional form. Despite its simplicity, the subtraction between the two non-negative definite matrices that comes naturally in Equation 23 is a potential hazard in terms of quantization errors. We must not forget, though, that the RLS algorithm is just one method for solving M simultaneous linear equations. If machine precision makes the inversion of an ill-conditioned $\hat{\mathbf{R}}(n)$ matrix inaccurate, other methods that have already been used with good results in off-line linear-algebra problems might be more appropriate. One method that has been known for its excellent numerical properties is the QR-decomposition method, which consists of modifying the problem such that the solution can be obtained through the inversion of a triangular matrix. This is possible since every $n \times M$ matrix $\mathbf{A}$ with linearly independent columns can be factored into $\mathbf{A} = \mathbf{Q}^T\mathbf{R}$, where $\mathbf{Q} \in \mathrm{IR}^{n \times n}$ is orthogonal and $\mathbf{R} \in \mathrm{IR}^{n \times M}$ is upper triangular.[26] No confusion should arise at this point between the QR nomenclature and the spectral decomposition of the autocorrelation matrix mentioned above. For the sake of consistency with the references, let $\mathbf{Q}(n)$ and $\mathbf{R}(n)$ be the QR-decomposition matrices of a data matrix $\mathbf{A}(n)$.

Now let $\mathbf{A}(n)$ be the weighted input-data matrix defined as

$$\mathbf{A}(n) = \begin{bmatrix} \lambda^{(n-1)/2} & 0 & \cdots & 0 \\ 0 & \ddots & & \vdots \\ \vdots & & \lambda^{1/2} & 0 \\ 0 & \cdots & 0 & 1 \end{bmatrix} \begin{bmatrix} u(1) & 0 & \cdots & 0 \\ u(2) & u(1) & & 0 \\ \vdots & \vdots & & \vdots \\ u(n) & u(n-1) & \cdots & u(n-M+1) \end{bmatrix}$$

We can easily verify that $\mathbf{A}^T(n)\mathbf{A}(n)$ is the autocorrelation matrix defined in Equation 18 and the requirement of linear independence of the columns of $\mathbf{A}(n)$ can be translated as the necessary condition for the positive definiteness of Equation 18 for a unique solution of the LS problems. Furthermore, let the weighted reference-data vector be defined as

$$\mathbf{d}(n) = \begin{bmatrix} \lambda^{(n-1)/2} & 0 & \cdots & 0 \\ 0 & \ddots & & \vdots \\ \vdots & & \lambda^{1/2} & 0 \\ 0 & \cdots & 0 & 1 \end{bmatrix} \begin{bmatrix} d(1) \\ \vdots \\ d(n-1) \\ d(n) \end{bmatrix}$$

The objective function to be minimized, Equation 16, can be rewritten as

$$\mathcal{E}(n) = \|\varepsilon(n)\|_2^2$$

where

$$\varepsilon(n) = \mathbf{d}(n) - \mathbf{A}(n)\mathbf{w}(n)$$

is the weighted *a posteriori* error vector. Since the two-norm is invariant under orthogonal transformations,[26]

$$\mathcal{E}(n) = \|\mathbf{Q}(n)\varepsilon(n)\|_2^2$$

can be minimized with respect to the elements of $\mathbf{w}(n)$ if $\mathbf{Q}(n)$ is orthogonal. Therefore, by factoring the weighted input-data matrix $\mathbf{A}(n)$ as

$$\mathbf{Q}(n)\mathbf{A}(n) = \begin{bmatrix} \mathbf{R}(n) \\ \mathbf{0} \end{bmatrix}$$

where $\mathbf{R}(n)$ is an $M \times M$ upper triangular matrix, the minimization problem becomes

$$\begin{aligned} \mathcal{E}_{\min}(n) &= \min\|\mathbf{Q}(n)\mathbf{d}(n) - \mathbf{Q}(n)\mathbf{A}(n)\mathbf{w}(n)\|_2^2 \\ &= \min\left\| \begin{bmatrix} \mathbf{p}(n) \\ \mathbf{q}(n) \end{bmatrix} - \begin{bmatrix} \mathbf{R}(n) \\ \mathbf{0} \end{bmatrix} \mathbf{w}(n) \right\|_2^2 \end{aligned}$$

where

$$\begin{bmatrix} \mathbf{p}(n) \\ \mathbf{q}(n) \end{bmatrix} = \mathbf{Q}(n)\mathbf{d}(n)$$

The solution is obtained for the vector $\mathbf{w}(n)$ satisfying

$$\mathbf{p}(n) = \mathbf{R}(n)\mathbf{w}(n) \tag{25}$$

for which $\mathcal{E}_{min}(n) = \|\mathbf{q}(n)\|_2^2$. For convenience, we have used $\mathbf{R}(n)$ and $\mathbf{p}(n)$ in Equation 25, although they do not denote the autocorrelation matrix of the input signal and the crosscorrelation vector between the input-signal vector and the reference signal. Actually, $\mathbf{R}(n)$, being upper triangular, could never denote an autocorrelation matrix. The solution $\mathbf{w}(n)$ can be easily obtained by applying back substitution to Equation 25.[26]

The next step toward a QR-RLS adaptation algorithm is to find a recursive procedure that can produce the solution with roughly the computational complexity of the RLS algorithm, i.e., $O(M^2)$.

Suppose that at instant n the solution obtained at instant $n - 1$ is available in the form of matrix $\mathbf{R}(n-1)$ and vector $\mathbf{p}(n-1)$. The updated weighted input-data matrix at instant n can be constructed from the previous one as

$$\mathbf{A}(n) = \begin{bmatrix} \lambda^{1/2}\mathbf{A}(n-1) \\ \mathbf{u}^T(n) \end{bmatrix}$$

Now let $\mathbf{Q}(n|n-1)$ be the order-updated orthogonal matrix constructed from $\mathbf{Q}(n-1)$ as

$$\mathbf{Q}(n \mid n-1) = \begin{bmatrix} \mathbf{Q}(n-1) & \mathbf{0} \\ \mathbf{0}^T & 1 \end{bmatrix}$$

By applying $\mathbf{Q}(n|n-1)$ to $\mathbf{A}(n)$, we have

$$\mathbf{Q}(n \mid n-1)\mathbf{A}(n) = \begin{bmatrix} \lambda^{1/2}\mathbf{Q}(n-1)\mathbf{A}(n-1) \\ \mathbf{u}^T(n) \end{bmatrix} = \begin{bmatrix} \lambda^{1/2}\mathbf{R}(n-1) \\ \mathbf{0} \\ \mathbf{u}^T(n) \end{bmatrix} \tag{26}$$

Similarly,

$$\mathbf{d}(n) = \begin{bmatrix} \lambda^{1/2}\mathbf{d}(n-1) \\ d(n) \end{bmatrix}$$

which leads to

$$\mathbf{Q}(n \mid n-1)\mathbf{d}(n) = \begin{bmatrix} \lambda^{1/2}\mathbf{p}(n-1) \\ \lambda^{1/2}\mathbf{q}(n-1) \\ d(n) \end{bmatrix} \tag{27}$$

From Equations 26 and 27 it is clear that to obtain $\mathbf{R}(n)$ and $\mathbf{p}(n)$ from their previous values we must find a transformation matrix $\mathbf{G}(n)$ that makes $\mathbf{Q}(n|n-1)\ \mathbf{A}(n)$ upper triangular, which would force the elements of the last row of Equation 26 to be zeros. The resulting upper triangular matrix is the updated matrix $\mathbf{R}(n)$ in the relation

$$\mathbf{G}(n)\mathbf{Q}(n \mid n-1)\mathbf{A}(n) = \mathbf{Q}(n)\mathbf{A}(n) = \begin{bmatrix} \mathbf{R}(n) \\ \mathbf{0} \end{bmatrix}$$

The vector $\mathbf{p}(n)$ is similarly updated as

$$\mathbf{Q}(n)\mathbf{d}(n) = \begin{bmatrix} \mathbf{p}(n) \\ \mathbf{q}(n) \end{bmatrix}$$

The task of finding $\mathbf{G}(n)$ is much simpler than that of finding $\mathbf{Q}(n)$ since $\mathbf{G}(n)$ need only cancel the last line of Equation 26. This can be done by M successive Givens rotations,[7] each of the form

$$\mathbf{G}_k(n) = \begin{bmatrix} 1 & 0 & \cdots & 0 & 0 & & \cdots & & 0 \\ 0 & 1 & \cdots & 0 & 0 & & \cdots & & 0 \\ \vdots & & \ddots & 0 & 0 & & & & \vdots \\ 0 & & \cdots & 1 & 0 & & \cdots & & 0 \\ 0 & & \cdots & & c_k & 0 & \cdots & 0 & s_k \\ 0 & & \cdots & & 0 & 1 & \cdots & 0 & 0 \\ \vdots & & \cdots & & \vdots & & \ddots & & \vdots \\ 0 & & \cdots & & 0 & & \cdots & 0 & 1 \\ 0 & & \cdots & & -s_k & 0 & \cdots & 0 & c_k \end{bmatrix}; \quad k = 1, \ldots, M$$

where c_k and s_k denote cosines and sines, calculated as

$$c_k = \frac{\beta_{kk}}{\sqrt{\beta_{kk}^2 + \beta_{nk}^2}}$$

$$s_k = \frac{\beta_{nk}}{\sqrt{\beta_{kk}^2 + \beta_{nk}^2}}$$

and β_{kk} is the (k, k)-element of matrix $\mathbf{G}_{k-1}(n) \ldots \mathbf{G}_1(n)\mathbf{Q}(n|n-1)\mathbf{A}(n)$ which has been modified by the previous rotations. Since $\mathbf{G}_k(n)$, $k = 1, \ldots, M$, is very sparse, only the cosines and sines must be calculated. It must be stressed here that although some of the matrices involved grow in size with n, only $M \times 1$ vectors and $M \times M$ matrices need to be stored and it is not necessary to compute $\mathbf{q}(n)$.

The last step in the development of the algorithm is to provide an initialization method for $n \le M$. In this case, an exact procedure can be developed such that $\mathcal{E}_{\min}(n) = 0$. The idea is to update $\mathbf{R}(n)$ and $\mathbf{p}(n)$ both in order and in time such that $\mathbf{d}(n) = \mathbf{A}(n)\mathbf{w}(n)$ is always a vector of zeros, starting with $\mathbf{R}(0) = 0$ and $\mathbf{p}(0) = 0$. Although unnecessary for the purposes of the algorithm, it can be easily shown that during this initialization period $\|\mathbf{q}(n)\|_2^2 = 0$.

A summary of the QR-RLS algorithm is given in Table 2.

6.3 FAST IMPLEMENTATIONS OF THE RLS ALGORITHM

As a final note on alternative implementations of the RLS algorithm, we briefly mention certain implementations that are computationally more efficient, requiring only $O(M)$ multiplications per iteration. The diversity of these implementations, individual properties, and origins are enough to

Table 2 QR-RLS Algorithm

Available at time instant n:

$\mathbf{u}(n) = [u(n)\ u(n-1)\ \ldots\ u(n-M+1)]^T \quad d(n)$

Initialize:

$\mathbf{R}(0) = \mathbf{0}_{(M\times M)};\quad \mathbf{p}(0) = \mathbf{0}_{(M\times 1)};\quad \mathbf{x} = \mathbf{0}_{(M\times 1)};\quad z = 0;\ n = 1, 2 \ldots$

{

if ($n <= M$) $N = n$; /* during exact init. */

else $N = M$;

$z = d(n);\quad \mathbf{x} = \mathbf{u}(n);$

for ($k = 1$; $k \le N$; k ++)

{

$$c = \frac{\lambda^{1/2}\mathbf{R}_{kk}(n-1)}{\sqrt{\lambda\mathbf{R}_{kk}^2(n-1) + \mathbf{x}_k^2}};$$

$$s = \frac{\mathbf{x}_k}{\sqrt{\lambda\mathbf{R}_{kk}^2(n-1) + \mathbf{x}_k^2}};$$

for ($j = k$; $j \le N$; j++)

{

$\mathbf{R}_{kj}(n) = c\lambda^{1/2}\mathbf{R}_{kj}(n-1) + s\mathbf{x}_j;$

$\mathbf{x}_j = -s\lambda^{1/2}\mathbf{R}_{kj}(n-1) + c\mathbf{x}_j;$

}

$\mathbf{p}_k(n) = c\lambda^{1/2}\mathbf{p}_k(n-1) + sz;$

$z = -s\lambda^{1/2}\mathbf{p}_k(n-1) + cz;$

}

for ($k = N$; $k >= 1$; k– –)

{

$\mathbf{w}_k(n) = \mathbf{p}_k(n)$

for ($j = k + 1$; $j <= N$; j+ +)

$\mathbf{w}_k(n) = \mathbf{w}_k(n-1) - \mathbf{R}_{kj}\mathbf{w}_j(n);$

$\mathbf{w}_k(n) = \mathbf{w}_k(n)/\mathbf{R}_{kk}(n);$

}

}

set them apart in a new class, called *fast* RLS algorithms in the literature. The drastic reduction in the number of multiplications in these algorithms can be achieved by updating the gain vector

$$\mathbf{k}(n) = \hat{\mathbf{R}}^{-1}(n)\mathbf{u}(n)$$

instead of updating $\hat{\mathbf{R}}^{-1}(n)$, using $\mathbf{k}(n-1)$ without the need of matrix multiplications. For detailed derivations of some of these algorithms, the reader is referred to References 7 and 27 through 30.

7. DISCUSSION

Numerous structures, algorithms, and implementations have been proposed in recent years for the different applications of adaptive filters, each with its own qualities and drawbacks. The choice of the best scheme for a given application must, therefore, be subjected to careful scrutiny. The LMS algorithm has been one of the most widely used, despite its slow convergence in some cases, largely because of its simplicity and robustness. When compared with filters based on the LMS algorithm, transform-domain adaptive filters can show improvement in convergence speed at the expense of a relatively small increase in computational complexity. They have gained popularity especially for processing voice in acoustic echo cancellation. With the development of more-robust and computationally efficient techniques and further advancements in VLSI technology, algorithms that employ second-order information are becoming more and more attractive, since they offer improved performance in terms of convergence speed and misadjustment.

REFERENCES

1. **K. F. Gauss,** *Theory of the Motion of the Heavenly Bodies Moving about the Sun in Conic Sections — A Translation of Theoria Motus.* New York: Dover, 1963.
2. **J. J. Shynk,** Adaptive IIR filtering, *IEEE Signal Process. Mag.* 6, 4–21, April 1989.
3. **S. L. Netto, P. S. R. Diniz, and P. Agathoklis,** Adaptive IIR filtering algorithms for system identification: a general framework, *IEEE Trans. Educ.* 38, 54–66, Feb. 1995.
4. **M. M. Sondhi and D. A. Berkley,** Silencing echoes on the telephone network, *Proc. IEEE,* 68, 948–963, Aug. 1980.
5. **E. Hansler,** The hands-free telephone problem — an annotated bibliography, *Signal Process.* 27, 259–271, June 1992.
6. **S. U. H. Qureshi,** Adaptive equalization, *Proc. IEEE,* 73, 1349–1385, Sept. 1985.
7. **S. Haykin,** *Adaptive Filter Theory,* 2nd ed., Englewood Cliffs, N.J., Prentice-Hall, 1991.
8. **T. Kailath,** A view of three decades of linear filtering theory, *IEEE Trans. Inf. Theor.,* IT-20, 145–181, Mar. 1974.
9. **A. Papoulis,** *Probability, Random Variables, and Stochastic Processes,* 3rd ed., New York: McGraw-Hill, 1991.
10. **B. Widrow and M. E. Hoff, Jr.,** Adaptive switching circuits, in *IRE WESCON Conv. Rec.,* 4, 96–104, 1960.
11. **B. Widrow and S. D. Stearns,** *Adaptive Signal Processing.* Englewood Cliffs, NJ: Prentice-Hall, 1985.
12. **B. Widrow, J. M. McCool, M. G. Larimore, and C. R. Johnson, Jr.,** Stationary and nonstationary learning characteristics of the LMS adaptive filter, *Proc. IEEE,* 64, 1151–1162, Aug. 1976.
13. **B. Widrow and E. Walach,** On the statistical efficiency of the LMS algorithm with nonstationary inputs, *IEEE Trans. Inf. Theor.,* IT-30, 211–221, Mar. 1984.
14. **F. F. Yassa,** Optimality in the choice of the convergence factor for gradient-based adaptive algorithms, *IEEE Trans. Acoust. Speech Signal Process.,* ASSP-35, 48–59, Jan. 1987.
15. **D. T. M. Slock,** On the convergence behaviour of the LMS and the normalized LMS algorithms, *IEEE Trans. Signal Process.,* 41, 2811–2825, Sept. 1993.
16. **B. Farhang-Boroujeny and S. Gazor,** Selection of orthonormal transforms for improving the performance of the transform domain normalized LMS algorithm, *IEEE Proc. Part F,* 139, 327–335, Oct. 1992.
17. **D. F. Marshall, W. K. Jenkins, and J. J. Murphy,** The use of orthogonal transforms for improving performance of adaptive filters, *IEEE Trans. Circuits Syst.,* 36, 474–483, Apr. 1989.
18. **H. Robbins and S. Monro,** A stochastic approximation method, *Ann. Math. Stat.,* 22, 400–407, 1951.
19. **G. C. Goodwin and S. K. Sin,** *Adaptive Filtering Prediction and Control.* Englewood Cliffs, NJ: Prentice-Hall, 1984.
20. **S. S. Reddi,** A time-domain adaptive algorithm for rapid convergence, *Proc. IEEE,* 72, 533–535, Apr. 1984.
21. **H. W. Sorenson,** Least-squares estimation: from Gauss to Kalman, *IEEE Spectrum,* 63–68, July 1970.
22. **C. Caraiscus and B. Liu,** A roundoff error analysis of the LMS adaptive algorithm, *IEEE Trans. Acoust. Speech Signal Process.,* ASSP-32, 34–41, Feb. 1984.
23. **S. Ljung and L. Ljung,** Error propagation properties of recursive least-squares adaptation algorithms, *Automatica,* 21,(2), 157–167, 1985.
24. **G. E. Bottomley and S. T. Alexander,** A novel approach for stabilizing recursive least squares filters, *IEEE Trans. Signal Process.,* 39, 1770–1779, Aug. 1991.
25. **J. M. Cioffi and T. Kailath,** Fast, recursive-least-squares transversal filters for adaptive filtering, *IEEE Trans. Acoust. Speech Signal Process.,* ASSP-32, 304–337, Apr. 1984.
26. **G. Golub and C. F. Van Loan,** *Matrix Computations,* 2nd ed., Baltimore: The Johns Hopkins University Press, 1989.
27. **D. D. Falconer and L. Ljung,** Application of fast Kalman estimation to adaptive equalization, *IEEE Trans. Commun.* COM-26, 1439–1446, Oct. 1978.
28. **G. Carayannis, D. G. Manolakis, and N. Kalouptsidis,** A fast sequential algorithm for least-squares filtering and prediction, *IEEE Trans. Acoust. Speech Signal Process.,* ASSP-31, 1394–1402, Dec. 1983.
29. **D. T. M. Slock and T. Kailath,** Numerically stable fast transversal filters for recursive least squares adaptive filtering, *IEEE Trans. Signal Process.,* 39, 92–114, Jan. 1991.
30. **S. T. Alexander,** Fast adaptive filters: a geometrical approach, *IEEE Signal Process. Mag.,* 3, 18–28, Oct. 1986.

Round-Off Errors and Limit Cycles in Digital Filters

P. Agathoklis and Ayman Tawfik

CONTENTS

1. INTRODUCTION

There are many different possibilities to implement digital filters depending on the speed and hardware requirements of the particular application. Such possibilities include software on a general-purpose computer, DSP chips either using fixed-point or floating-point arithmetic, or hardware implementations using VLSI and FPGAs. All these implementations use finite word lengths (FWL) which generate what is called FWL effects. The severity of these effects, however, can vary significantly. For applications which do not require high speeds, such as non-real-time applications, long word lengths can be used so that the resulting FWL effects are negligible. At the other end, applications where speed and/or hardware complexity are important factors, long word lengths are not always possible and the FWL effects need to be considered carefully. Between these two extremes there are many other cases where it is important to know that the filter implementation will perform as expected after implementation. Techniques to analyze and predict the performance of a digital filter implementation are therefore very useful. In a typical application, the filters to be implemented are obtained using design specifications implied from the particular application. Obviously, there are many different possible designs which satisfy these specifications, and these filters do not perform the same under FWL constraints. In fact, their performances can vary

0-8493-8951-8/97/$0.00+$.50
© 1997 by CRC Press LLC

significantly even for medium (16 bits) word lengths. Such differences come from changes in the frequency response of the filter due to the finite word length of the parameters or from a decrease in the signal-to-noise ratio in the output due to noise generated from rounding or truncation during arithmetic operations. Further, there is the possibility that the output of the filter exhibits limit cycles due to overflow and becomes independent of the input signal.

In this chapter a brief overview of the FWL effects in digital filter implementations is given. The fixed-point and floating-point number representations, the two representations used in most DSP chips, are briefly presented in Section 2. The finite word-lengths effects in fixed-point implementations are discussed in Section 3, and some of the approaches in the analysis of quantization errors in fixed-point arithmetic implementations are presented in Section 4. Finally in Section 5 some brief comments on quantization errors in floating-point implementations are included.

2. FIXED-POINT AND FLOATING-POINT ARITHMETIC

Numbers are stored in binary words, i.e., strings of bits containing 0s or 1s. In fixed-point representation a number is represented by $s\beta_{n-1} \ldots \beta_1\beta_{0\Delta}\beta_{-1}\beta_{-2} \ldots \beta_{-b}$, where s is the sign bit indicating whether this is a positive number ($s = 0$) or negative ($s = 1$), and Δ, the binary point, separates the integer part from the fraction part. The word length of such a representation is given by $n + b + 1$. For example, the six-bit binary number $0101_\Delta\ 11$ is equal to 5.75 in the decimal number system. The location of the binary point is fixed for an implementation, and, depending on its location, the numbers represented can be signed integers, signed fractions, or a combination of both. If all except the sign bit are on the right of Δ, the number represented is a fraction; if all the bits are on the left of Δ, the number represented is an integer. Integers are usually represented without showing explicitly the binary point. Because of the FWL, the numbers which can be represented have a finite range of values, called the *dynamic range* of the number representation. For example, the dynamic range for all four-bit positive integers ($n = 4$, $b = 0$, $s = 0$) is $0 \le x \le 15$. If the result of an arithmetic operation exceeds the dynamic range, it cannot be represented, and this is called *overflow*. An *overflow* occurs when the result is within the dynamic range, but cannot be exactly represented with the available precision.

There are three standard fixed-point formats; sign magnitude, one's complement, and two's complement. The representation of positive numbers ($s = 0$) is the same in all three formats. The difference consists in the representation of negative numbers. Consider a signed fraction representation as shown in Figure 1. For sign magnitude, the decimal value of the number is given by

$$x = (-1)^s \sum_{i=1}^{b} \beta_{-i} 2^{-i} \tag{1}$$

For one's complement the negative number is given by

$$x = -s\left(1 - 2^{-b}\right) + \sum_{i=1}^{b} \beta_i 2^{-i} \tag{2}$$

For two's complement representation, the negative number can be given by

$$x = -s + \sum_{i=1}^{b} \beta_i 2^{-i} \tag{3}$$

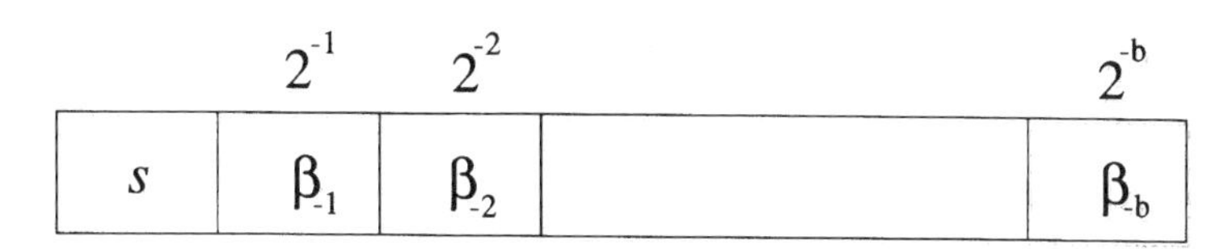

Figure 1

Each of these three representations has its advantages and disadvantages in the implementation of arithmetic operations. The one most commonly used in digital filter implementations is the two's complement.

When two fixed-point numbers are added, the result may exceed the dynamic range of the implementation; i.e., addition of fixed-point numbers may cause overflow. When two fixed-point numbers are multiplied, the result, in general, is twice the initial word length. If both numbers have an integer part, the results may overflow. If both numbers are fixed-point fractions, the result remains a fraction and cannot overflow. It is, therefore, preferable in a fixed-point implementation to scale all numbers so that they are fractions. Multiplication of two fixed-point fractions, even though free of overflow, does lead to underflow. In IIR digital filter implementations, a word-length reduction is necessary to prevent the word lengths of the signals from increasing. This reduction of the word length is called *signal quantization.*

A floating-point representation of a decimal number x is of the form $x\ M 2^c$, where c is called the *exponent* and M is the *mantissa*. The mantissa is a b-bit fixed-point number in the range $0.5 \leq M < 1$. Both M and c are stored in fixed-point form. Multiplication of $x_1 = M_1 2^{c_2}$ and $x_2 = M_2 2^{c_2}$ in floating-point form leads to $x = M_1 M_2 2^{c_1 + c_2}$ involving the multiplication of the mantissa parts and addition of the exponent parts. In practice, the resulting exponent is readjusted so that the resulting mantissa is brought back to the range $0.5 \leq M < 1$. Addition of x_1 and x_2 is a little more complicated. First, the exponent and mantissa of the number with the smaller magnitude are adjusted until the exponents match and then the mantissas can be added. Finally, the resulting representation is rescaled so that the mantissa is in the range [0.5,1). Floating-point multiplication and addition require more-complicated hardware and result in slower computational speed than fixed-point multiplication and addition. The floating-point arithmetic units found in most floating-point DSP chips are designed based on the IEEE standard.

In summary, fixed-point fractional implementations have the advantage of simplicity. Addition does not need quantization, but multiplication does and this produces round-off errors. Addition can lead to overflow but fractional multiplication cannot. Floating-point implementations, on the other hand, are more complicated, but permit a significantly larger dynamic range and overflow can be easily avoided. Both addition and multiplication need quantization and, therefore, produce round-off errors. For non-real-time applications on general-purpose computers, floting-point arithmetic is preferred since neither the cost of hardware nor the processing speed is a significant factor. For real-time applications, where speed and complexity of the dedicated hardware are significant factors, fixed-point arithmetic is preferred.

3. FWL EFFECTS IN FIXED-POINT IMPLEMENTATIONS

Digital filter implementations using FWL can suffer from some undesirable effects related to the finite accuracy of the implementation. For most implementations in general-purpose computers with long word lengths, these effects are negligible. For real-time applications, however, where the requirements on computation speed and hardware complexity imply short word lengths, these effects can have a significant impact in the performance of a filter implementation. There are three FWL effects in fixed-point IIR digital filters implementations. These effects are

1. Changes in the input/output descriptionof the filter due to finite accuracy of the filter coefficients representation;
2. Round-off noise and quantization limit cycles caused by the quantization of the signals within the filter realization;
3. Limit cycles due to overflow.

These FWL effects will be briefly described in the following subsections. Another source of quantization noise is the A/D conversion of the input signal when analog signals are processed using digital filters. These quantization effects, however, do not depend on the filter implementation and thus are not discussed here.

3.1 COEFFICIENT QUANTIZATION

The coefficients of a digital filter implementation are the rounded or truncated filter coefficients obtained during design. The quantization of the filter coefficients is manifested by a change in the input/output (I/O) characteristic of the filter transfer function $H(z)$. There are several techniques to characterize the sensitivity of the filter to coefficient quantization errors, some deterministic and some using statistical means. One popular approach is based on examining the movements of poles and zeros caused by coefficient quantization.[2] Consider the transfer function of an IIR filter implementation

$$\tilde{H}(z) = \frac{\sum_{k=0}^{M} \tilde{b}_k z^{-k}}{1 + \sum_{k=1}^{N} \tilde{a}_k z^{-k}} = \frac{\prod_{k=1}^{M} \left(1 - \tilde{z}_k z^{-1}\right)}{\prod_{k=1}^{N} \left(1 - \tilde{p}_k z^{-1}\right)} \tag{4}$$

where $\tilde{b}_k$ and $\tilde{a}_k$ are the quantized coefficients. $\tilde{p}_k$ and $\tilde{z}_k$ are the poles and zeros of $\tilde{H}(z)$, respectively. b_k, a_k, p_k, z_k, are the corresponding variables of $H(z)$ before quantization. The relation between the perturbation of the poles Δp_k due to the coefficient quantization error Δa_k has been shown to be

$$\Delta p_i = -\sum_{k=1}^{N} \frac{p_i^{N-k}}{\prod_{\substack{m=1 \\ m \neq i}}^{N} \left(p_i - p_m\right)} \Delta a_k \tag{5}$$

This expression provides a measure for the sensitivity of the ith pole to change in the coefficients Δa_k. If the poles are close (e.g., narrowband filters), the magnitude of the factors in the denominator will be small, and, hence, a small change in the coefficients Δa_k will result in a large perturbation of the poles. A similar result can be obtained for the sensitivity of the zeros to errors Δb_k. The main conclusion of this analysis is that, if the poles are close together, as is the case in narrow band filters, a small change in the denominator coefficients of $H(z)$ can cause a large change in the location of the poles (a similar argument can be made for the zeros of the filter). This may lead to a frequency response for the filter implementation which differs significantly from the response of the original design. Finding filter structures which are not sensitive to coefficient quantization errors and thus can be implemented with short coefficient word lengths is an active area of research, and an extensive literature on this subject has appeared. There are many different approaches to this problem, and many low-sensitivity structures have been proposed so far. One interesting property of these low-sensitivity structures is that they tend to have a reasonably good behavior with respect to the other FWL effects.

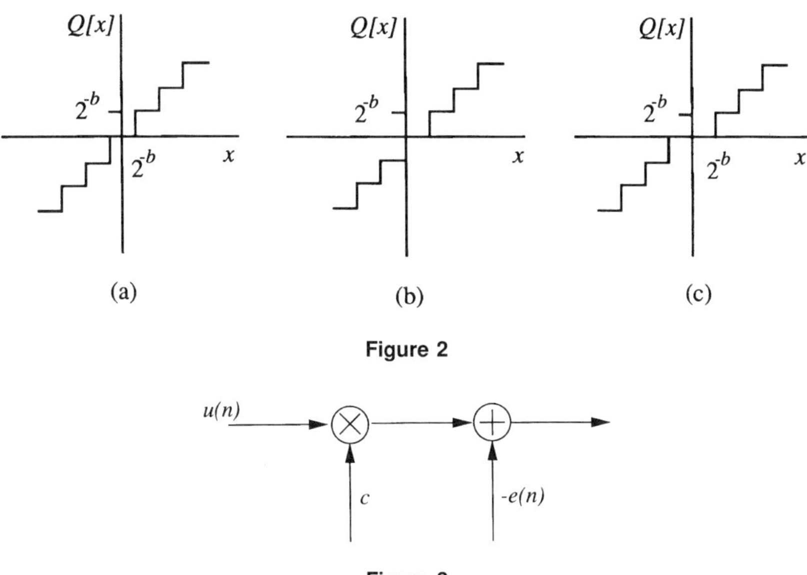

Figure 2

Figure 3

3.2 SIGNAL QUANTIZATION

In a fractional fixed-point digital filter implementation, the number of bits required for accurate representation of the result of a multiplication is given by the sum of the word lengths of the multiplicands. In an IIR filter, this would lead to continuous increase in the word length due to the feedback loop. It is therefore necessary to quantize the signals such that

$$Q|x| = x - e$$

where $Q[x]$ is the quantization characteristic and e is the quantization error. Rounding and truncation[1,3] are commonly used for $Q[\ ]$, and the resulting quantization errors are shown in Figure 2.

It is assumed that the signal word length after quantization is b bits which implies that the quantization step size is $q = 2^{-b}$. Rounding quantizes x to the nearest available b bits fraction (Figure 2a) and, depending on the implementation, rounds up or down when x is exactly halfway between two consecutive values. This leads to an error e which is in the interval ($-q/2$, $q/2$ or $-q/2$, $q/2$), depending on the implementation. In most cases when rounding errors are analyzed, e is considered to be in $[-q/2, q/2]$. Truncation quantizes x by discarding all bits on the right of the first b bits. For a negative number, the error e depends on the type of arithmetic used. In the case of sign magnitude and one's complement the rounding error e is in the interval $[-q, q]$, whereas for two's complement the error e is in $[q, 0]$ as shown in Figure 2.

The FWL multiplier can be modeled as in Figure 3. By assuming fractional representation with $1 \gg 2^{-b}$ and an input signal $u(n)$ which is sufficiently exciting, e can be considered as a random variable. The probability density function $p(e)$ for the three quantization characteristics presented in Figure 2 can be easily obtained. For rounding arithmetic the $p_{RD}(e)$ is $^1/_2q$ in the interval $-q/2 \le e \le q/2$ and zero otherwise, implying zero mean and a variance $\sigma_e^2 = q^2/12$ for the error e. For sign magnitude or one's complement truncation, $p_{T1}(e)$ is $1/2q$ in the interval $-q \le e \le q$ and zero otherwise, implying zero mean and a variance $\sigma_e^2 = q^2/3$ for the error e. For two's complement truncation, $p_{T2}(e)$ is $1/q$ in the interval $-q \le e \le 0$ implying a mean $m_e = -q/2$ and a variance of $\sigma_e^2 = q^2/12$ for the error e. This nonzero mean of the error in two's complement truncation is a disadvantage. In practice, the most commonly used quantization characteristic is rounding.

A common approach to the round-off noise analysis of a fractional fixed-point digital filter implementation consists in replacing all multiplications with the model of Figure 3 and modeling

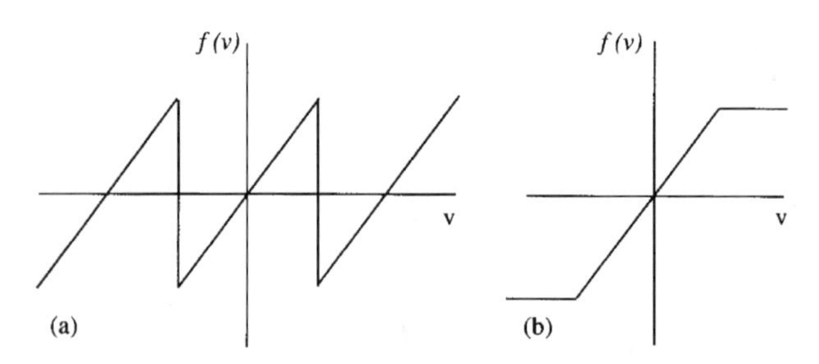

Figure 4

the error variable e as a random variable with the appropriate mean and variance depending on the arithmetic used. The round-off noise in the output is a measure indicating the sensitivity of a digital filter structure to quantization errors. It should be noted that this approach is vlid as long as the stochastic model for the quantization error is valid. If, for example, the input signal does not change much, i.e., is not sufficiently exciting, then this approach is not valid. For example, when the input signal is zero or constant, limit cycles can occur due to quantization. These limit cycles usually have small amplitude and are called quantization (or granularity) limit cycles.

3.3 OVERFLOW OSCILLATIONS

Overflow may occur when two fixed-point numbers are added. If the magnitude of the result exceeds the dynamic range available, then an overflow has occurred which causes errors in the computations. In IIR digital filters these errors can lead to limit cycles of significant amplitude called *overflow oscillations*.[5] They can be avoided by scaling the internal variables. However, completely avoiding overflow requires very conservative scaling which leads to significant decrease of the signal-to-noise ratio in the output due to quantization errors. In practice, less-conservative scaling is used which leads to reducing the probability of overflow, instead of eliminating it. Overflow can still take place, but its probability and thus the probability for overflow oscillations to occur are reduced. The various scaling rules give trade-offs between the reduction of the probability of overflow and the increase of the output round-off noise variance. A commonly used scaling rule is l_2 scaling which is less conservative than other scaling rules.

The scaling rules reduce the probability of overflow, but overflows still do occur; when they do, the result of the addition depends on the overflow characteristic used. The two's complement, Figure 4(a), and saturation, Figure 4(b), are two commonly used overflow characteristics. The two's complement charactristic arises when the overflow bits are ignored in a two's complement representation. It has the disadvantage that overflow can easily lead to limit cycles with large amplitudes. The saturation characteristic is more complicated to implement, since it requires that overflow be detected and then the variable set to the minimum or maximum possible value, but leads to better performance. For second-order sections, it can eliminate overflow oscillations and, if possible, should be used in higher-order filter implementations, too. A detailed technical analysis of the overflow performance of a filter structure requires that the digital filter is analyzed as a nonlinear system which can become quite complicated even for low-order filters.

4. ANALYSIS OF QUANTIZATION ERRORS IN FIXED-POINT IMPLEMENTATIONS

The quantization errors in an IIR filter implementation vary significantly depending on the filter structure. Several techniques have been developed to analyze the output round-off noise and sensitivity to parameter variations of a realization and to obtain structures with low output round-off noise and/or low parameter sensitivity (see Reference 9, Chapters 6 and 7). Work in this area was initially focused on direct forms, because of their simplicity, and on special filter structures. More recently, several techniques based on the state space approach have been developed.[6]

Direct realizations of IIR digital filters are some of the simplest realizations of such filters and use the coefficients of the transfer function directly. Their advantage is the low computational complexity (they require in general only $2N + 1$ multiplications), but they tend to perform poorly under FWL constraints. They have high output round-off noise, high coefficient sensitivity, and they are susceptible to overflow and quantization limit cycles.[4,5] Their bad performance becomes more apparent and more severe when narrowband IIR filters are considered. Cascade connection of second-order sections, where each second-order section can be realized as a direct form, improves the FWL performance to some extent, but is still not satisfactory for narrowband IIR filters. Several techniques have been proposed for improving the performance of the direct form, such as delay replacement[12] or error spectral shaping, which offer improvements at the cost of higher computational complexity depending on the approach. Many realizations which perform better than the direct form have been also proposed in the literature, such as lattice filters, wave digital filters, etc.

A powerful technique to study the output round-off noise of filter realizations and to obtain different realizations for the same transfer function is the state space approach. The minimal state space description $\{\mathbf{A}, \mathbf{b}, \mathbf{c}, d\}$ of the Nth-order stable single-input single-output digital filter is given by

$$\mathbf{x}(n+1) = \mathbf{A}\mathbf{x}(n) + \mathbf{b}u(n) \qquad \mathbf{x}(n) \in R^N \tag{6}$$

$$y(n) = \mathbf{c}\mathbf{x}(n) + du(n) \qquad u(n),\ y(n),\ d \in R \tag{7}$$

where $\mathbf{A}$, $\mathbf{b}$, $\mathbf{c}$ are matrices of dimension $N \times N$, $N \times 1$, and $1 \times N$, respectively, and d is a scalar. The corresponding transfer function is given by

$$H(z) = d + \mathbf{C}(z\mathbf{I} - \mathbf{A})^{-1}\mathbf{B} \tag{8}$$

Clearly, all realizations $\{\mathbf{A}, \mathbf{b}, \mathbf{c}, d\}$ of $H(z)$ have the same performance if infinite precision is assumed. In a FWL implementation this is not the case, and the effect of quantization errors needs to be investigated. Analysis of errors due to signal quantization is based on the assumptions that $\{\mathbf{A}, \mathbf{b}, \mathbf{c}, d\}$ are exactly represented and that the input signal $u(n)$ is sufficiently exciting.

The filter implementations can then be described by

$$\hat{\mathbf{x}}(n+1) = \mathbf{A}Q[\hat{\mathbf{x}}(n)] + \mathbf{b}u(n) \tag{9}$$

$$\hat{y}(n) = \mathbf{c}Q[\hat{\mathbf{x}}(n)] + du(n) \tag{10}$$

where the quantization process $Q[\cdot]$ is given by

$$Q[\hat{\mathbf{x}}(n)] = \hat{\mathbf{x}}(n) - \mathbf{e}(n) \tag{11}$$

$\mathbf{e}(n)$ is the round-off quantization vector with each element having a variance of σ_r^2 as discussed in Section 3.1. In Equations 9 and 10, the addition operations are assumed to be executed in double-precision accumulators.

The states of Models 9 and 10 can still overflow, which could lead to overflow oscillations as discussed earlier. Scaling is used to reduce the probability of overflow and consequently the probability of overflow oscillations. The commonly used l_2 scaling with equal probability of overflow for all states can be obtained using the following constraint

$$k_{ii} = 1 \qquad \forall i \tag{12}$$

where k_{ii} is the ith diagonal element of the covariance matrix $\mathbf{K}$ (also called controllability gramian) given by

$$\mathbf{K} = \frac{1}{2\pi j} \oint_{\Gamma} \mathbf{f}(z)\mathbf{f}^t\left(z^{-1}\right)z^{-1}dz \tag{13}$$

"t" denotes the transpose of a matrix and

$$\mathbf{f}(z) = \left[f_1(z) f_2(z) \cdots f_N(z)\right]^t = (z\mathbf{I} - \mathbf{A})^{-1}\mathbf{b} \tag{14}$$

where $f_i(z)$ is the transfer function from the input to the ith state. The covariance matrix $\mathbf{K}$ can be obtained also as the solution of

$$\mathbf{K} = \mathbf{AKA}^t + \mathbf{bb}^t \tag{15}$$

The variance σ^2 of the output noise due to quantization errors is used as a measure for the performance of the filter implementations. It is obtained using the error model, Figure 3, for the multiplications in Equations 9 and 10. The statistics of the quantization error e depend on the type of fixed-point arithmetic used as discussed Section 3.1. Assuming double-precision additions and no overflows, the variance of the output round-off noise σ^2 is obtained as

$$\sigma^2 = \sigma_r^2 \left[1 + \sum_{i=1}^{N} w_{ii} k_{ii}\right] \tag{16}$$

where w_{ii} is the diagonal elements of the noise matrix $\mathbf{W}$ (also called observability gramian), which is defined as

$$\mathbf{W} = \frac{1}{2\pi j} \oint_{\Gamma} \mathbf{g}^t(z)\mathbf{g}\left(z^{-1}\right)z^{-1}dz \tag{17}$$

with

$$\mathbf{g}(z) = \left\{g_1(z) g_2(z) \cdots g_N(z)\right\} = \mathbf{c}(z\mathbf{I} - \mathbf{A})^{-1} \tag{18}$$

where $g_i(z)$ denotes the transfer function from the ith state to the output. The noise matrix $\mathbf{W}$ can be also obtained as the solution of

$$\mathbf{W} = \mathbf{A}^t\mathbf{WA} + \mathbf{c}^t\mathbf{c} \tag{19}$$

The first term on the right-hand side of Equation 16 is due to rounding at the output node while the second term is due to the propagation of the round-off errors of the internal states. When the filter is l_2 scaled, i.e.,

$$\sigma^2 = \sigma_r^2 \left[1 + \text{tr}(\mathbf{W})\right] \tag{20}$$

where tr denotes the trace of a matrix. The analysis of the effects of parameter quantization is based on the sensitivity of the transfer function to changes in the parameters of the realization $\{\mathbf{A}, \mathbf{b}, \mathbf{c}, d\}$. A commonly used sensitivity measure is given by[8]

$$M = \left\|N_A(\omega)\right\|_1^2 + \left\|N_b(\omega)\right\|_2^2 \left\|N_c(\omega)\right\|_2^2 \tag{21}$$

where $\|\cdot\|_p$ denotes the l_p norm and

$$N_A(\omega) = \left(\sum_{i=1}^{N} \sum_{j=1}^{N} \left| \frac{\delta H(z)}{\delta a_{ij}} \right|^2 \right)^{1/2} \tag{22}$$

$$N_b(\omega) = \left(\sum_{i=1}^{N} \left| \frac{\delta H(z)}{\delta b_i} \right|^2 \right)^{1/2} \tag{23}$$

$$N_c(\omega) = \left(\sum_{i=1}^{N} \left| \frac{\delta H(z)}{\delta c_i} \right|^2 \right)^{1/2} \tag{24}$$

It has been shown that

$$M = \text{tr}(\mathbf{K})\text{tr}(\mathbf{W}) + \text{tr}(\mathbf{K}) + \text{tr}(\mathbf{W}) \tag{25}$$

and for l_2-scaled filter realizations it becomes

$$M = (N+1)\text{tr}(\mathbf{W}) + N \tag{26}$$

One of the advantages of the state space analysis is the use of similarity transformations to obtain realizations with minimum output round-off noise variance σ^2 and minimal sensitivity measure M. For a given realization $\{\mathbf{A}_0, \mathbf{b}_0, \mathbf{c}_0, d\}$ it is required to find a similarity transformation $\mathbf{T}$, such that the resulting realization

$$\{\mathbf{A}_T, \mathbf{b}_T, \mathbf{c}_T, d\} = \{\mathbf{T}^{-1}\mathbf{A}_0\mathbf{T}, \mathbf{T}^{-1}\mathbf{b}_0, \mathbf{c}_0\mathbf{T}, d\}$$

with

$$\mathbf{K}_T = \mathbf{T}^{-1}\mathbf{K}_0\mathbf{T}^{-1} \quad \text{and} \quad \mathbf{W}_T = \mathbf{T}^t\mathbf{W}_0\mathbf{T}$$

has minimal output round-off noise variance σ^2 and l_2 scaled. This problem has been solved by Roberts and Mullis[6] and Hwang,[7] and it was shown that minimum round-off noise variance is obtained when

$$\mathbf{W} = \rho^2 \mathbf{K} \tag{27}$$

for a real ρ. It was later shown that this condition also leads to a minimum sensitivity measure M. The technique proposed by Mullis, Roberts, and Hwang leads to implementations with minimum output round-off noise, minimum sensitivity, and freedom of zero input limit cycles. However, these advantages are offset by the large number of arithmetic operations required for the implementation, $(N + 1)^2$ multiplications and $N(N + 1)$ additions. One approach to reduce the computations is to implement the filter as a cascaded or parallel combination of first- and second-order minimum-noise sections. Although this solution significantly reduces the computations, the resulting structure loses some of the desirable features such as minimum output round-off noise and freedom of zero-input limit cycles. Furthermore, finding the optimal zero–pole pairing and section ordering is generally difficult, especially for high-order filters.

Another approach to improve the performance of a filter implementation is error spectral shaping, also called residue feedback (see Reference 1, Chapter 11). It has been effectively used to reduce the output quantization noise and/or to eliminate limit cycles of IIR digital filters in both direct forms and state space realizations. The residue feedback technique is implemented by extracting the quantization errors (also called residues) due to multiplication and feeding them back in the next iteration. In general, residues could be weighted, but this requires extra multiplication operations leading to increased computational complexity. Other residue feedback schemes have been proposed in which the residue coefficients are restricted to be either integer or power of two. Residue feedback, in general, leads to significant reduction in the output round-off noise at the price of higher computational complexity for the residue operations.

5. ROUND-OFF NOISE IN FLOATING-POINT IMPLEMENTATIONS

The introduction of floating-point DSP chips has made the implementation of digital filters using floating-point arithmetic feasible even for many real-time applications. One of the main advantages of floating-point implementation is the fact that overflow can be avoided. Further, since the mantissa is normalized, the resolution of the representation, i.e., the difference between consecutive levels, depends on the magnitude of the numbers. A disadvantage of floating-point implementation is the fact that quantization errors occur in both additions and multiplications.

The analysis of the quantization errors in a floating-point implementation is more complicated than the one for fixed-point implementations. Quantization errors are caused by both multiplication as well as addition, and these errors are relative to the signal magnitude. The results are also dependent on how the multiplications and additions are sequenced, which further complicates the analysis. Recently, the analysis of quantization errors in floating-point implementations has received attention in the literature, but many of the results presented deal with special or low-order cases.[9] A systematic approach to the floating-point round-off error analysis for various filter structures is given in Reference 10. With floating-point DSP chips becoming more affordable, this area is expected to receive more attention.

REFERENCES

1. **A., Antoniou,** *Digital Filters: Analysis, Design, and Applications,* McGraw-Hill, New York, 1993.
2. **F. F. Kuo and J. F. Kaiser,** *Systems by Digital Computers,* Wiley, New York, 1966.
3. **J. G. Proakis and D. G. Manolakis,** *An Introduction to Digital Signal Processing,* MacMillan Publishing Company, New York, 1988.
4. **L. B. Jackson,** *Digital Filters and Signal Processing,* Kluwer Academic, Boston, 1989.
5. **P. M. Ebert, J. E. Mazo, and M. G. Taylor,** Overflow oscillations in digital filters, *Bell Syst. Tech. J.,* 48, 2999–3020, Nov. 1969.
6. **R. A. Roberts and C. T. Mullis,** *Digital Signal Processing,* Addison Wesley, Reading, MA, 1987.
7. **S. Y. Hwang,** Minimum uncorrelated unit noise in state space digital filtering, *IEEE Trans. Acoust. Speech Signal Process.,* ASSP-25, 273–281, Aug. 1977.

8. **V. Tavsanoglu and L. Thiele,** Optimal design of state-space digital filters by simultaneous minimization of sensitivity and roundoff noise, *IEEE Trans. Circuits Syst.,* CAS-31, 884–888, Oct. 1984.
9. **L. M. Smith, B. W. Bomar, R. D. Joseph, and G. Yong,** Floating-point roundoff noise analysis of second-order state-space digital filter structures, *IEEE Trans. Circuits Syst. II: Analog Digital Signal Process.,* 39, 90–98, Feb. 1992.
10. **B. D. Rao,** Floating-point arithmetic and digital filters, *IEEE Trans. Signal Process.,* 40, 85–95, Jan. 1992.
11. **S. K. Mitra and J. F. Kaiser,** *Handbook for Digital Signal Processing,* John Wiley, New York, 1995.
12. **D. Williamson,** *Digital Control and Implementation, Finite Wordlength Considerations,* Prentice Hall, Englewood Cliffs, NJ, 1991.

Hardware Implementation of Digital Filters

Stuart S. Lawson

CONTENTS

1. INTRODUCTION

1.1 PREAMBLE

To consider the hardware implementation of a digital filter, we must first consider the particular application. The value of parameters such as order, sampling rate, output signal-to-noise ratio (SNR) will have a bearing on the hardware choices to be made. In addition, the filter may be part of a larger system which needs to be integrated. The main approaches that can be considered are as follows:

0-8493-8951-8/97/$0.00+$.50
© 1997 by CRC Press LLC

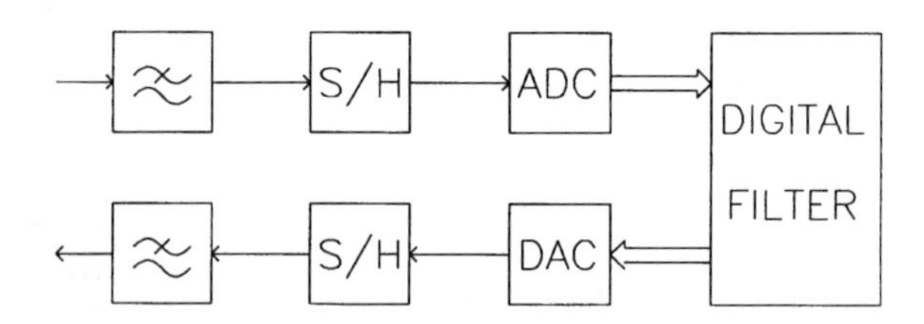

Figure 1 Block diagram of digital filtering system.

1. Board-level implementation using a mixture of off-the-shelf integrated circuits (i.c.) including, for example, 74 series TTL components, memories, microprocessors, multipliers, etc.
2. One or more ASICs (application-specific integrated circuits) such as gate arrays, semicustom or full-custom chips. ASICs may be appropriate in the board-level design, too.
3. General-purpose digital signal processing (DSP) microprocessors such as those from Texas Instruments, Motorola, AT&T, NEC, Analog Devices, etc.
4. Software implementation using a mainframe computer or workstation. Here, the principal design function is writing software, whereas in the other approaches hardware and possibly software will have to be designed.

In the space we have in this chapter, it is not possible to consider all these approaches in detail. We will concentrate on the third technique since there is now considerable support from the industry. The reader is referred to Reference 1 for a more general discussion on hardware design.

1.2 DIGITAL FILTER SYSTEMS

A block diagram of a complete filtering system is shown in Figure 1. If we assume that the source signal which is required to be filtered is continuous, then in order to be processed by a digital filter it needs to be sampled and digitized. An example in which Figure 1 would be relevant is the telephone modem, most of which are now fully digital. The input signal in this case would, of course, be human speech and would be sampled at around 8 kHz.

The very first stage, however, is a lowpass filter that is known as an antialiasing filter because its job is to remove any signal energy above half of the sampling frequency. A low-order analogue filter is sufficient for this task. The signal is then fed to a sample-and-hold amplifier (SHA) which samples the signal at a rate determined by a clock and will generally be greater than half the sampling frequency so as not to violate Nyquist's sampling theorem.[2] The SHA will aim to hold its output constant between clock pulses, thus allowing the analogue-to-digital converter (ADC) to perform an accurate conversion. ADC chips vary in speed and accuracy. Word lengths of 16 bits are now commonplace, and speeds in the video range, i.e., tens of megahertz, are easily obtained. There are several important types of ADCs, (1) successive approximation, (2) flash, (3) integrating. Sometimes the ADC and SHA are integrated into one device.

Having sampled and digitized the input signal, we can then carry out the digital filtering. We will look at this in a moment. First, let us consider the filter output which will be in digital form. If reconversion into analogue form is required, then the first stage will be a digital-to-analogue converter (DAC). The basic structure of a DAC is a weighted sum network, in which each output bit is scaled by a number related to its position in the output word. Its structure is simpler than that of an ADC, and word lengths and speeds compatible with those of ADCs have been achieved. The output of the DAC, being a discrete-time signal, is affected by sin x/x roll-off (Figure 2); this can be compensated for by a SHA. The final stage necessary to recover an analogue signal is lowpass filtering. The spectrum of a sampled signal contains the frequency information of an analogue signal together with reduced energy aliases of this information at multiples of the sampling frequency. So long as a sufficiently wide guard-band exists between the aliases and the baseband signal, it can be recovered by using an analogue lowpass reconstruction filter (Figure 3).[1]

We have seen how an analogue signal has to be prepared for digital processing and how it can be recovered after processing. We next look at the filter algorithm and what computational elements are needed.

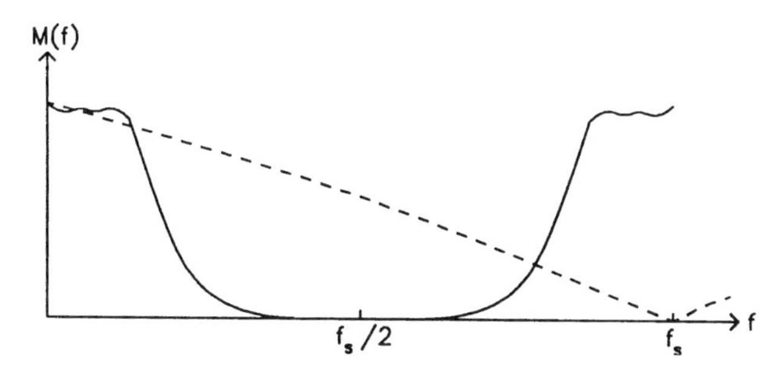

Figure 2 Typical DAC output showing sin *x*/*x* roll-off effect.

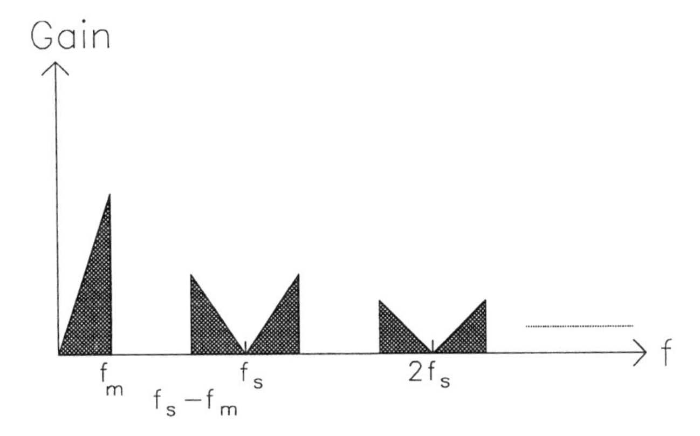

Figure 3 The need for a reconstruction filter.

1.3 REALIZING FILTER EQUATIONS

The ADC provides one n-bit sample every T seconds, where $T = 1/f_s$ is the sampling interval or period and f_s is the sampling frequency. We will assume for this discussion that all the sample bits are available at the same time. Serial data output from an ADC is possible and is required in some applications. Consider a simple second-order digital filter whose structure is illustrated in Figure 4. Note that it is direct form 2 and is canonic in delays.[3] The filter algorithm is essentially a set of linear difference equations, as follows:

$$\left.\begin{aligned} w_k &= Kx_k + \beta_1 w_{k-1} + \beta_2 w_{k-2} \\ y_k &= w_k + \alpha_1 w_{k-1} + \alpha_2 w_{k-2} \end{aligned}\right\} \tag{1}$$

The equations relate the output signal, y_k, to the input signal, x_k, and the filter coefficients $\{\alpha_k, \beta_k : k = 1,2\}$. K is an input scaling coefficient. In this case we also need an internal signal or state, w_k, which represents the value to be stored in a delay element. Note that x_k, is the same as $x(kT)$, i.e., the value of the signal x at the discrete-time point kT where k is any non-negative integer and T is the sampling period. x may have originated as a discrete-time signal, or it may be a sampled version of a continuous-time signal $x(t)$.

Equation 1 expresses the computations required for each new sample of the input signal x. Thus, we can easily deduce that, in general, five multiplications and four additions are required. In addition, some bookkeeping operations are required. For instance, at the end of the computation of Equation 1, delay storage must be updated. Thus w_{k-2} takes the value of w_{k-1} and w_{k-1} takes the value of w_k. Initially, i.e., when $k = 0$, the numbers stored in the two delay elements are assumed to be zero. Thus $w_{-1} = w_{-2} = 0$.

Let the time to carry out one multiplication be T_m and one addition be T_a and all the other operations have negligible delay. Then, the total time to process one sample will be $5T_m + 4T_a$.

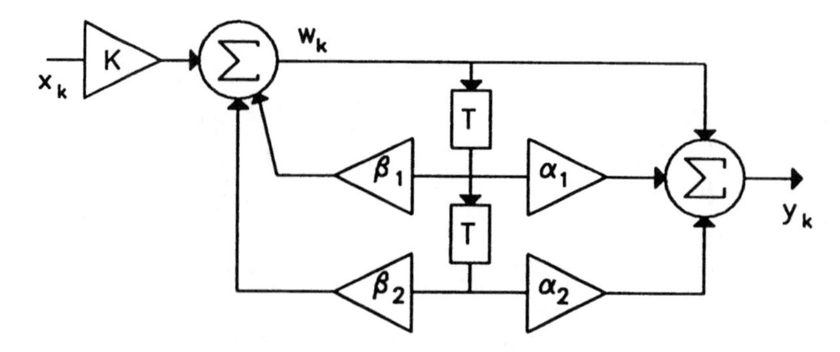

Figure 4 Block diagram of second-order recursive digital filter section.

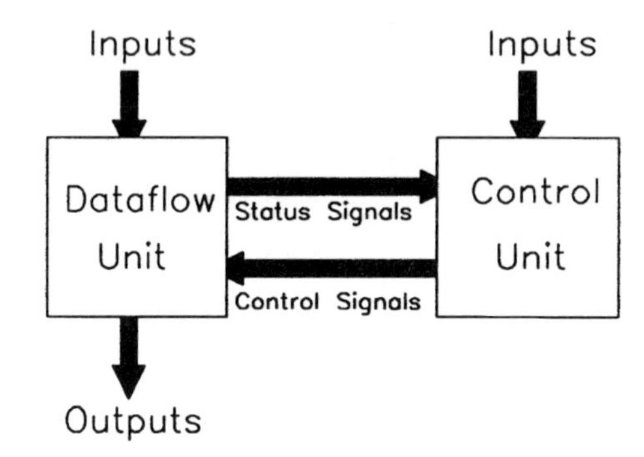

Figure 5 Partitioning of a digital system into a data flow unit and a control unit.

This will be a lower bound for the sampling period; its reciprocal will be an upper bound for the sampling frequency. To put some practical values to these quantities, suppose that $T_m = 100$ ns and $T_a = 20$ ns then $f_s \leq 1.8$ MHz. This has assumed that operations are performed sequentially. If several multipliers are available, then the maximum sampling frequency can be increased implying that signals with higher-frequency contents can be filtered without aliasing. The second-order block of Figure 4 can be used to realise higher-order filters, and this will imply a reduction in the maximum sampling rate possible. If there are N such blocks then the sampling rate will be reduced by approximately the same factor.

From the above discussion, it can be seen that a hardware digital filter will consist of memory for the coefficients, probably ROM (read-only memory) unless downloadable from a host computer in which case RAM (random access memory) would be used. In addition, an arithmetic unit would be needed with a conventional ALU (arithmetic logic unit) and a multiplier. Finally, the delays could be implemented with shift registers, but often a RAM is used. The various functional blocks would then be interconnected with data paths, and an overall control unit would supervise the filtering operations. Effectively, a digital filter is a special-purpose digital computer. This might sound like a very complex device, but, in practice, this need not be so. All the functional blocks are available as integrated circuits, and, because of the advances of modern technology, filters on a chip are now commonplace.

It is often simpler to consider a digital system as two communicating subsystems, the first being the data flow unit in which the desired computations are carried out while the second is the control unit which supervises the computations, ensuring the correct sequence is carried out. In addition, the data flow unit generates status information, e.g., overflow indication, which is then processed by the control unit. Figure 5 illustrates the model for this approach. There are numerous ways of implementing the control unit, and we briefly review them here. The classic technique is the design of a finite state machine which can be implemented using JK flip-flops to store the states. Alternatively, a single programmable logic device such as a PAL could realise the control functions. A more flexible approach is to use the concept of *microprogramming*.[1,4] Essentially, the control signals for the data processor are stored in a control memory which could be either RAM or ROM. Each *microinstruction* word in memory represents the state of every control signal in a particular clock interval. A *microprogram sequencer*, which is a sophisticated counter, is used to generate addresses to the microprogram memory. The sequencer allows normal sequential access to the memory, but it also allows jumps and subroutine calls. For example, a microprogram for a multiplication in a digital filter would need to be used many times and so it would be appropriate to treat the code as a subroutine. Microprogramming is expensive in terms of hardware, but it does allow flexibility both in the way that programs are written and the fact that a new control structure can be implemented by replacing the memory chips. The data flow unit can be loosely defined as an arrangement of storage registers (memories) and arithmetic/logic units together with the data paths between them. Thus, it can be thought of as a digital system which can take, for example, the contents of two registers, perform an arithmetical or logical operation on them, and return the result to another register.[4] As we will see, the assembly language instructions of DSP microprocessors perform very similar functions.

1.4 BRIEF HISTORY OF DIGITAL FILTER IMPLEMENTATION

The earliest implementations were on mainframe computers in software. In 1975, Pye TMC announced a N-type metal-oxide semiconductor (NMOS) chip incorporating two second-order sections, and at about the same time Rockwell developed a similar chip. Around 1979, Intel introduced their 2920 programmable analogue signal processor chip which contained an A/D, D/A, and processor. The sampling frequency was dependent on the length of the program, which was stored on on-board ROM. For example, a program of length 116 instructions would give a sampling frequency of around 22 kHz. British Telecom announced their bit-serial FAD (filter-and-detect) chip in 1981 which was implemented in NMOS technology.[5] It consisted of a second-order filter section with control circuitry that would allow multiplexing to any desired order. The FAD chip was specifically designed for multi-frequency 4 (MF4) signalling receivers as a tone detector running at a sampling frequency of 8 kHz. However, it could also be used as a general digital filtering system. Other contemporaneous developments included the American Microsystems, Inc., S2811 which was designed in 1978 but not released until 1982 because of technological problems and the NEC µPD7720 which was available from 1980.[6,7]

The next major advance came in 1982 when Texas Instruments developed the TMS32010, their first DSP microprocessor chip. Since that time, many other manufacturers have entered the field, e.g., Motorola, Philips, SGS/Thompson, Hitachi, Fujitsu, Zoran, and Analog Devices. AT&T was using its own DSP microprocessors from 1979, but the first to be commercially available, the DSP32, came out in 1984.[6,7] Most, if not all, of the available devices are general purpose and so can be used to implement all the most widely used DSP algorithms, such as signal detection, signal recovery, spectral analysis, correlation, adaptive filtering, etc. The DSP chips today are available in fixed-point or floating-point form.

One last development in DSP chips is worth mentioning. In 1986 Inmos brought out its CMOS A100 chip which was designed to be used as an FIR filter. It consisted of 32 16×16 multipliers which could be configured as a 32-tap filter. Chips were cascadable so higher filter orders were possible, too. Sampling frequencies of between 2.5 and 10 MHz were reported.[8]

The DSP microprocessor was developed specifically to increase data throughput for digital signal-processing algorithms. These are characterized by large numbers of multiply–add operations per unit time. Multiplication in a conventional microprocessor such as the Intel 80386 would be roughly ten times slower than with a Texas 320C25, for example.

The technology used today in DSP devices is complementary metal-oxide semiconductor (CMOS); there were some early designs fabricated in NMOS. Apart from the A100, sampling rates of around 300 kHz are possible for a 32-tap FIR filter. The large difference is due to the fact that the A100 is pure hardware; it is not software driven. A very recent development in the search for a faster and faster DSP is the Texas TM320C40. This device is configured in such a way that it can form part of a processing array (Figure 6). Potentially, the speed can be increased indefinitely,

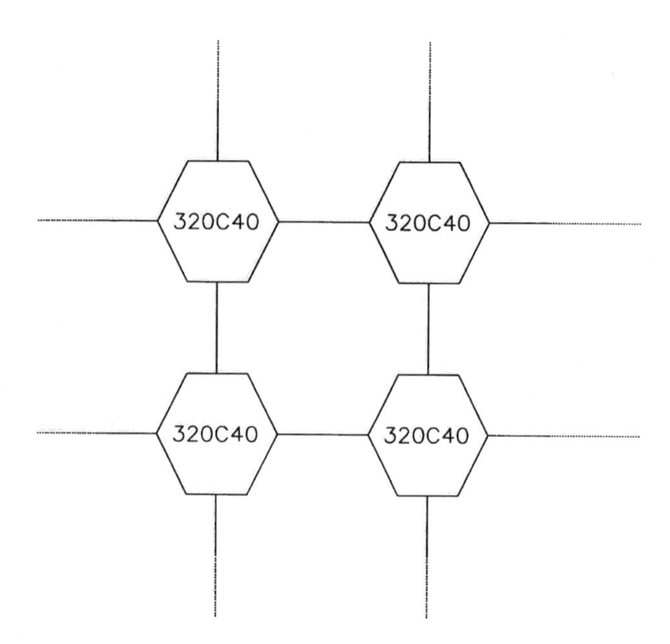

Figure 6 Use of Texas TMS320C40 in processor array.

although other factors would come into play, such as communication overheads, cost, power consumption, etc. For even higher performance a special-purpose chip must be designed from scratch. This can be a particularly expensive option.

The discussion in this section has been predominantly concerned with devices that are available off-the-shelf. In addition, many custom digital filter i.c.s have been designed. Unfortunately, there is no space here to discuss these further.

2. DESIGN ISSUES

2.1 OBJECTIVES

Before embarking on a full hardware design of a digital filter, some important questions need to be asked. Given the filter specification, we need, of course, to apply one of the various design methods described earlier to determine the coefficients. However, the final target technology may determine whether an IIR or FIR filter is chosen because of the different complexities involved. There are many other interrelated factors that need to be taken into account. Power consumption may be a key parameter which will determine whether MOS or bipolar technology is used. Component density on-chip is higher for MOS than for bipolar so this will affect the complexity of the final design and whether or not it can be integrated onto a single chip.

In addition to the standard specification of passband and stopband edges and passband ripple and minimum stopband attenuation, there is the question of the required output SNR. The value of SNR will determine signal and coefficient word length within the filter, and this, again, will affect the choice of technology. In addition, the type of filter will influence internal word length. It may be difficult to realize filters with very sharp cutoffs or narrow passbands with particular structures or with short word lengths. Where the Q of a second-order section in a cascade of such sections is high, the output could potentially be large enough to overflow. Thus, it will be necessary to scale down the input to that section sufficiently to prevent it. Scaling is normally accomplished with simple binary shifts. In order to maintain the best dynamic range within the filter and hence keep SNR high, it will be necessary to scale up at the section output, too. Normally, decisions about signal and coefficient word length can only be made after extensive modeling and simulation of the digital filter in question.

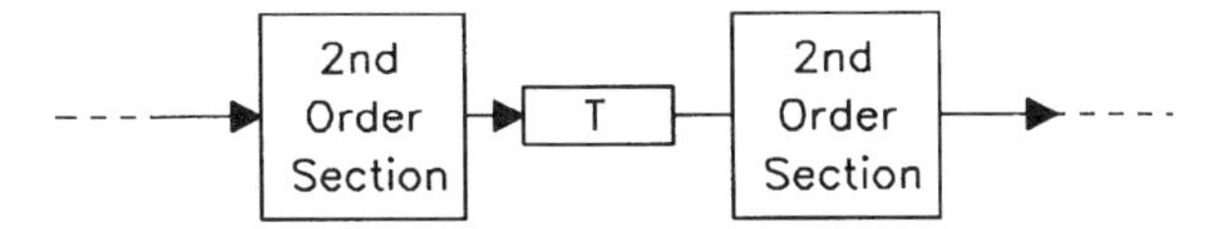

Figure 7 Pipelining second-order filter sections.

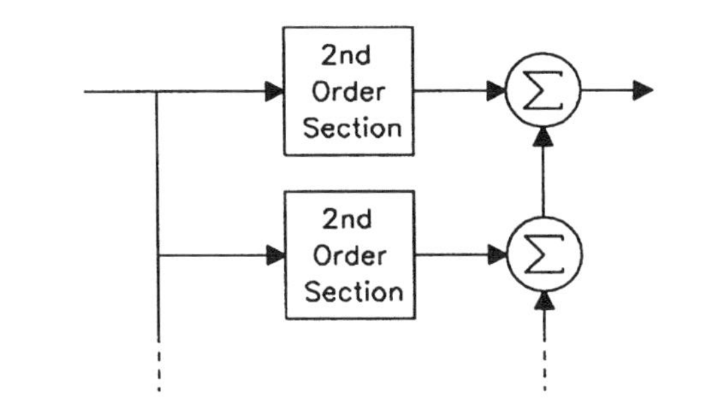

Figure 8 Second order sections in a parallel architecture.

Signal bandwidth must be brought into the equation, too. The sampling frequency must be greater than twice the bandwidth to allow for realizable reconstruction filtering. The value of f_s will, in turn, affect the choice of system clock rate.

Finally, requirements of cost, flexibility, and robustness will have an effect on the hardware design. By flexibility, we mean the programmability of coefficient values. A fixed filter will be more area efficient than a variable filter because the multipliers can be optimized as one of their inputs will be known. Robustness is the ability of the filter to operate correctly in real time even when sudden violent changes occur in the input signal and also under finite word-length conditions.

2.2 HARDWARE CHOICES

Given that the filter specification has been met and questions of technology have been answered, we need to look at the hardware architecture.

We will review four important hardware design approaches: pipelining, paralleling, multiplexing, and systolicizing.[1,9-12] *Pipelining* in an algorithm is the partitioning into smaller subalgorithms or processes. Each subprocess can be implemented as a separate piece of hardware or as a separate area on a chip. When running the system, data flows into the hardware for the first subprocess and, after processing, into the next, and so on to the output of the filter. The purpose of this arrangement is to maximize the data rate or throughput. Hardware for every subprocess must be able to operate simultaneously. Pipelining can be applied to digital filters realized as cascades of second-order sections by placing T-second delays between sections, effectively decoupling adjacent sections (Figure 7). The sampling rate is a function only of the delay through one second-order section. Note that it is not necessary that each subprocess be identical.

Paralleling is the partitioning of an algorithm so that data flows to each of a set of subprocesses simultaneously. The parallel arrangement of second-order filter sections shown in Figure 8 is a good example, although, again, the subprocesses do not have to possess the same structure. The sampling frequency will depend only on the delay through one section. Clearly, if the subprocesses are different, then we must use the section with the most computation to calculate the sampling frequency.

In the previous two approaches, high sampling frequencies are possible at the expense of greater hardware. Where overall speed is not a prime consideration, resources can be shared. In *multiplexing*, one piece of hardware is used sequentially to perform the computations for several subprocesses. Here, the order of subprocesses may be important. For example, in the cascade of second-order sections, we must perform the computations on the first so as to find the next output, which will then become the input for the next section, and so on. It is assumed that the pipelining delays

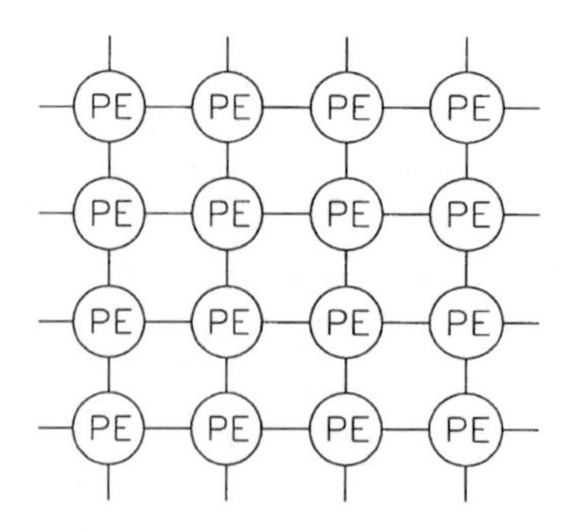

Figure 9 Systolic array architecture.

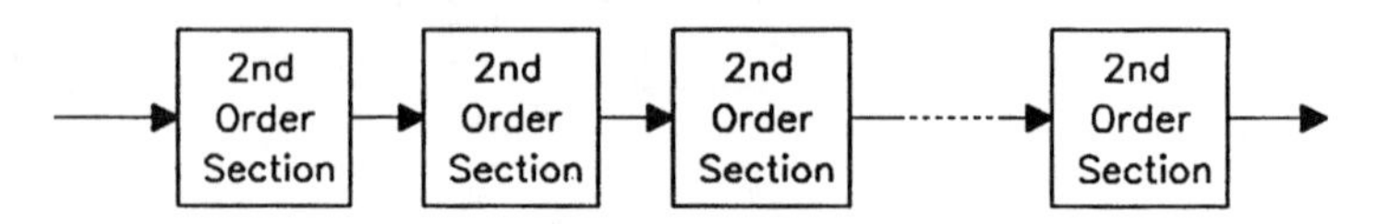

Figure 10 Cascade of second-order sections.

have been removed, see Figure 7. However, in the parallel arrangement of Figure 8, section computation can be carried out in any order.

In more-general cases where we have, say, M subprocesses (the tasks) and N computational units or processors (the resources) and $M < N$ then the problem is one of *scheduling,* which is an important area of research.[13]

The final architectural approach is *systolicizing* and can be thought of as an extension of both pipelining and paralleling in which, normally, the computational units are all identical. *Systolic arrays* are structures in which processing elements (PE) are interconnected to each other in an array configuration (Figure 9). Thus, a PE can communicate only with adjacent PEs although they may be a global clock for synchronization purposes. A PE can be a full processor such as a transputer or a simple circuit of gates and flip-flops.[14]

2.3 SOFTWARE DESIGN

In this section we will briefly describe an algorithm and its corresponding Pascal code to implement a digital filter based on a cascade of second-order sections (Figure 10). Let us suppose we have a filter which can be factorized into N second-order sections. Let $y_k^{(n)}$ represent the input to the kth section at time $t = nT$, where T is the sampling period. Note that the output of the kth section is also the input to the k + 1th section so the final filter output is $y_{N+1}^{(n)}$. In each second-order section there are two delay elements so that the total requirement for the filter is $2N$ delays. If we let the output of the first adder of section k at time $t = nT$ be $w_k^{(n)}$, then we can write the following pair of equations

$$\left.\begin{aligned} w_k^{(n)} &= \alpha_k y_k^{(n)} + b_{lk} w_k^{(n-1)} + b_{2k}^{(n-2)} w_k^{(n-2)} \\ y_{k+1}^{(n)} &= w_k^{(n)} + a_{lk} w_k^{(n-1)} + a_{2k} w_k^{(n-2)} \end{aligned}\right\} \tag{2}$$

where $\{a_{ik} b_{ik}: i = 1 \ldots 2, k = 1 \ldots N\}$ are the feedforward and feedback coefficients, respectively, and $\{\alpha_k : k = 1 \ldots N\}$ are the scaling factors. A Pascal procedure for this algorithm is given in Figure 11.

3. DSP PROCESSOR: ARCHITECTURE

3.1 INTRODUCTION

In this section, we will look at one particular DSP processor, the Texas TMS320C25. Its architecture will be described and its special features discussed. We have chosen this processor because it is

```
PROCEDURE Filter_Cascade(alpha:Ary1;a,b:Ary2;N:INTEGER ;
                        VAR w:Ary2;Xin,Xout:REAL);

{a and b are coefficient arrays of dimension Nx2. alpha is an array of dimension Nx1 to
store scaling constants. w is an array of dimension Nx2 to store past values of state
variables. N is the number of sections and Xin and Xout are the filter input and output
respectively. Ary1 and Ary2 are array types and have to be declared in the calling
program as follows:

TYPE
        Ary1 = ARRAY[1..N] OF REAL;
        Ary2 = ARRAY[1..N,1..2] OF REAL;
}

VAR
        x1,x2  : REAL;
        k      :INTEGER;
BEGIN
        x1 := Xin;
        FOR k := 1 TO N DO
        BEGIN
                x2 := alpha[k]*x1 + b[k,1]*w[k,1] + b[k,2]*w[k,2];
                x1 := x2 + a[k,1]*w[k,1] + a[k,2]*w[k,2];
                w[k,2] := w[k,1];
                w[k,1] := x2;
        END;
        Xout := x1;
END;
```

Figure 11 Pascal routine to implement digital filter based on cascade of second-order sections.

widely used, low-priced, and easily available. However, much that is said here also applies to other fixed-point processors, including the latest-generation chips such as the TMS320C50. By describing the chip in detail, it is hoped that its potential for high-performance digital signal processing can be appreciated. However, this section is not meant to replace the large amount of information already available from Texas Instruments and others.[15,16]

We will examine the hardware structure of the chip and its interface with the outside world. The TMS320C25 is a fixed-point processor with 16-bit program and data word length. On-chip is a 32-bit ALU/accumulator and a 16×16 parallel multiplier which provides the full 32-bit result. In addition, several parallel shifters are present to allow data scaling within one machine cycle. All the usual facilities that are seen in conventional microprocessors are available for program control such as a hardware stack which is used in interrupts and subroutine calls. In summary, the architecture is designed to optimize arithmetic operations of the form

$$S = S + A_i B_j \tag{3}$$

i.e., sums of products, which are used extensively in FIR and IIR filtering.

3.2 INTERNAL STRUCTURE

The TMS320C25 has what is known as a *Harvard architecture*; that is, the program and data have their own independent busses for communication. However, there are instructions that allow, for example, the instruction register to be loaded from data memory. This will be made clearer by observing the block diagram of Figure 12. This shows the internal structure in much detail. Let us look at the program area first. A program memory, ROM or EPROM, contains the application software. A program counter, PC, indicates the address in ROM of the next instruction to be executed. This instruction is loaded into the instruction register and decoded. Depending on the type of instruction, the controller then sets up its execution. This may include obtaining two numbers stored in data RAM and applying them to the ALU and then storing the result back in RAM. The

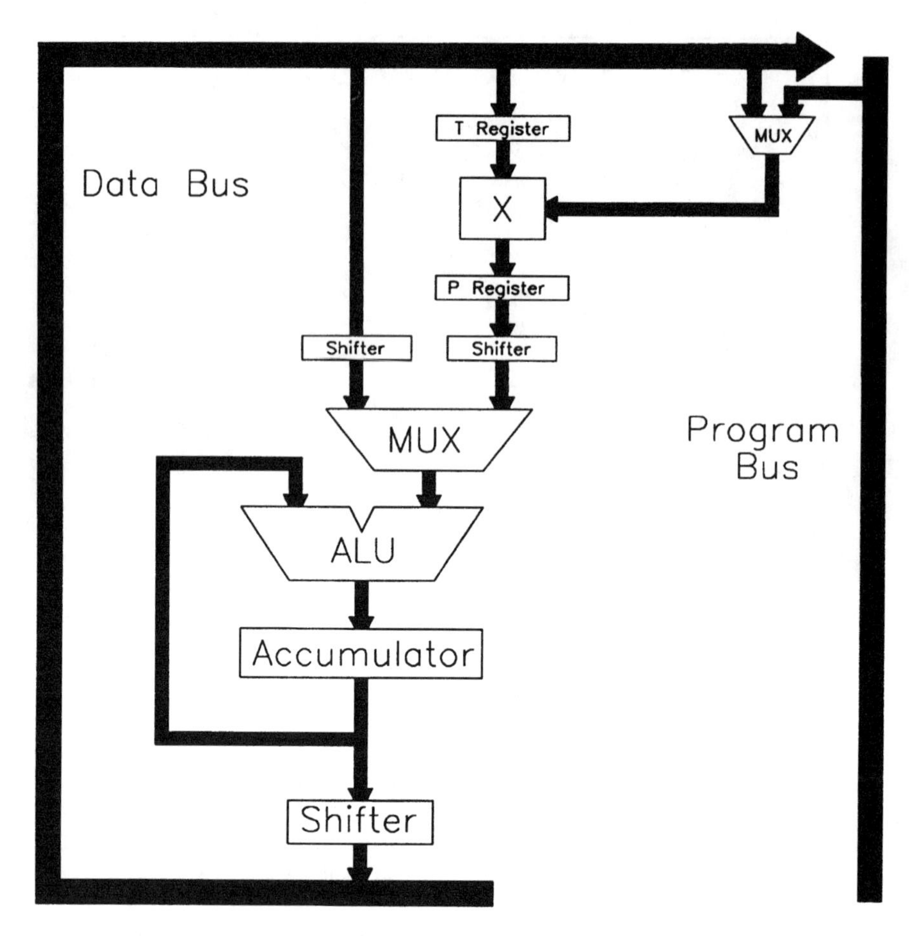

Figure 12 Architecture of TMS320C25 showing the data processor section.

program area also contains several other registers that are required to add flexibility to program writing. An eight-level hardware stack is provided to store the current value of PC so as to allow external interrupts and subroutine calls to be made. Various other registers are available for interrupt control, instruction pre-fetching, instruction repeating, block data moves, and timing. Finally, in the program area, there is a 4096-word mask-programmable ROM for storing program code.

In the data area, we find a 32-bit ALU which allows various arithmetical and logical operations to be performed on at most two operands. The result is stored in a 32-bit accumulator register whose output feeds to the one of the ALU inputs, thus enabling the accumulation operation defined in Equation 3. Various shifters are provided to allow data scaling before an ALU operation. In addition, a shifter is positioned at the output of the accumulator prior to placing the number on the program or data busses or in data RAM. The data area also has an auxiliary register file consisting of eight 16-bit registers. This may be used for addressing data memory, temporary storage, and in integer arithmetic processing via the auxiliary register arithmetic unit, ARAU. This last feature allows various standard forms of addressing memory, such as autoincrement- or decrement-indexed addressing. The TMS320C25 has three RAM areas, B0, B1, and B2. B0, which has 256 words, can be used both by data and by program code. B1 and B2 are available for data storage only and contain 256 words and 32 words, respectively.

The TMS320C25 is designed so that for many DSP applications, a single-chip solution is possible. Of course, there will have to be support and interface chips, but the algorithm will be fully implemented on-chip. For larger problems both in a program or data sense, addressing off-chip is possible. As the word length is 16 bits, the maximum program and data memory size is 64K, i.e., 65,536 words.

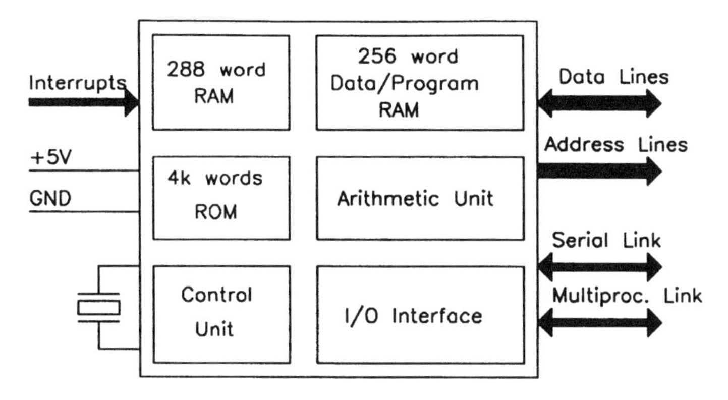

Figure 13 Block diagram of TMS320C25 showing external connections.

3.3 EXTERNAL CONNECTIONS

A simplified block diagram of the chip is shown in Figure 13. It can be seen that both the program and data busses are accessible to the external world. Apart from power and clock connections, two other features deserve mention. The first is the serial interface. This allows communications with serial devices such as Codecs. The second feature facilitates systems in which there may be more than one processor. For example, the TMS320C25 could be a slave processor to a mainframe host sharing some global memory space. Alternatively, several DSP chips may be configured to work in parallel. Also supported is *direct memory access* (DMA) to the external program or data memory of the chip.

4. DSP PROCESSOR: INSTRUCTION SET

In the small amount of space we have here, it is not possible to discuss each instruction in detail. However, we aim to present an overview showing the main features including the various addressing modes and the different classes of instructions.

First, let us review the three addressing modes:

1. **Direct** — We can consider the data memory space as 512 pages, each of 128 words. The nine-bit data memory page pointer (DP) register stores the current page pointer. A direct addressing instruction such as ADD 9,5 will take (DP)+9, where ( ) denotes the contents of the memory location contained within the brackets, left-shifted by five bits, and add it to the contents of the accumulator register. Direct addressing can be used with all instructions except CALL, branch type, immediate operand, and those with no operands.
2. **Indirect** — In this mode, use is made of the auxillary registers (AR), AR0 to AR7 and the ARAU. In addition, the auxillary register pointer (ARP) stores a pointer to one of the eight registers. Various types of indirect addressing are possible. The simplest uses (AR(ARP)) as the data memory address. An extension of this allows the address to be incremented or decremented after memory access. A further refinement allows addition or subtraction of (AR0) after access, i.e., an offset. Finally, bit-reversed (BR) addressing is possible, which is useful for programming fast Fourier transform algorithms.
3. **Immediate** — In immediate addressing, the instruction has a field which contains the actual value to be used in the particular operation. For example, the ADDK 213,2 instruction will add the eight-bit constant 213 to the accumulator with a shift to the left of 2, effectively adding 852 to the contents of the accumulator.

The instruction set can be grouped into six different classes, as follows:

1. Accumulator Memory Reference — various arithmetic and logic instructions involving the accumulator and memory;
2. Auxiliary Registers and Data Page Pointer — various instructions to load or modify values in these registers;

3. T and P Registers and Multiply — various instructions for multiplications and for setting and modifying the multiplicand (T) and product (P) registers;
4. Branch/Call — various conditional and unconditional branch instructions, subroutine calls, and software interrupts;
5. Input/Output and Data Memory — various instructions related to the serial port and parallel ports, also block data transfer instructions;
6. Control — various instructions related to the stack, status bits, interrupts, and repeat instruction.

Overall, the instruction set is similar to those for standard microprocessors except that there is a much larger subset devoted to multiplication and accumulation, operations specific to DSP algorithms. In a later section, we will compare the performance of various DSP chips and some standard microprocessors.

5. DSP PROCESSOR: HARDWARE AND SOFTWARE SUPPORT

Let us suppose that it has been decided to implement a digital filter using a DSP chip together with some support circuits, such as analogue signal acquisition and output, external memory, clock, etc. Using the instruction set reviewed in the previous section, we may write the program or, more precisely, *Assembly Code* module. This module would then be used as input to a program called an *Assembler* which translates it into a sequence of binary words. This is normally achieved in two passes through the code because the exact location of each instruction is required before jumps can be fully coded into binary. In addition, it may be that even after assembly, some subroutine call addresses and constants cannot be evaluated because they are defined in other modules. For this reason, a final process is required before the final executable code is produced. This process is known as *linking* in which all the relevant modules are combined. The final assembled and linked binary code can then be electrically stored in the program ROM on-chip. It is also possible to write the filter algorithm using a high-level language such as C or Pascal and then automatically compile into DSP code. Such a procedure is known as *cross-compilation* in which the application program is compiled on a computer different from the target processor.

For small programs the above approach may be acceptable, but a more elaborate design technique is required for larger programs, particularly where the digital filter is part of a larger DSP system. Figure 14 shows a block diagram summarizing a complete top-down design cycle. At each stage, software and hardware support is available.

In the first stage the overall system which includes the digital filter is modeled. This may be achieved using graphical entry of the system block diagram, using a high-level description language, or with the aid of a high-level language such as C, Fortran 90, or Pascal. Many commercial computer-aided design (CAD) systems for DSP incorporate graphical entry and supply libraries of standard blocks such as arithmetic and logic functions, memories, and delays. They also allow vector operations. Users can build up their own library of functions using the standard blocks. In addition, suppliers offer their own higher-level building blocks, such as signal generators, modulators, demodulators, filters, adaptive systems, encoders and decoders, and various mathematical functions. Once modeled, the system can be simulated with test data to verify its correctness.

The second stage involves the generation of DSP processor code. This is achieved automatically from the graphical data. The code can then be downloaded to the target DSP chip and run. Alternatively, the code can be loaded into a hardware simulator which is a program capable of accurately modeling the DSP chip. From this simulation, information about the implemented algorithm can be gauged, such as robustness, accuracy, and overall timing, as well as monitoring changes of value in registers, data memory, and input/output ports.

If the previous stages have been successfully completed, then the final hardware implementation can be carried out. The discussion has assumed that the target is a DSP chip. However, it may be that we wish to implement the DSP system on an ASIC in which case the design can be translated into the industry standard design language VHSIC hardware description language (VHDL), which

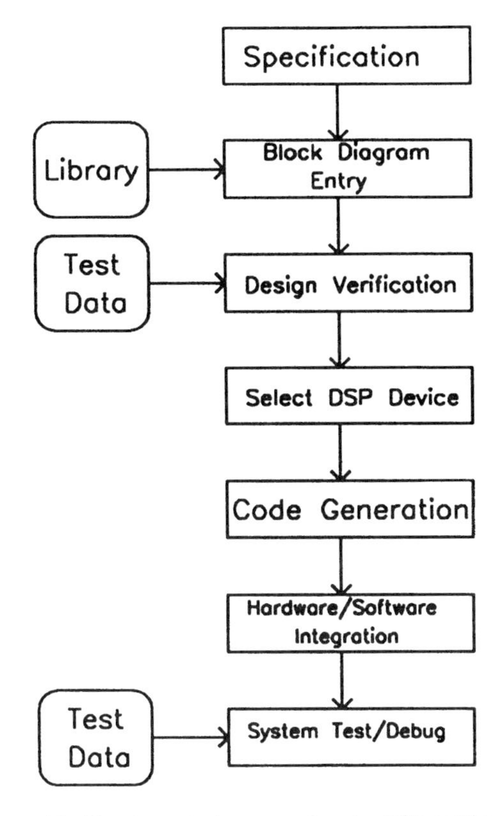

Figure 14 Top-down design procedure for DSP applications.

is an accepted input form for many very large scale integration (VLSI) CAD systems. A guide to engineering software including CAD for DSP is to be found in Reference 17.

Before we leave this section, a little should be said about hardware for DSP and digital filtering, in particular. Ultimately, once software for a particular application has been designed and thoroughly tested, it will be stored in internal and/or external ROM and the DSP chip will be embedded in the application on a purpose-built printed circuit board (PCB). For development purposes, many manufacturers offer various PC-compatible PCBs based on the most popular DSP chips. These boards typically include multichannel analogue input and output interfaces, RAM as well as the DSP processor and PC interface and, together with Assembler and Linker, a cross-compiler, and a suitable monitor package, provide a low-cost development environment.

6. REVIEW OF DSP CHIPS CURRENTLY AVAILABLE

We have centred our discussion so far on the Texas TMS320C25 which is a 16-bit fixed-point processor. It is useful to compare the various chips available from the point of view of speed, power consumption, memory size, accuracy, software support, etc. Table 1 shows some basic facts about five reasonably well-known fixed-point processors. The power column represents 5 V supply requirements. Cycle time is the time to perform one instruction. However, in that time, most processors can take two operands, multiply them together, and add the result to the accumulator. In addition, they can perform various addressing operations including fetching the next instruction. The word length figures are, respectively, that of the ALU and the size of the accumulator register. The figures for on-chip memory and external addressing are self-explanatory except that the Motorola DSP 560001 has two independent areas for both RAM and ROM on-chip.

No table can do full justice to each processor. For example, there are several variants in the Motorola 56000 family, including a low-voltage (3.3 V) DSP56L002 with a power requirement of

Table 1 Comparison of Basic Features of DSP Fixed-Point Processors

Processor	Technology (μm CMOS)	Power (mW)	Cycle Time (ns)	ALU, Acc. Word Lengths (bits)	On-Chip ROM	On-Chip RAM	Maximum Addressable ROM	Maximum Addressable RAM
Texas TMS320C25	1.8	125	100	32,32	4K × 16	544 × 16	64K × 16	64K × 16
Texas TMS320C50	0.8	315	50	32,32	2K × 16	8K × 16 544 × 16	64K × 16	64K × 16
AT&T DSP16	1	350	75	32,36	2K × 16	512 × 16	64K × 16	
Motorola DSP56001	1.5	450	50	56,56	256 × 4(2)	256 × 24(2) 512 × 24	64K × 24	128K × 24
Analog Devices ADSP2100	1.5	475	80	40,40	—	—	32K × 24	16K × 16

Table 2 Comparison of Basic Features of DSP Floating-Point Processors

Processor	Technology	Power (W)	Cycle Time (ns)	ALU, Acc. Word Lengths (bits)	On-Chip ROM	On-Chip RAM	Maximum Addressable Memory
Texas TMS320C30	1 μm CMOS	1.0	60	32,40	4K × 32	1K × 32(2)	16M × 32
Texas TMS320C40	0.8 μm CMOS	1.75	40	32,40		4K × 32	4G × 32
AT&T DSP32C	0.75 μm CMOS		80	32,40	2K × 32	1K × 32	16M × 32
Motorola DSP 96002	CMOS	1.5	50	32,96	1K × 32(2)	512 × 32(2) 1K × 32	4G × 32
Analog Devices ADSP-21060	0.6 μm CMOS	2.1	30	32,80	—	128K × 32 80K × 48	4G × 32

Table 3 Performance of DSP and Other Processors on Filter Benchmarks

			FIR Filter		IIR Filter	
Processor	Introduced	Cycle Time (ns)	50-Tap Filter Delay (μs)	100-Tap Filter Delay (μs)	7th-Order Filter Delay (μs)	13th-Order Filter Delay (μs)
Texas TMS320C25	1987	100	5.8	10.8	1.4	2.6
Texas TMS320C30	1988	60	3.3	6.3	0.84	1.56
Motorola DSP56001	1987	50	2.85	5.35	0.7	1.3
Motorola DSP96002	1990	50	2.85	5.35	0.7	1.3
Analog Devices 21060	1994	30	1.65	3.15	0.42	0.78

165 mW. Further information can be found in Reference 18 as well as manufacturers' data books. In addition, many books have been recently published which discuss applications in much detail.[19-23]

Table 2 summarizes similar properties for floating-point processors, and Table 3 compares processor performance on FIR and IIR filter benchmarks. The IIR filter referred to is based on a cascade of second-order sections. The delays given in the table are roughly proportional to the filter orders, although each processor has unique features concerning memory access and arithmetical calculations. So, for instance, the latest DSP chip, the ADSP-21060, is the fastest predominantly because its instruction cycle time is only 30 ns. Generally, we can say that each new generation is faster, has more on-chip memory and additional or improved features compared with the previous generation.

We have concentrated here on general-purpose digital signal processors. At least two manufacturers supply building-block chips for FIR filters. Motorola has developed a cascadable device, DSP56200, capable of implementing up to a 256-tap filter with a speed of 100 ns per tap. The chip has 16-bit signal accuracy with 24-bit coefficients and 40-bit accumulation. It can operate as a single or dual FIR filter or as an adaptive FIR filter. Harris Semiconductors has developed several filter chips. The HSP43168 has features very similar to those of the Motorola DSP56200 except that signal and coefficient precision are both 10 bits. It has a power dissipation of 2.4 W at 5 V. The HSP43891 is a cascadable video-speed FIR filter with sampling rates of up to 30 MHz. It consists of eight filter cells cascaded internally with a shift-and-add output stage. Each cell contains a 9×9 multiplier and a 26-bit accumulator, together with three decimation registers. Its power consumption is 2.4 W at 5 V.

7. CONCLUSIONS

In this section, we have considered hardware aspects of digital filter implementation. As in many design problems, the solution depends on the total specification, i.e., the filtering requirements and the environment in which the filter will operate. Parameters, such as sampling rate, SNR, power dissipation, accuracy, robustness, and so on, must be considered. For many applications the DSP processor represents a very efficient solution which, in some circumstances, can be a single-chip solution. The range of DSP devices is very large, and performance is greatly enhanced in each new generation. However, there will be some applications for which an ASIC chip or a PCB containing very high performance LSI and VLSI components will be more appropriate.

REFERENCES

1. **Lawson, S. and Mirzai, A.R.,** *Wave Digital Filters*, Ellis Horwood, New York, 1990.
2. **Taylor, F.J.,** *Principles of Signals and Systems*, McGraw-Hill, New York, 1994.
3. **Antoniou, A.,** *Digital Filters: Analysis, Design, and Applications*, 2nd. ed., McGraw-Hill, New York, 1993.
4. **Hill, F.J. and Peterson, G.R.,** *Digital Systems: Hardware Organization and Design*, 3rd ed., Wiley, New York, 1987.
5. **Challener, P.,** FAD — Flexibility in digital signal processing, *Microprocessors Microsyst.*, 7(10), 475–481, 1983.
6. **Lee, E.A.,** Programmable DSP architectures: part I, *IEEE Acoust. Speech, Signal Process. Mag.*, 5(4), 4–19, 1988.
7. **Lee, E.A.,** Programmable DSP architectures: part II, *IEEE Acoust. Speech, Signal Process. Mag.*, 6(1), 4–14, 1989.
8. *Inmos Transputer Development and iq Systems Databook*, Inmos Limited, Bristol, 1989.
9. **Bowen, B.A. and Brown, W.R.,** *VLSI Systems Design for Digital Signal Processing*, Vol. 1, Prentice-Hall, Englewood Cliffs, NJ, 1982.
10. **Bowen, B.A. and Brown, W.R.,** *VLSI Systems Design for Digital Signal Processing*, Vol. 2, Prentice-Hall, Englewood Cliffs, NJ, 1985.
11. **Hwang, K. and Briggs, F.A.,** *Computer Architecture and Parallel Processing*, McGraw-Hill, New York, 1985.
12. **DeFatta, D.J., Lucas, J.G., and Hodgkiss, W.S.,** *Digital Signal Processing: A System Design Approach*, Wiley, New York, 1988.
13. **Hoang, P.D. and Rabaey, J.M.,** Scheduling of DSP programs onto multiprocessors for maximum throughput, *IEEE Trans. Signal Process.*, 41(6), 2225–2235, 1993.
14. **Fortes, J.A.B. and B.W.Wah (Eds.),** Special issue on systolic arrays, *IEEE Comp.*, 20(7), 12–103, 1987.
15. *TMS320C2x User's Guide*, Texas Instruments, Dallas, 1993.
16. **Marven, C. and Ewers, G.,** *A Simple Approach to Digital Signal Processing*, Texas Instruments, Bedford, UK, 1993.
17. **Kaplan, G. (Ed.),** Guide to engineering and scientific software, *IEEE Spectrum*, 29(11), 28–83, November 1992; 31(11), 33–83, November 1994.
18. **Smith, M.R.,** How RISCy is DSP?, *IEEE Micro.*, 12, 10–22, December 1992.
19. **Ingle, V. and Proakis, J.,** *Digital Signal Processing Laboratory Using the ADSP-2101 Microcomputer*, Prentice-Hall, Englewood Cliffs, NJ, 1993.
20. **Hutchins, B.A. and Parks, T.W.,** *Digital Signal Processing Laboratory Using the TMS320C25*, Prentice-Hall, Englewood Cliffs, NJ, 1990.
21. **Motorola, Inc.,** *Real Time Digital Signal Processing Applications with the Motorola DSP 56001*, Prentice-Hall, Englewood Cliffs, NJ, 1990.
22. **Papamichalis, P.,** *Digital Signal Processing Applications with the TMS320 Family*, Vol. 1–3, Prentice-Hall, Englewood Cliffs, NJ, 1988 and 1991.
23. **Chassaing, R.,** *Digital Signal Processing with C and the TMS320C30*, Wiley, New York, 1993.

Chapter 7

Switched-Capacitor Filters

Cascade and Coupled-Biquad Switched-Capacitor Filters

Ezz I. El-Masry

CONTENTS

1. INTRODUCTION

Filters perform one of the most important functions of present day electronics systems. Their implementations have evolved drastically over the years as a result of the development in technology. Originally, filters had been implemented using passive components, such as resistors, capacitors, and inductors. Because inductors are bulky, noisy, lossy, and unsuitable for miniaturization,

0-8493-8951-8/97/$0.00+$.50

© 1997 by CRC Press LLC

tremendous efforts to simulate and/or replace them by circuits containing R, C, and operational amplifiers (op amp) had begun following the development of an inexpensive monolithic op amp by Widlar in 1967.[1] Most of the resulting analog RC–op amp filters were constructed in hybrid form with monolithic op amps and chip capacitors soldered on a board containing thick-film resistors. The deposited resistors are trimmed using a laser in order to tune the filter after fabrication. This process has resulted in high unit cost.

Practically, to achieve full integration in silicon chips, the designer would have to choose between the bipolar or MOS (*M*etal *O*xide *S*emiconductor) integrated circuit (IC) technologies.[2,3] Each technology has advantages and disadvantages. The bipolar technology is faster, offers higher transconductance values, and has lower turn-on values (0.5 V for silicon). On the other hand, the MOS technology has the ability to store signal-carrying charges at a node for a relatively long period of time, to move the charge between different nodes (under the control of a digital clock), and to continuously, nondestructively sense these charges. Prior to about 1977, there existed a clear separation of the bipolar and the MOS technologies, according to the function required. MOS technology with its superior device density, was used mostly for digital logic and memory applications; on the other hand, bipolar ICs such as op amps were used to implement analog filters.

A strong motivation for analog filter designers to develop circuit techniques that are compatible with MOS technology came with the development of the MOS LSI (*L*arge *S*cale *I*ntegration) microprocessor and other low-cost digital ICs. Currently, the MOS technology is the most widely used IC technology for digital VLSI (*V*ery *L*arge *S*cale *I*ntegration) circuits. In addition, new developments occurred in communication technology (such as digital telephony, data transmission via telephone lines, etc.) which require mixed analog and digital signal processing circuitry. Therefore, problems of compatibility with digital systems, cost, reliability, weight, and size added more incentives to the development of analog circuits that can be fabricated using MOS technologies.

There has been a long-standing difficulty in producing fully integrated MOS RC–op amp filters with precise characteristics. This is because the MOS fabrication process does not result in a sufficiently precise control over the absolute values of resistance and capacitance (and, hence, the RC product). In addition, MOS integrated (diffused) resistors have poor temperature and linearity characteristics, as well as requiring a large silicon area. Ordinarily, the RC values that determine the filter characteristics cannot be controlled to better than 20%. This limitation was soon overcome by the development of circuit techniques, whereas the resistor is simulated by capacitor and MOS switches. Hence the frequency response of the filter depends on the ratios of MOS capacitors. It was shown that MOS technology has the capability of realizing capacitor ratios within 0.1% accuracy. This class of filters is known as switched-capacitor (SC) filters, and they have received considerable attention recently. They are very accurate and do not require tuning or laser trimming. Besides, filter characteristics can be controlled by a precision digital clock. In addition, absolute capacitance values can be sufficiently reduced while maintaining the same capacitance ratios, resulting in circuits that consume a very small chip area.

The principle of SC is not new; it is relatively old. In fact, it was mentioned in 1873 by Maxwell[4] in his fundamental book on electricity and magnetism. Many years later, Fudim[5,6] pointed out that the SC approach to circuit synthesis was introduced by him in 1962 and was documented in his Ph.D. thesis in 1965.* The development of the resonant transfer principle[7,8] had a great impact on the SC network theory. The use of capacitor and MOS switches to realize digital filters was demonstrated by Kuntz[9] in 1968. More recently (1972), Fried[10] demonstrated the way to implement active filters using only capacitors, MOS transistors, and a pulse generator. Later (1977), the idea of replacing the resistor in RC–op amp filters by capacitor and MOS switches was fully investigated and implemented to design SC filters by Caves et al.[11] and Hosticka et al.[12] Since then, SC circuit design techniques have made great and rapid progress and are widely spreading, with great success, among many communications and signal processing applications. For almost the last two decades

* He is the first to use the name "capacitor-switch" in his Ph.D. thesis (1965).

the most successful realizations of analog filters in MOS technologies have been based on the well-established SC circuit techniques.[13-18]

1.1 SWITCHED-CAPACITOR VS. DIGITAL SIGNAL PROCESSORS

SC filters can be classified as analog signal processors since they process signal samples directly, without quantization and digitization processes. Therefore, they do not suffer from the quantization and round-off errors that are associated with such processes. In addition, SC circuits required to perform such basic operations as addition, multiplication, and delay are much simpler than the digital counterparts. This results in faster signal processing, higher density of operation on a chip, lower DC power supply requirement, and less power dissipation than digital signal processors (DSP). On the other hand, SC filters have limited accuracy as a result of the capability of MOS technology to realize ratios of capacitors not better than 0.1% (which corresponds to 10-bit floating-point accuracy in digital terms). Also, because of the relatively high noise level associated with SC components (op amps and switches), clock, supply lines, etc., the dynamic range of SC filters is lower than that of DSP. Furthermore, SC filters suffer from the effect of the nonidealities, such as parasitic capacitors, the clock feedthrough, finite open-loop DC gain and bandwidth of the op amps, finite on-resistance of the MOS switch, etc. Another important advantage of DSPs is due to the fact that they utilize the standard microprocessor technology, taking advantage of the immense digital cell libraries and information available for such devices (such as basic circuits and accurate device models for computer simulation).

From the above discussion one concludes that there is no general rule for choosing one signal processing technique over the other. This actually was not the intention of the above comparison. Moreover, it is becoming more cumbersome to partition electronic systems into wholly separate analog and digital LSI chips. It is rather clear that the two techniques rarely compete but, on the contrary, serve complementary roles. In fact, there are well-established techniques that permit combination of analog and digital blocks on the same chip to implement advanced signal processors that meet the requirements of present day communication systems. In this case the main issue is the assignment of different tasks to the most appropriate (digital or SC) functional block.

2. SWITCHED-CAPACITOR BUILDING BLOCKS

In this section we will provide more insight into the fundamental principles of operation of SC circuits and introduce the basic building blocks that are required for the design of parasitic-insensitive SC filters.

2.1 SWITCHED-CAPACITOR SIMULATION OF A RESISTOR

The recognition that a resistor could be simulated with a combination of a capacitor and MOS switches was the basic theory underlying the operation of SC filters. To illustrate this principle, let us consider the SC configuration shown in Figure 1, where the switch is moving back and forth between positions 1 and 2 with frequency f_c Hz. In the analysis to follow, the switch position 1 or 2 will be referred to it as switch-1 or switch-2, respectively. Assume that v_1 and v_2 are voltage sources (or virtual ground or op amp's output); that is, the values of v_1 and v_2 are not affected by the switch actions. When the switch is in position 1 (switch-1 is on) as shown above, the capacitor is charging to q_1 given by

$$q_1 = Cv_1 \tag{1}$$

It is assumed that the switch stays at that position long enough to fully charge the capacitor. Now if the switch moves to position 2 (switch-2 is on), the capacitor will be charging (discharging) if $v_2 > v_1$ ($v_2 < v_1$) at voltage v_2; then the new charge, q_2, is given by

$$q_2 = Cv_2 \tag{2}$$

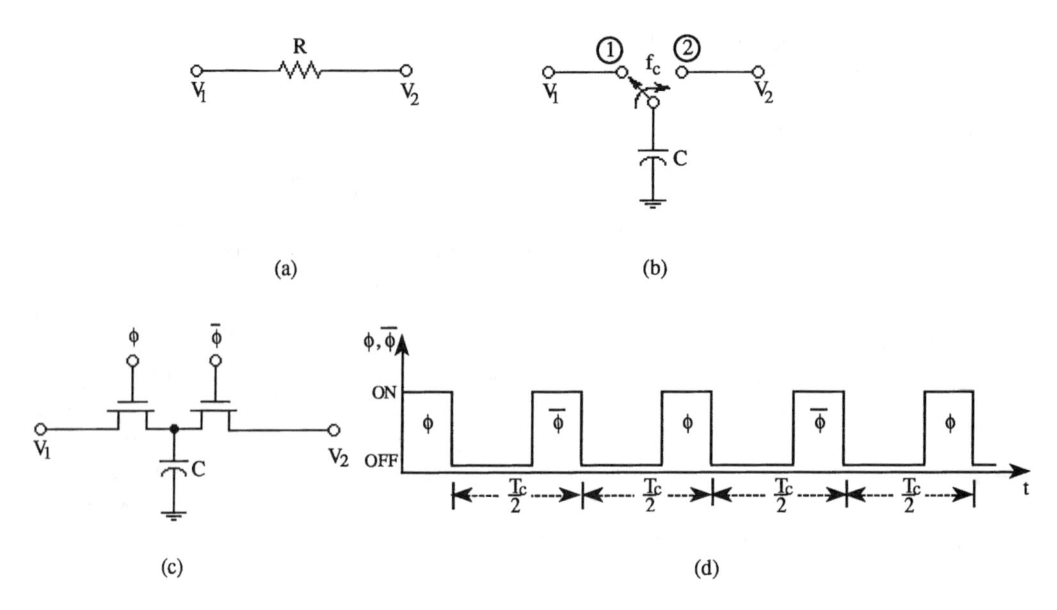

Figure 1 (a) Resistor to be simulated by SC; (b) SC simulation of R; (c) MOS implementation of (b); (d) waveforms of the clock signals.

Therefore, the transfer of charge Δq is

$$\Delta q = q_1 - q_2 = C(v_1 - v_2) \tag{3}$$

and this will be accomplished in $T_c = 1/f_c$ s. The electric current will be on the average

$$i(t) = \frac{\partial q}{\partial t} = \lim_{\Delta t \to 0} \frac{\Delta q}{\Delta t} \cong \frac{C(V_1 - V_2)}{T_c}, \qquad T_c << 1 \tag{4}$$

The size of an equivalent resistor that yields the same value of current is then

$$R = \frac{(V_1 - V_2)}{i} \cong \frac{T_c}{C} = \frac{1}{Cf_c} \tag{5}$$

For the above approximation to be valid, it is necessary that the switching frequency, f_c, be much larger than the frequencies of v_1 and v_2. This can be easily achieved in the case of audio-frequency filtering, and the SC may be regarded as a direct replacement for the resistor in RC–op amp circuits. However, this is only valid under the assumption that v_1 and v_2 are not affected by the switch closures. While this is in effect true in RC–op amp filters which are based on integrators, it is not true in more general circuits. In these cases simple replacement of the resistor by an SC can yield wrong results. The practicality and technological design considerations of the above approach have been discussed in detail by many authors.[19,20] MOS realization of the SC resistor, together with the pertinent clock signals, is shown in Figure 1c and d.

2.2 SWITCHED-CAPACITOR INTEGRATORS

SC integrators are the key elements in the realization of SC filters. A straightforward realization of an SC integrator, Figure 2, is obtained by applying the SC simulation of a resistor, Figure 1, to the RC–op amp inverting integrator. Analysis of this circuit using the previous approach yields approximate results. Exact results, however, can be achieved by applying a sampled-data analysis

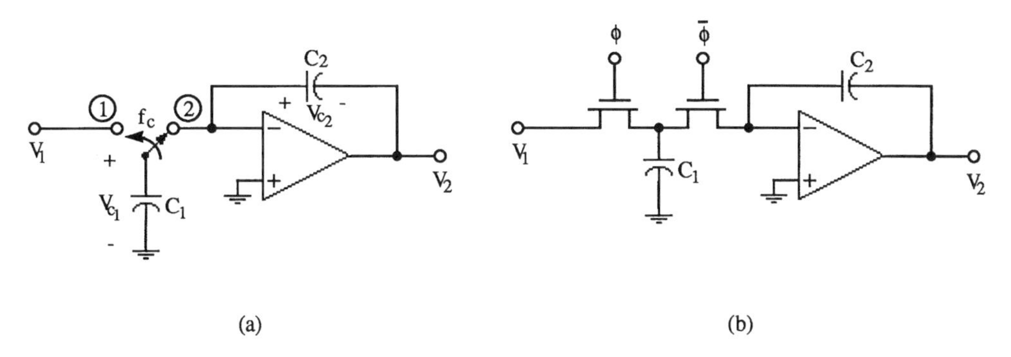

Figure 2 (a) SC integrator; (b) MOS implementation of (a).

technique. Because of the sampled-data nature of SC filters, therefore, they are more accurately described, as well as designed, by the use of the z transformation.[21] In this formulation the charge distribution in the SC circuit, for one clock cycle, is described by a difference equation. This difference equation is then converted into an algebraic equation employing the z transformation. Moreover, the application of this transformation yields a rational function in z which describes the behavior of the SC circuit. Begin by assuming that switch-1, in Figure 2, is closed (ϕ is high)* and C_1 is charged to a value

$$q_1 = C_1 v_{c_i} = C_1 v_1 \tag{6}$$

The concept of discrete time is introduced at this point. The simplest way of looking at this concept is to visualize the input voltage as being sampled at some discrete point in time. Mathematically, one can state that the signal is uniformly sampled at times $t + nT_c$ where T_c is the clock period and n is an integer. The choice of values for n is arbitrary, and thus the value $n = -1$ is chosen for this first phase. Equation 6 can now be rewritten as

$$q_1(t - T_c) = C_1 v_1(t - T_c) \tag{7}$$

Assume that the input voltage remains constant during this clock phase and C_1 charges instantly to the value in Equation 7 and holds that value for the duration of that phase. This assumption is valid as long as the frequency of the input voltage is smaller than the clock frequency. During the same time interval, C_2 is holding a charge

$$q_2(t - T_c) = C_2 v_2(t - T_c) \tag{8}$$

Since both switch-2 and the inverting input of the op amp present infinite impedance, no current is allowed to flow through C_2. This requires that the voltage across C_2 remain constant, or equivalently, that any charge stored in C_2 during the previous clock phase is held there during this phase. Therefore, v_2 remains constant, due to the virtual ground at the inverting input terminal of the op amp. After half clock cycle, at the instant $t - (T_c/2)$, switch-1 is opened and switch-2 is closed (ϕ is low), connecting the top plate of C_1 directly to the inverting input of the op amp and disconnecting the input from the rest of the circuit. Because of the virtual ground, C_1 is completely discharged and all the charge is transferred to C_2. The net effect of this charge transfer is most easily understood with the aid of the following charge equations for this time instant:

* The clock signals are depicted in Figure 1d.

$$q_1\left(t-\frac{T_c}{2}\right)=0$$

$$q_2\left(t-\frac{T_c}{2}\right)=q_2\left(t-T_c\right)+C_1\left\{v_{c_1}\left(t-\frac{T_c}{2}\right)-v_{c_1}\left(t-T_c\right)\right\} \tag{9a}$$

Thus,

$$C_2v_2\left(t-\frac{T_c}{2}\right)=C_2v_2\left(t-T_c\right)+C_1\left\{v_{c_1}\left(t-\frac{T_c}{2}\right)-v_{c_1}\left(t-T_c\right)\right\} \tag{9b}$$

with v_{c_1} $[t-(T_c/2)]=0$ and $v_{c_1}(t-T_c)=v_1(t-T_c)$.

Equations 9 make use of the charge conservation principle, which states that charge is neither created nor destroyed in a closed system. Once the charge transfer has taken place, all currents go to zero, and no further voltage changes occur.

After another half clock cycle (at the instant t) switch-2 is opened (ϕ is high); therefore, no change will occur in the voltage across C_2, as explained earlier. Thus,

$$v_2(t)=v_2\left(t-\frac{T_c}{2}\right) \tag{10}$$

Equations 9 and 10 may then be combined to form the complete difference equation

$$C_2v_2(t)=C_2v_2\left(t-T_c\right)-C_1v_1\left(t-T_c\right) \tag{11}$$

Implicit in Equation 11 is the assumption that the output of the circuit is sampled at the instants $t+nT_c$ of operation (when ϕ is high). Now applying the z transformation to Equation 11 yields

$$C_2V_2(z)=C_2z^{-1}V_2(z)-C_1z^{-1}V_1(z) \tag{12}$$

The transfer function of the integrator is then obtained as

$$\frac{V_2}{V_1}(z)=-\left(\frac{C_1}{C_2}\right)\frac{z^{-1}}{1-z^{-1}} \tag{13}$$

Readers familiar with numerical integration methods can easily recognize that Equation 13 is the forward-Euler numerical integration scheme. Therefore, the integrator in Figure 2 with the output signal sampled when switch-1 is closed (ϕ is high) is known as the *F*orward-*E*ular *D*iscrete *I*ntegrator (FEDI).

2.3 PARASITIC CAPACITORS

A major problem with the integrator of Figure 2 is that it suffers from the effect of the inevitable parasitic capacitors which are associated with the switches and capacitors due to the MOS fabrication process. These parasitic capacitors have considerable effect on the frequency response of SC filters. It is difficult, if not impossible, to compensate for these parasitics since they do not have specific values, but rather they assume a range of values that are poorly controlled. The sources of these parasitic are briefly described next.[16,17]

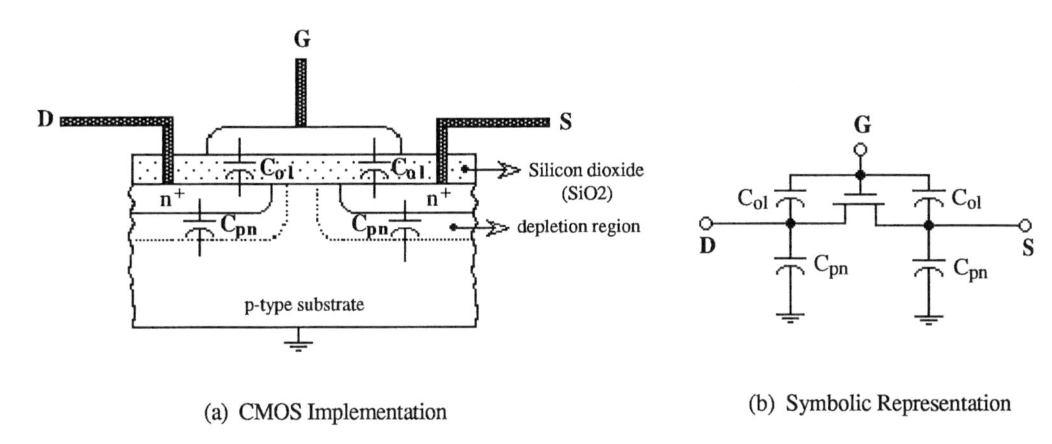

Figure 3 NMOS switch and the associated parasitic capacitances. (a) CMOS implementation; (b) symbolic representation.

2.3.1 Parasitic Capacitors Associated with the MOS Switch

Associated with every MOS transistor, on a chip, are some undesired capacitance values. This is inherent in the MOS technology itself. Figure 3 shows an NMOS switch together with the major parasitic capacitors associated with it. C_{ol} are the gate-diffusion overlap capacitance values due to errors in the mask alignment. They are about 0.005 pF each and couple the clock signals into the filter. This clock feedthrough appears at the output as an additional DC offset. These parasitic capacitors can be greatly reduced by using self-alignment MOS technologies that result in very small overlaps between the gate and both source and drain of a transistor. C_{pn} are the source (drain)-substrate capacitance values due to the reversed-biased *pn* junctions between the source (drain) and the substrate. They are about 0.02 pF each.

2.3.2 Parasitic Capacitors Associated with the MOS Capacitor

Figure 4 shows a cross section of a typical MOS capacitor using the double polysilicon MOS technology together with the associated parasitic capacitors and an equivalent circuit model. C_{p1} is a sizable parasitic capacitance that exists from the bottom plate of the MOS capacitor, C, to the substrate. This is the capacitance of the silicon dioxide layer under the first layer of polysilicon as shown in Figure 4. Typically, C_{p1} can have a value in the range of 5 to 20% of C. C_{p2} is a small capacitance from the top plate of C to the substrate, and it is due to the interconnection of C to other circuitry. Typically, values of C_{p2} can be from 0.1 to 1% of C, depending on the layout technique.

Recognizing that the range of capacitance values of the capacitors used as passive components in SC filters are typically from 0.01 to 100 pF, therefore, the effects of the above parasitic on the frequency response should not be ignored. The design of SC filters must be done in such a way that the parasitic capacitors do not degrade the performance of the filter as will be seen later.

The effect of the parasitic capacitors is illustrated in Figure 5 for the FEDI. Most of the top and bottom plate parasitic capacitors, shown in Figure 5a, have no effect on V_2 since they do not deliver any charge to C_2 (C_{p12}, C_{pn22}, and C_{p21} are shorted assuming that the open-loop gain of the op amp is sufficiently high, C_{pn11} and C_{p22} are voltage driven, and, therefore, they are insignificant). The overlap capacitance (C_{ol1} and C_{ol2}) can be drastically reduced in the fabrication process (using the self-alignment techniques); therefore, they are irrelevant. The parasitic capacitors that have a direct effect on the output of the integrator are C_{pn12}, C_{p11}, and C_{pn21}. Their parallel combination is represented by C_p as shown in Figure 5b. C_p is poorly controlled and may be as large as 0.05 pF, making the net value of C_1 uncertain. If, for example, a 1% accuracy is required for C_1, it should have a value greater than 5 pF. Since, for practical reasons, the integrating capacitance $C_2 >> C_1$, the integrator will, therefore, consume a large chip area.

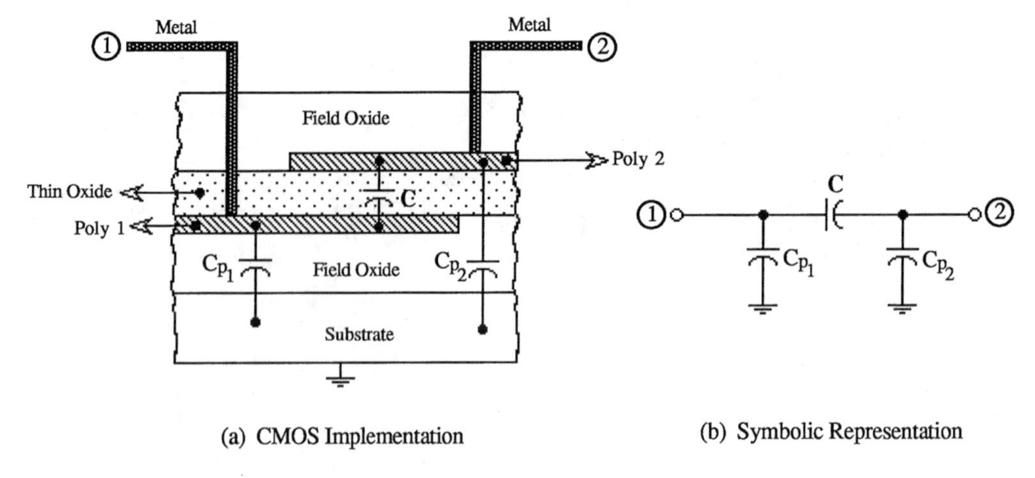

Figure 4 Double polysilicon CMOS capacitor and the associated parasitic capacitances. (a) CMOS implementation; (b) symbolic representation.

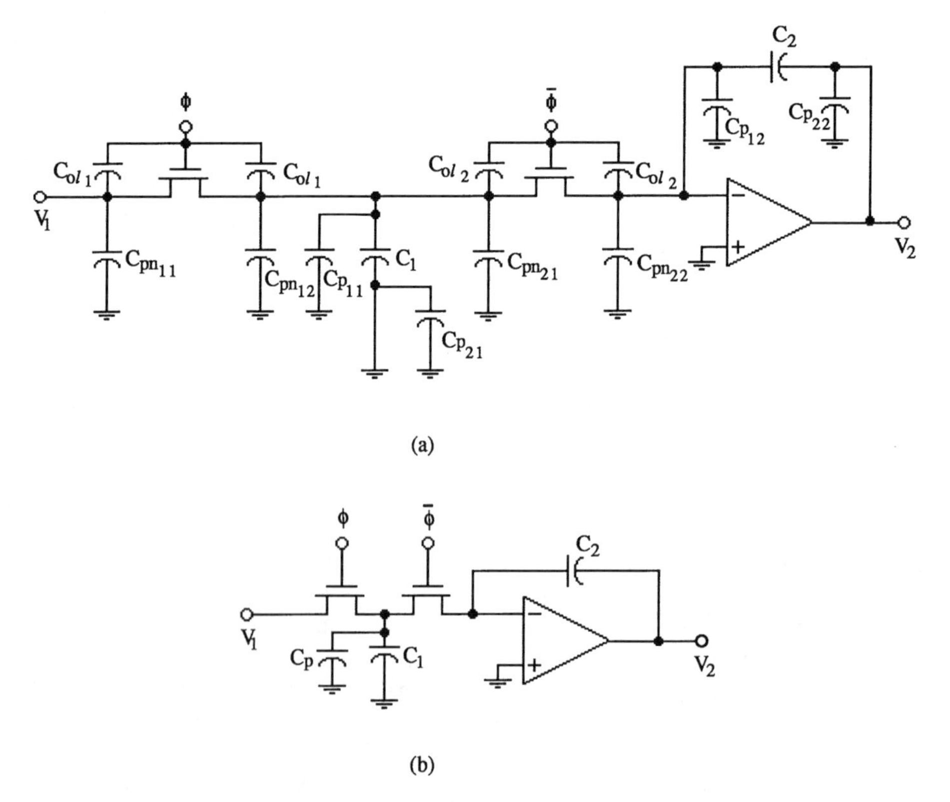

Figure 5 (a) SC integrator and the associated parasitic capacitances; (b) effect of the parasitics on the performance of SC integrator.

2.4 PARASITIC-INSENSITIVE SWITCHED-CAPACITOR INTEGRATORS

The effect of the parasitic capacitors can be eliminated by the use of the parasitic-insensitive integrator configurations shown in Figure 6.[12,22] The circuit in Figure 6a is the parasitic-insensitive version of the FEDI, where C_1 is charging through the top plate and discharging through the bottom plate. Therefore, the transfer function of this integrator, with v_2 sampled when ϕ is high, is a noninverting one. It should be noted that if v_1 is sampled and held for a full clock cycle, then v_2 will also be held constant for a full cycle. The integrator shown in Figure 6c has an inverting

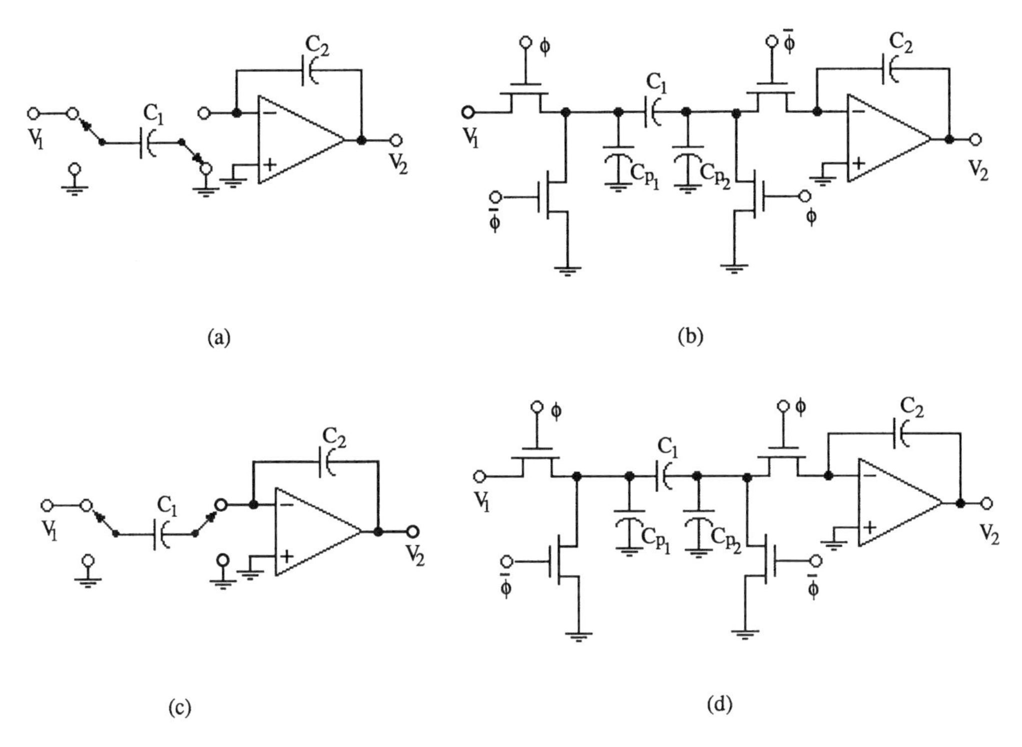

Figure 6 (a) Parasitic-insensitive FEDI; (b) MOS realization of (a); (c) parasitic-insensitive BEDI; (d) MOS realization of (c).

transfer function. Applying the charge equation method, previously described in Section 2.2, to the integrator of Figure 6c and assuming v_2 is sampled when ϕ is high, one can easily derive the following transfer function:

$$\frac{V_2}{V_1}(z) = -\left(\frac{C_1}{C_2}\right)\frac{1}{1-z^{-1}} \tag{14}$$

It should be noted that the numerical integration in Equation 14 is similar to the backward-Euler numerical integration scheme. Therefore, the integrator in Figure 6c, with the output signal sampled when ϕ is high, is known as the *B*ackward-*E*uler *D*iscrete *I*ntegrator (BEDI).

The MOS implementations along with the relevant parasitic capacitors, C_{p1} and C_{p2}, are also given in Figures 6b and d. Note that C_{p1} and C_{p2} are periodically switched between v_1-ground and the op amp virtual ground–ground, respectively. Therefore, they do not deliver any charge to C_2. Thus, the integrator output voltage, v_2, is insensitive to these parasitic capacitors. Also, as a result of this insensitivity, the capacitance values C_1 and C_2 need not be much larger than the parasitic. Therefore, the size of the integrator can be made smaller than that of the parasitic-sensitive realization.

2.5 SWITCHED-CAPACITOR BIQUAD

SC circuits that realize the discrete-time (DT) biquadratic transfer function of Equation 15 are known in the literature as SC biquad. They are the most useful circuits to filter designers because they form a fundamental building block for the construction of higher-order filters in the form of cascade or multiple-loop feedback topologies. In this section we will introduce an SC biquad that will be used later to realize higher-order filters. The SC circuit shown in Figure 7 realizes the general DT biquadratic transfer function[23]

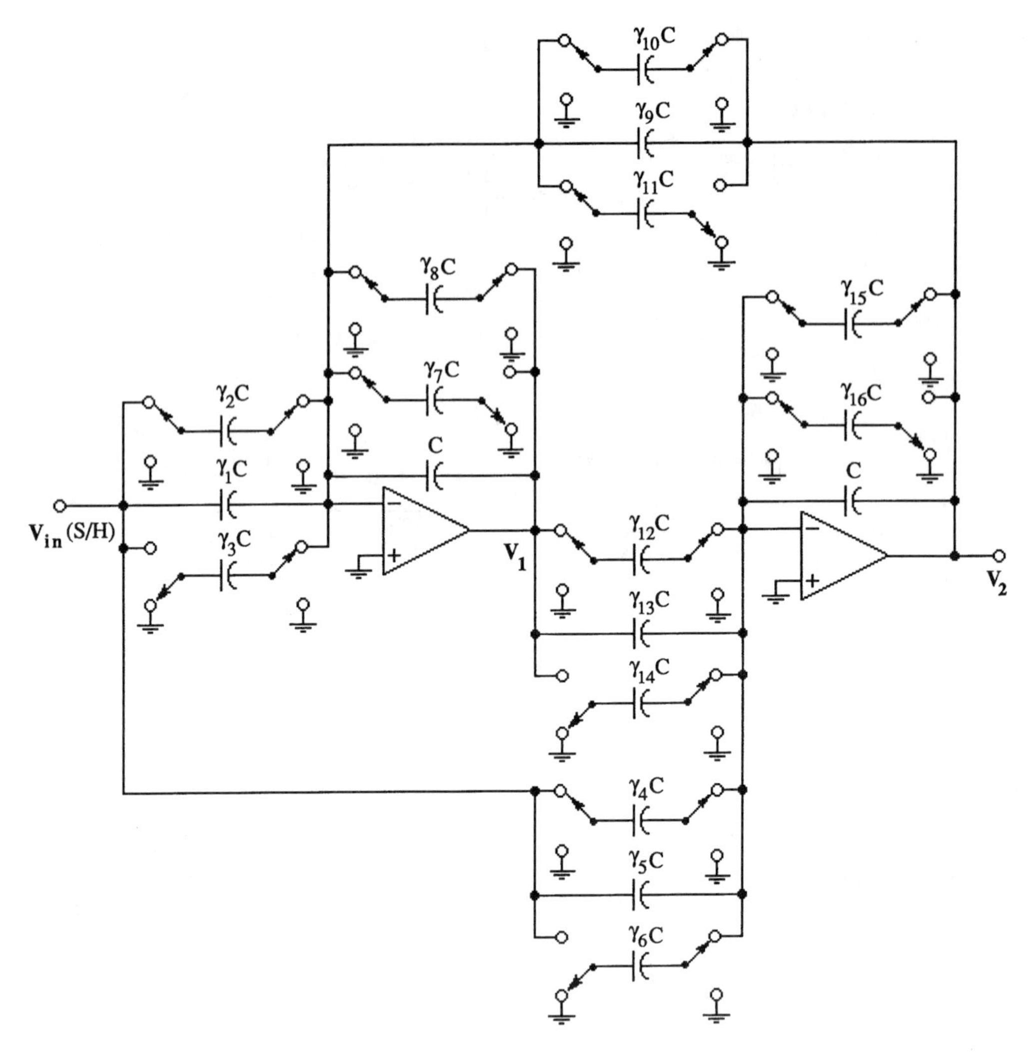

Figure 7 General SC biquad.

$$T(z) = \frac{\alpha_o + \alpha_1 z^{-1} + \alpha_2 z^{-2}}{1 + \beta_1 z^{-1} + \beta_2 z^{-2}} \tag{15}$$

with no exceptions. The circuit is based on the two-integrator loop topology and implements the parasitic-insensitive inverting and noninverting BEDI and FEDI, respectively.

Damping is provided, in general, by the SCs $\gamma_8 C$ and $\gamma_{15} C$ (resistive damping) and by the SC $\gamma_{10} C$ and $\gamma_{12} C$ (capacitive damping). Usually only one of these four SCs is used. The SC $\gamma_7 C$, $\gamma_{11} C$, and $\gamma_{16} C$ realize positive feedback and the SCs $\gamma_7 C$ and $\gamma_{16} C$ provide local positive feedback. Typically, only one of these capacitors is used. Transmission zeros are realized via the feedforward approach; that is, the input signal is fed to the input terminal of both integrators through capacitors $\gamma_1 C$ and $\gamma_5 C$ and SCs $\gamma_2 C$, $\gamma_3 C$, $\gamma_4 C$, and $\gamma_6 C$. Assuming that the switches are operated by nonoverlapping clocks and the input signal is sampled and held for the full period, T_c, then the voltage-ratio transfer functions of the circuit are obtained as[23]

$$T_1(z) = \frac{V_1}{V_{\text{in}}}(z) = \frac{a_o + a_1 z^{-1} + a_2 z^{-2}}{c_o + c_1 z^{-1} + c_2 z^{-2}} \tag{16a}$$

and

$$T_2(z) = \frac{V_2}{V_{\text{in}}}(z) = \frac{b_o + b_1 z^{-1} + b_2 z^{-2}}{c_o + c_1 z^{-1} + c_2 z^{-2}} \tag{16b}$$

where

$$\begin{aligned}
a_o &= (\gamma_4 + \gamma_5)(\gamma_9 + \gamma_{10}) - (\gamma_1 + \gamma_2)(1 + \gamma_{15}),\\
a_1 &= (\gamma_1 + \gamma_2)(1 + \gamma_{16}) + (\gamma_1 + \gamma_3)(1 + \gamma_{15}) - (\gamma_4 + \gamma_5)(\gamma_9 + \gamma_{11}) - (\gamma_5 + \gamma_6)(\gamma_9 + \gamma_{10}),\\
a_2 &= (\gamma_5 + \gamma_6)(\gamma_9 + \gamma_{11}) - (\gamma_1 + \gamma_3)(1 + \gamma_{16}),\\
b_o &= (\gamma_1 + \gamma_2)(\gamma_{12} + \gamma_{13}) - (\gamma_4 + \gamma_5)(1 + \gamma_8),\\
b_1 &= (\gamma_4 + \gamma_5)(1 + \gamma_7) + (\gamma_5 + \gamma_6)(1 + \gamma_8) - (\gamma_1 + \gamma_2)(\gamma_{13} + \gamma_{14}) - (\gamma_1 + \gamma_3)(\gamma_{12} + \gamma_{13}),\\
b_2 &= (\gamma_1 + \gamma_3)(\gamma_{13} + \gamma_{14}) - (\gamma_5 + \gamma_6)(1 + \gamma_7),\\
c_o &= (1 + \gamma_8)(1 + \gamma_{15}) - (\gamma_9 + \gamma_{10})(\gamma_{12} + \gamma_{13}),\\
c_1 &= (\gamma_9 + \gamma_{11})(\gamma_{12} + \gamma_{13}) + (\gamma_9 + \gamma_{10})(\gamma_{13} + \gamma_{14}) - (1 + \gamma_7)(1 + \gamma_{15}) - (1 + \gamma_8)(1 + \gamma_{16}),
\end{aligned} \tag{16c}$$

and

$$c_2 = (1 + \gamma_7)(1 + \gamma_{16}) - (\gamma_9 + \gamma_{11})(\gamma_{13} + \gamma_{14})$$

By using the bilinear z transformation

$$S = 2f_c \frac{1 - z^{-1}}{1 + z^{-1}} \tag{17}$$

the equivalent analog pole frequency, ω_o, and its Q-factor are obtained as

$$\omega_o = 2f_c \sqrt{\frac{c_o + c_1 + c_2}{c_o - c_1 - c_2}} \quad \text{and} \quad Q = \frac{\sqrt{(c_o + c_1 + c_2)(c_o - c_1 + c_2)}}{2(c_o - c_1)} \tag{18}$$

It should be noted that as a result of the sampled and held input the output voltages V_1 and V_2 will also be held constant for the full period T_c. This feature is very useful for the realization of higher-order SC filters in cascade or multiple-loop feedback topologies.

Other popular SC biquads based on the two-integrator loop topology can easily be generated from the circuit of Figure 7. For example, the biquad introduced by Fleischer and Laker[24] can be obtained from the circuit of Figure 7 by letting $\gamma_1 = \gamma_5 = \gamma_7 = \gamma_8 = \gamma_{11} = \gamma_{12} = \gamma_{13} = \gamma_{16} = 0$. Also, the biquad by Martin and Sedra[25] can be obtained from Figure 7 by letting $\gamma_3 = \gamma_4 = \gamma_6 = \gamma_7 = \gamma_8 = \gamma_{11} = \gamma_{12} = \gamma_{13} = \gamma_{15} = \gamma_{16} = 0$.

3. CASCADE AND COUPLED-BIQUAD SWITCHED-CAPACITOR FILTERS

SC filters making use of the bilinear z transformation and implementing SC biquads have been reported in the literature in different structures, such as the cascade,[16,26,27] the leapfrog and nodal simulation,[25] the follow-the-leader feedback (FLF),[28] the inverse FLF (IFLF),[29] and the generalized multiple-loop feedback structures.[30] In this section we will demonstrate the cascade and the coupled-biquad SC filter design strategy.

3.1 CASCADE SC FILTER DESIGN

The cascade method is by far the simplest and the most popular for filter realization. In this approach the filter transfer function $T(z)$ is realized as the voltage transfer ratio of a cascade of filter sections (modules), each having a second-order (biquadratic) transfer function. The realization involves two major steps:

Step l: Decompose the nth-order transfer function $T(z)$ into the product of biquadratic transfer functions

$$T_j(z) = \frac{\alpha_{0j} + \alpha_{1j}z^{-1} + \alpha_{2j}z^{-2}}{1 + \beta_{1j}z^{-1} + \beta_{2j}z^{-2}}$$

i. e.,

$$T(z) = \frac{\sum_{i=0}^{i=n} \alpha_i z^{-i}}{1 + \sum_{k=1}^{k=n} \beta_k z^{-k}} = \prod_{j=1}^{N} T_j(z), \qquad N = \begin{cases} \dfrac{n}{2}, & n \text{ is even} \\ \dfrac{n+1}{2}, & n \text{ is odd} \end{cases}$$

It is assumed that $T(z)$ is obtained by applying the bilinear z transformation, Equation 17, to the continuous-time (CT) transfer function $T(s)$ that meets the prewarped filter specification. Note that for n is odd, we let $m = n + 1$ (an even number) and set $\alpha_m = \beta_m = 0$ in the realization.

Step (2): Realize each biquadratic transfer function T_j using the basic SC circuit of Figure 8.

The SC biquad of Figure 8 is obtained from the one of Figure 7 after removing many of the unnecessary capacitors and switches. The transfer functions T' and T are easily obtained using Equation 16) as

$$T'(z) = \frac{V'}{V_{\text{in}}}(z) = \frac{(CF + CG - AI) + (BI + AI - DF - DG - CG)Z^{-1} + (DG - BI)Z^{-2}}{HI + (EF + EG - 2HI)Z^{-1} + (HI - EG)Z^{-2}} \tag{19a}$$

$$T(z) = \frac{V}{V_{\text{in}}}(z) = -\frac{CH + (AE - CH - DH)Z^{-1} + (DH - BE)Z^{-2}}{HI + (EF + EG - 2HI)Z^{-1} + (HI - EG)Z^{-2}} \tag{19b}$$

The first step (the decomposition problem) involves the following issues:[31]

1. Pole–zero pairing (there are $N!$ possible different combinations to choose from).
2. Cascade sequence (there are $N!$ possible different sequences).
3. Gain distribution.

The solution of the decomposition problem affects the filter dynamic range more than any other performance measure. Dynamic range maximization techniques will be discussed in Section 3.4.

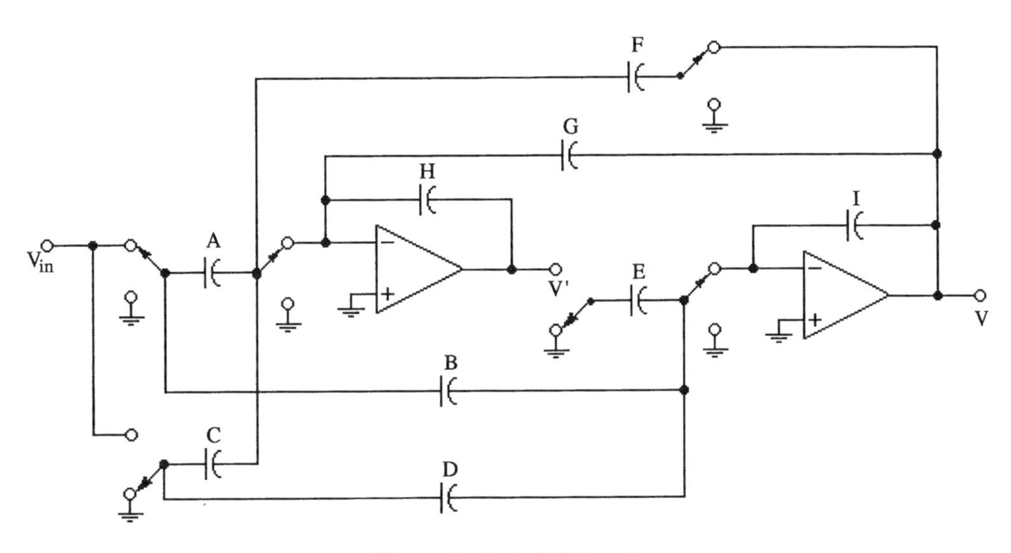

Figure 8 Basic SC biquad.

Cascade Design Example

The cascade design is demonstrated by the design of a stagger-tuned six-order Chebyshev bandpass filter. The filter has a maximum of 1 dB attenuation in the passband that extends between 800 and 1250 Hz.

Using the bilinear z transformation with clock frequency of 32 kHz, the cascade design is accomplished by the following step-by-step procedure:

Step 1: Obtain the CT specifications from the given one by applying the prewarping equation

$$\frac{\Omega}{2f_c} = \tan\left(\frac{\omega}{2f_c}\right) \tag{20}$$

where Ω and ω are the CT and the DT frequency variables, respectively, and f_c is the clock frequency.

Step 2: Obtain the CT transfer function, $T(s)$, by consulting the tables available for CT filters.

$$T(s) = \left(\frac{2231.05s}{s^2 + 1397.23s + 3.95 \times 10^7}\right)\left(\frac{2231.05s}{s^2 + 550.03s + 2.56 \times 10^7}\right)\left(\frac{2231.05s}{s^2 + 847.21s + 6.08 \times 10^7}\right)$$

Step 3: Obtain the DT transfer function, $T(z)$, by applying the bilinear z transformation to $T(s)$; that is,

$$T(z) = T(s)\Big|_{g=\frac{2z-1}{f_c z+1}} = \left(\frac{0.034(1 - z^{-2})}{1 - 1.92z^{-1} + 0.958z^{-2}}\right)\left(\frac{0.034(1 - z^{-2})}{1 - 1.958z^{-1} + 0.983z^{-2}}\right)$$

$$\left(\frac{0.034(1 - z^{-2})}{1 - 1.917z^{-1} + 0.974z^{-2}}\right) = T_1T_2T_3$$

Step 4: Use the biquad circuit of Figure 8 to realize T_1, T_2, and T_3; then connect them in cascade as shown in the circuit of Figure 9 with the capacitance values given in Table 1. The magnitude response of the circuit of Figure 9 is depicted in Figure 10.

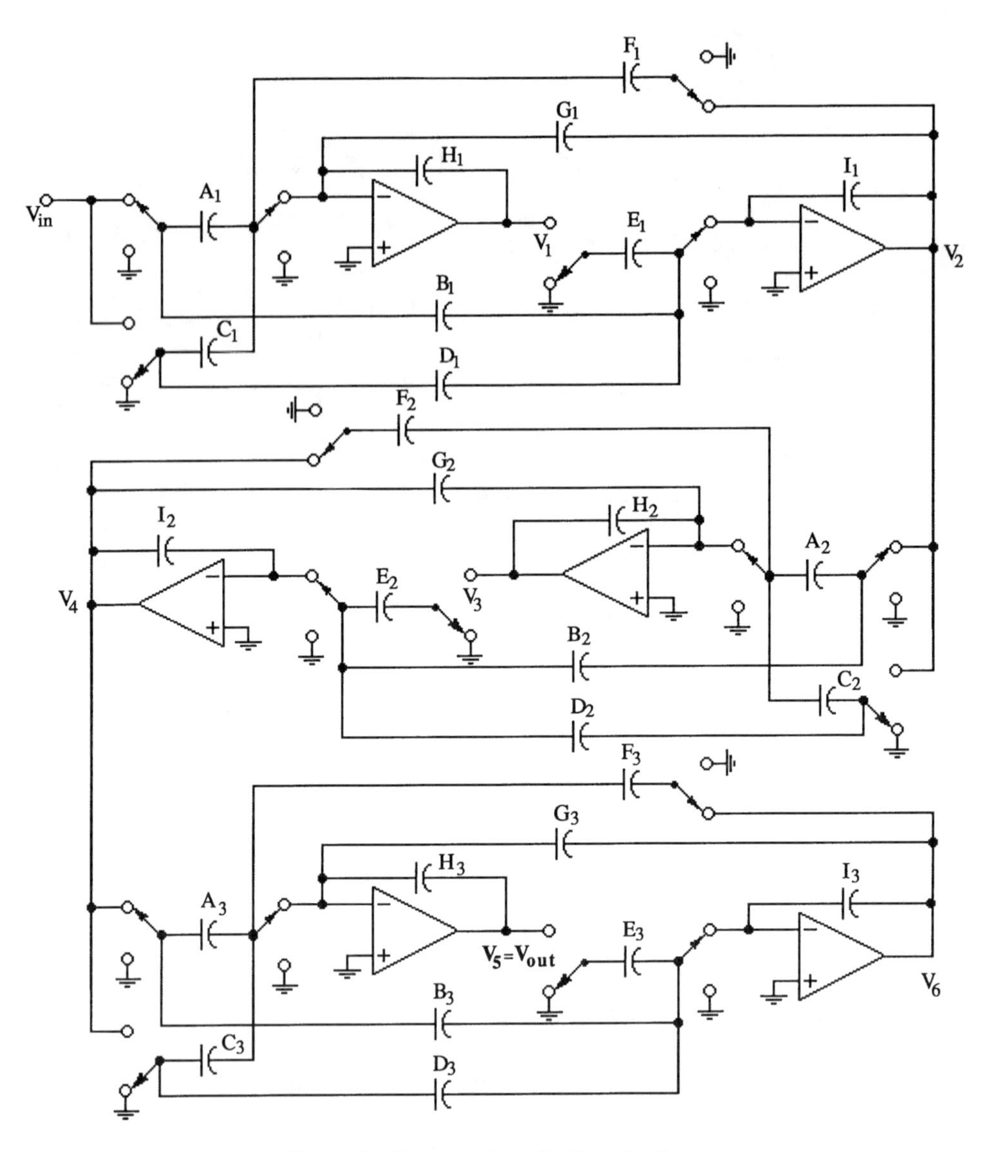

Figure 9 Six-order cascade filter structure.

Table 1 Capacitance Values for the Circuit of Figure 9

	$i = 1$	$i = 2$	$i = 3$
A_i	0.611	0.2303	0.2056
B_i	0.807	1.0	0.6666
C_i	0.2	0.1319	0.1
D_i	0.807	1.0	0.6666
E_i	0.206	0.1572	0.2602
F_i	0.593	0.1572	0.2602
G_i	0.206	0.1075	0.09916
H_i	1.0	1.0	1.0
I_i	1.0	1.0	1.0

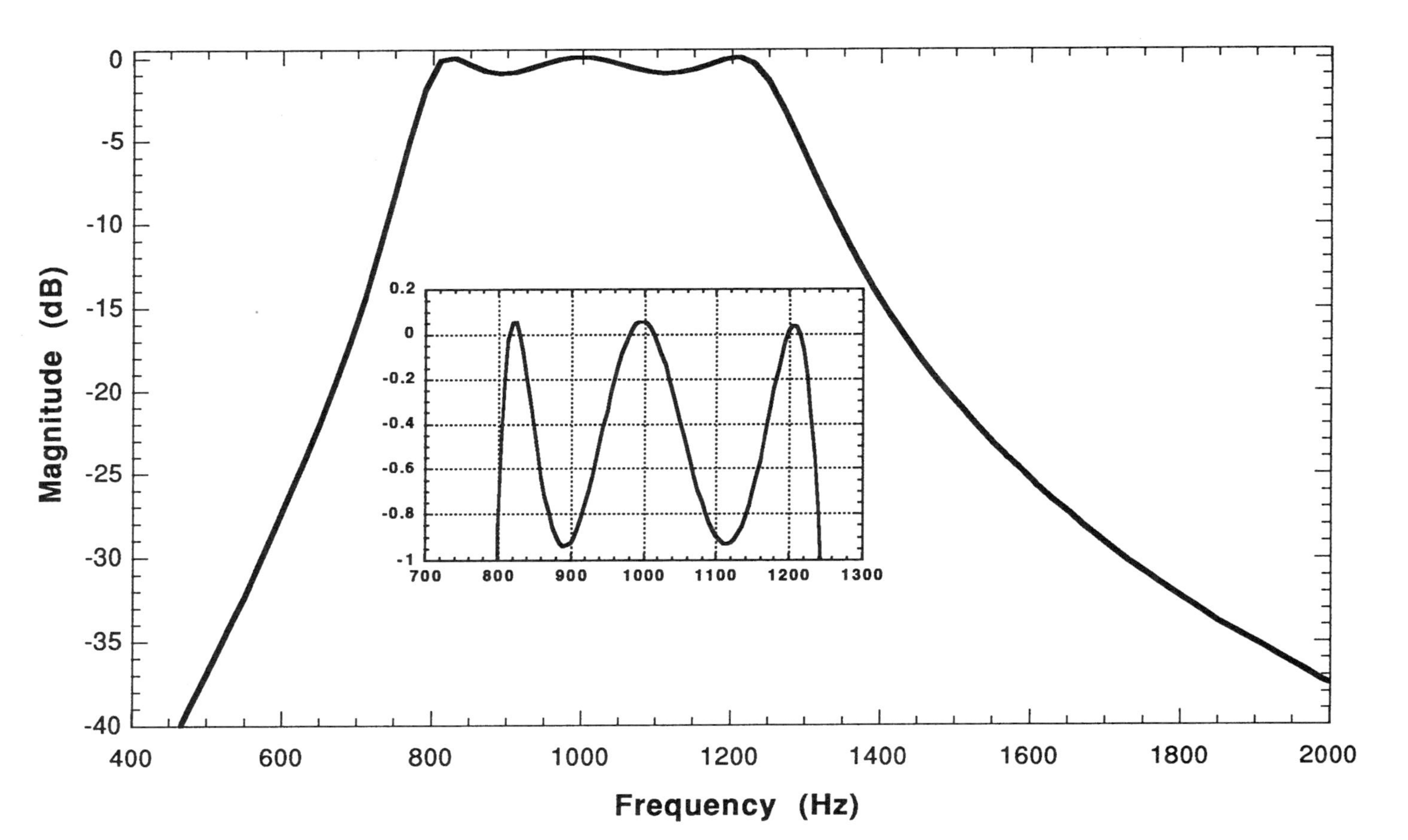

Figure 10 Magnitude response of the six-order Chebyshev bandpass filter.

3.2 COUPLED-BIQUAD SC FILTER DESIGN

The cascade realization, though simple, is often (especially for high-Q narrowband filters) found to be quite sensitive to component variations in the passband and, therefore, unacceptable in practice. On the other hand, realizations that are based on multiple-loop feedback topologies can have better passband sensitivities if couplings between sections are carefully chosen. Two of the multiple-loop feedback SC filter structures that have been found useful in practice, namely, the FLF and the IFLF, will be presented.

3.2.1 Biquadratic State-Space Representation

In the following we will present state-space formulations that represent the general nth-order DT transfer function, $T(z)$, that meets the specifications of the SC filter. These representations result in simple topologies (e.g., FLF and IFLF) with all the feedback and feedforward coefficients determined directly from the coefficients of $T(z)$. Furthermore, it would be possible to realize SC filters implementing the modular parasitic-insensitive biquad. In general $T(z)$ is given as

$$T(z) = \frac{V_{\text{out}}}{V_{\text{in}}}(z) = \frac{\sum_{j=0}^{n} \alpha_j z^{-j}}{1 + \sum_{k=1}^{n} \beta_k z^{-k}} \tag{21}$$

State-space representation of Equation 21 is obtained as[32]

$$\begin{aligned} z\left[\frac{x_o}{x_e}\right] &= \left[\begin{array}{c|c} A_{11} & A_{12} \\ \hline A_{21} & A_{22} \end{array}\right]\left[\frac{x_o}{x_e}\right] + \left[\frac{B_o}{B_e}\right]V_{\text{in}} \\ V_{\text{out}} &= \left[C_o \,|\, C_e\right]\left[\frac{x_o}{x_e}\right] + dV_{\text{in}} \end{aligned} \tag{22}$$

where

$$x_o = \left[x_1 \;\; x_3 \;\; \ldots \;\; x_{n-3} \;\; x_{n-1}\right]^t$$

$$x_e = \left[x_2 \;\; x_4 \;\; \ldots \;\; x_{n-2} \;\; x_n\right]^t$$

and

$$d = \alpha_o$$

The submatrices $\mathbf{A_{ij}}$ and the vectors $\mathbf{B_o}$, $\mathbf{B_e}$, $\mathbf{C_o}$, and $\mathbf{C_e}$ are defined based on the desired topology. Here we will consider two well-known structures, namely, the FLF and the IFLF.

3.2.2 Synthesis of FLF SC Filters

For the FLF topology the submatrices and the vectors in Equation 22 are given as[28]

$$A_{11} = \begin{bmatrix} -\beta_1 & -\beta_3 & \cdots & -\beta_{n-3} & -\beta_{n-1} \\ \hline 0 & 0 & \cdots & 0 & 0 \\ 0 & 0 & \cdots & 0 & 0 \\ \vdots & \vdots & \ddots & \vdots & \vdots \\ 0 & 0 & \cdots & 0 & 0 \end{bmatrix}, \quad A_{12} = \left[\begin{array}{cccc|c} -\beta_2 & -\beta_4 & \cdots & -\beta_{n-2} & -\beta_n \\ \hline 1 & 0 & \cdots & 0 & 0 \\ 0 & 1 & \cdots & 0 & 0 \\ \vdots & \vdots & \ddots & \vdots & \vdots \\ 0 & 0 & \cdots & 1 & 0 \end{array}\right]$$

$$A_{21} \equiv \frac{n}{2} \times \frac{n}{2} \ \text{Identity Matrix}, \ A_{22} \equiv 0$$

$$B_o = \left[b_1 \ b_3 \ \ldots \ b_{n-3} \ b_{n-1}\right]^t$$

$$B_e = \left[b_2 \ b_4 \ \ldots \ b_{n-4} \ b_{n-2}\right]^t \tag{23}$$

$$b_i = \alpha_{n-i} - \sum_{\substack{k=1 \\ i \neq n}}^{n-i} \beta_k b_{i+k}, \quad i = 1, 2, \ldots, n$$

$$C_o \equiv 0 \quad \text{and} \quad C_e = [0 \ 0 \ \ldots \ 1]$$

Eliminating $\mathbf{x_0}$ from Equation 22, making use of Equations 23 and letting $x_e = W$, yields the following biquadratic representation

$$W = \left(z^{-1}A_{11} + z^{-2}A_{12}\right)W + \left(z^{-1}B_o + z^{-2}\hat{B}\right)V_{\text{in}}$$
$$V_{\text{out}} = w_{n/2} + \alpha_o V_{\text{in}} \tag{24}$$

where

$$\mathbf{W} = \left[\mathbf{w}_1 \ \mathbf{w}_2 \ \cdots \ \mathbf{w}_{n/2}\right]^t$$

$$\hat{\mathbf{B}} = \left[\left(\mathbf{b_0} + \sum_{j=1}^{n/2} \beta_{2j-1} b_{2j-1}\right) \ b_2 \ b_4 \ \ldots \ b_{n-4} \ b_{n-2}\right]^t$$

In Equations 22 and 24, it is assumed that n is an even number; they still hold, however, if n is odd. In this case we use $m = n + 1$ (an even number) and set $\alpha_m = \beta_m = 0$. A flow diagram representation of Equation 24 is shown in Figure 11. A modified version of Figure 11 is shown in Figure 12, where the output summer has been eliminated, and, therefore, it becomes more suitable for realization using SC biquads. According to Figure 12, we have

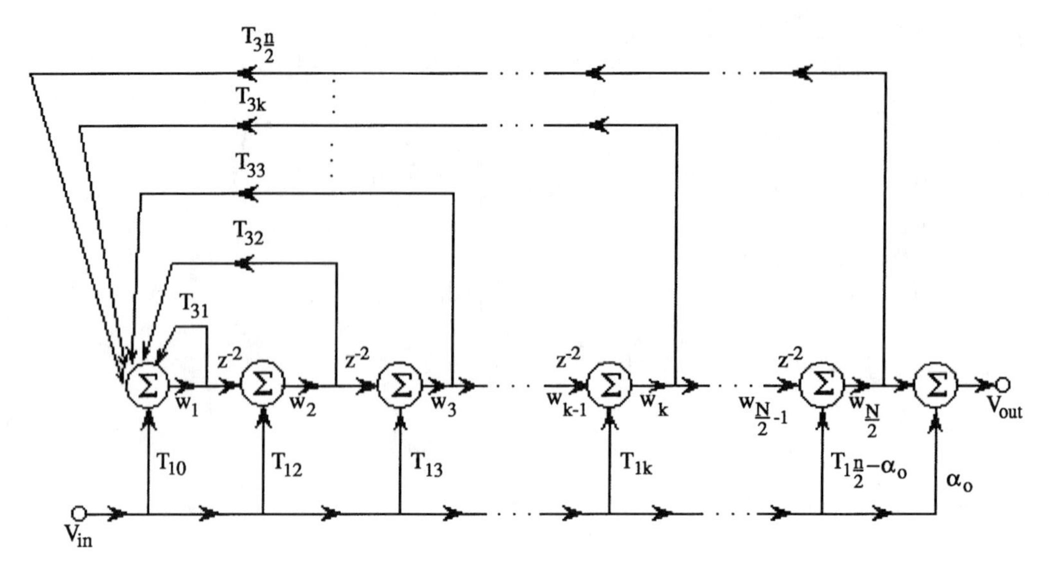

Figure 11 Signal flow graph representation of Equation 24.

$$
\begin{aligned}
w_1 &= T_{11}(z)V_{\text{in}} + \sum_{k=1}^{n/2} T_{3k}(z)w_k \\
w_k &= T_{1k}(z)V_{\text{in}} + T_{2k}(z)w_{k-1} \quad \text{for} \quad k = 2, 3, \ldots, n/2
\end{aligned}
\tag{25}
$$

where the transfer functions, $T_{ij}(z)$, are defined as

$$
T_{11}(z) = \left(b_1 + \alpha_o \beta_{n-1}\right)z^{-1} + \left(\alpha_n - \sum_{i=1}^{n/2} b_{2i}\beta_{2i}\right)z^{-2}
$$

and

$$
\begin{aligned}
T_{1k}(z) &= b_{2k-1}z^{-1} + b_{2k-2}z^{-2} \quad \text{for} \quad k = 2, 3, \ldots, (n/2) - 1 \\
T_{1n/2}(z) &= \alpha_o + b_{n-1}z^{-1} + b_{n-2}z^{-2} \\
T_{2k}(z) &= z^{-2} \quad \text{for} \quad k = 2, 3, \ldots, n/2
\end{aligned}
\tag{26}
$$

and

$$
T_{3k}(z) = -\beta_{2k-1}z^{-1} - \beta_{2k}z^{-2} \quad \text{for} \quad k = 1, 2, \ldots, n/2
$$

The individual sections of Figure 12 can be realized by the SC biquad circuits depicted in Figures 13 and 14. These circuits are generated from the general SC biquad in Figure 7, with the number of switches minimized and with additional input nodes to accommodate more than one input signal. Figure 13 is used to realize section #1, while Figure 14 is used to realize the remaining sections, Sec. # 2 to Sec. # *n*/2).

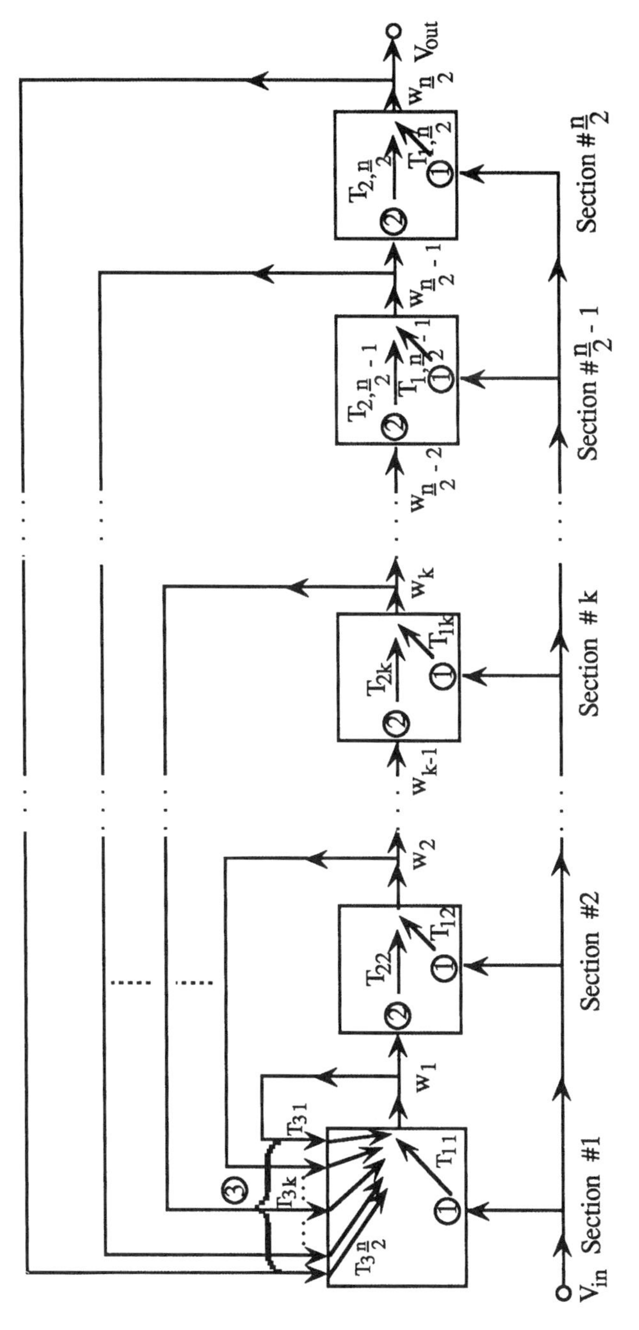

Figure 12 Modified version of Figure 11.

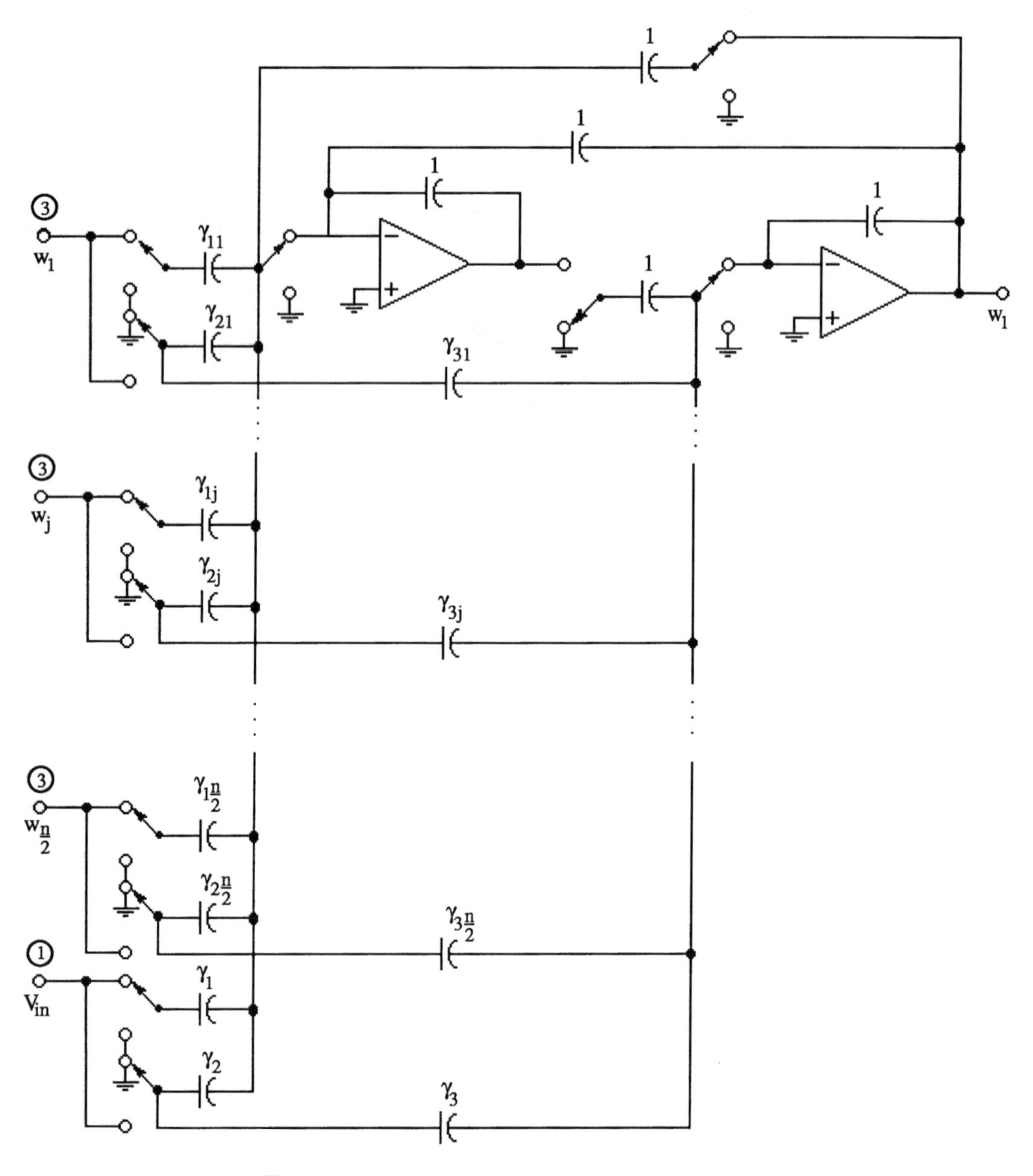

Figure 13 SC realization of section #1 in Figure 12.

The output voltages of the circuits of Figures 13 and 14 are obtained from Equations 16 as

$$w_1 = \left[(\gamma_3 - \gamma_1)z^{-1} + (\gamma_2 - \gamma_3)z^{-2}\right]V_{in} + \sum_{k=1}^{n/2}\left[(\gamma_{3k} - \gamma_{1k})z^{-1} + (\gamma_{2k} - \gamma_{3k})z^{-2}\right]w_k$$

$$w_k = \left[-D_k + (C_k + D_k - A_k)z^{-1} + (B_k - C_k)z^{-2}\right]V_{in} + z^{-2}w_{k-1} \quad \text{for} \quad k = 2, 3, \ldots, n/2 \tag{27}$$

and

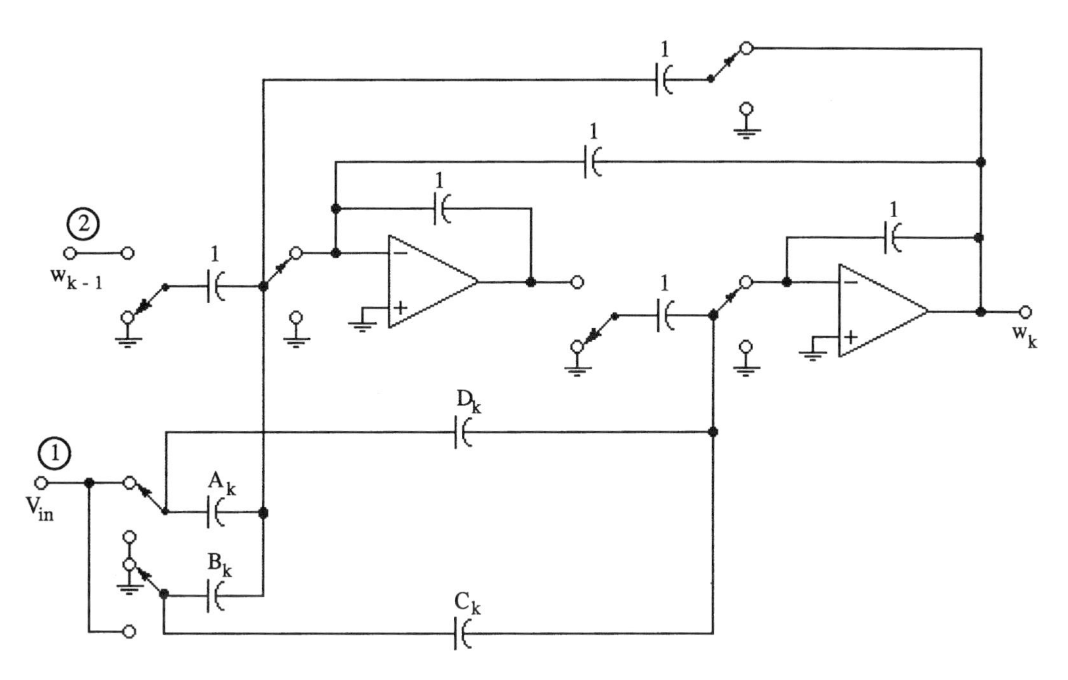

Figure 14 SC realization of the *k*th section in Figure 12.

$$D_k = 0 \quad \text{for} \quad k = 2, 3, \ldots (n/2) - 1$$

From Equations 25 through 27 one can easily obtain the following sets of design equations:

$$\left.\begin{aligned} \gamma_3 - \gamma_1 &= b_1 + \alpha_o B_{n-1} \\ \gamma_2 - \gamma_3 &= \alpha_n - \sum_{i=1}^{(n/2)-1} b_{2i}\beta_{2i} \end{aligned}\right\} \tag{28a}$$

$$\left.\begin{aligned} \gamma_{3k} - \gamma_{1k} &= -\beta_{2k-1} \\ \gamma_{2k} - \gamma_{3k} &= -\beta_{2k} \end{aligned}\right\} \quad \text{for} \quad k = 1, 2, \ldots n/2 \tag{28b}$$

$$\left.\begin{aligned} C_k - A_k &= b_{2k-1} \\ B_k - C_k &= b_{2k-2} \end{aligned}\right\} \quad \text{for} \quad k = 2, 3, \ldots (n/2) - 1 \tag{28c}$$

$$\left.\begin{aligned} D_{n/2} &= -\alpha_o \\ C_{n/2} + D_{n/2} - A_{n/2} &= b_{n-1} \\ B_{n/2} - C_{n/2} &= b_{n-2} \end{aligned}\right\} \tag{28d}$$

In case α_o is a positive number, we may multiply the transfer function to be realized, $T(z)$, by -1, to obtain a positive value for $D_{n/2}$; that is, we realize $-T(z)$. Each set of the above design equations has one degree of freedom; that is, a free design parameter. The choice of these free design parameters can be made to enhance the performance of the realization (e.g., maximizing the dynamic range, minimizing the sensitivity, minimizing the capacitance ratios and/or total capacitance, etc.). This task may be carried out by implementing an optimization routine.[33,34] As far as the sensitivity is concerned, it is shown[32] that the transfer function sensitivities with respect to the capacitance ratios, γ values, are directly proportional to those ratios. That is, the smaller the γ values (closer to zero), the smaller the sensitivities would be. On the other hand, for the filter to consume smaller chip area, γ values should be closer to unity. Therefore, a trade-off between low sensitivity and practical realization must be exercised.

3.2.3 Synthesis of IFLF SC Filters

For the IFLF topology[28] the submatrices $\mathbf{A_{ij}}$ are simply the transpose of those of the FLF topology; that is,

$$A_{11} = \left[\begin{array}{c|cccc} -\beta_1 & 0 & 0 & \dots & 0 \\ -\beta_3 & 0 & 0 & \dots & 0 \\ \vdots & \vdots & \vdots & \ddots & \vdots \\ -\beta_{n-3} & 0 & 0 & \dots & 0 \\ -\beta_{n-1} & 0 & 0 & \dots & 0 \end{array}\right], \quad A_{21} = \left[\begin{array}{c|cccc} -\beta_2 & 1 & 0 & \dots & 0 \\ -\beta_4 & 0 & 1 & \dots & 0 \\ \vdots & \vdots & \vdots & \ddots & \vdots \\ -\beta_{n-2} & 0 & 0 & \dots & 1 \\ \hline -\beta_n & 0 & 0 & \dots & 0 \end{array}\right]$$

$$A_{12} \equiv \frac{n}{2} \times \frac{n}{2} \quad \text{Indentity Matrix}, \quad A_{22} \equiv 0 \tag{29}$$

$$B_o = \left[\alpha_1 - \alpha_o\beta_1 \quad \alpha_3 - \alpha_o\beta_3 \quad \dots \quad \alpha_{n-1} - \alpha_o\beta_{n-1}\right]^t$$

$$B_e = \left[\alpha_2 - \alpha_o\beta_2 \quad \alpha_4 - \alpha_o\beta_4 \quad \dots \quad \alpha_n - \alpha_o\beta_n\right]^t$$

and

$$C_e \equiv 0 \quad \text{and} \quad C_o = [1 \quad 0 \quad \dots \quad 0]$$

In this case, the biquadratic state-space representation can be obtained by eliminating x_e from Equations 22, making use of Equations 29, and letting $x_o = W$, thus, yielding

$$\begin{aligned} W &= \left(z^{-1}A_{11} + z^{-2}A_{21}\right)W + \left(z^{-1}B_o + z^{-2}B_e\right)V_{\text{in}} \\ V_{\text{out}} &= w_1 + \alpha_o V_{\text{in}} \end{aligned} \tag{30}$$

A flow diagram representation of Equations 30 is shown in Figure 15. The output of the kth block shown in Figure 15 is obtained as

$$V_4^k(z) = \sum_{j=1}^{3} H_{jk}(z) V_j^k(z) \tag{31a}$$

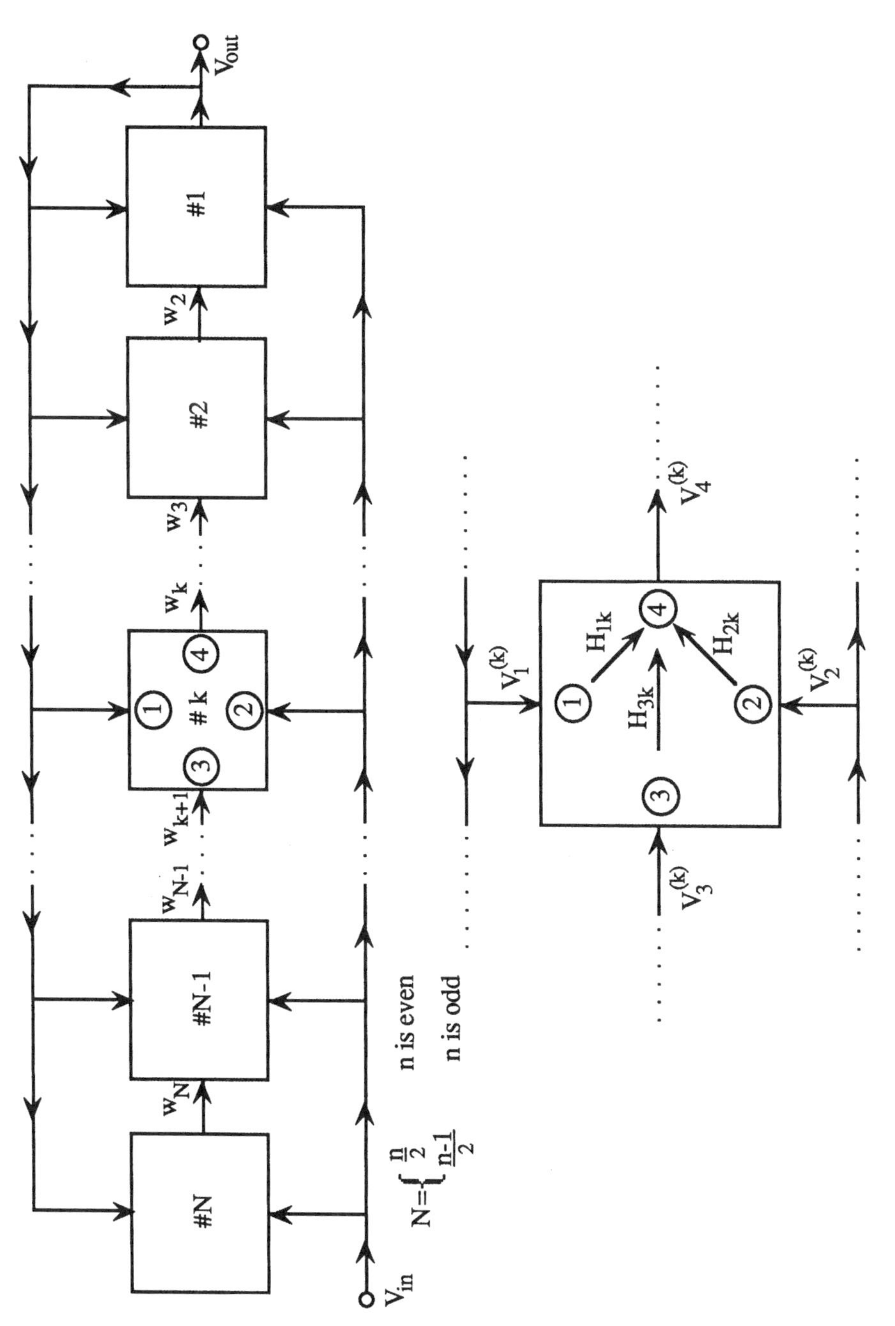

Figure 15 Block diagram representation of Equation 30.

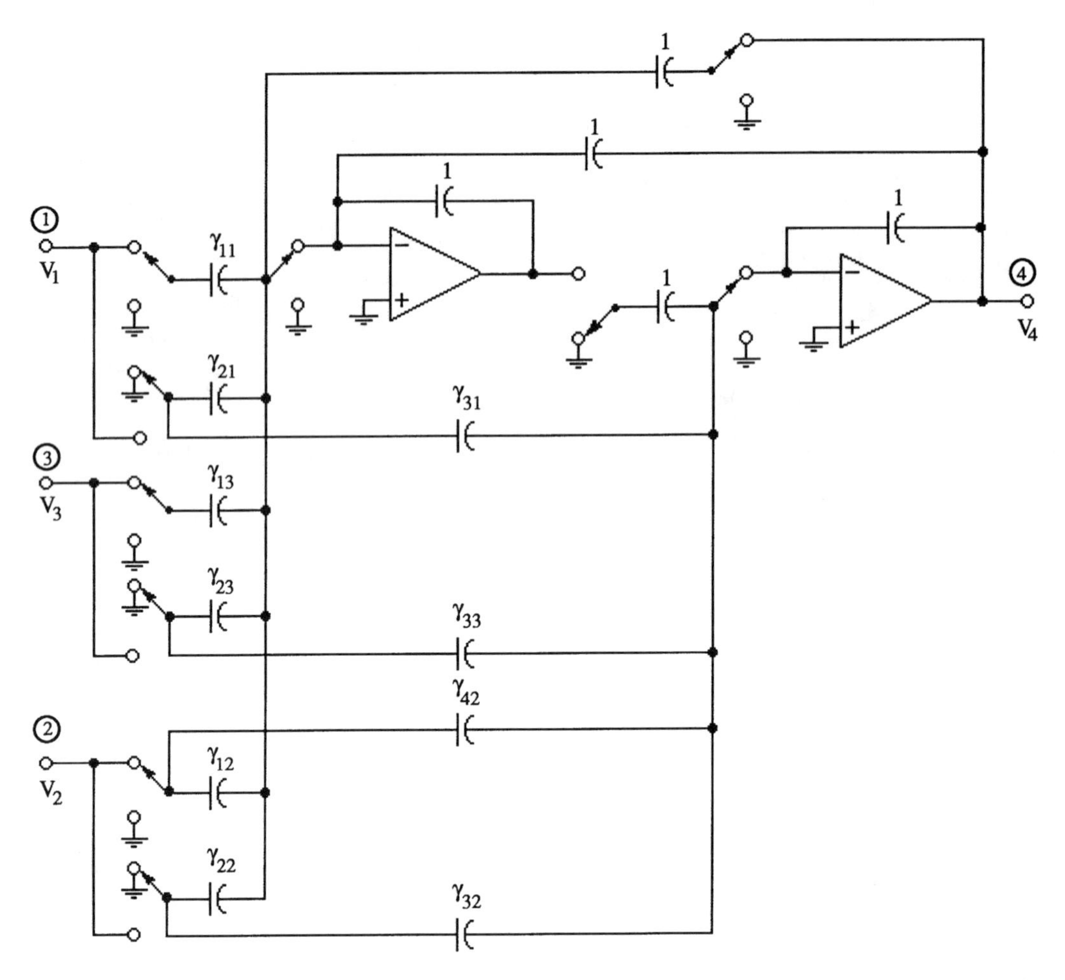

Figure 16 SC realization of the kth section in Figure 15.

where

$$H_{1k}(z) = \left.\frac{V_4^k}{V_1^k}\right|_{V_2^k = V_3^k = 0} = -\beta_{2k-1} z^{-1} - \beta_{2k} z^{-2}, \quad k = 1, 2, \ldots, N$$

$$H_{21}(z) = \left.\frac{V_4^1}{V_2^1}\right|_{V_1^1 = V_3^1 = 0} = \alpha_o + \alpha_1 z^{-1} + \alpha_2 z^{-2}$$

$$H_{2k}(z) = \left.\frac{V_4^k}{V_2^k}\right|_{V_1^k = V_3^k = 0} = \alpha_{2k-1} z^{-1} + \alpha_{2k} z^{-2}, \quad k = 2, 3, \ldots, N \tag{31b}$$

$$H_{3k}(z) = \left.\frac{V_4^k}{V_3^k}\right|_{V_1^k = V_2^k = 0} = z^{-2}, \quad k = 1, 2, \ldots, N-1$$

An SC module that realizes the kth section in Figure 15 is shown in Figure 16. This module is obtained from the general biquad of Figure 7 with the number of switches minimized and with two additional input nodes to accommodate three input signals. By using Equations 16, the output of the SC circuit of Figure 16 is obtained as

$$V_4(z) = \sum_{j=1}^{3} \left[-\gamma_{4j} + \left(\gamma_{3j} + \gamma_{4j} - \gamma_3\gamma_{1j}\right)z^{-1} + \left(-\gamma_3\gamma_{2j} - \gamma_{3j}\right)z^{-2} \right] V_j(z) \tag{32}$$

where $\gamma_{41} = \gamma_{43} = 0$.

From Equations 31 and 32 one can easily obtain the following design equations for the kth section:

$$\begin{aligned} \gamma_{31} - \gamma_{11} &= -\beta_{2k-1} \\ \gamma_{21} - \gamma_{31} &= -\beta_{2k} \end{aligned} \tag{33a}$$

$$\begin{aligned} \gamma_{33} &= \gamma_{13} \\ \gamma_{23} - \gamma_{33} &= 1 \quad k \neq N \end{aligned} \tag{33b}$$

$$\gamma_{42} = \begin{cases} \pm\alpha_o & k = 1 \\ 0 & k = 2, 3, \ldots, N \end{cases} \tag{33c}$$

$$\begin{aligned} \gamma_{42} + \gamma_{32} - \gamma_{12} &= \pm\alpha_{2k-1} \\ \gamma_{22} - \gamma_{32} &= \pm\alpha_{2k} \end{aligned} \tag{33d}$$

If $\alpha_o < 0$ ($\alpha_o > 0$), then the negative (positive) sign in Equation 33c and the positive (negative) sign in Equation 33d are applicable.

Coupled-Biquad Filter Design Example

To illustrate the above coupled-biquad synthesis techniques, we consider the design of a fourth-order FLF highpass elliptic filter with the following specifications: the passband edge frequency is 3141.6 rad/s, the stopband edge frequency is 1570.8 rad/s, the minimum attenuation in the stopband is 47 dB, and the maximum attenuation in the passband is 0.5 dB. By using a clock frequency of 8 kHz, the above attenuation specifications are prewarped to the CT specifications (using Equation 20) and then the CT transfer function, $T(s)$, is obtained by consulting the tables available for CT filters. $T(z)$ is then obtained by applying the bilinear z transformation of Equation 17 to $T(s)$ as

$$T(z) = T(s)\Big|_{s=\frac{2z-1}{f_c z+1}} = \frac{-0.57368 + 2.27037z^{-1} - 3.39352z^{-2} + 2.27037z^{-3} - 0.57368z^{-4}}{1 - 2.94852z^{-1} + 3.42788z^{-2} - 1.83922z^{-3} + 0.39421z^{-4}}$$

By using the FLF topology, the design equations of the required two biquads are obtained from Equations 28 as

1. *For Biquad I:*

$$\gamma_3 - \gamma_1 = 1.11102$$

$$\gamma_2 - \gamma_3 = -1.55273$$

$$\gamma_{31} - \gamma_{11} = 2.94852$$

$$\gamma_{21} - \gamma_{31} = 3.42788$$

$$\gamma_{32} - \gamma_{12} = 1.83922$$

$$\gamma_{22} - \gamma_{32} = 0.39321$$

2. *For Biquad II:*

$$D_2 = 0.57368$$

$$C_2 + D_2 - A_2 = 0.57887$$

$$B_2 - C_2 = 0.27978$$

A solution to the above equations yields the capacitance values given in Table 2. The corresponding SC circuit is shown in Figure 17. The circuit was analyzed using SC analysis program and the magnitude response is shown in Figure 18.

Table 2 Capacitance Values for the Fourth-Order Elliptic HP Filter Circuit of Figure 17

Capacitor	Value	Capacitor	Value
γ_1	0.42171	γ_{31}	3.42788
γ_2	0.0	γ_{32}	1.83922
γ_3	1.53273	A_2	0.99481
γ_{11}	0.47936	B_2	1.27978
γ_{12}	0.0	C_2	1.0
γ_{21}	0.0	D_2	0.57368
γ_{22}	1.4451		

3.3 DYNAMIC RANGE MAXIMIZATION

In SC filter design, the output of the op amps is limited to within a few volts of the supply voltage. Overdriving these op amps will make them saturate and will result in "clipping" which distorts the signal. For cascade or coupled-biquad structures, each section input level must be adjusted accordingly so as to restrict operation to the linear region. It is a good practice to observe the output of each op amp using SC analysis program, while sweeping the filter input over the frequency range of interest at the maximum operating level to ensure linear operation.

In order to ensure that the SC filters have the widest possible signal range–handling capability, the dynamic range must be maximized. Without appropriately adjusting the dynamic range, the output level of the op amps may be overloaded (output level too high) or very low which contributes to some additional noise. In this section we will demonstrate the use of a scaling technique to maximize the dynamic range of the SC filters.

In order to insure maximum dynamic range it is required that the voltage maxima at all the op amp outputs be equal. This requirement can be mathematically written as

$$\max_{\omega}\left|\frac{V_k(\omega)}{V_{\text{in}}(\omega)}\right| = \max_{\omega}\left|\frac{V_{\text{out}}(\omega)}{V_{\text{in}}(\omega)}\right| \qquad k = 1, 2, \ldots, n \tag{34}$$

where $V_k(\omega)$ are the output of the kth op amps.

Let us consider the SC biquad shown in Figure 8 with the transfer functions T' and T are obtained in Equation 19. This circuit is the basic building block for the SC filters discussed in this chapter. Note that to adjust the voltage level V', without affecting V, only the capacitors connected to node V' (E and H) need to be scaled. More precisely, if it is desired to modify the gain constant associated with V' according to $T' \rightarrow \mu T'$, then it is only necessary to scale E and H by $1/\mu$. That is, according to Equation 19a,

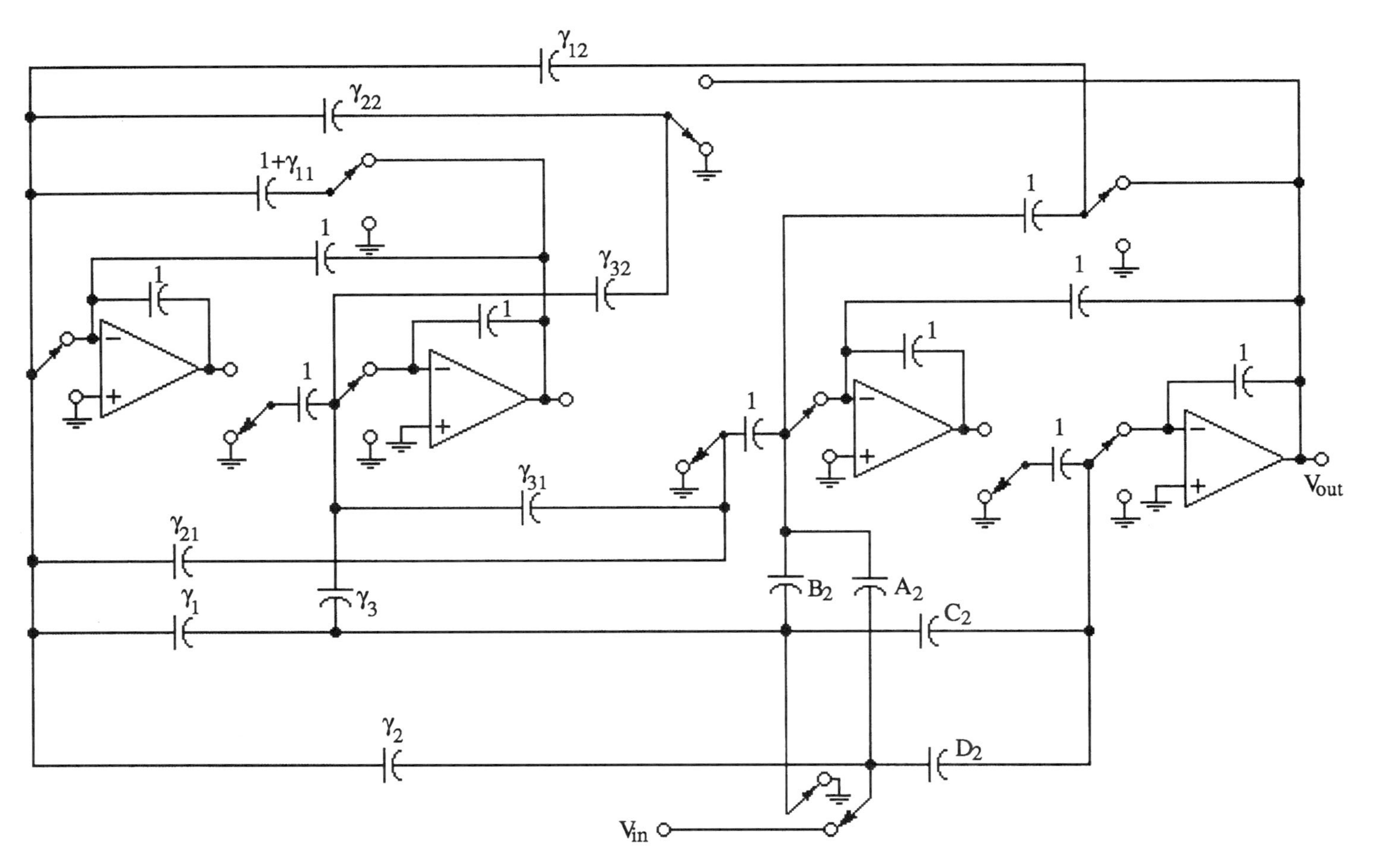

Figure 17 SC realization of the fourth-order elliptic highpass filter.

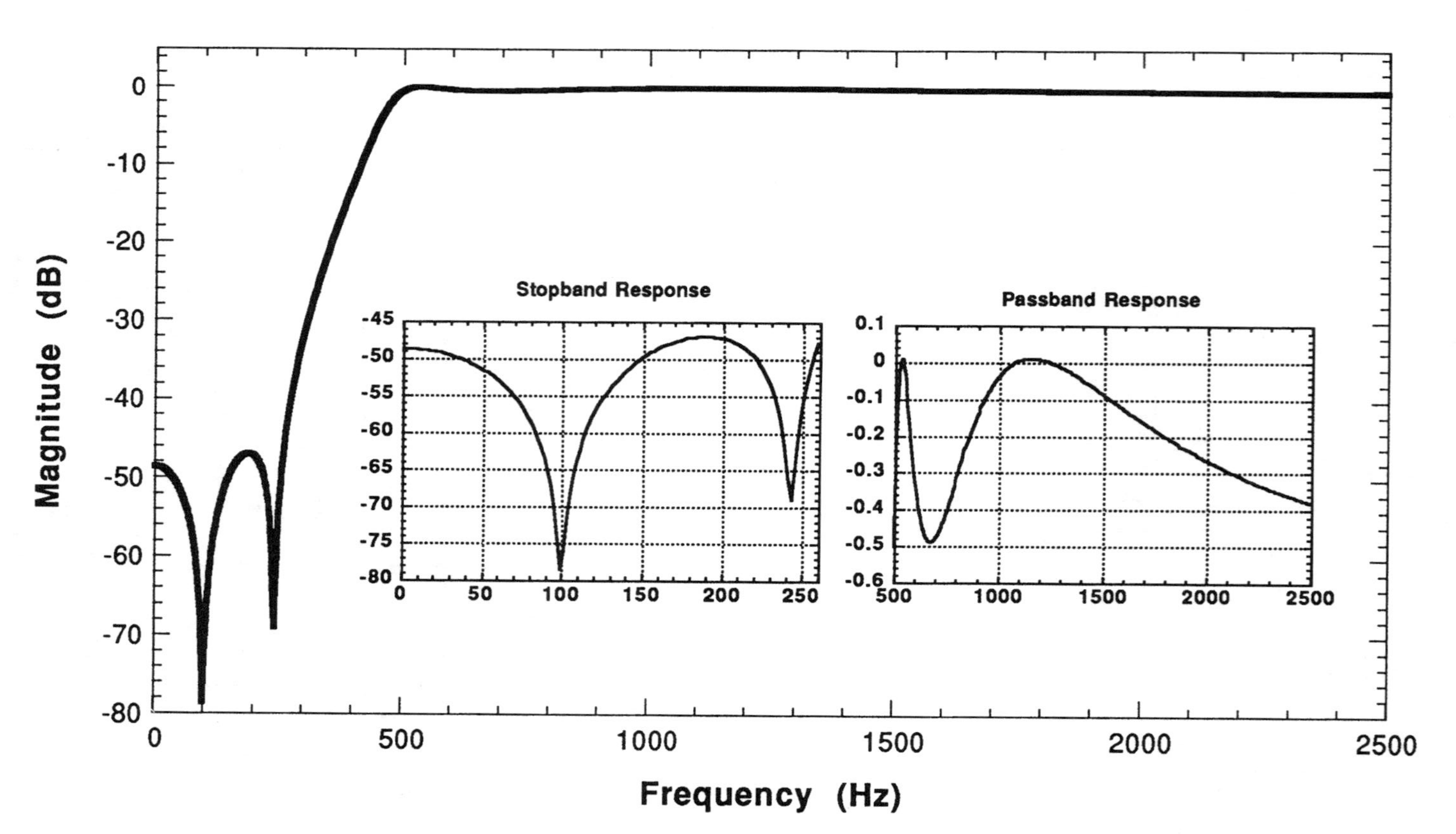

Figure 18 Magnitude response of the fourth-order elliptic highpass filter.

$$T'(z) = \mu \frac{V'}{V_{\text{in}}}(z) = \frac{N'(z)}{\frac{1}{\mu}D(z)} = \mu \frac{N'(z)}{D(z)}$$

Observe that scaling E and H does not affect T according to Equation 19b; hence, the gain constant associated with T remains invariant under this scaling. In a similar manner, if the flat gain associated with V is to be modified by ν, say, that is, $T \to \nu T$, then the capacitors connected to node V (F, G, and I) need to be scaled by 1/ν. Again, according to Equation 19a this does not affect T'. This procedure is readily generalized for n op-amps. The correctness of this procedure follows directly from signal flow graph theory.[35]

Thus, the dynamic range can be maximize by the following step-by-step procedure.

Step 1 Determine the voltage at the output of each op amp (i.e., V_1, V_2, ..., V_n) using an SC analysis program.

Step 2 Calculate the required scaling factor

$$a_k = \frac{\max_{\omega} \left| \dfrac{V_{\text{out}}(\omega)}{V_{\text{in}}(\omega)} \right|}{\max_{\omega} \left| \dfrac{V_k(\omega)}{V_{\text{in}}(\omega)} \right|} \qquad k = 1, 2, \ldots, n$$

Step 3 Scale all the capacitors connected (directly or through switches) to the output of the kth op amp ($k = 1, 2, ..., n$) by the factor $1/a_k$. This, in effect, scales the output voltage of the kth op amp by the factor a_k.

3.3.1 Normalization of Capacitor Values

After scaling the capacitor for maximum dynamic range, it is desirable to scale the admittance associated with each stage so that the minimum capacitance value in the circuit becomes unity. This makes it easier to observe the maximum capacitance ratios required to realize a given circuit and also serves to "standardize" different designs so that the total capacitance required can be readily observed. The two groups of capacitors that are to be scaled together are listed as (see Figure 8)

Group 1: (A,C, F,G, H)
Group 2: (B, D, E,I)

Note that capacitors in each group are distinguished by the fact that they are all incident on the same input node (virtual ground) of the op amp.

Thus, normalizing the capacitance values is achieved by the following procedure:

1. Group the capacitors connected to the input of each op amp.
2. Identify the minimum capacitance value in each group, say, g_k.
3. Scale the capacitors in each group by $1/g_k$.

To demonstrate the above procedure let us consider the design of a stagger-tuned six-order bandpass Butterworth filter. The filter has a center frequency of 225 Hz and a passband that extends from 80 to 305 Hz. The passband attenuation should not exceed 0.5 dB. The filter has been designed following the procedure of Section 3.1. The final circuit is shown in Figure 19 with the capacitance values are given in Table 3. The circuit was simulated using an SC analysis program and the magnitude response is depicted in Figure 20. The following maximum voltages at each of the six op amps and the corresponding scaling factor a_k were obtained as

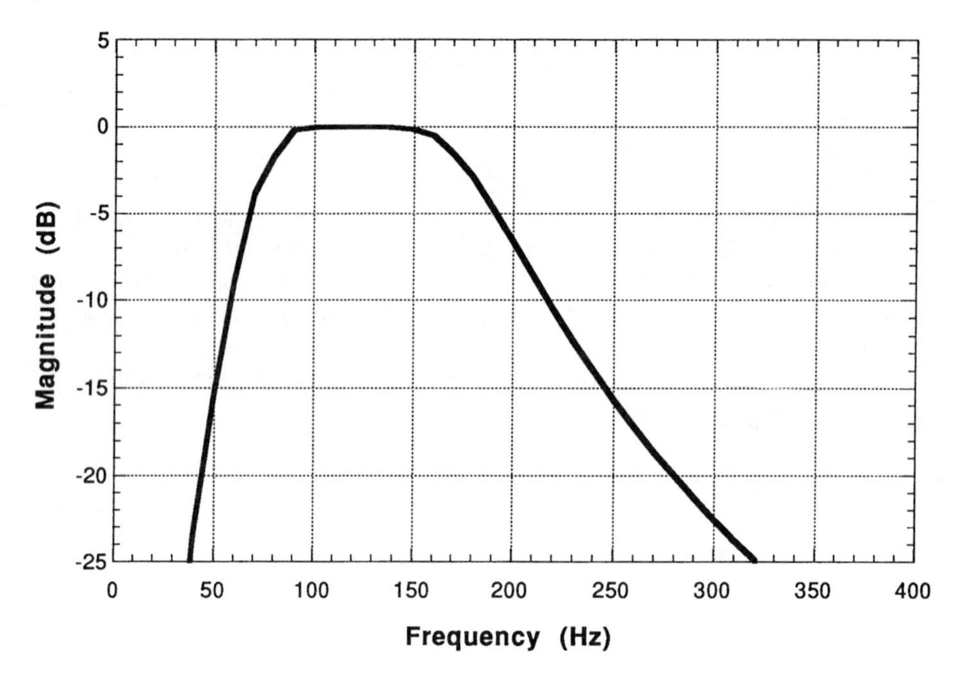

Figure 19 Magnitude response of the six-order Butterworth bandpass filter.

Table 3 Capacitance Values for the Six-Order Butterworth BP Filter Circuit of Figure 9

	$i = 1$			$i = 2$			$i = 3$		
	Unscaled Value	**Adjusted Value[a]**	**Normalized Value**	**Unscaled Value**	**Adjusted Value[a]**	**Normalized Value**	**Unscaled Value**	**Adjusted Value[a]**	**Normalized Value**
A_i	1.1336	1.1336	3.2527	1.172	1.1646	2.6071	1.694	3.8559	4.1068
B_i	1.0	1.0	4.7916	1.0	0.9937	6.2379	1.0	2.2762	16.1662
C_i	1.1045	1.1045	3.1702	1.1137	1.1066	2.4773	1.1060	2.5175	2.6813
D_i	1.0	1.0	4.7916	1.0	0.9937	6.2379	1.0	2.2762	16.1662
E_i	0.21	0.2087	1.0	0.07	0.1593	1.0	0.15	0.1408	1.0
F_i	0.2381	0.3483	1.0	0.2857	0.4464	1.0	0.8	1.291	1.375
G_i	1.0	1.4634	4.2	1.0	1.5634	3.4999	1.0	1.6137	1.7187
H_i	1.0	0.9937	2.8522	1.0	2.2762	5.0956	1.0	0.9389	1.0
I_i	1.0	1.4634	7.02	1.0	1.5634	9.8142	1.0	1.6137	11.4609

[a] Value adjusted for maximum dynamic range.

$$V_1 = -0.05530 \text{ dB (at 110 Hz)}, \ a_1 = 9.899$$

$$V_2 = 3.30729 \text{ dB (at 130 Hz)}, \ a_2 = 0.166$$

$$V_3 = 7.14423 \text{ dB (at 70 Hz)}, \ a_3 = 0.0766$$

$$V_4 = 3.88149 \text{ dB (at 80 Hz)}, \ a_4 = 0.0141$$

$$V_5 = V_{out} = -0.54745 \text{ dB (at 140 Hz)}, \ a_5 = 1$$

$$V_6 = 4.15658 \text{ dB (at 70 Hz)}, \ a_6 = 0.1317$$

The output of the filter is also scaled in this case to be 0 dB. The capacitor values were also normalized so that the minimum capacitor value in the filter design becomes unity. Both the scaled and normalized capacitance values are also included in Table 3.

REFERENCES

1. **Graeme, J. G.,** *Operational Amplifier: Third-Generation Technologies*, McGraw-Hill, New York, 1973.
2. **Gray, P. R. and Meyer, R. G.,** *Analysis and Design of Analog Integrated Circuits*, John Wiley & Sons, New York, 1984.
3. **Grebene, A. B.,** *Bipolar and MOS Analog Integrated Circuits Design*, John Wiley & Sons, New York, 1984.
4. **Maxwell, J. C.,** *A Treatise of Electricity and Magnetism*, Clarendon Press, Oxford, 1873, 347.
5. **Fudim, E. F.,** Fundamentals of the switched-capacitor approach to circuit synthesis, *IEEE Circuits Syst. Mag.*, 6, 12, 1984.
6. **Fudim, E. F.,** Capacitor-switch approach and its use in pneumatics, Candidate of Technical Sciences (Ph.D.) Dissertation, Institute of Control Sciences, Moscow, USSR, April 8, 1965.
7. **Boite, R. and Thiran, J. P. V.,** Synthesis of filters with capacitances switches, and regenerating devices, *IEEE Trans. Circuits Theor.*, CT-15, 447, 1968.
8. **Fettweis, A.,** Theory of resonant-transfer circuits and their application to the realization of frequency selective networks, *Proc. Summer School on Circuit Theory,* Czechoslovak Academy of Science, Prague 1968.
9. **Kuntz, W.,** A new sample and hold device and its application to the realization of digital filters, *IEEE Proc.,* 56, 2092, 1968.
10. **Fried, D. L.,** Analog sampled-data filters, *IEEE J. Solid-State Circuits,* SC-7, 302–304, August 1972.
11. **Caves, J. T., Copeland, M. A., Rahim, C. F., and Rosenbaum, S. D.,** Sampling analog filtering using switched capacitors as resistor equivalents, *IEEE J. Solid-State Circuits,* SC-12, 592, 1977.
12. **Hosticka, B. J., Brodersen, R. W., and Gray, P. R.,** MOS sampled-data recursive filters using switched capacitor integrators, *IEEE J. Solid-State Circuits,* SC-12, 600, 1977.
13. **Allan, P. E. and Sanchez-Sinencio, E.,** *Switched Capacitor Circuits*, Van Nostrand Reinhold, New York, 1984.
14. **Unbenhauen, R. and Cichocki, A.,** *MOS Switched-Capacitor and Continuous-Time Integrated Circuits and Systems*, Springer-Verlag, New York, 1989.
15. **Moschytz, G. S.,** Ed., *MOS Switched-Capacitor Filters: Analysis and Design*, IEEE Press, 1984.
16. **Ghausi, M. S. and Laker, K. R.,** *Modern Filter Design*, Prentice-Hall, Englewood Cliffs, NJ, 1981, Chap. 6.
17. **Gregorian, R. and Temes, G. C.,** *Analog MOS Integrated Circuits for Signal Processing*, John Wiley & Sons, New York, 1986, Chap. 5.
18. **Nakayama, K. and Kuraishi, Y.,** Present and future applications of switched-capacitor circuits, *IEEE Circuits Devices Mag.,* 10, September 1987
19. **Brodersen, R. W., Gray, P. R., and Hodges, D. A.,** MOS switched-capacitor filters, *IEEE Proc.,* 67, 61, 1979.
20. **Allstot, D. J. and Black, W. C., Jr.,** Technological design considerations for monolithic MOS switched-capacitor filtering systems, *IEEE Proc.,* 71, 967, 1983.
21. **Jury, E. I.,** *Theory and Application of the z Transform Method*, John Wiley & Sons, New York, 1964.
22. **Martin, K. and Sedra, A. S.,** Strays-insensitive switched-capacitor filters based on the bilinear z transformation, *Elec. Lett.,* 15, 365, 1979.
23. **El-Masry, E. I.,** Strays-insensitive active switched-capacitor biquad, *Elec. Lett.,* 16, 480, 1980.
24. **Fleischer, P. E. and Laker, K. R.,** A family of active switched-capacitor building blocks, *Bell Syst. Tech. J.,* 58, 2235, 1979.
25. **Martin, K. and Sedra, A. S.,** Exact design of switched-capacitor band-pass filters using coupled-biquad structures, *IEEE Trans. Circuits Syst.,* CAS-27, 469, 1980.
26. **Fleischer, P. E., Laker, K. R., March, D. G., Ballantyne, J. P., Yiannoulos, A. A., and Fraser, D. L., Jr.,** An NMOS analog building block for telecommunication application, *IEEE Trans. Circuits Syst.,* CAS-27, 552, 1980.
27. **Gregorian, R. G.,** Switched-capacitor filter design using cascaded sections, *IEEE Trans. Circuits Syst.,* CAS-27, 515, 1980.
28. **El-Masry, E. I.,** Synthesis of a follow-the-leader feedback switched-capacitor structure, *IEE Proc. Part G,* 132, 18, 1985.
29. **El-Masry, E. I.,** Strays-insensitive state-space switched-capacitor filters, *IEEE Trans. Circuits Syst.,* CAS-30, 474, 1983.
30. **El-Masry, E. I. and Lee, H. L.,** Low-sensitivity realization of switched-capacitor filters, *IEEE Trans. Circuits Syst.,* CAS-34, 510, 1987.
31. **Sedra, A. and Brackett, P.** *Filter Theory and Design: Active and Passive,* Matrix Publisher, 1978, Chap. 10.
32. **El-Masry, E. I.,** Design of switched-capacitor filters in the biquadratic state-space form, *Proc. 1981 IEEE Int. Symp. Circuits and Systems,* 1981, 179.
33. **El-Masry, E. I.,** State-space switched-capacitor structures, *Proc. 14th Asilomar Conference on Circuits, Systems and Computers,* CA, 1980.
34. **Attaie, N. and El-Masry, E. I.,** Multiple-loop feedback switched-capacitor structures, *IEEE Trans. Circuits Syst.,* CAS-30, 865, 1983.
35. **Perry, D. J.,** Improvement of dynamic range and reduction of limiting effects in multiple feedback filters, in *Electronic Filters, IEE Conf. Publ.* 167, 36, 1978.

Design of Switched-Capacitor Ladder Filters

John T. Taylor

CONTENTS

1. INTRODUCTION

The importance and relevance of the switched-capacitor (SC) method of filter design is now well established[1-3] and has been described elsewhere in this chapter. Practical design methods based on *biquadratic sections* have also been described.[4] The simplest realization of a high-order filter employs a cascade of second-order sections to realize the chosen transfer function. These realizations are very simple and allow the filter designer maximum freedom and flexibilty in the choice of possible transfer functions since there are no restrictions placed on the permitted locations of poles and zeros. In addition, a particular advantage of the biquad cascade is that the parameters of such a system can be readily scaled for optimum dynamic range. On the other hand, transfer functions realized by such structures are very likely to be highly sensitive to parameter changes, rendering them unsuitable for very demanding applications where component tuning is not available. In order to overcome the parameter sensitivity problem, a variety of techniques are available. For

0-8493-8951-8/97/$0.00+$.50
© 1997 by CRC Press LLC

example, *multiple-loop feedback* (MLF) can be employed to form *coupled biquad* structures.[5] These methods were described in the Chapter 7.1 and allow considerable reductions in sensitivity to be obtained in exchange for some loss of flexibility in other areas.

The optimum solution to the parameter sensitivity problem is provided by the *ladder realization* method, which maps the elements of an LCR prototype filter into the active domain, in this case into the SC domain. This special feature derives from a well-known property of doubly terminated reactance ladder filters first described by Orchard[6] in 1966 and subsequently extended.[7] If one assumes that such a prototype is designed to transmit maximum power to its load at a certain frequency (or set of frequencies), then, at that frequency, the filter response will have zero sensitivity (i.e., zero first-order differential sensitivity) to parameter changes. Given a well-chosen approximating function, it is usually possible to arrange for the parameter sensitivity to be negligibly small across a band of frequencies, usually chosen to be the filter passband. Notice, however, that the existence of an LCR prototype required by realization as a ladder implies some loss of the design flexibility enjoyed by the cascade family of realizations.

There are two general approaches for designing active filters based on LCR ladder prototypes. The first is inductor replacement using an impedance transformation device of which the gyrator and negative impedance converter (NIC) are perhaps the best-known examples.[8] The second method simulates the electrical terminal properties of the elements of the prototype, which for this purpose is usually represented by its *signal flow graph* (SFG). Although both methods have been employed successfully in the design of practical active RC filters,[8,9] in practice only the ladder simulation method is frequently used in the design of SC filters. This is because it is very difficult to design a suitable impedance transformer which can be made insensitive to parasitic capacitance (stray-free). Since we wish to scale the components in the integrated realization to minimize die size while at the same time preserving a high level of transfer function accuracy, insensitivity to parasitic capacitance is clearly crucial. In fact, this requirement tends to dominate practical SC filter design, since only a fairly small subset of SC subcircuits can be made entirely stray-free.[2,3]

In this section we begin by reviewing briefly the SFG method, since this is such a frequently used technique in the design of active filters based on ladder prototypes, including SC filters. We then consider the frequency transformations which are available to map the continuous-time ladder filter into the discrete-time domain, and we show how the choice of transformation is constrained by the availability of suitable SC building blocks. An illustrative design example is then given which is the design of an all-pole lowpass ladder SC filter using the lossless discrete integrator (LDI) method. This example deals with the complete design process from the choice of an LCR prototype to circuit schematic including frequency warping and nodal voltage scaling. This realization is entirely stray-free and performs well in highly oversampled applications, i.e., where the clock frequency is many times greater than the baseband edge frequency. More-advanced techniques, such as the exact design of ladder filters using the bilinear transformation, are beyond the scope of this chapter and are referred to in the bibliography.

2. SIGNAL FLOW GRAPH METHOD

2.1 BASIC PRINCIPLES

Electrical networks are made up of components characterized by a scalar parameter such as L, C, or R. However, the voltage–current relationships which characterize the components represent a mathematical operation in which the component value generally appears as a scaling factor. So, for example, the series inductor shown as an elementary one-port network in Figure 1a has the following time-domain voltage–current relationship:

$$v(t) = L\frac{di(t)}{dt} \tag{1a}$$

Or, alternatively, in the s domain where s is the Laplace frequency variable:

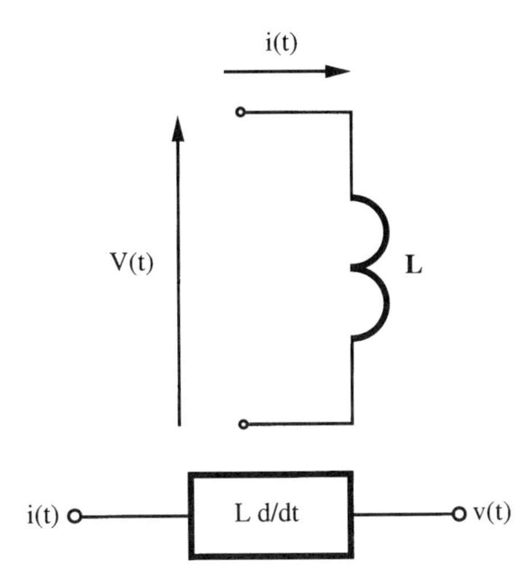

Figure 1 (a) One-port representation of an inductor; (b) the circuit of (a) interpreted as an operational (SFG) element.

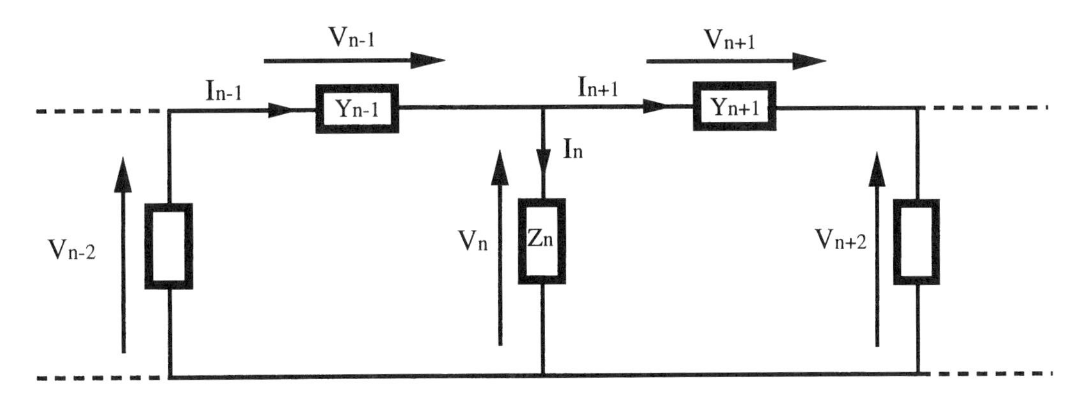

Figure 2 Section of a general immittance ladder network.

$$V(s) = Ls \tag{1b}$$

If we consider this operation to be described by differentiation with respect to time as shown in Equation 1a, the process can be represented by the operational diagram shown in Figure 1b. This *operational representation* is the essence of the SFG method. By applying Kirchoff's voltage and current laws to the general immittance ladder of which a three-node section is shown in Figure 2, the following set of simultaneous linear equations can be developed:

$$\begin{gathered} \vdots \\ I_{n-1} + I_n + I_{n+1} = 0 \\ \cdot \\ ---- \\ V_{n-2} + V_{n-1} - V_n = 0 \\ V_n + V_{n+1} - V_{n+2} = 0 \\ \cdot \\ ---- \\ \cdot \end{gathered} \tag{2}$$

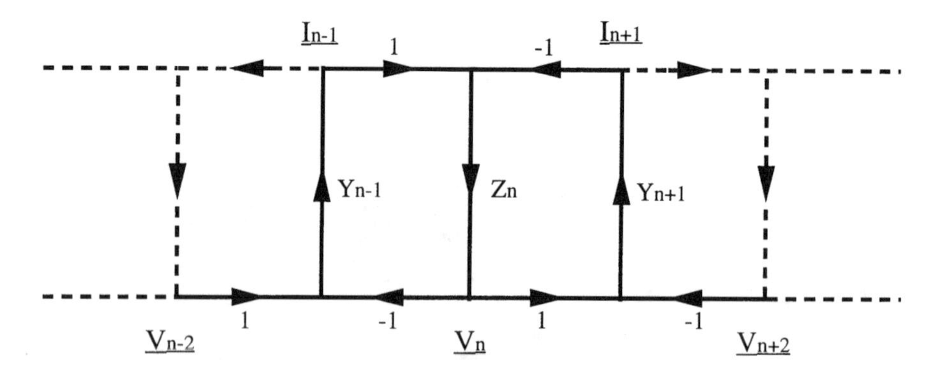

Figure 3 SFG representation of the network of Figure 2.

together with the following *branch relationships*:

$$
\begin{gathered}
\vdots \\
I_{n-1} + Y_{n-1} \cdot V_{n-1} = 0 \\
-Z_n \cdot I_n + V_n = 0 \\
I_{n+1} + Y_{n+1} \cdot V_{n+1} = 0 \\
\vdots
\end{gathered}
\tag{3}
$$

Combining these expressions, we derive the following nonunique set of *state equations* for the network (Equation 4).

$$
\begin{gathered}
\vdots \\
I_{n-1} + Y_{n-1}\left(V_{n-2} - V_n\right) \\
V_n = Z_n\left(I_{n-1} - I_{n+1}\right) \\
I_{n+1} + Y_{n+1}\left(V_n - V_{n+2}\right) \\
\vdots
\end{gathered}
\tag{4}
$$

These are in the operational form required for SFG representation, the operator being an immittance function in each case. The resulting SFG is shown in Figure 3. In the representation given in Equation 4, the dependent or state variables on the left-hand side of the equations have been chosen to be the currents flowing through the series arms of the network and the voltages developed across the shunt arms. Although other choices are possible (i.e., by eliminating different variables between Equations 2 and 3, etc.) the form given in Equation 4 is especially important in active filter (including SC) design since the immittance operators correspond to integration in the time domain. As we shall see shortly, the most effective SC building block is the LDI integrator. This type of realization is referred to as a *leapfrog* filter.[8]

2.2 SIMPLIFICATION AND SCALING

Having constructed the SFG, it is frequently necessary to simplify it by eliminating and/or combining certain branches. This is carried out by means of standard operations on the graph which correspond to manipulating the underlying linear equations. The most-elementary SFG operations are given in Table 1, and a more comprehensive treatment of the SFG method is given in References 10 and 11.

Another important SFG operation is *scaling for maximum signal-handling capability.* This enables certain state variables within the system to be adjusted locally, i.e., without affecting the

Table 1 Fundamental SFG Operations

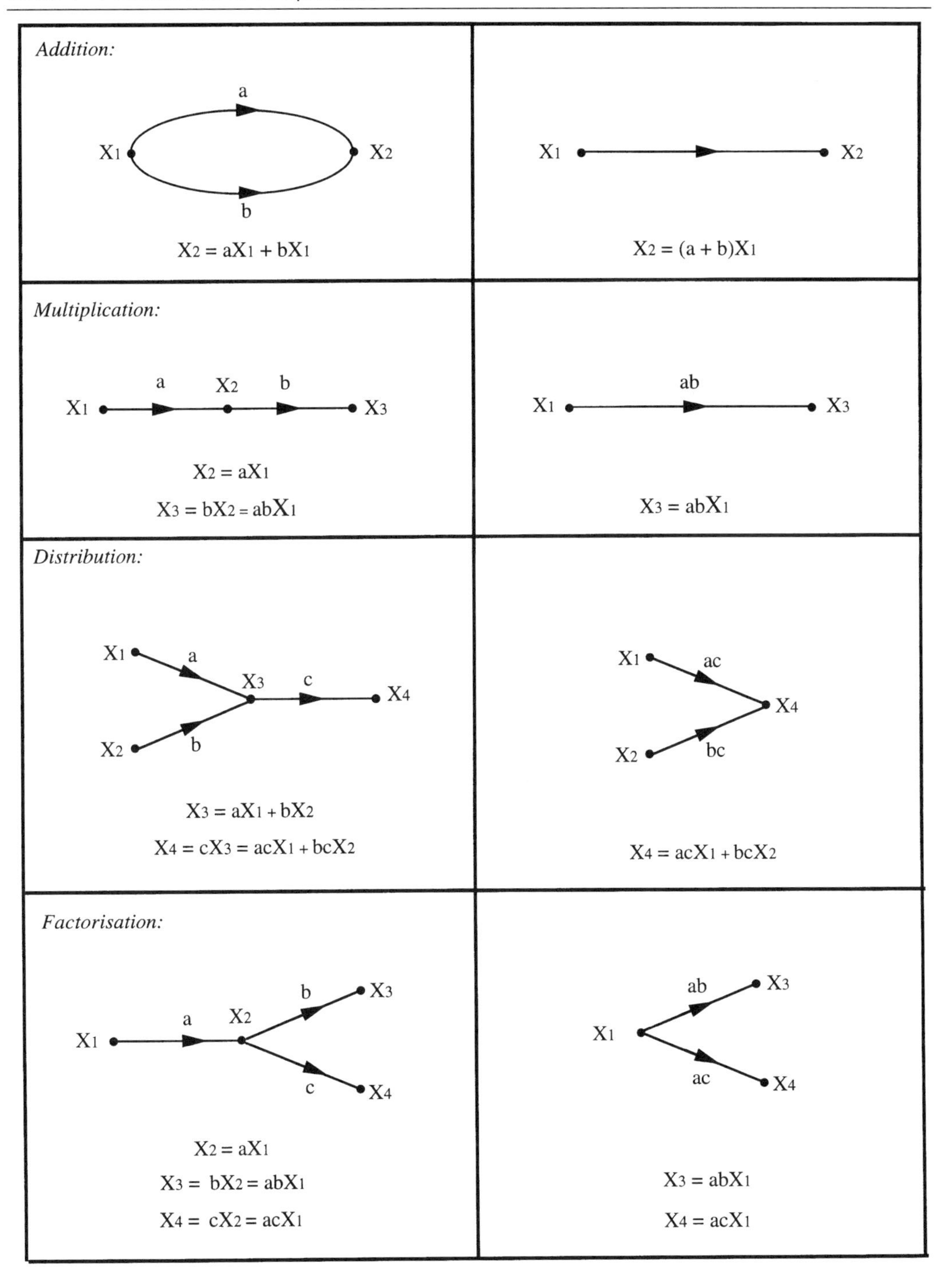

overall system transfer function. Consider a *cutset* of SFG branches, i.e., any minimal set of branches which, when deleted, separates the graph into exactly two pieces (this procedure is sometimes called *contour scaling*). Then, if the branches entering the cutset are multiplied by a scaling factor λ (where λ can be a real or complex factor) and those leaving the cutset are multiplied by λ^{-1} (or vice versa), the transfer function H(s) is unchanged. The only exception to this rule occurs when the cutset separates the input node from the output node, in which case H(s) is scaled by λ.[12] For

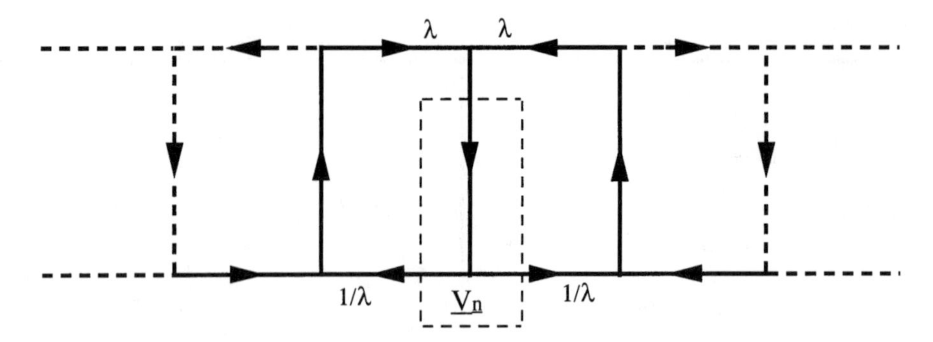

Figure 4 Section of an SFG illustrating scaling.

example, consider the section of an SFG shown in Figure 4. We wish to scale the signal level represented by the variable V_n by a constant factor λ. A cutset is drawn on the SFG (shown by the dashed box) which contains the branches connected to the node V_n, and the scaling transformation is carried out as above.[12] This process is a generalized form of scaling transformations proposed elsewhere in the literature (see, e.g., Reference 13).

Other scaling operations enable the total capacitance (and hence the die area) of the system to be minimized. This is a relatively straightforward process and is described in Reference 13.

2.3 EXAMPLE 1: THIRD-ORDER ALL-POLE LOWPASS FILTER

Given the LCR prototype filter shown in Figure 5, construct the SFG with the state variables chosen as in Equation 3 above.

Solution

The three state equations are as follows:

$$I_{L1} = \frac{V_{in} - V_{c2}}{R + sL_1}$$

$$V_{c2} = \frac{I_{L1} - I_{L3}}{sC_2}$$

$$I_{L3} = \frac{V_{C2} - V_{out}}{sL_3}$$

and, additionally:

$$V_{out} = I_{L3} \cdot R$$

where the state variables have been chosen to be the inductor currents and capacitor voltages. It can be readily shown that these equations are represented by the SFG of Figure 6a, where the terminations are realized by the additional feedback connections. In an active realization, we generally require all state variables, whether voltage or current, to be represented by voltages. In the circuit of Figure 6a, this can be achieved by scaling the impedance levels in the circuit by a factor R^{-1}, using cutsets around the appropriate nodes as indicated by the dashed boxes. The resulting impedance-scaled SFG is shown in Figure 6b.

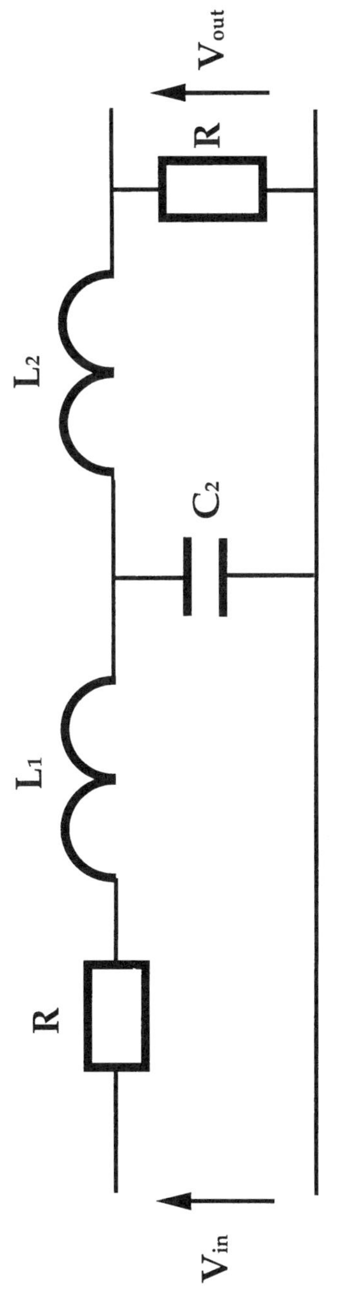

Figure 5 Third-order all-pole lowpass LCR prototype filter (see example 1).

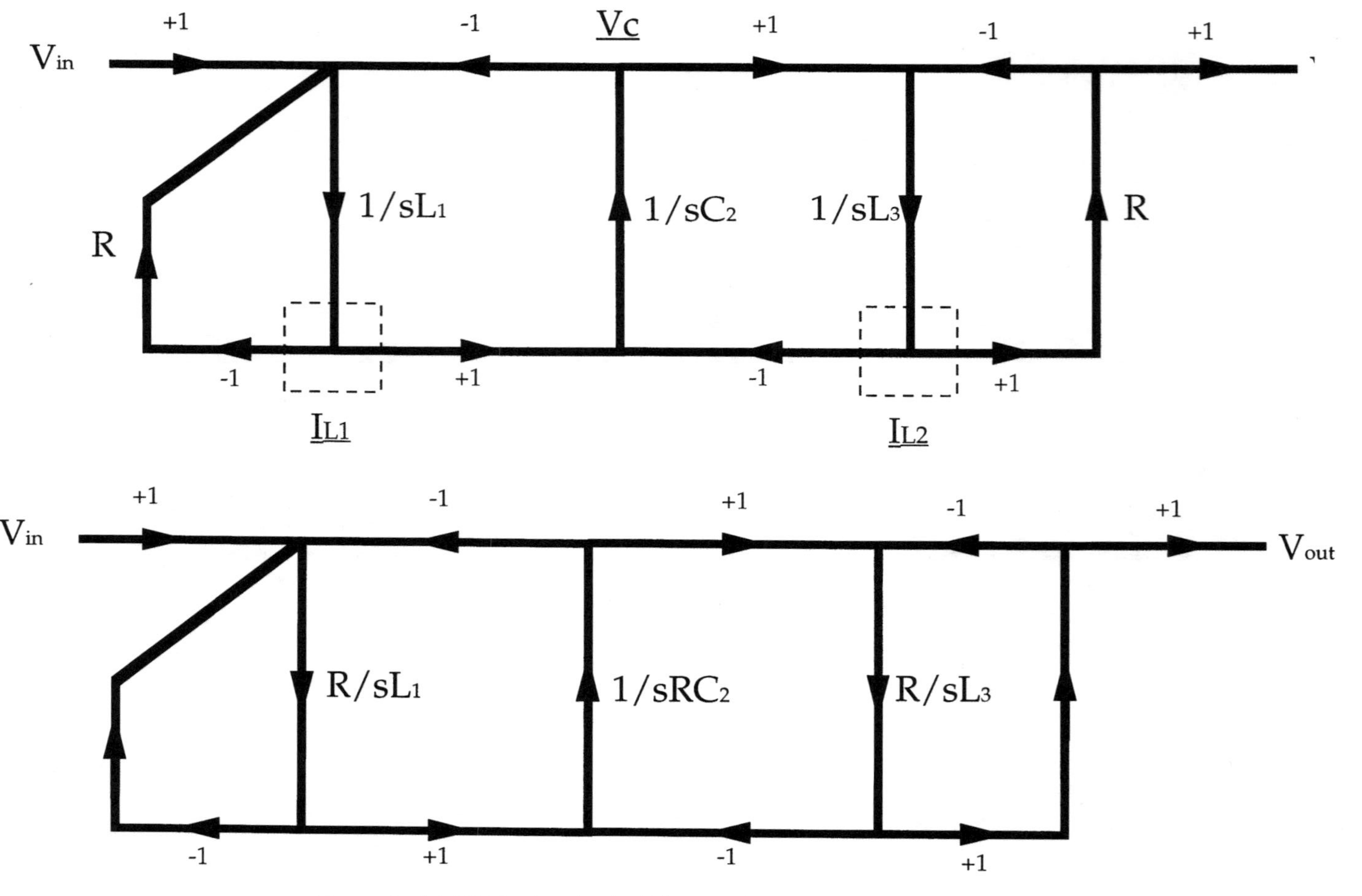

Figure 6 (a) SFG for the circuit of Figure 4; (b) impedance-scaled version.

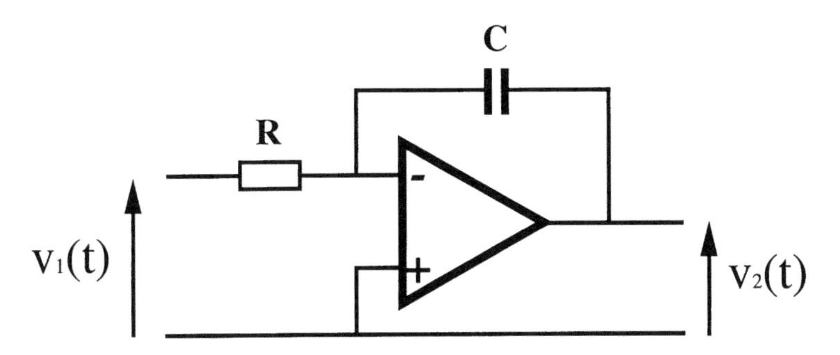

Figure 7 Active RC integrator.

2.4 OTHER TYPES OF FILTERS: ADDITION OF FINITE TRANSMISSION ZEROS

Using the methods described above in conjunction with standard transformations, it is possible to construct SFGs for bandpass, highpass, and bandstop filters, beginning with an all-pole lowpass prototype. Frequently, however, we wish to realize filters with finite transmission zeros, such as elliptic types, and there are various ways of modifying the SFG to do this; and these are described in References 10 and 11 and in various citations contained in Reference 1.

3. FREQUENCY TRANSFORMATIONS AND SC SUBCIRCUITS

3.1 BASIC PRINCIPLES

Although the operation of the SC integrator has been widely discussed, a simplified explanatory analysis is presented here. Consider the continuous-time active RC integrator shown in Figure 7. With ideal components, the transfer function, H(s), is as follows:

$$H(s) = \frac{V_2(s)}{V_1(s)} = \frac{-1}{sRC} \tag{5}$$

Since the SC equivalent of an active RC integrator is a discrete-time system, it is appropriate to begin by analyzing the circuit of Figure 7 on a discrete-time basis. For a time interval from $(\tau - T)$ to τ, where T is the sample period and τ is an arbitrary point on the time axis, the charge transfer through R is given by

$$\Delta Q = \frac{1}{R}\int_{\tau-T}^{\tau} v_1(t)dt \tag{6}$$

ΔQ is therefore represented by the area under that portion of the graph of $v_1(t)$ against t which falls in the range:

$$(\tau - T) \leq t \leq \tau$$

SC circuits must be sought which adequately approximate the integral of Equation 6, since a discrete-time system cannot compute this exactly. Some examples of numerical forms which are important in this context are given in Figure 8. The following approximate integrals result:

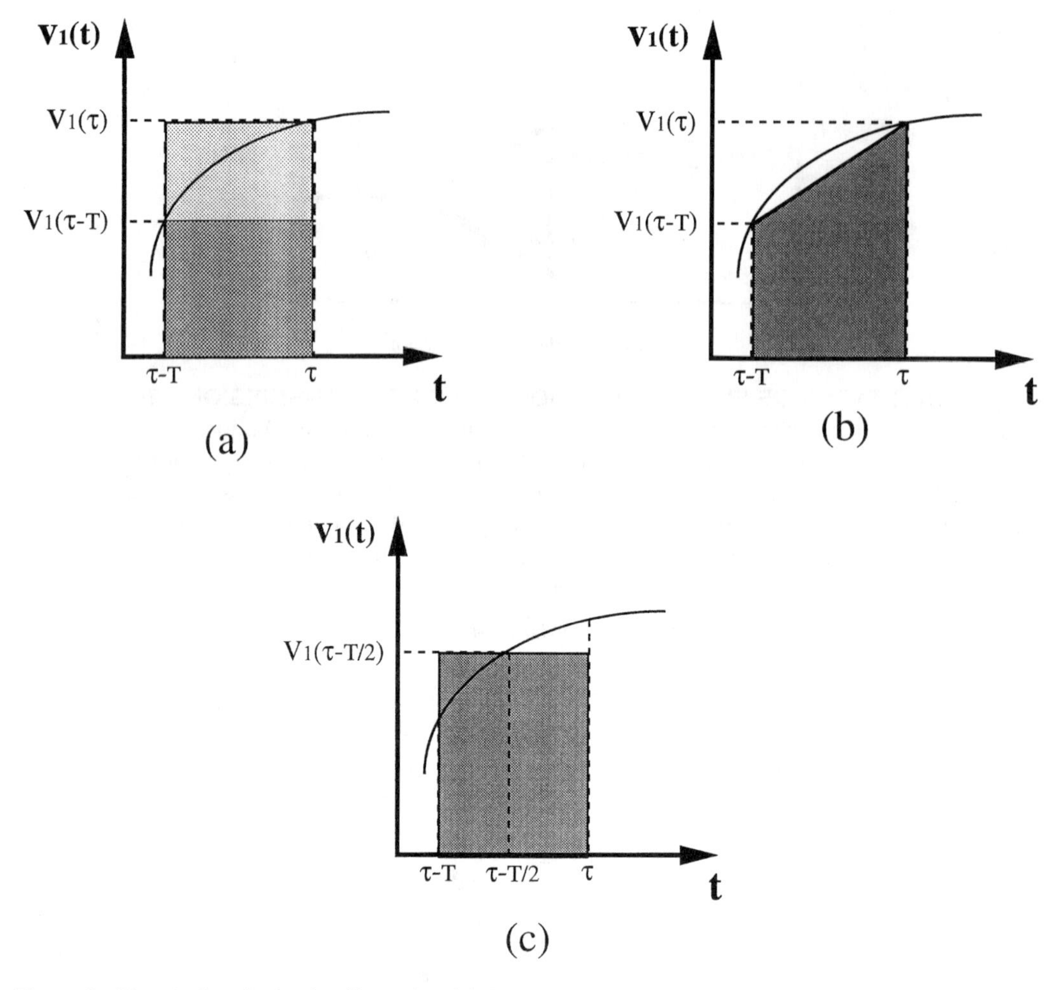

Figure 8 Discrete-time integration illustrating: (a) Euler, (b) trapezoidal (bilinear), and (c) intermediate rectangular (LDI) integration.

Forward Euler integration: $$\Delta Q = \frac{T}{R} \cdot v_1(\tau - T) \tag{7a}$$

Backward Euler integration: $$\Delta Q = \frac{T}{R} \cdot v_1(\tau) \tag{7b}$$

Intermediate rectangular integration (LDI): $$\Delta Q = \frac{T}{R} \cdot v_1(\tau - T/2) \tag{7c}$$

Trapezoidal (bilinear) integration: $$\Delta Q = \frac{T}{2R} \cdot \left[v_1(\tau - T) + v_1(\tau)\right] \tag{7d}$$

In all cases, ΔQ is transferred onto the integrating capacitor, C, resulting in a transition in the output voltage $v_2(t)$:

$$\Delta Q = -C\left[v_2(\tau) - v_2(\tau - T)\right] \tag{8}$$

Combining Equation 6 with one of the forms of Equation 5 and taking z transforms,* we obtain the following voltage transfer ratios:

$$\frac{V_2(z)}{V_1(z)} = -\frac{T}{CR} \cdot \frac{z^{-1}}{1 - z^{-1}} = -\frac{T}{2CR} \cdot \frac{z^{-1/2}}{\sinh(sT/2)} \tag{9a}$$

$$\frac{V_2(z)}{V_1(z)} = -\frac{T}{CR} \cdot \frac{1}{1 - z^{-1}} = -\frac{T}{2CR} \cdot \frac{z^{1/2}}{\sinh(sT/2)} \tag{9b}$$

$$\frac{V_2(z)}{V_1(z)} = -\frac{T}{CR} \cdot \frac{1}{z^{1/2} - z^{-1/2}} = -\frac{T}{2CR} \cdot \frac{1}{\sinh(sT/2)} \tag{9c}$$

$$\frac{V_2(z)}{V_1(z)} = -\frac{T}{2CR} \cdot \frac{1 + z^{-1}}{1 - z^{-1}} = -\frac{T}{2CR} \cdot \frac{1}{\tanh(sT/2)} \tag{9d}$$

To obtain the frequency response, set $z = e^{j\omega T}$:

$$\frac{V_2(j\omega)}{V_1(j\omega)} = -\frac{T}{j2CR} \cdot \frac{e^{-j\omega T/2}}{\sin(\omega T/2)} \tag{10a}$$

$$\frac{V_2(j\omega)}{V_1(j\omega)} = -\frac{T}{j2CR} \cdot \frac{e^{j\omega T/2}}{\sin(\omega T/2)} \tag{10b}$$

$$\frac{V_2(j\omega)}{V_1(j\omega)} = -\frac{T}{j2CR} \cdot \frac{1}{\sin(\omega T/2)} \tag{10c}$$

$$\frac{V_2(j\omega)}{V_1(j\omega)} = -\frac{T}{j2CR} \cdot \frac{1}{\tan(\omega T/2)} \tag{10d}$$

Equations 9c and d (and 10c and d) implement *integration in discrete time*, the only significant difference from the continuous-time case being that the frequency variable is a trigonometric function of the corresponding continuous-time variable. This turns out to be a relatively minor point in practice and can be corrected, if necessary, by *prewarping* the continuous-time frequency variable. By contrast, the forms shown in Equations 9a and b (and 10a and b) contain an extra numerator term which adds an extra half-cycle of delay which is not present in the continuous-time integrator transfer function. If such circuits were used to construct a ladder filter, the additional phase contributions resulting from the numerator terms would introduce high levels of distortion into the filter characteristic. This point was recognized in the early development of SC filters, and practical systems implementing the LDI or bilinear transformations, i.e., Equations 9c and d (and 10c and d), have been described.[14] For the rest of this chapter, we will deal exclusively with these cases.

* The z transform is dealt with extensively in Chapter 6.1.

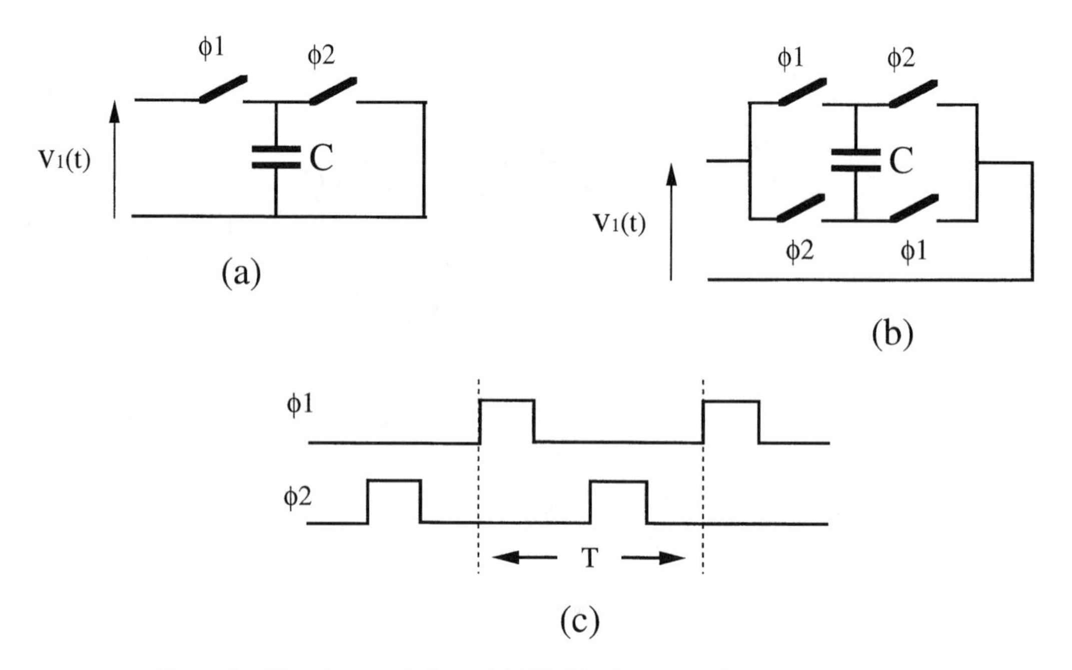

Figure 9 SC resistor equivalents: (a) LDI, (b) bilinear, and (c) clock waveforms.

3.2 SC RESISTOR EQUIVALENTS

We next consider practical SC circuits to replace the input resistance R in Figure 7. Since, with ideal components, R (and its SC equivalent) is connected to a virtual ground it may be treated as a grounded one-port network. Consider the arrangement shown in Figure 9a. In this circuit, the continuous-time input voltage $v_1(t)$ is sampled by the switch ϕ_1, held on the capacitor C_1, and then discharged to ground by the second switch, ϕ_2. The switching waveform, in which the ON phases are less than 50% of the overall clock period, T, and do not overlap, is shown on the figure. The voltage on C_1 tracks $v_1(t)$ until ϕ_1 turns OFF, and, with the notation used above, let us assume that this occurs at time $(\tau - T/2)$. The amount of charge stored and dumped per period is given by

$$\Delta Q(t) = C_1 v_1(\tau - T/2) \tag{11}$$

Combining Equations 11 and 8, where the integrating capacitor is now relabeled C_2, we get

$$C_1 v_1(\tau - T/2) = -C_2\left[v_2(\tau) - v_2(\tau - T)\right]$$

and taking z transforms, we get

$$\frac{V_2(z)}{V_1(z)} = -\frac{C_1}{C_2} \cdot \frac{1}{z^{1/2} - z^{-1/2}} = -\frac{C_1}{2C_2} \cdot \frac{1}{\sinh(sT/2)} \tag{12}$$

which, apart from the scaling constant, realizes the LDI transformation of Equation 9c. Comparing Equations 9c and 12, we see that the one-port circuit shown in Figure 9a simulates an SC equivalent grounded resistor with effective resistance:

$$R_{eff} = \frac{T}{C_1}$$

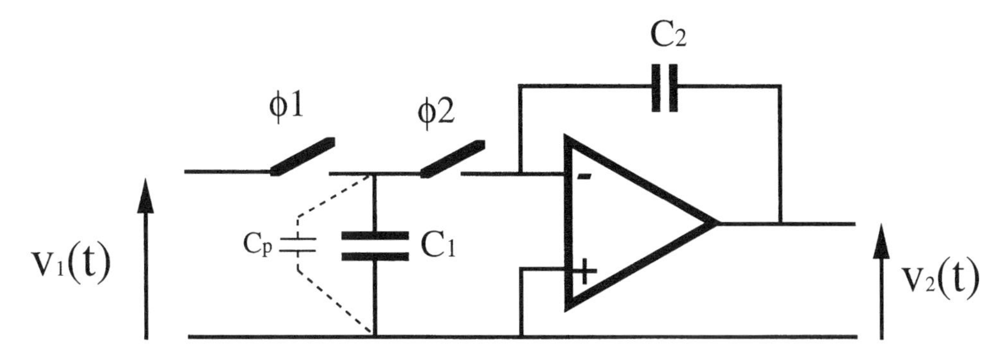

Figure 10 Parasitic sensitive LDI SC integrator.

By similar arguments, it can be shown that the one-port circuit shown in Figure 9b realizes the bilinear transformation (Equation 9d), also with equivalent resistance T/C_1.

3.3 SC INTEGRATORS AND PARASITIC SENSITIVITY

Replacing R in Figure 7 by an SC resistor equivalent results in a complete SC integrator, of which the LDI type is shown in Figure 10. Note, however, that the transfer function of this circuit is sensitive to parasitic capacitance associated with both the analogue switches and with the MOS capacitor, C_1. The presence of parasitics affects the minimum size of the capacitors which must be realized (i.e., to dominate the parasitics in order to obtain the required level of transfer function accuracy) and hence the required silicon area. In addition, since certain parasitics contain nonlinear voltage-dependent elements, their presence can also introduce considerable signal distortion.

If one assumes that the operational amplifier is ideal and that the circuit is fed from an ideal voltage source, the main parasitics which affect the operation of the circuit are associated with the top and bottom plates of C_1. The top plate component consists of the voltage-dependent source-drain capacitances of the switches and a fixed contribution from the capacitance between the aluminium track interconnections and the substrate with the field (thick) oxide as dielectric. The bottom plate parasitic is a large fixed capacitance between the bottom plate of C_1 and the substrate and can amount to as much as 20% of the value of C_1. The effect of the bottom plate parasitic is generally avoided by connecting it to an independent voltage source (or an op amp output). However, the top plate capacitance, which can amount to as much as 5% of C_1, cannot be easily reduced using the circuit of Figure 10, which represents the residual parasitics as a single, lumped capacitor, C_p. Fortunately, the effect of C_p can be very much reduced by means of the modified circuit shown in Figure 11[15] (see also Reference 1), where the function G(z) is given by

$$G(z) = \frac{1}{z^{1/2} - z^{-1/2}} = \frac{1}{2\sinh(sT/2)}$$

It is easily shown that this is a *parasitic insensitive* noninverting integrator, insensitive to both top and bottom plate parasitics.

In addition to providing the circuit schematic of the parasitic insensitive integrator, Figure 11 shows the equivalent SFG. Note that although the integrator has one input it has, in effect, two outputs, since the transfer function is different depending on the clock phase in which the output sample is taken. It is of the greatest importance, therefore, that the interconnection of each integrator to its neighbours should be phased correctly so as to implement the LDI transformation. The integrators a and b shown in Table 2 have been described as *canonic* integrators since they use the minimum possible number of components and the input branch weights are constrained to be equal.[16] There are distinct advantages to using these integrators in terms of die area and also because there are no errors in the realization of the input branches. On the other hand, the

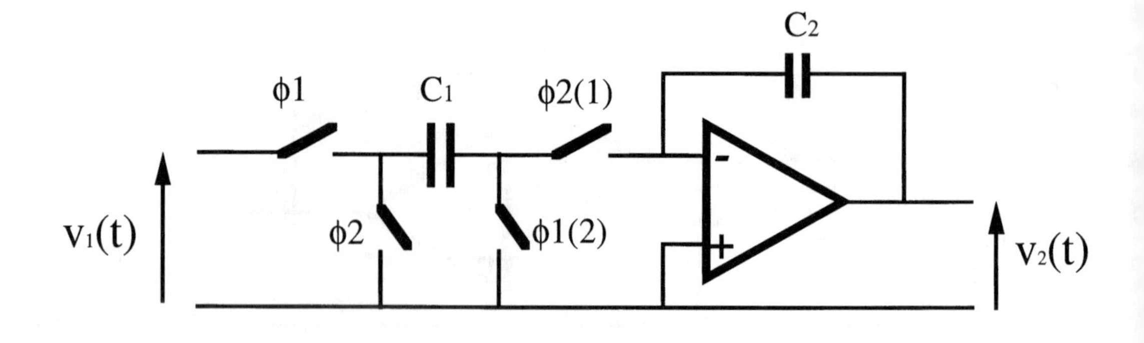

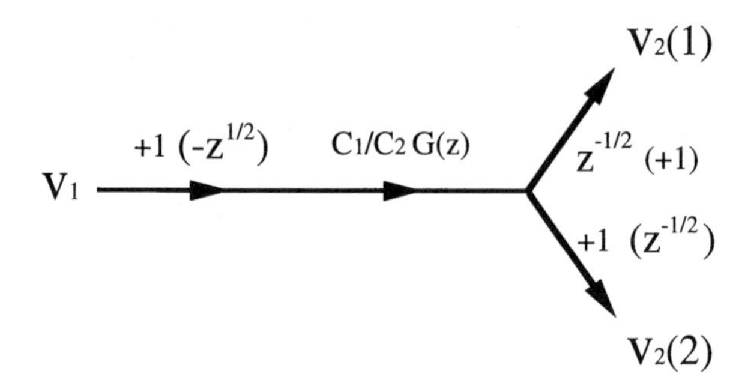

Figure 11 Parasitic insensitive noninverting (inverting) LDI SC integrator and its SFG.

use of canonic integrators is often inconvenient as the fixed input weights makes the application of contour scaling much more difficult. In the case of noncanonic integrators (see Table 2c and d), the input branch weights are set by separate input capacitors (C1 and C2) and so can be scaled independently, but at the expense of additional capacitor ratio errors. It should be noted that since such tolerances do not correspond to elements in the LCR prototype, the zero sensitivity property[6] does not apply.

Figure 9b shows an SC equivalent resistor which realizes the bilinear transformation. Although there are several crucial advantages to designing filter systems which implement the bilinear transformation (see below), unlike the LDI case, no simple, entirely parasitic-free integrator based on the circuit of Figure 9b has been discovered. It is possible, however, to realize a bilinear SC filter using LDI integrators by means of special scaling transformations.[17]

3.4 LDI AND BILINEAR TRANSFORMATIONS

These s-to-z transformations were alluded to above and since their properties are well understood,[2,3] they will not be discussed in detail here. There are, however, two very important differences between the transformations, which can be appreciated by returning to Equations 10c and d. Firstly, in the bilinear case consider the following property:

$$\tan\frac{\omega T}{2} \to \infty \quad \text{as} \quad \omega \to \frac{\pi}{T}$$

i.e., as ω approaches the Nyquist frequency (π/T), the bilinear frequency variable approaches infinity, resulting in a zero of transmission in the frequency response of the integrator of Equation 10d. Filters realized using the bilinear transform method therefore possess transmission zeros at the Nyquist frequency if their continuous-time prototypes possess such zeros as ω approaches infinity. Such filters can therefore realize stopband responses with high levels of attenuation. In the LDI case, however,

Table 2 Differential LDI SC Integrators and Their SFGs

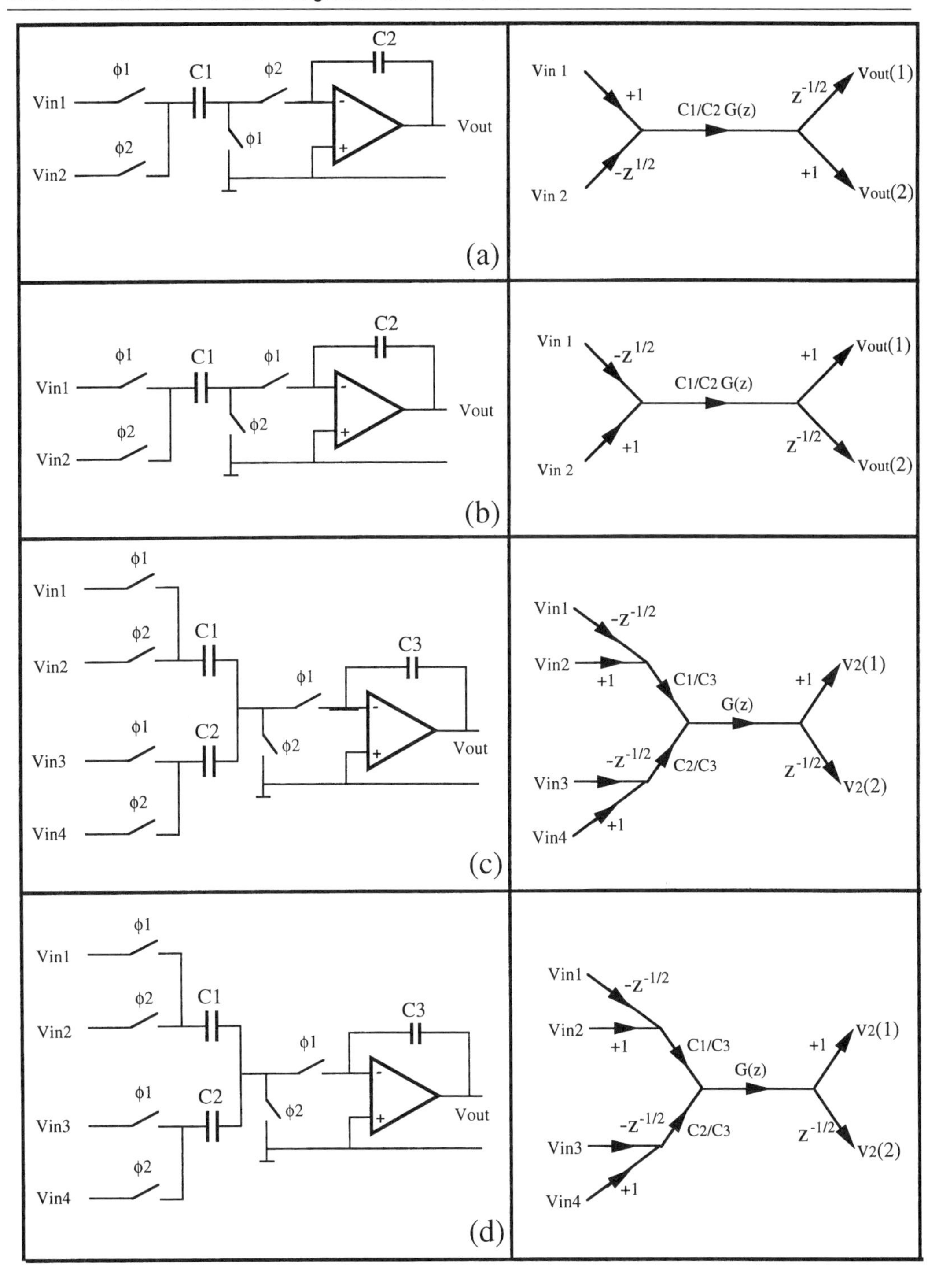

$$\sin\frac{\omega T}{2} \rightarrow 1 \quad \text{as} \quad \omega \rightarrow \frac{\pi}{T}$$

corresponding to a finite level of attenuation at the Nyquist frequency. If a high level of stopband attenuation is required, the order of the filter can be increased, or, alternatively, the ratio of sampling frequency to bandedge frequency (the oversampling ratio) can be increased. However, for practical reasons, neither of these solutions may be very attractive.

The second major difference between the bilinear and LDI transformations relates to the manner in which resistive elements, such as terminations, are realized. In the bilinear case, a resistor, R, transforms exactly into the discrete-time domain, but in the LDI case, it becomes a frequency-dependent element:

$$R \rightarrow R \cdot z^{-1/2} = R\left(\cosh\frac{sT}{2} - \sinh\frac{sT}{2}\right)$$

The sinh(sT/2) part of this expression can generally be absorbed into the reactive elements of the filter, but the cosinusoidal term remains, resulting in errors in the realized frequency response and a departure from the zero-sensitivity property, except, perhaps at DC.[16]

The effect of the termination error is reduced for high levels of oversampling, or, alternatively, exact design methods have been described which synthesize an LDI ladder including its terminations. However, even in these cases, the stopband attenuation problem remains. It is for these reasons that in practice, in most cases, the bilinear transformation is the preferred method of realizing SC filters, with the important caveat that the building blocks of such systems are LDI integrators. The design of such filters is rather complex, however, and beyond the scope of this text.[17]

3.5 FREQUENCY WARPING

We have seen that the frequency variable in an SC filter is a trigonometric function of the continuous-time variable, ω. However, trigonometric functions are actually dimensionless and so cannot be compared directly with ω. Let us formalize the relationship between ω and the discrete-time frequency variable, Ω. In the LDI case:

$$\sin\frac{\omega T}{2} \rightarrow \frac{\omega T}{2} \quad \text{as} \quad \frac{\omega T}{2} \rightarrow 0$$

Define the LDI frequency variable as follows:

$$\Omega_{LDI} = \frac{2}{T}\sin\frac{\omega T}{2} \tag{13}$$

where we note that

$$\Omega \rightarrow \omega \quad \text{as} \quad \frac{\omega T}{2} \rightarrow 0$$

This is a nonlinear frequency mapping (warping) which is exact only at zero frequency. At other frequencies, the continuous-time frequency scale is warped as a result of its mapping into the discrete-time domain. So, for example, the cutoff frequency of an SC lowpass filter will not, in general, coincide with the cutoff frequency of its LCR prototype. The same comments apply to the bilinear case, where the corresponding variable is given by

$$\Omega_{BI} = \frac{2}{T}\tan\frac{\omega T}{2}$$

The frequency response of the LDI integrator is therefore

$$\frac{V_2(j\omega)}{V_1(j\omega)} = \frac{-1}{j\left[T\frac{C_2}{C_1}\right]\left[\frac{2}{T}\sin(\omega T/2)\right]}$$

and the effective integrator time constant, τ, is given by:

$$\tau = T(C_2/C_1) \tag{14}$$

(similar comments apply for the bilinear case). The mapping can be made exact (i.e., $\Omega = \omega$) at one finite frequency in addition to zero by modifying Equation 13 as follows:

$$\Omega_{LDI} = \frac{2x}{T}\sin\frac{\omega T}{2} \tag{15}$$

where x is a dimensionless parameter and the integrator time constant is therefore scaled by x^{-1}. It is easy to show that for $\Omega = \omega$, we require:

$$x = \frac{\frac{\omega T}{2}}{\sin\frac{\omega T}{2}} = \frac{\frac{\omega\pi}{\omega_s}}{\sin\frac{\omega\pi}{\omega_s}} \tag{16}$$

where ω_s is the sampling frequency. For example, suppose we wish to design an LDI SC lowpass filter with a cutoff frequency of 10 kHz and a sample frequency of 100 kHz. We require that the frequency mapping is exact at the cutoff frequency. Substituting in Equation 16, we have

$$x = \frac{\frac{\pi}{10}}{\sin\frac{\pi}{10}} = 1.0166$$

and the integrator time constant is scaled by x^{-1}.

3.6 EXAMPLE 2: THIRD-ORDER ALL-POLE LOWPASS SC FILTER USING CANONIC LDI INTEGRATORS

The appropriate LCR prototype filter is shown in Figure 5, and a Chebyshev response with passband ripple amplitude of 0.28 dB is required. In addition, a cutoff frequency of 1 kHz and a switching frequency of 10 kHz are specified. This example is intended to illustrate the simplest possible type of practical SC ladder filter design.

The normalized prototype values are as follows (Saal: C0325):[18]

$$R_S = R_T = 1\Omega$$

$$L_1 = L_3 = 1.3453$$

$$C_2 = 1.1414$$

The L and C values obtained from tables are next denormalized in the conventional manner (see, e.g., Reference 9) for the required bandedge frequency of 1 kHz:

$$L_1 = L_3 \rightarrow 1.3453/(2\pi \cdot 10^3) = 2.1411.10^{-4}$$

$$C_2 \rightarrow 1.1414/(2\pi \cdot 10^3) = 1.8166.10^{-4}$$

Following the procedure of Example 1 above, the *leapfrog* SFG shown in Figure 6b is constructed and the continuous-time integrators are repaced by the canonic LDI units shown in Table 2a and b. As described above, the switch phases are chosen so that the forward path of the SFG contains no additional terms in $z^{-1/2}$ which would introduce high levels of phase distortion into the filter response. The only exception to this occurs in the realization of the terminations which is done by applying feedback around integrators #1 and #3 to realize *damped* (or *lossy* integrators). From the integrators in Table 2, it is easy to show that the frequency response of integrator #1 is

$$\frac{V_2(j\omega)}{V_1(j\omega)} = \frac{-1}{j\sin(\omega T/2)\left[2\dfrac{C_2}{C_1} + \dfrac{C_3}{C_1}\right] + \left(\dfrac{C_3}{C_1}\right)\cos(\omega T/2)} \tag{17}$$

and that the transfer function of integrator #3 is the same as integrator #1 with $C_3 = C_1$.

The resulting SFG is shown in Figure 12 and its SC realization in Figure 13, where we note once again that realization of the terminations under the LDI transformation involves extra terms in $z^{-1/2}$ with the consequences discussed above.

By using Equations 14 and 17, the capacitor ratios for the three integrators can be calculated as follows:

$$\left(\frac{C_2}{C_1}\right)^{\#1} = \left(\frac{C_2}{C_1}\right)^{\#3} = 1.6411$$

$$\left(\frac{C_2}{C_1}\right)^{\#2} = 1.8166$$

The frequency response of this filter was simulated using the package SWITCAP and the amplitude of this response is shown in Figure 14a and b. Note that the amplitude at the passband edge (1 kHz) peaks by approximately 0.35 dB (Figure 14a). This is due partly to the frequency-warping effect described above, but is mainly caused by inaccuracies in the realization of the LDI terminations. Notice also that the stopband attenuation of the LDI SC filter levels off at about –35 dB.

Figure 15 shows the passband behaviour of the three state variables: I_{L1}, V_{C2}, and I_{L3} (= V_o), which are represented in this active realization by the outputs of the three operational amplifiers. Note that the output of integrator #2, corresponding to V_{C2}, peaks at about 4.6 dB above the output, which in an active realization corresponds to 4.6 dB loss of signal-handling capacity. Similarly, the output of integrator #1 peaks at about 3.3 dB above the output. Ideally, we would like to equalize all the nodal voltage maxima, including the input, in order to maximize signal-handling capacity and to remove the –6 dB insertion loss produced by the prototype circuit of Figure 13. However, examination of Figure 13 reveals that the contour-scaling technique described above cannot be employed here, because the differential inputs to canonic integrators are fixed and equal. However, if *noncanonic* integrators (see Table 2c and d) are used, contour scaling can be readily applied. A modified version of the design using noncanonic intergrators is shown in Figure 16 (all nominal capacitor values = 1.00, except for the three integrating capacitors as shown).

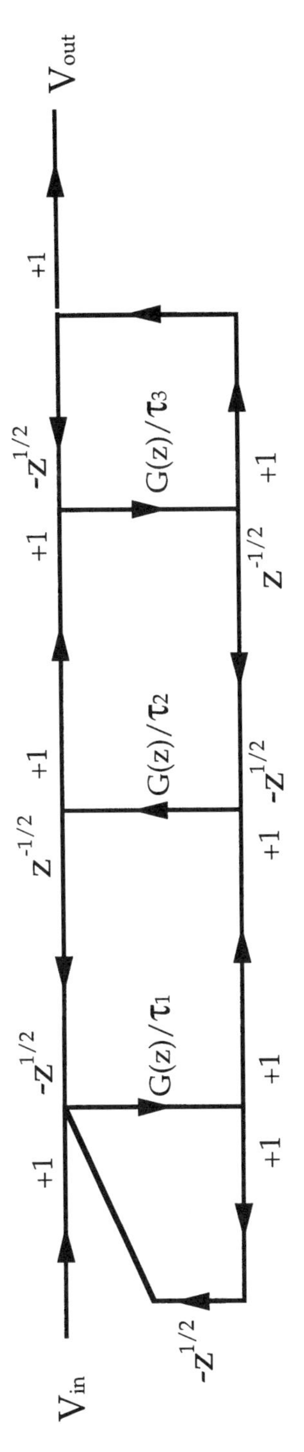

Figure 12 SFG of Figure 6b with SC integrators (see Example 2).

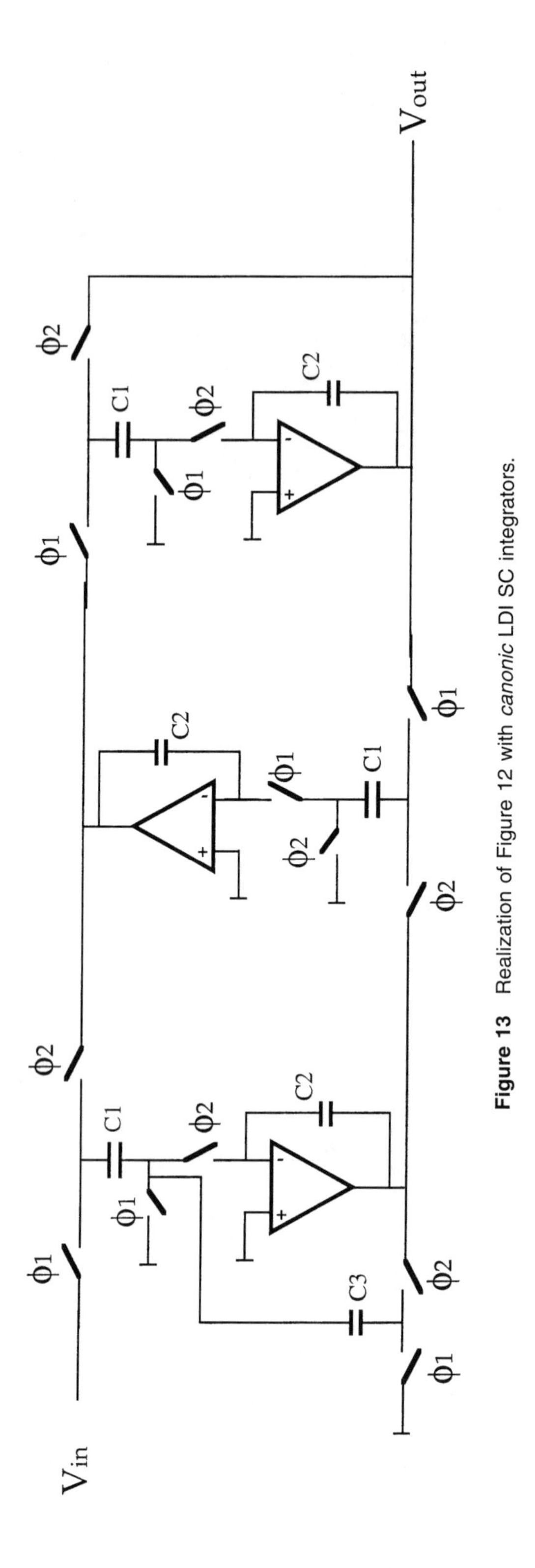

Figure 13 Realization of Figure 12 with *canonic* LDI SC integrators.

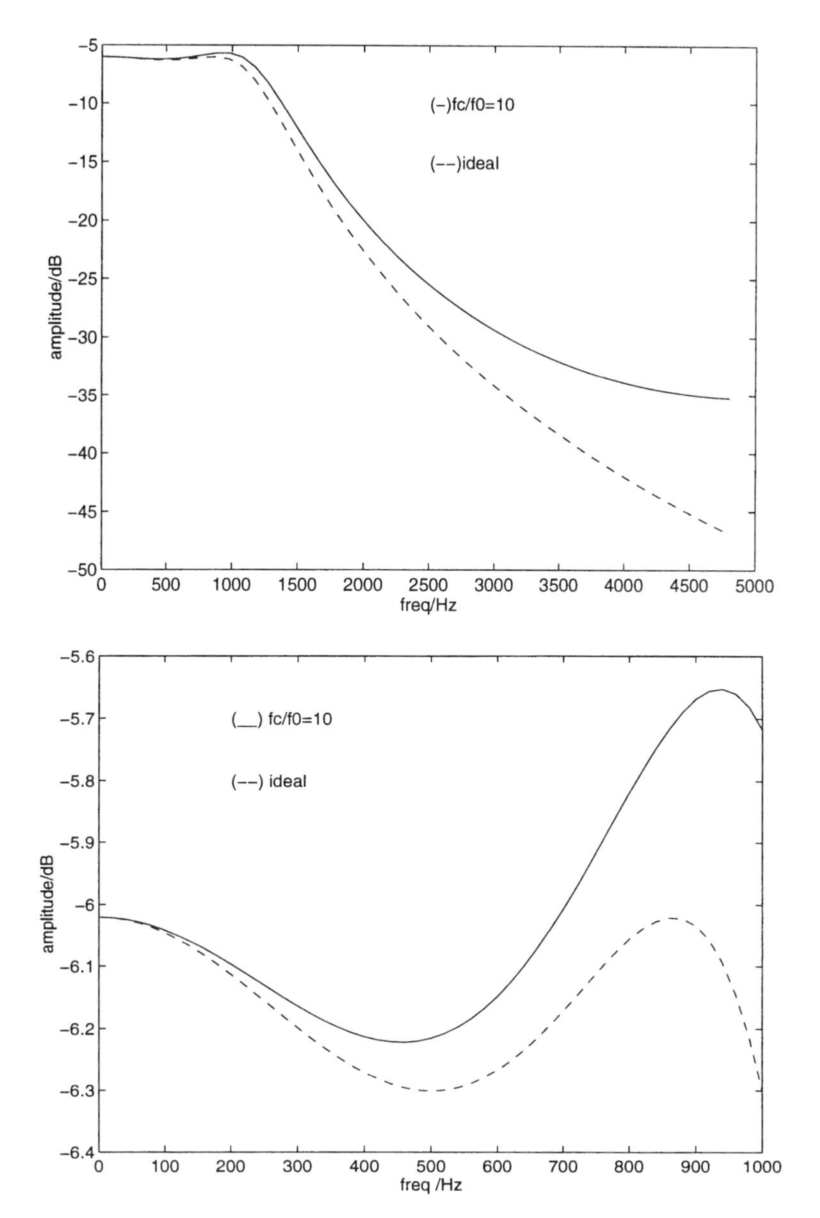

Figure 14 (a) Response of the filter shown in Figure 13; (b) passband response of the filter shown in Figure 13.

Applying the contour-scaling technique, we scale the inputs of the integrators by constant factors λ_i ($\lambda_i < 1$) and the outputs by λ_i^{-1}, where the index i refers to the integrators, numbering from the input. In this instance, using the excess voltage maxima quoted above, we have:

$$\lambda_1 = 0.68$$

$$\lambda_2 = 0.59$$

The capacitors C_{a1} and C_{a2} in Figure 16 are therefore scaled by λ_2 (0.59) while the input capacitors C_{b1} and C_{b2} are scaled by λ_2^{-1} (1.69). Similarly, since C_{b1} is connected to the output of integrator #1, it is also scaled by λ_1 and hence:

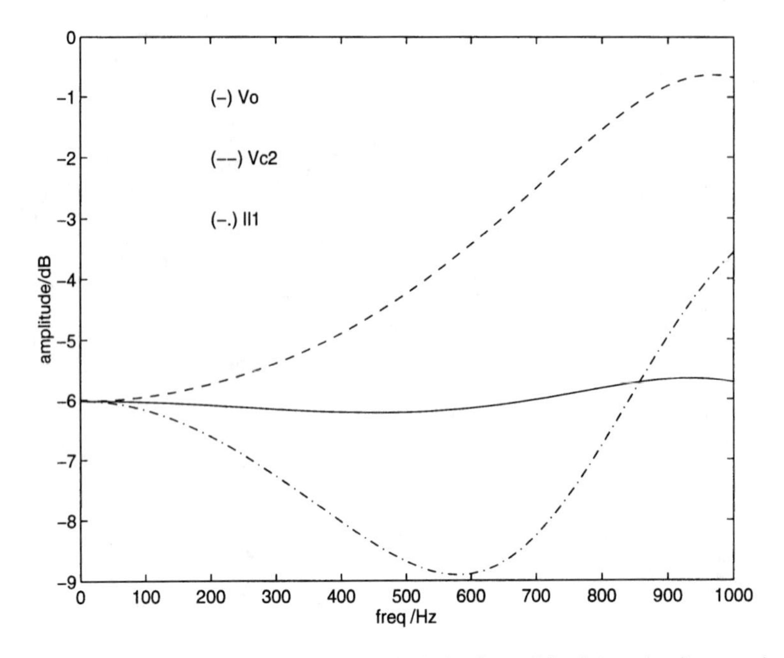

Figure 15 Passband response showing the behaviour of the internal voltage nodes.

$$C_{b1} \rightarrow C_{b1} \cdot \lambda_1 \lambda_2^{-1}$$

and so on for all the other capacitors in the circuit. Finally, the –6 dB insertion loss can be conveniently removed by doubling the value of the input capacitor. The scaled version of the circuit is shown in Figure 17 and its responses (not shown) all peak at 0 dB. Notice that the input branches to the individual integrators are no longer of equal weight and, hence, the integrators do not correspond exactly to elements in the LCR prototype filter. Any errors in realizing these branch weights will not, therefore, be minimized by the zero-sensitivity property of ladder filters.

3.7 DESIGN OF LADDER FILTERS USING THE BILINEAR TRANSFORMATION

From what has been written above, it is clear that bilinear transformation is the method of choice in the design of SC ladder filters, but, unfortunately, no simple entirely parasitic insensitive bilinear integrator circuit has yet been discovered. A solution to this problem was suggested by Datar and Sedra,[17] who began with a bilinearly transformed prototype ladder and then applied *frequency-dependent impedance scaling* to transform the elements of the ladder into units realizable by LDI integrators. Although this approach succeeds in obtaining the best of most possible worlds, a detailed account is beyond the scope of this section.

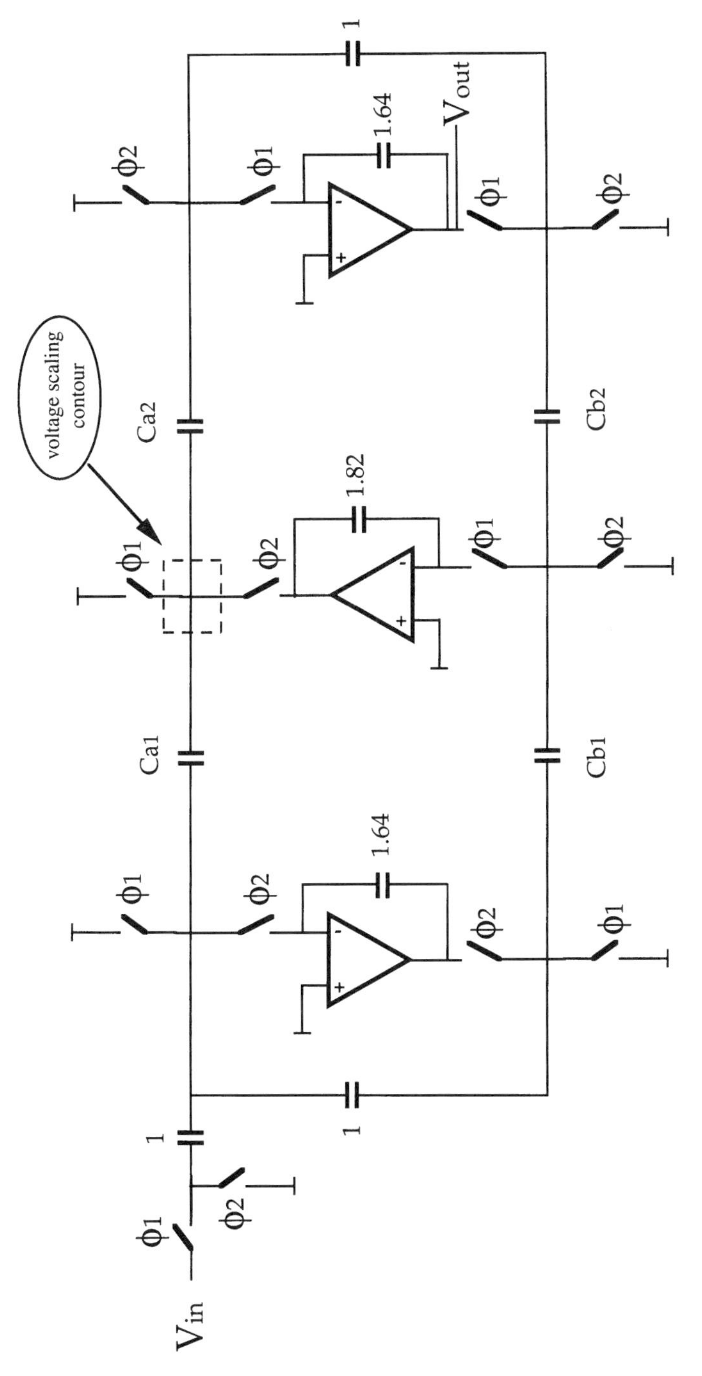

Figure 16 Realization of Figure 12 with *noncanonic* LDI SC Integrators.

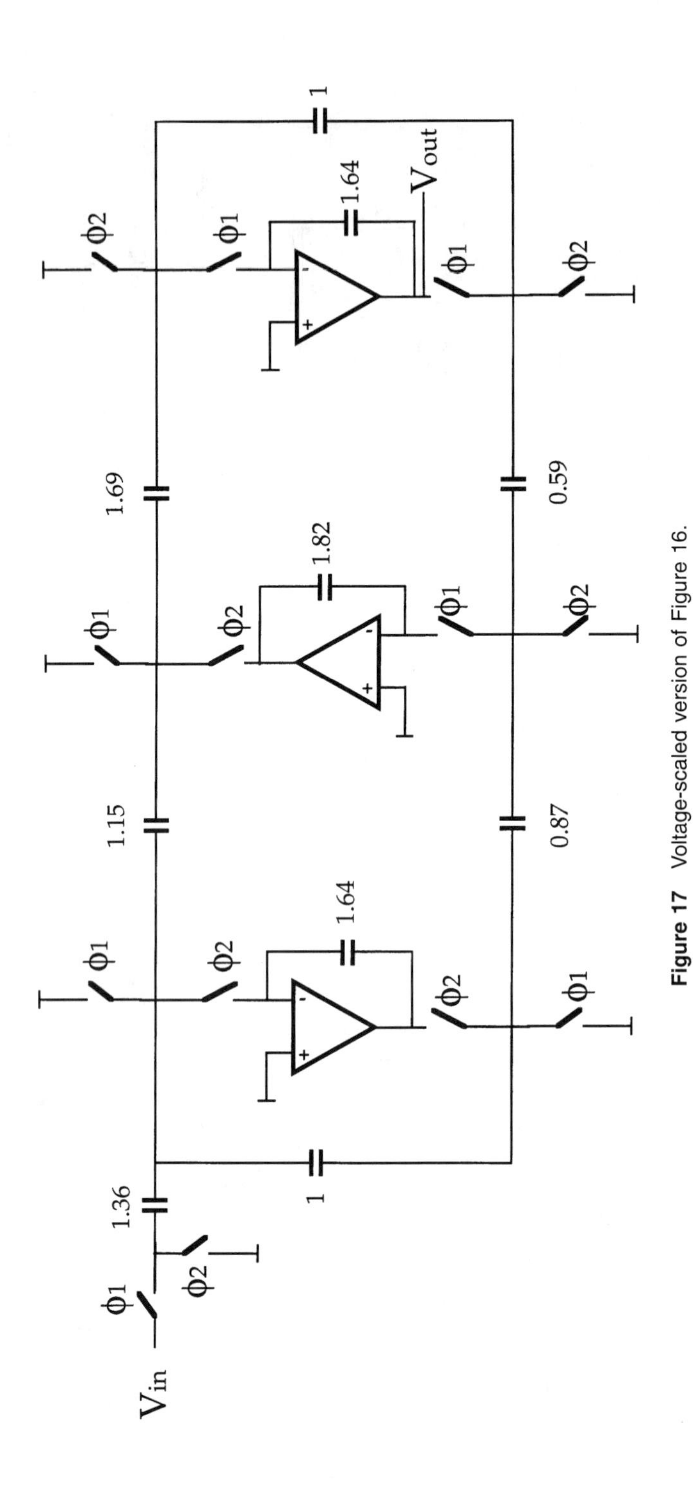

Figure 17 Voltage-scaled version of Figure 16.

REFERENCES

1. **G. Moschytz (ed):** *MOS Switched Capacitor Filters*, New York, IEEE Press, selected reprint series/Wiley, 1986.
2. **R. Schaumann, M. Ghausi, and K. Laker:** *Design of Analog Filters*, Prentice-Hall, Englewood Cliffs, NJ, 1990, chapter 8.
3. **R. Unbehauen and A. Cichocki:** *MOS Switched-Capacitor and Continuous-Time Integrated Circuits and Systems,* Springer-Verlag, Berlin, 1989, chapter 5.
4. **P. Fleischer and K. Laker:** A family of active switched-capacitor building blocks, *Bell Syst. Tech. J.* (B.S.T.J.), 58, 2235–2269, 1979.
5. **K. Martin and A. Sedra:** Exact design of switched-capacitor bandpass filters using coupled-biquad structures, *IEEE Trans. Circuits Syst.*, CAS-27, 469–475, 1980.
6. **H. Orchard:** Inductorless filters, *Electron. Lett.*, 2, 224–225, September, 1966.
7. **H. Orchard, G. Temes, and T. Cataltepe,** Sensitivity formulas for terminated lossless two-ports, *IEEE Trans. Circuits Syst.*, CAS-32, 459–466, 1985.
8. **W. Heinlein and H. Holmes:** Active filters for integrated circuits, Oldenbourg Verlag, Munich, 1974, chapter 8.
9. **G. Daryanani:** *Principles of Active Network Synthesis and Design*, John Wiley, New York, 1976, chapter 11.
10. **N. Balabanian and A. Bickart:** *Electrical Network Theory*, John Wiley, New York, 1969, chapter 9.
11. **P. Brackett and A. Sedra:** Direct SFG simulation of LC ladder networks with applications to active filter design, *IEEE Trans. Circuits Syst.*, CAS-23, 61–67, 1976.
12. **D. Perry:** Scaling transformation of multiple-feedback filters, *IEE Proc. (Part G),* 4, 176–179, August, 1981.
13. **R. Gregorian and G. Temes:** *Analog MOS Integrated Circuits for Signal Processing,* John Wiley, New York, 1986, chapter 5.
14. **G. Jacobs, D. Allstot, R. Brodersen, and P. Gray:** Design techniques for MOS switched-capacitor ladder filters, *IEEE Trans. Circuits Syst.*, CAS-25, 1014–1021, 1978.
15. **K. Martin:** Improved circuits for the realization of switched capacitor filters, *IEEE Trans. Circuits Syst.*, CAS-27, 237–244, 1980.
16. **D. Haigh, B. Singh, and J. Taylor:** Continuous time and switched capacitor monolithic filters based on current and charge simulation, *Proc. IEE (Part G)*, 137(2), 147–155, 1990.
17. **R. Datar and A. Sedra:** Exact design of strays-insensitive switched-capacitor ladder filters, *IEEE Trans. Circuits Syst.*, CAS-30, 888–898, 1983.
18. **R. Saal:** *Handbook of Filter Design*, AEG-Telefunken, Berlin, 1979.

7.3

Practical Design Considerations for Switched-Capacitor Filters

Qiuting Huang

CONTENTS

1. INTRODUCTION

The previous sub-chapters have been concerned with synthesizing discrete-time transfer functions and signal flow graphs, and implementing them with switched-capacitor (SC) building blocks such as stray-insensitive SC integrators. Under the assumption that the switches and op amps used are ideal, the description of an SC filter is nearly the same as that of a digital filter in the sense that signal processing takes place in discrete time which can be described by difference equations. Since each sample in an SC filter is held constant for the clock period, the transfer function of an SC filter is similar to an A/D converter, digital filter plus D/A converter combination. The first-order hold function, which an SC filter has in common with a D/A converter, introduces a 2 $\sin(\omega T/2)/\omega T$ modification to the exact transfer function. This should be taken into account during the filter synthesis if the clock-to-bandedge-frequency ratio is not very high.[1] Capacitance ratio inaccuracies were mentioned, reflecting the analog nature of SC circuits, to justify the use of ladder structures.

In practical implementations, however, SC filter performance not only depends on the accuracy of the ratios of capacitors which determine the coefficients in the transfer function, but also on how well the op amps and switches in the filter are implemented. Performance limitations of an

0-8493-8951-8/97/$0.00+$.50
© 1997 by CRC Press LLC

op amp, such as finite DC gain, finite gain-bandwidth product, and finite slew rate, and their influence on the performance of an SC filter, will be discussed in this chapter. In addition to causing the filter response to deviate from an ideal staircase shape of a sampled and held waveform, the switches and op amps used in an SC filter also generate noise and introduce undesirable paths which couple interferences from the power supply to the output. Such noise and interferences degrade the signal quality and limit the dynamic range of the filter. Circuit noise, power supply noise coupling, clock feedthrough, as well as circuit techniques which minimize those spurious signals, will be discussed in this section.

As the topics we are going to discuss are mostly technology oriented, the conclusions of the discussions are not easy to tabulate into tables, as one would expect from a handbook. Instead, the intention here is to compile the key ideas of the discussions that are scattered in research papers into a concise reference, where the guidelines to optimize each aspect of practical SC filter design can be found in a concise form. To enable the reader to understand the key steps leading to a conclusion, mathematical formulas, sometimes complex-looking ones, are included for reference. For the readers who are only interested in looking up the conclusion itself, the latter is designed to be understood without the derivations.

2. CAPACITANCE RATIO ACCURACY

The coefficients in an SC filter are realized by ratios of capacitances.[2] The accuracy of such ratios directly influences the locations of the filter poles and zeros. In CMOS technologies, a capacitor is usually implemented as two layers of conducting material sandwiching a thin layer of silicon dioxide which serves as high-quality dielectric. The two layers of conducting material may be metal and polysilicon, two layers of polysilicon, or polysilicon and a diffusion layer in the substrate. The value of the capacitance is determined by the area of the capacitor and the thickness of the dielectric. Fabrication errors such as underetching, which causes the capacitor plate area to be smaller by a fixed percentage, may cause certain systematic errors in the value of a capacitor. Such errors can be eliminated by laying capacitors out in a parallel combination of smaller, unit capacitors with constant area-to-perimeter ratio. The percentage error will then cancel out in the ratio of capacitances that define a coefficient, if the two capacitors are placed close together in a concentric way. In addition to systematic errors, there are also random errors which differ from capacitor to capacitor, which must be taken into consideration when we design capacitors. Due to granularity of the materials that make up the layers of a capacitor, the definition of the edges of the capacitor plates and the thickness of the dielectric are never perfect. Figure 1 illustrates the random edge variation of a square capacitor plate. Assuming the random variation in the direction perpendicular to the ideal edge, $l(x)$, is a zero mean, stationary variable, then the area of the capacitor will be $A = L^2 + \Delta A$, where

$$\Delta A = \int_0^{4L} l(x)dx \tag{1}$$

is the deviation of the area from its nominal value L^2. The mean value of ΔA is zero. Its variance is given by

$$\sigma_{\Delta A}^2 = \sigma_A^2 = E\left\{\int_0^{4L} l(x_1)\int_0^{4L} l(x_2)dx_1dx_2\right\} = \int_0^{4L}\int_0^{4L} E\{l(x_1)l(x_2)\}dx_1dx_2 = \int_0^{4L}\int_0^{4L} R(x_1 - x_2)dx_1dx_2 \tag{2}$$

where $R(x)$ is the autocorrelation function of $l(x)$. Substituting $z = x_1 - x_2$ into Equation 2, it can be shown that

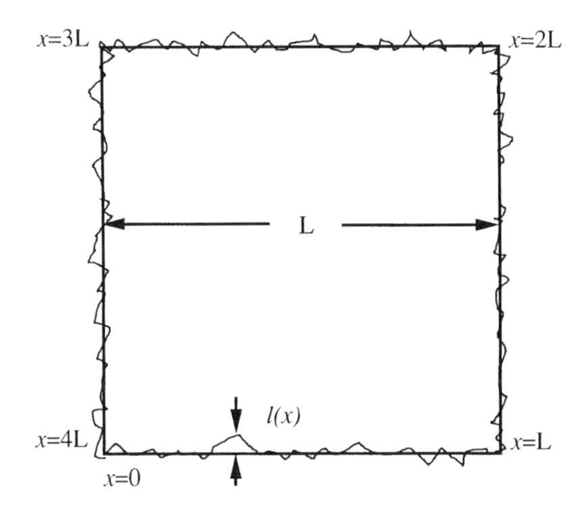

Figure 1 A capacitor with inaccurate edge definition.

$$\sigma_A^2 = \int_{-4L}^{4L} (4L - |z|) R_l(z) dz \approx \int_{-d}^{d} (4L - |d|) R_l(z) dz \approx 4L \int_{-d}^{d} R_l(z) dz \propto L \tag{3}$$

where d is the correlation radius beyond which the autocorrelation function becomes negligible. This radius d should be of the same order of magnitude as the granularity of the material, which is much less than the perimeter of the capacitor, $4L$. The relative error in capacitance is

$$\frac{\sigma_c}{C} \propto \frac{\sigma_A}{A} \propto \frac{1}{L^{1.5}} \tag{4}$$

Following a similar analysis, it can be shown that the relative capacitance error due to random variation of the oxide thickness and permittivity is proportional to $1/L$. For designers, it is thus useful to know that given a capacitance ratio, the accuracy improves with larger size. As we mentioned earlier, the systematic errors in capacitance due to underetching and oxide thickness and permittivity gradient along a certain direction on the wafer should be minimized by constructing each capacitor with smaller unit capacitors of identical layout. For a given capacitance, the more unit capacitors it is split into (hence, smaller L for the unit capacitor), the higher the error due to random variations, but the lower the error due to systematic variations. The choice of the size of the unit capacitor is therefore a careful trade-off based on knowledge of the quality of processing technology in terms of systematic and random variations. For a given unit capacitance, on the other hand, the more such units are connected in parallel, the smaller the relative capacitance error, as the total capacitance increases linearly with the number of parallel unit capacitors, whereas the random capacitance error increases as the square root of the number. If a filter coefficient is realized by the ratio of two capacitances of very different magnitude, then the larger capacitance has much lower relative error than the smaller capacitance. Ensuring the accuracy of the filter coefficient is therefore often a question of ensuring the size of the smallest capacitances in the circuit. In practical circuits at the time of writing, typical minimum capacitances in SC circuits range between 0.25 to 1 pF. If an SC filter design results in high spread of capacitances, that is, high ratio of the largest capacitance and the minimum capacitance in the circuit, then the total capacitance is dominated by the largest capacitance. Capacitance spread is therefore a measure of how much silicon area the circuit consumes, which is one of the important cost measures. The choice of the minimum capacitance is a trade-off between cost and accuracy. The former includes both area and power, as higher power is required for higher capacitance to maintain speed, and the latter includes both coefficient accuracy and noise. The influence of capacitance on noise performance will be discussed at the end of this chapter.

3. OPERATIONAL TRANSCONDUCTANCE AMPLIFIERS[3,4]

A conventional operational amplifier is defined as an amplifier that has very high (infinite) input impedance, very high gain (more than 10^5) and very low (kilohms or less) output impedance. Most such amplifiers are implemented in bipolar technologies. For reasons of cost and compatibility with the rest of the digital circuits in the system, SC circuits are typically implemented in CMOS technologies. The input resistance of an MOS transistor is nearly infinite, which makes it ideal for the input characteristic of an amplifier in an SC circuit, whose operating principle is based on charge conservation. It is, however, much harder for MOS transistor-based amplifiers to achieve very high gain and very low output impedance. This is primarily due to the fact that unlike bipolar transistors, whose I-V characteristic is exponential, the I-V relationship for an MOS transistor is quadratic. This gives the latter lower transconductance for practical levels of bias current than their bipolar counterpart. The output resistance of an MOS transistor is also typically lower than that of a bipolar transistor, making it harder to achieve high gain. When using technologies available at the time of writing, the achievable low-frequency gain for a CMOS amplifier is typically between 70 and 90 dB, depending on the speed of the amplifier. Typically, the output impedance of an amplifier is related to the inverse of the transconductance of the output transistor, if a source-follower circuit is used to buffer the amplifier. Even if higher power consumption is not a major concern and low output resistance is achieved by higher level of bias current, the loss of output voltage swing, due to the MOS transistor threshold voltage, in a push–pull type of output stage is usually unacceptable. On the other hand, in SC circuits there is no resistive load to be driven by the amplifier. Only transient currents are required to change the voltage across a capacitor from one clock phase to the next. There is therefore also no need for amplifiers to have very low output resistance in an SC filter. On the contrary, amplifiers currently used in most SC circuits have no buffers so that their output resistances are very high. They are generally referred to as operational transconductance amplifiers (OTAs). The most important requirements on such amplifiers are DC gain, slew rate, and gain-bandwidth product, which we shall discuss below.

Two typical OTA circuits used in SC filters are shown in Figure 2. Although they are different in their ability to provide gain, speed, slew rate, and voltage swing, for the purpose of modeling a practical SC circuit they can both be represented by the same first-order small signal equivalent circuit model shown in Figure 2c, where g_m represents the total transconductance of the amplifier, g_o represents the output conductance, and C_L represents the compensation capacitance that is required for the stability of the amplifier. It is assumed here that the phase margin of the amplifier is close to 90° so that the amplifier has single pole characteristic, to simplify the analysis. We shall see in the following how the values of those model parameters affect the charge transfer accuracy of an SC integrator.

Figure 3a redraws a noninverting SC integrator with an explicit load capacitor C_L, which may also include any capacitors connected or switched to the OTA output from other parts of an SC circuit during a particular clock phase. If we replace the OTA by its equivalent circuit model in Figure 2c, the SC integrator can be modeled by Figure 3b. The charge transfer in an SC integrator, which carries the information of the input to the integrator output, takes place in three steps in a noninverting integrator. In the first step the input capacitor C_{in} is charged to the input voltage V_{in} during one of the clock phases, $\phi 1$ (Figure 4a). This charge is transferred to the integrating capacitor C in the next clock phase, $\phi 2$, in two steps. At the very beginning of $\phi 2$, which we denote $t = 0^+$, the three capacitors connected to the OTA form a closed loop (Figure 4b). Although the OTA has no time yet to supply current to any of the capacitors, an instant charge redistribution takes place among the three capacitors so that charge conservation and Kirchhoff's voltage law are held true simultaneously. This will cause a step change of both the voltage at the inverting input node V_1 of the amplifier and that at its output node, V_O. The step change of voltage V_1 causes the OTA to deliver current to the capacitors, until the input capacitor C_{in} is completely discharged and V_1 returns to virtual ground (Figure 4c).

Referring to Figure 4b, the following equalities hold for $t = 0^+$:

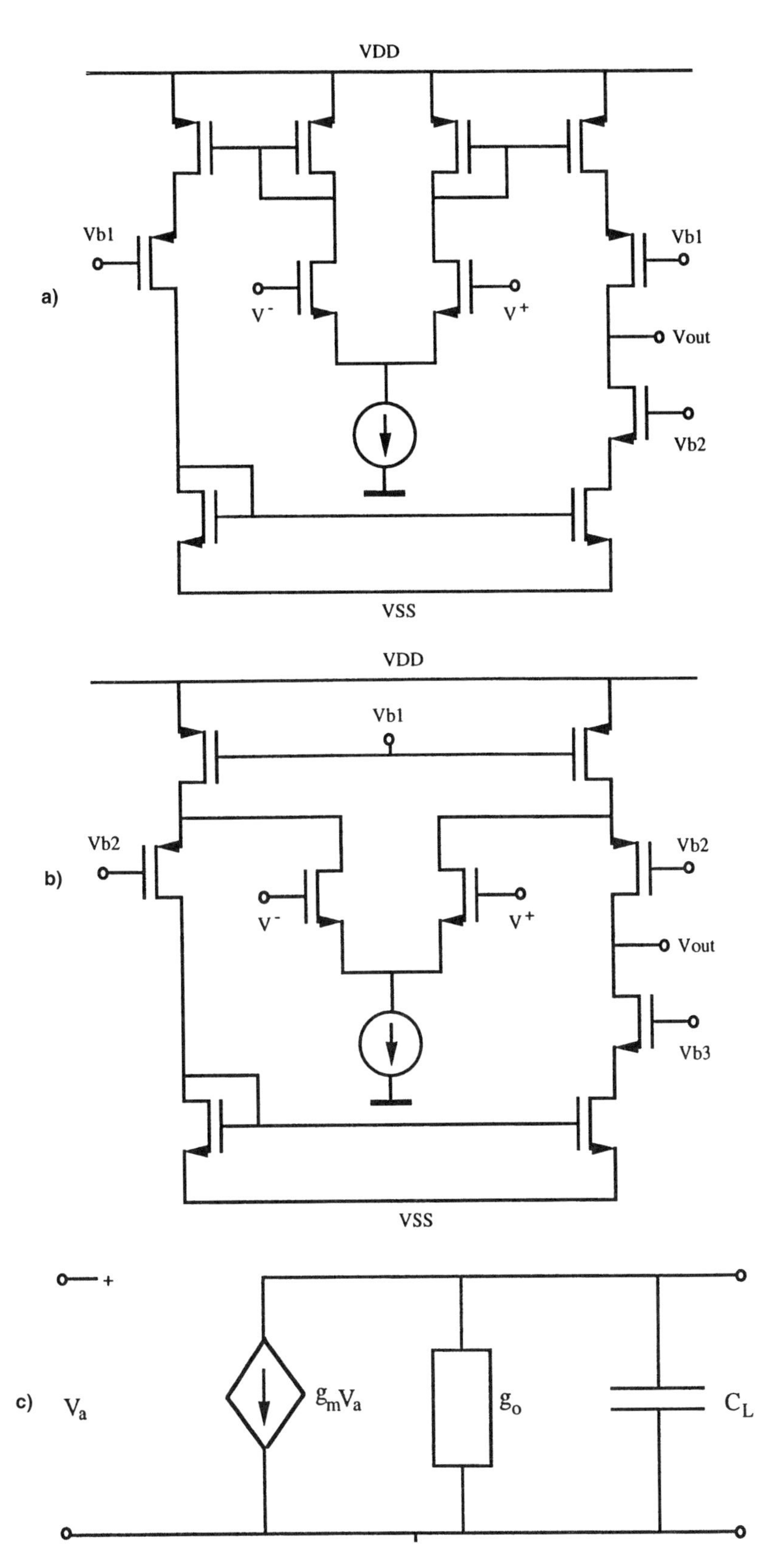

Figure 2 (a) A cascode CMOS OTA; (b) a folded cascode CMOS OTA; (c) small signal equivalent circuit model of an OTA.

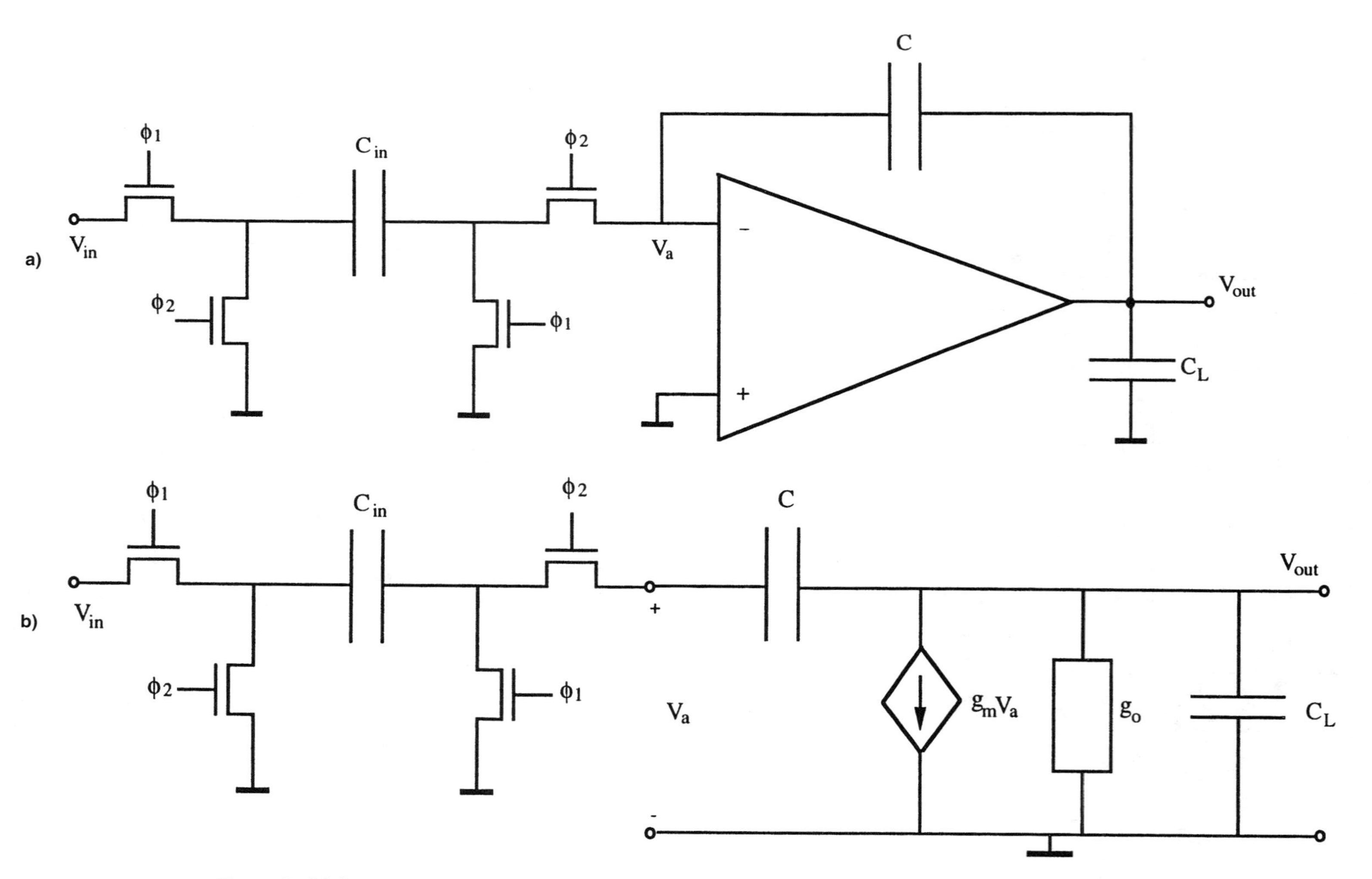

Figure 3 (a) A noninverting SC integrator; (b) a noninverting SC integrator with equivalent model of OTA.

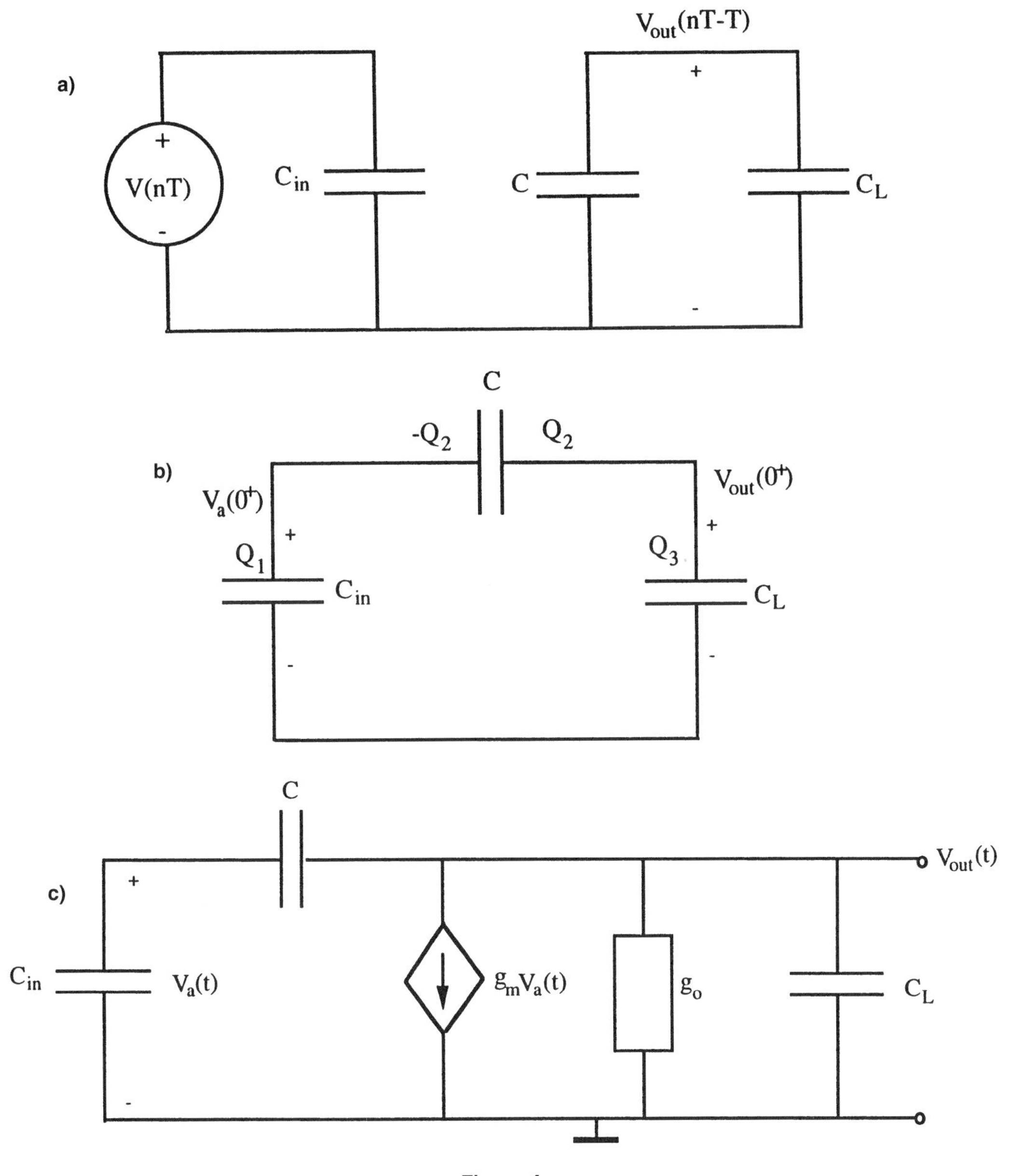

Figure 4

$$\frac{Q_1}{C_{\text{in}}} + \frac{Q_2}{C} = \frac{Q_3}{C_L} \tag{5}$$

$$V_a\left(0^+\right) = \frac{Q_1}{C_{\text{in}}} \tag{6}$$

$$V_{\text{out}}\left(0^+\right) = \frac{Q_3}{C_L} \tag{7}$$

$$Q_1 - Q_2 = -C_{\text{in}} \cdot V_{\text{in}}(nT) - C \cdot V_{\text{out}}(nT - T) \tag{8}$$

$$Q_2 + Q_3 = (C + C_L) \cdot V_{\text{out}}(nT - T) \tag{9}$$

Solving Equations 5 through 9 for the initial voltage steps $V_a(0^+)$ and $V_{\text{out}}(0^+)$, we find

$$V_a(0^+) = -\frac{(C + C_L) \cdot C_{\text{in}}}{C_{\text{in}} \cdot C + C_L \cdot C + C_{in} \cdot C_L} \cdot V_{\text{in}}(nT) \tag{10}$$

$$V_{\text{out}}(0^+) = -\frac{C \cdot C_{\text{in}}}{C_{\text{in}} \cdot C + C_L \cdot C + C_{\text{in}} \cdot C_L} \cdot V_{\text{in}}(nT) + V_{\text{out}}(nT - T) \tag{11}$$

The nonzero value of V_a causes the OTA to deliver current to the capacitors, which transfers the charge remaining on C_{in} to the integrating capacitor C during clock phase $\phi 2$, so that $V_a(t)$ approaches zero and $V_{\text{out}}(t)$ approaches the final value determined by the transfer function of the integrator (Figure 5). Referring to Figure 4c, the charge transfer transient is described by differential Equations 12 and 13:

$$C_{\text{in}} \cdot \frac{dV_a(t)}{dt} = C \cdot \frac{d}{dt}\left[V_{\text{out}}(t) - V_a(t)\right] \tag{12}$$

$$C \cdot \frac{d}{dt}\left[V_{\text{out}}(t) - V_a(t)\right] + C_L \cdot \frac{dV_{\text{out}}(t)}{dt} + g_o \cdot V_{\text{out}}(t) + g_m V_a(t) = 0 \tag{13}$$

By using the initial conditions given by Equations 10 and 11, solutions to Equations 12 and 13 can be found:

$$V_a(t) = \left\{ -\frac{C_{\text{in}} \cdot (C + C_L)}{C_{\text{in}} C + C_L C + C_{\text{in}} C_L} V_{\text{in}}(nT) + \frac{V_{\text{out}}(nT - T) + \frac{C_{\text{in}}}{C} V_{\text{in}}(nT)}{A} \right\}$$

$$\cdot \exp\left[-\frac{g_m + g_o\left(1 + \frac{C_{\text{in}}}{C}\right)}{C_{\text{eff}}} \cdot t \right] - \frac{V_{\text{out}}(nT - T) + \frac{C_{\text{in}}}{C} V_{\text{in}}(nT)}{A}$$

$$\approx -\frac{C_{\text{in}} \cdot (C + C_L)}{C_{\text{in}} \cdot C + C_L \cdot C_{\text{in}} \cdot C_L} \cdot V_{\text{in}}(nT) \cdot \exp\left[-\frac{g_m}{C_{\text{eff}}} \cdot t \right] - \frac{V_{\text{out}}(nT - T) + \frac{C_{\text{in}}}{C} V_{\text{in}}(nT)}{A} \tag{14}$$

$$V_{\text{out}}(t) = \left(1 + \frac{C_{\text{in}}}{C}\right) V_a(t) + V_{\text{out}}(nT - T) + \frac{C_{\text{in}}}{C} V_{\text{in}}(nT) \tag{15}$$

where the effective load capacitance C_{eff} and the amplifier DC gain A are defined as

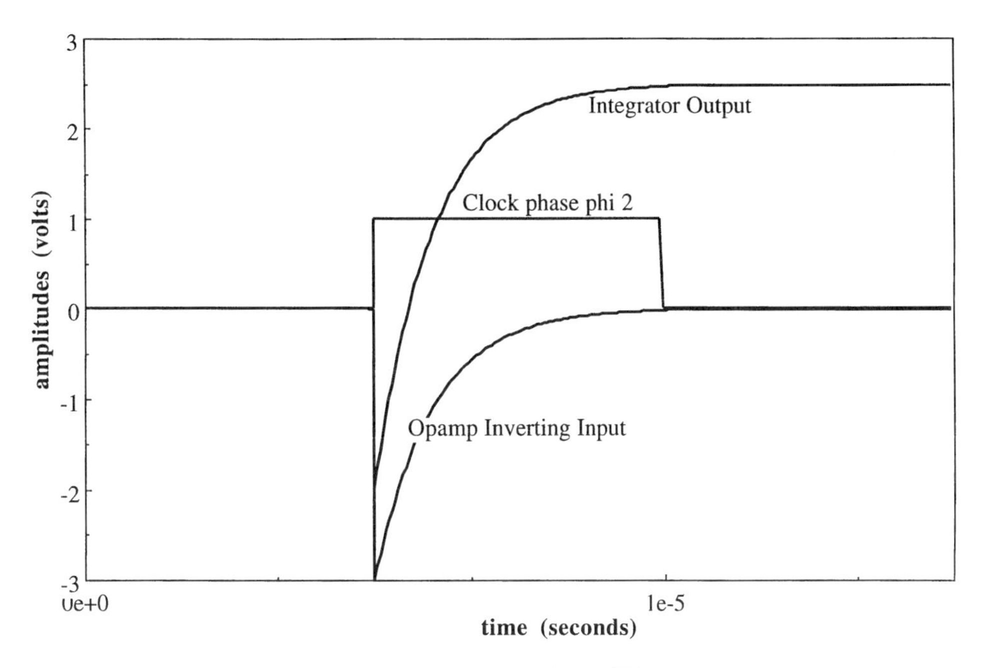

Figure 5 Switching transient in an SC integrator.

$$C_{\text{eff}} = C_{\text{in}} + C_L + \frac{C_{\text{in}} \cdot C_L}{C} \tag{16}$$

$$A = \frac{g_m + g_o\left(1 + \dfrac{C_{\text{in}}}{C}\right)}{g_o} \approx \frac{g_m}{g_o} \tag{17}$$

The last two terms in Equation 15 represent the output of an ideal integrator for an input signal $V_{\text{in}}(nT)$, assuming infinite DC gain and infinite transconductance for the amplifier. For practical amplifiers, the DC gain and the transconductance are finite and the charge transfer is incomplete. In most two-phase SC circuits, the duty cycle of each clock is close to 50%, so that the time available for the charge transfer in clock phase $\phi 2$ is about $T/2$ for a clock period of T. At the end of $\phi 2$, the output voltage of the SC integrator is given by

$$\begin{aligned} V_{\text{out}}(nT) &= \left(1 + \frac{C_{\text{in}}}{C}\right) V_a\left(\frac{T}{2}\right) + V_{\text{out}}(nT - T) + \frac{C_{\text{in}}}{C} V_{\text{in}}(nT) \approx \\ &- \frac{C_{\text{in}}}{C}\left(1 + \frac{C}{C_{\text{eff}}}\right) \cdot V_{\text{in}}(nT) \cdot \exp\left[-\frac{g_m}{C_{\text{eff}}} \cdot \frac{T}{2}\right] - \left(1 + \frac{C_{\text{in}}}{C}\right) \frac{V_{\text{out}}(nT)}{A} \\ &+ V_{\text{out}}(nT - T) + \frac{C_{\text{in}}}{C} V_{\text{in}}(nT) = -\beta V_{\text{out}}(nT) + V_{\text{out}}(nT - T) + \frac{C_{\text{in}}}{C}(1 - \alpha) \cdot V_{\text{in}}(nT) \end{aligned} \tag{18}$$

where

$$\alpha = \left(1 + \frac{C}{C_{\text{eff}}}\right) \cdot \exp\left[-\frac{g_m}{C_{\text{eff}}} \cdot \frac{T}{2}\right] \tag{19}$$

and

$$\beta = \frac{\left(1 + \frac{C_{\text{in}}}{C}\right)}{A} \tag{20}$$

If we take the z transform of Equation 18, the transfer function of an SC integrator with a nonideal OTA is given by

$$\frac{V_{\text{out}}(z)}{V_{\text{in}}(z)} \approx (1-\alpha)\frac{C_{\text{in}}}{C}\frac{1}{\left(1-z^{-1}\right)+\beta} \tag{21}$$

Compared with the ideal discrete-time integrator, we see that insufficient transient settling introduces an error in the effective capacitance ratio, thereby filter coefficient, in the form of α. Insufficient DC gain A, on the other hand, causes the transfer function to deviate from that of an ideal integrator in the form of the β term in the denominator.

In Equation 19, α is an exponential function of the unity-gain frequency (gain-bandwidth product) of the OTA, g_m/C_{eff} (in rad/s). Considering the integrator alone, the unity-gain frequency needs only to be two or three times the clock frequency to reduce the coefficient error due to α to less than 0.1%. When many integrators are interconnected into a loop in an SC filter, the settling is slower so that higher unity-gain frequencies are required for the amplifiers. Op amp slewing, which has been ignored in the analysis above but which occurs in most practical situations, also reduces the time available for linear settling.[4] Another simplification in the above analysis is the 90° phase margin. In practice, amplifiers all have nondominant poles which contribute to additional phase shift, so that phase margins are between 50 and 70°. This means that practical settling behavior of an SC circuit is better described by a ringing exponential settling, which can be much slower for poorer phase margins. Thus, in practice, designers usually aim to achieve an op amp unity-gain frequency that is five to seven times the clock frequency of the filter, with a phase margin of about 60°.

The effect of β, on the other hand, is to introduce a leakage path into an integrator. As the majority of integrators in a filter ought to be as close to ideal integrators as possible, β must be minimized. The effect of β becomes even more serious when the clock frequency is high. For a sine wave input of angular frequency ω, Equation 21 can be rewritten as

$$\frac{V_{\text{out}}(j\omega)}{V_{\text{in}}(j\omega)} \approx (1-\alpha)\frac{C_{\text{in}}}{C}\frac{1}{\left(1-e^{-j\omega T}\right)+\beta} \approx (1-\alpha)\frac{C_{\text{in}}}{CT}\frac{1}{j\omega+\frac{\beta}{T}} = (1-\alpha)\frac{C_{\text{in}}}{CT}\frac{1}{j\omega+\beta\cdot f_c} \tag{22}$$

From Equation 22 we see that due to the presence of β an ideal integrator is turned into a lowpass filter with a cutoff frequency of βf_c. The higher the clock frequency, therefore, the less ideal an integrator will become, and more stringent requirements may have to be placed on the op amp DC gain A. As higher clock frequency also requires a high bandwidth of the op amp, and as high DC gain and high bandwidth are conflicting requirements in an op amp design, high-frequency SC filters tend to be limited by the amplifiers. In practice, lower than 60 dB DC gain is considered unacceptable for most applications. Designers usually aim at a gain between 70 and 85 dB for CMOS technologies. Even with special circuit techniques, such as the regulated cascade OTA configurations,[5] it is quite difficult to exceed a 90 dB gain. Readers are also warned that simulations by computer programs such as SPICE often predict several tens of decibel higher gain (due to inaccurate transistor models), but not many CMOS amplifiers have been measured with more than 90 dB gain!

4. SWITCH NONIDEALITIES

Switches in CMOS SC circuits are implemented with either single MOS transistors or pairs of NMOS–PMOS transistors. Nonidealities of such switches include nonzero on-resistance, clock feedthrough, and noise.

When an MOS switch transistor is turned on, its I-V characteristic is determined by the following equation:

$$I_d = \mu \cdot C'_{ox} \frac{W}{L}\left[\left(V_{gs} - V_T\right)V_{ds} - \frac{V_{ds}^2}{2}\right] \tag{23}$$

where μ is the surface mobility of the conducting channel of the transistor, C'_{ox} is the capacitance per unit area under the transistor gate, and W and L are the transistor gate width and length, respectively. The threshold voltage of the transistor, which is usually between 0.5 and 1 V, is denoted by V_T. The on-resistance of the switch is given by

$$r = \left(\frac{dI_d}{dV_{ds}}\right)^{-1}_{V_{ds}\to 0} = \frac{L}{W}\frac{1}{\mu \cdot C'_{ox}\left(V_{gs} - V_T\right)} \tag{24}$$

For a 1-μm CMOS technology the product of $\mu C'_{ox}$ is about 100 μA/V², so that for a clock swing of 5 V, the on-resistance of a minimum-sized switch ($W = L$) is of the order of 2.5 kΩ. This introduces additional time delay in the charge transfer transients in an SC integrator. For a 0.5 pF capacitance, the time constant associated with a minimum-sized switch is about 1 ns. For an accurate settling, four to five times the time constant is required so that the clock frequency is limited to 100 MHz. This limit is usually much higher than that imposed by the settling time of an OTA. In some SC circuits for multirate signal processing, however, the input switches are clocked at a higher rate than those associated with the op amps. The on-resistance of the switch is then a limiting factor and must be taken into consideration. As Equation 24 suggests, the on-resistance of a switch can be reduced by using wider transistors (higher W). The drawback of doing this, however, is increased charge injection from the switch, also known as clock feedthrough, into the inverting input of the op amp.

When a switch transistor is in the on state, its conducting channel consists of a mobile layer of charge carriers, such as electrons. The amount of charge under the transistor gate is given by

$$Q_c = C'_{ox}WL\left(V_{gs} - V_T\right) \tag{25}$$

When the switch is turned on, the required channel charge is supplied by the signal source, which is usually the output of an amplifier. When the switch is suddenly turned off, however, the channel charge has to be removed. For very sharp clock edges half of the channel charge is injected into each terminal that the switch was connected to.[6] The switches whose clock feedthrough affects the signal processing of an SC filter are those connected to the inverting input terminal of each op amp, as the injected charge affects the charge conservation equation on that node. Since both terminals of such a switch were at the virtual ground before the switch is turned off, the amount of charge injected to the op amp input node is a fixed amount proportional to the high voltage of the clock signal (which is usually the power supply), and independent of the signals being processed by the SC circuit. The effect of such a charge injection is to cause an offset voltage equivalent to several millivolts at the input of the op amp. Such an offset usually cannot be eliminated by offset cancellation schemes based on switching, such as the so-called correlated double sampling. Clock feedthrough, therefore, places a limit on the amount of low-frequency gain that can be realized by an SC circuit.

5. INTERFERENCE COUPLING THROUGH THE POWER SUPPLY[7]

In most applications, an SC circuit finds itself very close to the analog/digital and digital/analog interface. At least for integrated circuit applications, SC circuits are often placed on the same chip as the other digital circuits of the system, for reasons of economy. One characteristic of CMOS digital circuits is that they draw a large amount of current from the power supply at the switching clock transitions, to charge and discharge parasitic capacitors. The peak currents at such transitions can amount to hundreds of milliamps. The impulsive nature of supply currents causes ringing of the power supply voltage on the chip, because of the inductance associated with the bond wires used to connect the power supply bonding pad to the power supply pin of the package. Such inductances, which are of the order of a few nanohenrys, form resonant circuits with parasitic capacitances associated with the source and drain junctions of MOS transistors in the digital circuit. For different switching states the amount of parasitic capacitances shorted to the power supply varies widely, so that the spectrum of the power supply ringing is very wideband, between 100 kHz to hundreds of megahertz. Such wideband interference can be coupled to the sensitive nodes in several different ways, which degrades the signal quality in an SC circuit. As an SC circuit contains many switches which perform sample-and-hold operations, high-frequency ringing above the clock frequency will be aliased into the baseband where it cannot be filtered out by the filter itself.

The first obvious way of digital switching noise coupling is through a common power supply. When a switch connected to the inverting input of an op amp is turned on, the power supply noise modulates the amount of charge in the conducting channel through the gate voltage. When the switch is suddenly turned off, the amount of channel charge is frozen (sampled) and half of that is dumped onto the inverting input of the op amp. In addition to causing offset, sampled supply noise is integrated. Amplifiers also couple supply noise into an SC circuit. A well-known type of amplifier, the two-stage amplifier internally compensated by a Miller capacitor (the 741-type of op amp), is not particularly good in this regard and deserves special consideration. An SC integrator using a two-stage op amp can be modeled as in Figure 6a, where the negative power supply forms the "ground," or noninverting input, of the second amplification stage. Since we are only interested here in the influence of V_{ss} on the output, the input signal source can be assumed to be zero. Capacitor C_{in} is the input capacitor, C the integrating capacitor, whereas capacitor C_M is the internal compensation capacitor of the op amp. As many capacitors (C_{in} in Figure 6) in an SC circuit are switched, the transfer function from V_{ss} to the output of the circuit is that of a periodically time-varying circuit. This is equivalent to two time-invariant topologies being multiplexed at the clock frequency, or each output being multiplied by a 1–0 sequence before added together. If the two topologies are very different, multiplexing them aliases high-frequency signals down to the baseband. In our integrator example the two different topologies are modeled in Figure 6b and c, for $\phi 2 = 1$ and $\phi 1 = 1$, respectively. Since the power supply coupling is a high-frequency phenomenon, the load resistances $r1$ and $r2$, which account for the DC gain of the amplifier have been assumed infinite. Referring to Figure 6b,

$$V_{out}(s)\frac{C}{C+C_{in}}g_{m1}+\left(V_{out}(s)-V_1(s)\right)sC_M=0 \tag{26}$$

$$V_{out}(s)s\frac{C_{in}C}{C_{in}+C}\left(V_{out}(s)-V_1(s)\right)sC_M=\left(V_{ss}(s)-V_1(s)\right)g_{m2} \tag{27}$$

Solving for the transfer function from V_{ss} to V_{out}, we find,

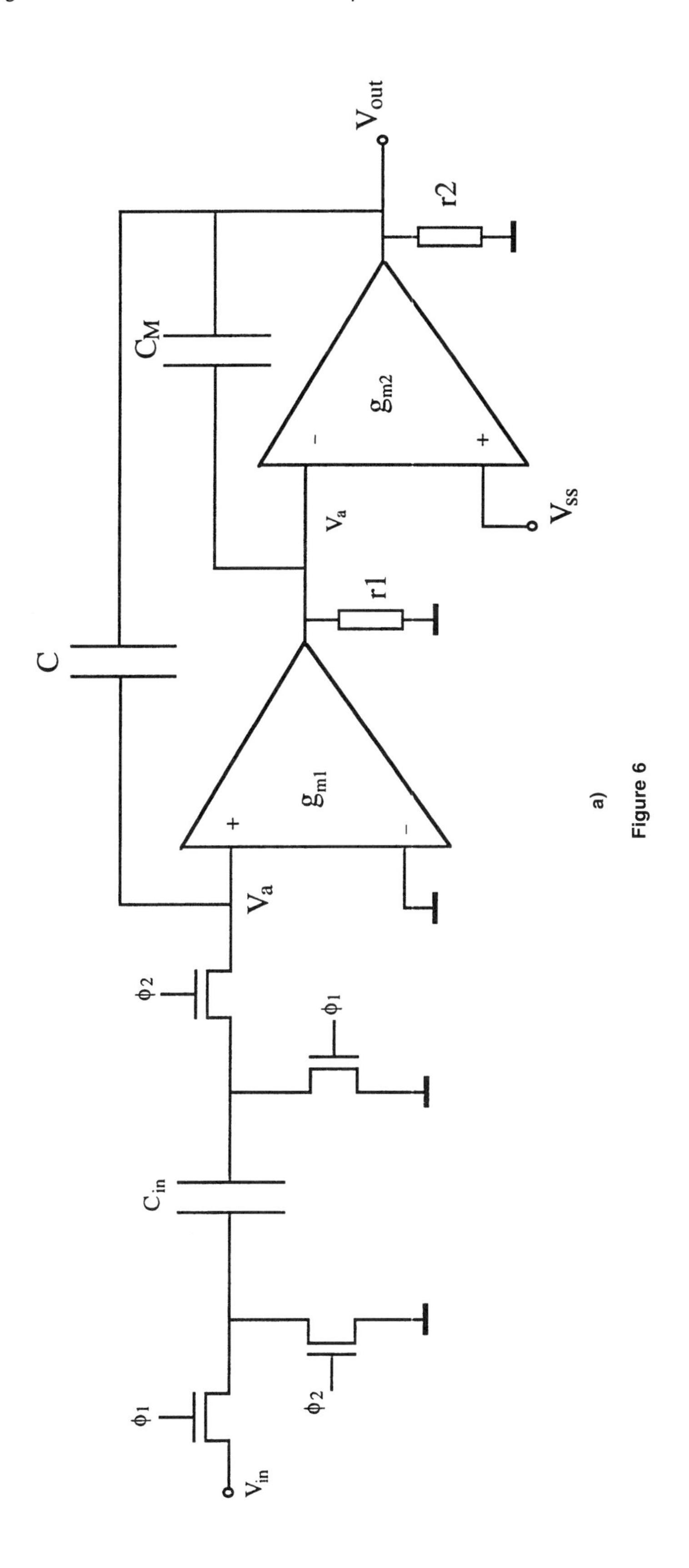

a)

Figure 6

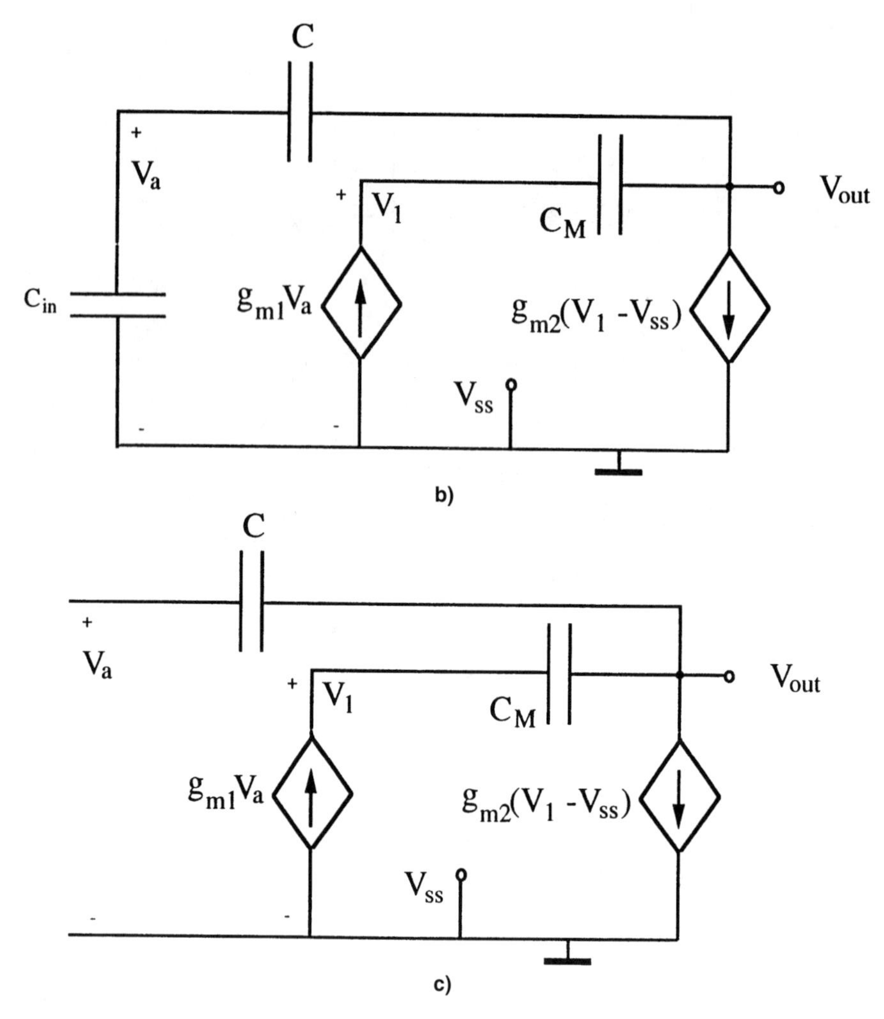

Figure 6 (continued)

$$H_{\phi 2}(s) = \frac{V_{\text{out}}(s)}{V_{ss}(s)} = \frac{s\dfrac{(C+C_{\text{in}})g_{m2}}{C_{\text{in}}C}}{s^2 + s\left[\dfrac{(C+C_{\text{in}})g_{m2}}{C_{\text{in}}C} - \dfrac{g_{m1}}{C_{\text{in}}}\right] + \dfrac{g_{m1}g_{m2}}{C_{\text{in}}C_M}} \tag{28}$$

As Figure 6c only differs from Figure 6b in C_{in}, the transfer function for Figure 6c is obtained by setting C_{in} to zero in Equation 28.

$$H_{\phi 1}(s) = \frac{V_{\text{out}}(s)}{V_{ss}(s)} = \frac{s}{s + \dfrac{g_{m1}}{\left(1 - \dfrac{g_{m1}}{g_{m2}}\right)C_M}} \tag{29}$$

$H_{\phi 1}(s)$ is a first-order highpass filter with a corner frequency somewhat higher than the op amp unity-gain frequency, $g_{m1}/2\pi C_M$. At high frequencies, interference from V_{ss} goes through to the output with little attenuation. $H_{\phi 2}(s)$, on the other hand, is a second-order bandpass filter that peaks at

$$\omega_o = \frac{1}{2\pi}\sqrt{\frac{g_{m1}g_{m2}}{C_{\text{in}}C_M}} \tag{30}$$

The peak amplitude is

$$\left|H_{\phi 2}(\omega_o)\right| = \frac{1}{\left(1 - \dfrac{g_{m1}}{g_{m2}}\dfrac{C}{C + C_{\text{in}}}\right)} \tag{31}$$

Interference signals at around ω_o, which is usually higher than the op amp unity-gain frequency, will again pass through to the output with no attenuation. As the clock frequency is always lower than the op amp unity-gain frequency, high-frequency interferences will be aliased down to the baseband with very little attenuation, by the sampling action of the SC circuit stages that follow the above integrator. Many op amp specifications quote very high numbers for the power supply rejection ratio (PSRR, defined as the ratio between the normal gain of the op amp and the gain between the power supply and output) at low frequencies. The reader must not be misled into believing that, because low-frequency interferences are well rejected (as also by the transfer functions in Equations 30 and 31), the baseband will be clear of supply interferences. What contaminates the baseband are unattenuated high-frequency interference signals aliased down to the baseband by the sampling operation and by the multiplexing of two different transfer functions, as mentioned above. Ordinary transfer function measurement, in which the instrument examines the output of a circuit at the same frequency as that of the test signal it supplies to the input (V_{ss}) of the circuit, will not reveal baseband signal degradation due to high-frequency interference.

One of the effective ways to reduce the influence of supply noise in an SC circuit is to use differential structures, both at the amplifier level by using fully differential input/differential output OTAs and at the filter level by duplicating all capacitors and switches for a negative input path. The output is the difference between the positive signal path and the negative signal path. Interferences coming from the power supply appear as common mode signals which are rejected by the differential structure. If a single-ended SC filter is preferred, then the amplifiers shown in Figure 2 tend to have better supply noise immunity than the 741-style two-stage amplifiers.

6. NOISE IN SC FILTERS

One of the basic limitations on achievable dynamic range in any electronic circuit is noise. In CMOS circuits, the dominant noise sources include thermal noise associated with random thermal motions of carriers in the conducting channel of the transistor, which is wideband, and flicker noise, whose power spectral density is inversely proportional to frequency. In continuous-time circuits, thermal noise is usually less serious in low-frequency, narrowband applications than flicker noise, as the spectral density of the latter is much higher at low frequencies. In SC circuits, however, the sample-and-hold operation aliases a very substantial amount of high-frequency thermal noise into the baseband, so that thermal noise is as important as flicker noise, given the high number of analog switches that contribute to the overall noise. The complexity of noise analysis in an SC filter makes it mainly a task for computer simulations. In this text, we shall limit the mathematical derivation to the minimum needed to understand the general principle and limit the examples to the simplest cases, as they will be sufficient for illustrating the kind of trade-off a designer has to make in a low-noise SC filter design.

We start with a brief review of the aliasing effect of a sample-and-hold circuit for wideband signals.[8] When a continuous-time signal $x(t)$ is sampled every T_s seconds and held for a period T, resulting in a waveform as shown in Figure 7, the latter can be expressed mathematically as

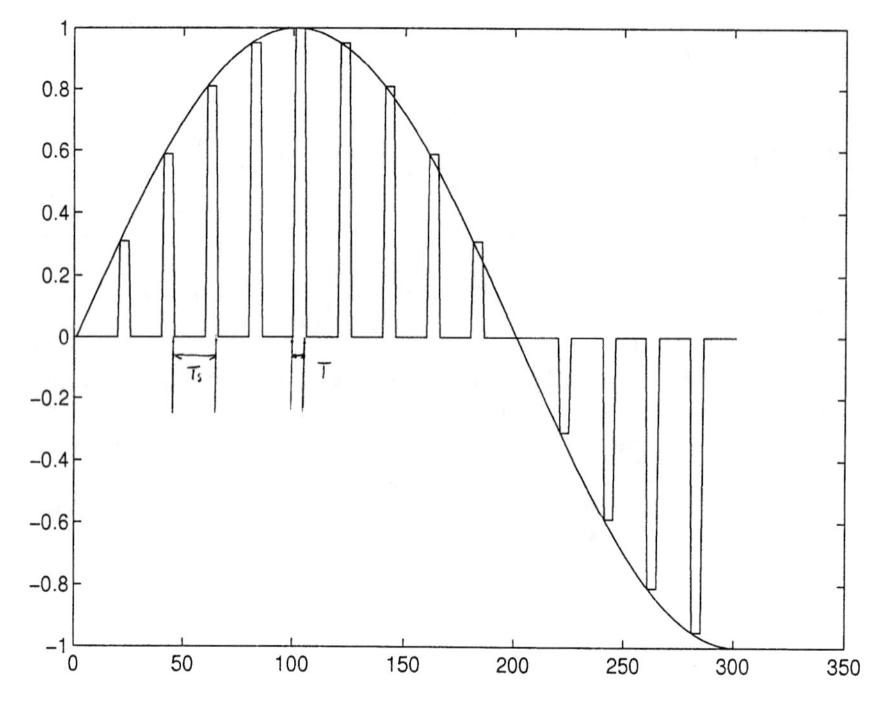

Figure 7 Flat-top sample-and-hold waveform.

$$\begin{aligned} x_s(t) &= \sum_{n=-\infty}^{\infty} x(nT_s)h(t-nT_s) = \int_{-\infty}^{\infty} \sum_{n=-\infty}^{\infty} (x(\tau)h(t-\tau))\delta(\tau-nT_s)d\tau \\ &= \int_{-\infty}^{\infty}\left(\sum_{n=-\infty}^{\infty} x(\tau)\delta(\tau-nT_s)\right)h(t-\tau)d\tau = \int_{-\infty}^{\infty} x_\delta(\tau)h(t-\tau)d\tau \end{aligned} \tag{32}$$

where $h(t)$ is a rectangular pulse of unit amplitude and duration T, defined as

$$h(t) = \begin{cases} 1, & 0 \le t \le T \\ 0, & \text{otherwise} \end{cases} \tag{33}$$

The infinite sum in Equation 32, represented by $x_\delta(t)$, is the instantaneously sampled version of the analog input $x(t)$. This means that a sampled-and-held waveform is equivalent to the instantaneously sampled version of the input convolved with the rectangular pulse $h(t)$. According to Parseval's theorem, the Fourier transform of $x_s(t)$ ($S(\omega)$), is then the product of the Fourier transform of $R_\delta(t)$ ($S_\delta(\omega)$) and that of $h(t)$ ($H(\omega)$).

$$S(\omega) = X_\delta(\omega)H(\omega) = \left(\frac{1}{T_s}\sum_{m=-\infty}^{\infty} X\left(\omega - m\frac{2\pi}{T_s}\right)\right)\left(T\frac{\sin\left(\frac{\omega T}{2}\right)}{\frac{\omega T}{2}}\exp\left(-j\frac{\omega T}{2}\right)\right) \tag{34}$$

Equation 34 shows that a sample-and-hold circuit replicates the input spectrum at frequency intervals equal to the sampling frequency, before filtering the replicated spectrum with a crude lowpass function, $\sin(x)/x$. Signals at frequencies higher than half of the sampling frequency will appear in the baseband and degrade the quality of signals there. This is called aliasing. To prevent aliasing of input signals, the latter is made bandlimited by a continuous-time antialiasing lowpass filter, and the sampling frequency is chosen to be higher than twice the resulting signal bandwidth.

While it is possible to prevent aliasing of input signals by proper filtering and sufficiently high sampling frequency, it is not possible to do the same with the noise signals introduced by the switches and op amps in an SC circuit. The reason for this is that the bandwidth associated with those noise-contributing elements must necessarily be several times higher than the sampling frequency to ensure sufficient charge transfer accuracy in an SC filter, as explained earlier. The noise spectral density of the output of an SC filter due to a particular noise source is therefore usually higher than the spectral density of the source itself due to aliasing. The spectral density of noise, which is a random signal, is defined as the Fourier transform of the autocorrelation function of the noise, which has the dimension of voltage squared per hertz. An estimate of the output noise spectrum ($S_N(\omega)$), consistent with the way such a spectrum is usually measured in the laboratory, can be obtained by the average of an estimate of the squared spectrum ($N_\delta(\omega)$) of the noise itself (rather than its autocorrelation) over a finite observation period.[9]

$$S_N(\omega) = N_\delta(\omega)H^2(\omega) = \left(\frac{1}{T_s^2}\sum_{m=-\infty}^{\infty} N\left(\omega - m\frac{2\pi}{T_s}\right)\right)\left(T\frac{2}{\omega T}\sin\left(\frac{\omega T}{2}\right)\right)^2 \tag{35}$$

The effect of aliasing, as described by Equation 35, on the noise spectral density in the signal baseband can be appreciated by the graphical example in Figure 8, where a noise source is sampled at a frequency three times less than its bandwidth. The result in the baseband output is that the noise spectral density (or rather, its estimate) has increased by a factor of two times the undersampling ratio (3 in this case), before being scaled by the $\sin(x)/x$ function. The latter is also plotted in Figure 8 with a duty cycle of 50%. In an SC filter, noise sources associated with switches tend to see a track and hold circuit with 50% hold time. In this case the tracking mode can be modeled as a transmission gate which multiplies the input by zero and one alternately. The track-and-hold circuit can then be represented as the parallel connection of a sample-and-hold circuit and a transmission gate. To include the influence of the track mode on the output noise spectrum in the baseband, Equation 35 can be modified to become

$$\begin{aligned} S_N(\omega) &= S_n\left[\left(\frac{T}{T_s}\right)^2\left(\frac{2}{\omega T}\sin\left(\frac{\omega T}{2}\right)\right)^2 + \left(1-\frac{T}{T_s}\right)^2\right] \qquad f_s > 2BW_n \\ S_N(\omega) &\approx S_n\left[2\frac{BW_n}{f_s}\left(\frac{T}{T_s}\right)^2\left(\frac{2}{\omega T}\sin\left(\frac{\omega T}{2}\right)\right)^2 + \left(1-\frac{T}{T_s}\right)^\alpha\right] \qquad 1 < \alpha \le 2,\ \frac{\omega}{\pi} < f_s \le 2BW_n \end{aligned} \tag{36}$$

where the noise source is assumed to be of uniform spectral density S_n and a finite (single side) bandwidth of BW_n. For $BW_n/f_s \geq 10$, $\alpha \approx 1$.

To associate noise bandwidth, undersampling, and aliasing with design parameters in an SC circuit, let us return to the basic circuit building block shown in Figure 3a, the SC integrator. Considering only the noise contribution of the two switches controlled by $\phi 2$, we can model the integrator as an equivalent resistor R_{on} (determined by Equation 24) in series with the input capacitor C_{in} during the track mode when $\phi 2$ is on. When $\phi 2$ turns off, the noise sampling takes place and the result held on the integrating capacitor C for the remaining half clock cycle. The equivalent noise is modeled as a voltage source with a uniform, double-sided power density of $v_n^2 = 2kTR_{\text{on}}$, where $k = 1.38 \times 10^{-23}$ is the Boltzmann constant and T is the absolute temperature in degrees Kelvin. This is shown in Figure 9. Assuming virtual ground for the inverting input node of the op amp (thus neglecting the influence of finite op amp bandwidth), the noise voltage across capacitor C_{in} is a lowpassed version of the noise source:

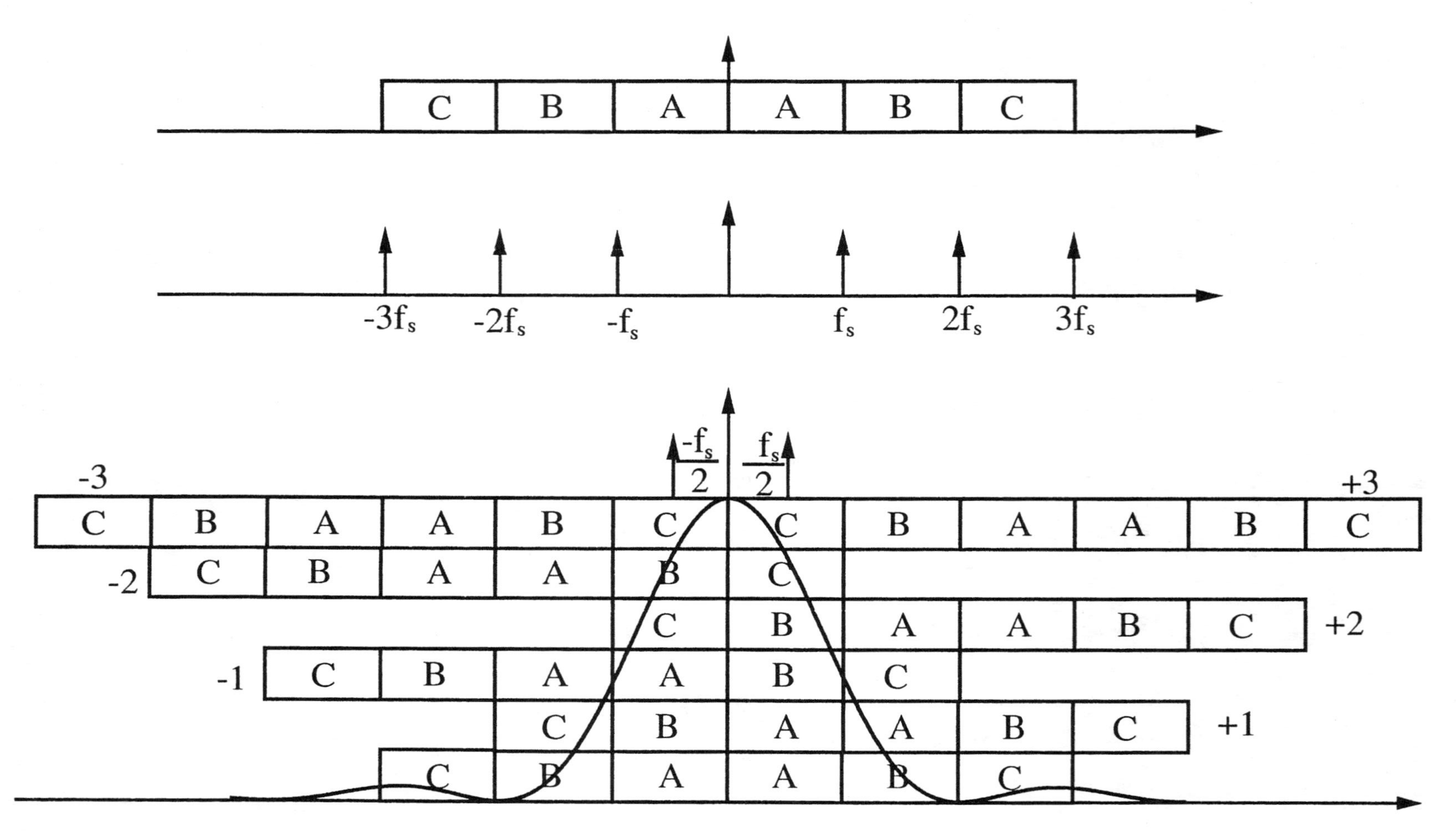

Figure 8 Sampled-and-held noise spectrum.

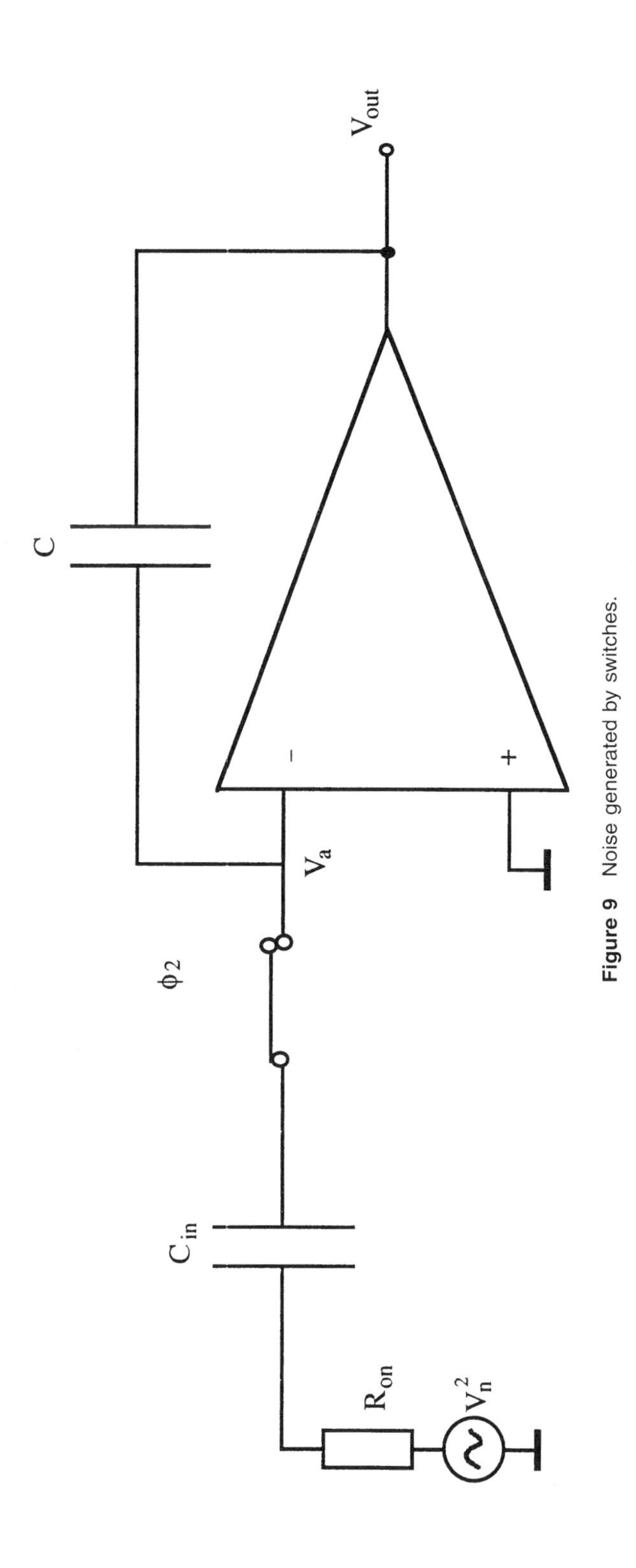

Figure 9 Noise generated by switches.

$$v_n^2(\omega) = 2kTR_{on}H^2(\omega) = 2kTR_{on}\frac{1}{1+(\omega R_{on}C)^2} \quad (37)$$

It can be shown that the integration of $v_n^2(\omega)$ over the entire frequency axis yields an integrated noise power of exactly kT/C_{in}. This result is independent of the value of the noise resistance and is referred to as kT/C noise. The same noise power may result from the original uniform noise source with a brick wall cutoff frequency of exactly $1/(2\pi R_{on}C_{in})$. If we approximate the noise density given by Equation 37 with a uniform noise source with such a brick wall cutoff frequency, then the effect of the sampling switch on noise voltage (which participates in the charge transfer equation at the inverting input of the op amp) can be appreciated by applying Equation 36:

$$S_N(\omega) \approx \left[\frac{4kT}{2\pi f_s C_{in}}\left(\frac{T}{T_s}\right)^2\left(\frac{2}{\omega T}\sin\left(\frac{\omega T}{2}\right)\right)^2 + 2kTR_{on}\left(1-\frac{T}{T_s}\right)^\alpha\right] \quad 1<\alpha\le 2 \quad \frac{\omega}{\pi}<f_s \quad (38)$$

The first term in Equation 38, which is usually much higher than the second term, shows that a sampled-and-held noise spectral density is also inversely proportional to capacitance. Similarly, it can be shown that although different noise sources, including those of the op amps referred to the input terminal, may have different transfer functions to the output of the filter, their output spectral density is usually dominated by the term which is of the form kT/C, due to the sampling operation of the SC filter. Thus, although capacitors are noiseless elements in SC filters, the choice of their values is very important in low-noise designs.

On the whole, increasing the value of capacitors in an SC filter reduces the noise spectral density. As higher capacitance costs more silicon area and causes the circuit to slow down and consume more power, the total capacitance of a circuit must be limited. An SC circuit optimized for noise should be one that minimizes output noise for a given total capacitance area. This means if one increases capacitance in one part of the circuit, it has to be reduced in other parts of the circuit so that the sum stays the same. The values of capacitors in an SC circuit, however, are linked by design equations to filter coefficients and maximum signal swing requirements, as explained in early chapters. The only way left to scale capacitors is therefore to scale them in groups that are independent of each other. Such groups exist around op amp–based integrators. Applying the knowledge from the early part of this chapter, we know that if we scale all the capacitors connected and switched to the input of an op amp, the transfer function of the filter remains unchanged. Such a group of capacitors can therefore be scaled together for noise minimization. As there are typically n op amps in an nth-order SC filter, there are n such groups of capacitors which can be scaled relative to each other to maintain a constant total capacitance. Since the integrators are located at different parts of the filter, the transfer functions from the noise sources associated with each scalable capacitor group are different and the weighting of the contribution of each group of noise sources in the output noise is different. The basic idea of noise optimization is to increase the capacitances associated with the noise sources with a higher weighting and to reduce the capacitances associated with those with a lower weighting. The total output noise is a function of n variables, each being the total capacitance of a scalable group. The n variables are confined by the total capacitance of the filter. Since the noise function is a sum of positive terms inversely proportional to capacitance, the optimization can be easily solved analytically with respect to capacitance assignment. An additional constraint, however, is that the spread of capacitance of the final assignment must not exceed the maximum capacitance spread determined purely by filter coefficients. Otherwise, although the noise is optimized, the coefficient accuracy may suffer as a result of the minimum capacitance being too small. In other words, if the minimum and maximum capacitances are determined by the group of scalable capacitors with the maximum spread prior to noise optimization, the capacitors in other groups must not have lower minimum capacitance or higher maximum capacitance than the first group after optimization. An "active set" method can be used to solve the noise optimization problem with this additional minimum spread criterion which is expressed by a set of $n - 1$ inequalities.[10] In a biquad design, the ratio between the two scalable capacitor groups can be obtained with a straightforward search with the help of noise simulation by computer.

REFERENCES

1. **J. Taylor and D. Haigh,** A stray-free first order circuit for the correction of sample and hold amplitude in switched capacitor filters, *Electron. Lett.* 23, 177–178, Feb. 1987.
2. **J. Shyu, G. Temes, and K. Yao,** Random errors in MOS capacitors, *IEEE J. Solid State Circuits,* SC-17(6), 1070–1076, 1982.
3. **K. Martin and A. S. Sedra,** Effects of the op amp finite gain and bandwidth on the performance of switched capacitor filters, *IEEE Trans. Circuits Syst.,* CAS-28, 822–829, August 1981.
4. **W. Sansen, Q. Huang, and K. Halonen,** Transient analysis of charge transfer in SC filters — gain error and distortion, *IEEE J. Solid-State Circuits*, SC-22, 268–276, April 1987.
5. **K. Bult and G. Geelen,** A fast settling CMOS op amp for SC circuits with 90 dB DC gain, *IEEE J. Solid State Circuits,* 25(6), 1379–1384, 1990.
6. **G. Wegmann, E. Vittoz, and F. Rahali,** Charge injection in analog MOS switches, *IEEE J. Solid State Circuits*, SC-22(6), 1091–1097, 1987.
7. **M. Steyaert and W. Sansen,** Power Supply rejection ratio in operational transconductance amplifiers, *IEEE Trans. Circuits Syst.,* 37(9), 1077–1086, 1990.
8. **S. Haykin,** *An Introduction to Analog and Digital Communications*, Wiley, New York, 1989.
9. **J. Fischer,** Noise sources and calculation techniques for switched capacitor filters, *J. Solid State Circuits*, SC-17, 742–752, Aug. 1982.
10. **A. Kaelin, J. Goette, W. Guggenbühl, and G. S. Moschytz,** A novel capacitance assignment procedure for the design of sensitivity and noise-optimized SC filters, *IEEE Trans. Circuits Syst.,* 38(11), 1255–1268, 1991.

7.4

Computer-Aided Design Methods for Switched-Compacitor Systems

Jiri Vlach

CONTENTS

1. INTRODUCTION

Switched-capacitor (SC) networks are normally analyzed by a specialized program. If it is not available, programming in Fortran or C is quite difficult. The situation has changed recently with the arrival of such mathematically oriented packages as MATLAB, MATHCAD, etc. They remove the burden of programming, and the user can analyze larger networks, if he knows the theory. This chapter gives information about how to set up the necessary equations in matrix form. Both time-domain and frequency-domain analyses are covered.

2. TIME-DOMAIN ANALYSIS

Consider the network in Figure 1a, with phase 1 shown in Figure 1b and Phase 2 in Figure 1c. To distinguish between switching phases and nodes of the network, we will use two subscripts. The first one will denote the phase, the second one, if needed, will give the node number (in parentheses).

In phase 1, at the instant of switching, each capacitor will have a charge, defined by the voltages at the end of the previous phase. We will denote time-dependent voltages by lowercase letters, initial voltages from the previous phase by uppercase letters. The initial charges are drawn in Figure 1 as independent charge sources using intersecting circles. Writing the balance of charges at node 2 and 3,

0-8493-8951-8/97/$0.00+$.50
© 1997 by CRC Press LLC

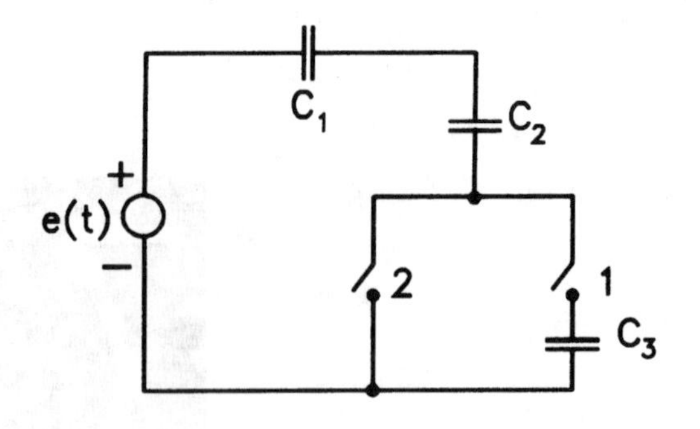

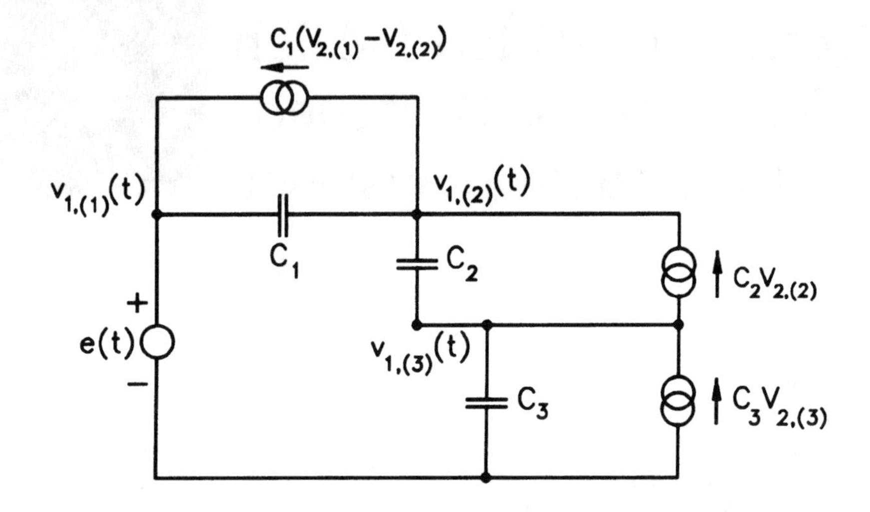

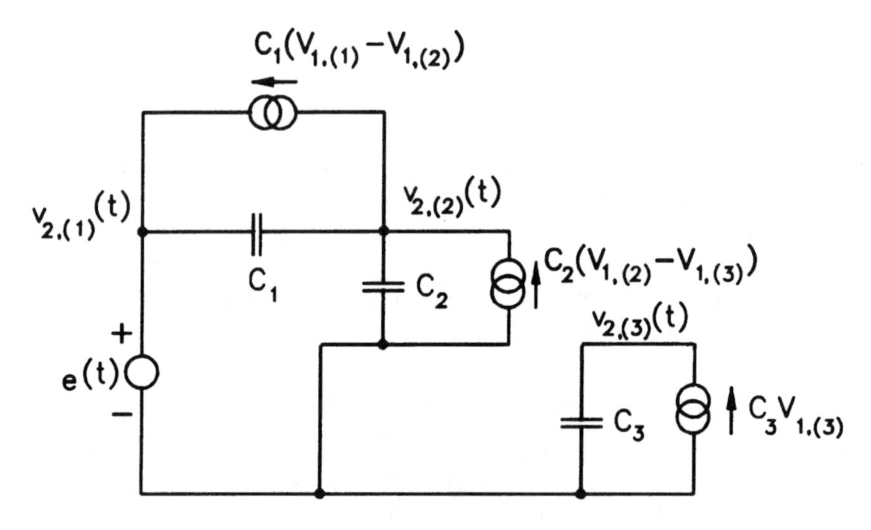

Figure 1 Introductory SC network.

$$-C_1 v_{1,(1)}(t) + (C_1 + C_2) v_{1,(2)}(t) - C_2 v_{1,(3)}(t) = -C_1\left(V_{2,(1)} - V_{2,(2)}\right) + C_2 V_{2,(2)}$$

$$-C_2 v_{1,(2)}(t) + (C_2 + C_3) v_{1,(3)}(t) = -C_2 V_{2,(2)} + C_3 V_{2,(3)}$$

We must also add the equation for the voltage source, $v_{1,(1)}(t) = e(t)$. In matrix form

$$\begin{bmatrix} -C_1 & C_1+C_2 & -C_2 \\ 0 & -C_2 & C_2+C_3 \\ 1 & 0 & 0 \end{bmatrix} \begin{bmatrix} v_{1,(1)}(t) \\ v_{1,(2)}(t) \\ v_{1,(3)}(t) \end{bmatrix} = \begin{bmatrix} -C_1 & C_1+C_2 & 0 \\ 0 & -C_2 & C_3 \\ 0 & 0 & 0 \end{bmatrix} \begin{bmatrix} V_{2,(1)} \\ V_{2,(2)} \\ V_{2,(3)} \end{bmatrix} + \begin{bmatrix} 0 \\ 0 \\ 1 \end{bmatrix} e(t)$$

For phase 2 we write at node 2 and 3

$$(C_1+C_2)v_{2,(2)}(t) - C_1 v_{2,(1)}(t) = -C_1\left(V_{1,(1)} - V_{1,(2)}\right) + C_2\left(V_{1,(2)} - V_{1,(3)}\right)$$

$$C_3 v_{2,(3)}(t) = C_3 V_{1,(3)}$$

and the source is taken into account by $v_{2,(1)}(t) = e(t)$. In matrix form

$$\begin{bmatrix} -C_1 & C_1+C_2 & 0 \\ 0 & 0 & C_3 \\ 1 & 0 & 0 \end{bmatrix} \begin{bmatrix} v_{2,(1)}(t) \\ v_{2,(2)}(t) \\ v_{2,(3)}(t) \end{bmatrix} = \begin{bmatrix} -C_1 & C_1+C_2 & -C_2 \\ 0 & 0 & C_3 \\ 0 & 0 & 0 \end{bmatrix} \begin{bmatrix} V_{1,(1)} \\ V_{1,(2)} \\ V_{1,(3)} \end{bmatrix} + \begin{bmatrix} 0 \\ 0 \\ 1 \end{bmatrix} e(t)$$

Another network with an operational amplifier is shown in Figure 2. Using similar steps as above, we get for phase 1

$$-C_2 v_{1,(2)}(t) = -C_2 V_{2,(2)}$$

$$v_{1,(1)}(t) = e(t)$$

leading to the matrix equation

$$\begin{bmatrix} 0 & -C_2 \\ 1 & 0 \end{bmatrix} \begin{bmatrix} v_{1,(1)}(t) \\ v_{1,(2)}(t) \end{bmatrix} = \begin{bmatrix} 0 & -C_2 \\ 0 & 0 \end{bmatrix} \begin{bmatrix} V_{2,(1)} \\ V_{2,(2)} \end{bmatrix} + \begin{bmatrix} 0 \\ 1 \end{bmatrix} e(t)$$

In the second phase we write

$$-C_1 v_{2,(2)}(t) = C_1 V_{1,(1)} - C_2 V_{1,(2)}$$

$$v_{2,(1)}(t) = e(t)$$

In matrix form

$$\begin{bmatrix} 0 & -C_2 \\ 1 & 0 \end{bmatrix} \begin{bmatrix} v_{2,(1)}(t) \\ v_{2,(2)}(t) \end{bmatrix} = \begin{bmatrix} C_1 & -C_2 \\ 0 & 0 \end{bmatrix} \begin{bmatrix} V_{1,(1)} \\ V_{1,(2)} \end{bmatrix} = \begin{bmatrix} 0 \\ 1 \end{bmatrix} e(t)$$

It is not difficult to write the above equations once the charges have been properly identified. Unfortunately, it is very easy to make a mistake. We need a method which can supply the matrices in a simple and easily programmable manner. Instructions on how to write the matrices are in the next section.

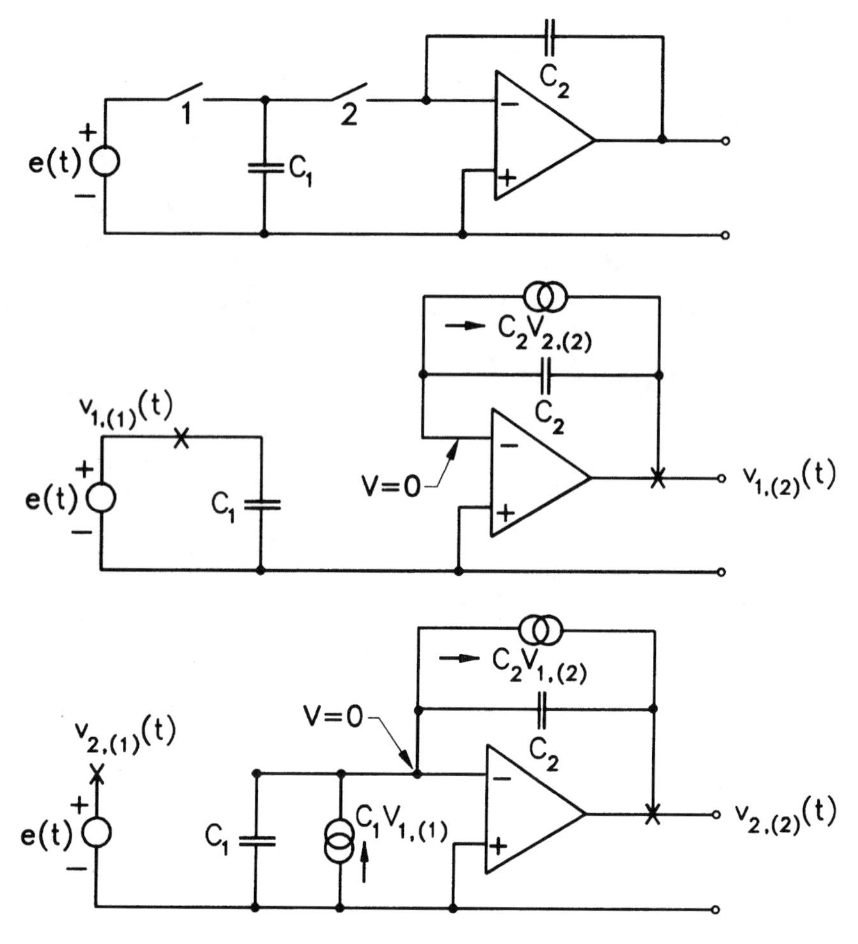

Figure 2 SC integrator.

3. SYSTEM MATRICES

The theory used for analysis of SC networks allows the use of only a very few network elements: ideal capacitors, ideal switches, ideal voltage sources, and voltage-controlled voltage sources. This restriction permits us to use voltages and charges as the network variables and avoids the need to use numerical integration in time-domain analysis.

To identify the network for a computer we need to give each node a name or number. Let the number of ungrounded nodes be n. When setting up the equations, we can keep the switches as elements of the network, but then each switch will increase the size of the system matrix by one row and one column. If we have m switches, our system matrix will be at least the size $m + n$. This can be avoided once we realize that an open switch (open circuit) need not be considered at all. Moreover, each closed switch (short circuit) collapses two nodes into one. If we are prepared to go into the trouble of properly renumbering the nodes, we can actually *reduce* the size of the system matrix below n. There are additional possibilities. An ideal operational amplifier (which is in fact a voltage-controlled voltage source) has the same voltages at its input nodes. Using this fact, we can *further reduce* the system matrix by one row and column for each operational amplifier. This will be our way to set up the matrices.

The method was originally developed using separate graphs for currents and voltages.[4] We give here a condensed version which also eliminates currents of all voltage sources. Because the theory requires the use of two graphs, each node of the network must be given two numbers. We will use

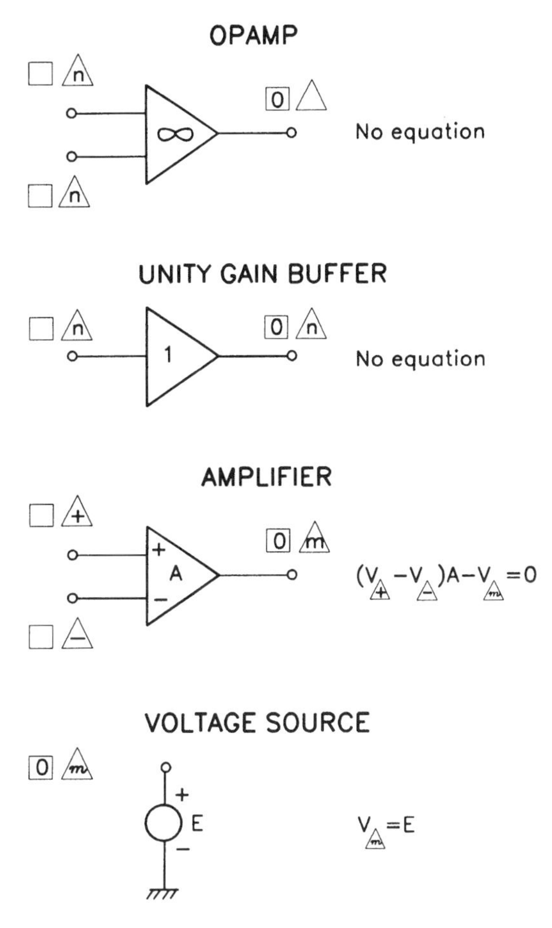

Figure 3 Node numbering for voltage sources.

1. Numbers in triangles to indicate node numbers on the voltage graph.
2. Numbers in squares to indicate node numbers on the current graph.

The method sets up the equivalent of a nodal matrix, but renumbering may make this matrix rectangular. An amplifier (voltage-controlled voltage source) with the gain A is described by the equation $(V_+ - V_-)A - V_m = 0$, and this equation must added to the system. An independent voltage source is described by the equation $V_m = E$ and this equation must also be attached to the system. Eventually, with all such additions, the system matrix will be square and thus will make solution possible.

In Figure 3, we summarize all voltage sources which we can use in the analysis of SC networks. The figure gives the necessary numbering. The symbols +, –, and m for the amplifier will be replaced by integers. If some terminals are at the same potential (unity-gain buffer, operational amplifier), the same letter n indicates that we must use the same voltage subscript at these nodes. Empty squares or triangles are filled with consecutive numbers, preferably going from one end of the network to the other.

We show how to use Figure 3 on the network in Figure 1. Its two phases are redrawn in Figure 4. First, we label the node of the voltage source; the remaining nodes are labeled by consecutive integers. The numbering of both systems is independent. The next step is to prepare the connectivity table. In all phases we must always give the *same* node as the first one (for instance the left or the upper node). The connectivity table for phase 1 is

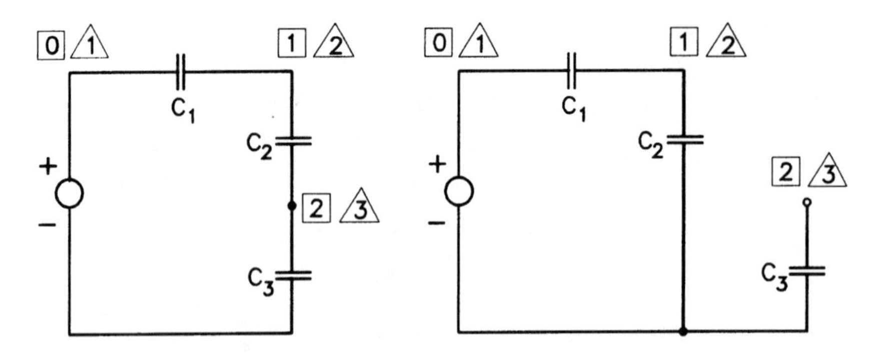

Figure 4 Node numbering for the network in Figure 2.

	$\mathbf{A}_1$	C_1	C_2	C_3
I-1	From node	0	1	2
	To node	1	2	0
V-1	From node	1	2	3
	To node	2	3	0

Here the set marked I-1 gives the *row* of the matrix where the element will appear, the set marked V-1 gives the *column.* The information also determines the sign. If the directions in the I and V table are the same, a positive sign is used; otherwise the sign is negative. Consider only the capacitor C_2. The following information provides the matrix entries:

I-1: from node 1, V-1: from node 2. Enter $+C_2$ in position (1,2)
I-1: from node 1, V-1: to node 3. Enter $-C_2$ in position (1,3)
I-1: to node 2, V-1: from node 2. Enter $-C_2$ in position (2,2)
I-1: to node 2, V-1: to node 3. Enter $+C_2$ in position (2,3)

If one of the node numbers is zero, the entry does not exist (we do not have zeroth row or column of the matrix). Proceeding this way for the other capacitors, we get the upper two rows of the matrix $\mathbf{A}_1$. We must also add the equation for the voltage source. Since it is $V_1 = E$ and columns are multiplied by the voltages, we enter a unit in the (3,1) position. The complete matrix becomes

$$\mathbf{A}_1 = \begin{bmatrix} -C_1 & C_1 + C_2 & -C_2 \\ 0 & -C_2 & C_2 + C_3 \\ 1 & 0 & 0 \end{bmatrix}$$

For the voltage source we must also prepare the right-hand side vector. Since the properties of the source appeared in the third row, the vector will be

$$\mathbf{g}_1 = [0 \quad 0 \quad 1]^T$$

Writing the connectivity table for the second phase we have

	$\mathbf{A}_2$	C_1	C_2	C_3
I-2	From node	0	1	2
	To node	1	0	0
V-2	From node	1	2	3
	To node	2	0	0

and using the above rules we get

$$\mathbf{A}_2 = \begin{bmatrix} -C_1 & C_1 + C_2 & 0 \\ 0 & 0 & C_3 \\ 1 & 0 & 0 \end{bmatrix}$$

where the third row again takes care of the voltage source. The right-hand side vector for the source is $\mathbf{g}_2 = \mathbf{g}_1$.

The matrices $\mathbf{B}_k$ take care of charges of capacitors from the previous phase. We use the I-k information of the same phase, but the V-(k – 1) information of the phase preceding in time. The remaining rules on the entries into the matrix remain unchanged. Copying the appropriate information from the above, we get the connectivity table for $\mathbf{B}_1$

	$\mathbf{B}_1$	C_1	C_2	C_3
I-1	From node	0	1	2
	To node	1	2	0
V-2	From node	1	2	3
	To node	2	0	0

and the matrix is

$$\mathbf{B}_1 = \begin{bmatrix} -C_1 & C_1 + C_2 & 0 \\ 0 & -C_2 & C_3 \\ 0 & 0 & 0 \end{bmatrix}$$

Since this matrix takes into account only the charges of capacitors, there will be zeros in the row that belongs to the description of the voltage source. For $\mathbf{B}_2$ we have the connectivity table

	$\mathbf{B}_2$	C_1	C_2	C_3
I-2	From node	0	1	2
	To node	1	0	0
V-1	From node	1	2	3
	To node	2	3	0

and the matrix

$$\mathbf{B}_2 = \begin{bmatrix} -C_1 & C_1 + C_2 & -C_2 \\ 0 & 0 & C_3 \\ 0 & 0 & 0 \end{bmatrix}$$

Another example is shown in Figure 5, complete with node numbering. The connectivity tables and matrices are in Figure 6.

4. FREQUENCY-DOMAIN ANALYSIS

In each phase we can write

$$\mathbf{A}_k v_k(t) = \mathbf{B}_k \mathbf{V}_{k-1} + \mathbf{g}_k e(t) \tag{1}$$

If the network has N phases, then in this expression $\mathbf{V}_0$ is to be replaced by $\mathbf{V}_N$. It is convenient to prepare

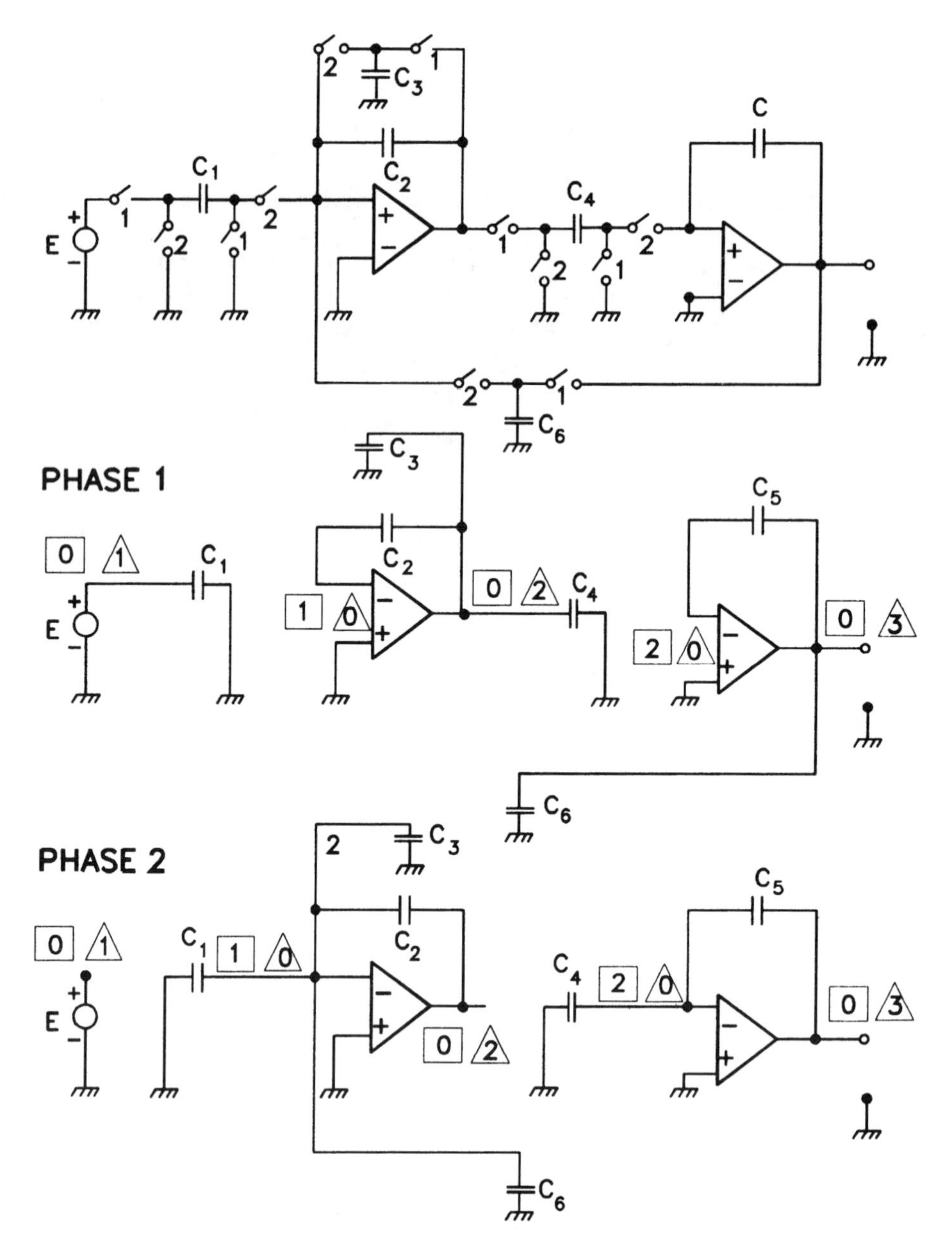

Figure 5 SC network with two operational amplifiers.

$$\begin{aligned} \mathbf{C}_k &= \mathbf{A}_k^{-1}\mathbf{B}_k \\ \mathbf{p}_k &= \mathbf{A}_k^{-1}\mathbf{g}_k \end{aligned} \tag{2}$$

and

$$\mathbf{v}_k(t) = \mathbf{C}_k\mathbf{V}_{k-1} + \mathbf{p}_k e(t) \tag{3}$$

Algebraic manipulations can discover an important property. In Equation 1, add and subtract simultaneously the product $\mathbf{A}_k\mathbf{V}_k$ on the left side. On the right side do the same with $\mathbf{g}_k E_k$, where E_k is the source voltage at the instant of switching. This gives

	$\mathbf{A}_1$	C_1	C_2	C_3	C_4	C_5	C_6
I-1	from node	0	1	0	0	2	0
	to node	0	0	0	0	0	0
V-1	from node	1	0	2	2	0	3
	to node	0	2	0	0	3	0

	$\mathbf{A}_2$	C_1	C_2	C_3	C_4	C_5	C_6
I-1	from node	0	1	1	0	2	1
	to node	1	0	0	2	0	0
V-1	from node	0	0	0	0	0	0
	to node	0	2	0	0	3	0

$$\mathbf{A}_1 = \begin{bmatrix} 0 & -C_2 & 0 \\ 0 & 0 & -C_5 \\ 1 & 0 & 0 \end{bmatrix} \qquad \mathbf{A}_2 = \begin{bmatrix} 0 & -C_2 & 0 \\ 0 & 0 & -C_5 \\ 1 & 0 & 0 \end{bmatrix} \qquad \mathbf{g}_1 = \begin{bmatrix} 0 \\ 0 \\ 1 \end{bmatrix}$$

	$\mathbf{B}_1$	C_1	C_2	C_3	C_4	C_5	C_6
I-1	from node	0	1	0	0	2	0
	to node	0	0	0	0	0	0
V-2	from node	0	0	0	0	0	0
	to node	0	2	0	0	3	0

	$\mathbf{B}_2$	C_1	C_2	C_3	C_4	C_5	C_6
I-1	from node	0	1	1	0	2	1
	to node	1	0	0	2	0	0
V-1	from node	1	0	2	2	0	3
	to node	0	2	0	0	3	0

$$\mathbf{B}_1 = \begin{bmatrix} 0 & -C_2 & 0 \\ 0 & 0 & -C_5 \\ 0 & 0 & 0 \end{bmatrix} \qquad \mathbf{B}_2 = \begin{bmatrix} -C_1 & -C_2+C_3 & C_6 \\ 0 & -C_4 & -C_5 \\ 0 & 0 & 0 \end{bmatrix} \qquad \mathbf{g}_2 = \begin{bmatrix} 0 \\ 0 \\ 1 \end{bmatrix}$$

Figure 6 Connectivity tables and matrices for the network in Figure 5.

$$\mathbf{A}_k\left[\mathbf{v}_k(t)-\mathbf{V}_k\right]+\mathbf{A}_k\mathbf{V}_k=\mathbf{B}_k\mathbf{V}_{k-1}+\mathbf{g}_k\left[e(t)-E_k\right]+\mathbf{g}_kE_k$$

At the switching instants, the terms in square brackets are zero and we can partition the above equation into two set of equations: the discrete set

$$\mathbf{A}_k\mathbf{V}_k=\mathbf{B}_k\mathbf{V}_{k-1}+\mathbf{g}_kE_k \tag{4}$$

valid only at the instances of switching, and the algebraic set

$$\mathbf{A}_k\left[\mathbf{v}_k(t)-\mathbf{V}_k\right]=\mathbf{g}_k\left[e(t)-E_k\right] \tag{5}$$

valid everywhere. The benefit of such splitting comes in frequency-domain analysis where we can consider the digital system separately.

We will describe the theory for a network with *equal* phases. Expressions for arbitrary phase durations are in References 1 and 4. Let the number of phases be N, the switching period T, and let all phase intervals be equal,

$$\tau = \frac{T}{N} \tag{6}$$

Define the z-domain variable

$$z = e^{j\omega T} \tag{7}$$

Here ω is the frequency under consideration in radians per second. In the following we assume that only one source having $E = 1V$ is present. The matrix equation of the system is

$$\begin{bmatrix} \mathbf{A}_1 & \mathbf{0} & \mathbf{0} & \cdots & \mathbf{0} & -\mathbf{B}_1 \\ -\mathbf{B}_2 & \mathbf{A}_2 & \mathbf{0} & \cdots & \mathbf{0} & \mathbf{0} \\ \mathbf{0} & -\mathbf{B}_3 & \mathbf{A}_3 & \cdots & \mathbf{0} & \mathbf{0} \\ \cdots & \cdots & \cdots & \cdots & \cdots & \cdots \\ \mathbf{0} & \mathbf{0} & \mathbf{0} & \cdots & -\mathbf{B}_N & z\mathbf{A}_N \end{bmatrix} \begin{bmatrix} \mathbf{V}_1 \\ \mathbf{V}_2 \\ \mathbf{V}_3 \\ \cdots \\ \mathbf{V}_N \end{bmatrix} = \begin{bmatrix} \mathbf{g}_1 z^{0/N} \\ \mathbf{g}_2 z^{1/N} \\ \mathbf{g}_3 z^{2/N} \\ \cdots \\ \mathbf{g}_N z^{(n-1)/N} \end{bmatrix} \tag{8}$$

Solve the system and select the output $V_{k,\text{out}}$ in each phase. Define

$$\begin{aligned} a_k &= z^{(1-k)/N} \qquad k = 1, 2 \ldots (N-1) \\ a_N &= z^{+1/N} \end{aligned} \tag{9}$$

and

$$W = \frac{1 - z^{-1/N}}{\ln z} \tag{10}$$

The complete output of the digital system is

$$V_{\text{out,digital}} = \sum_{k=1}^{N} a_k V_{\text{k,out}} \tag{11}$$

This is the desired transfer function in the z variable. It should be stressed that this is *not* the output of the SC network. If the signal is sampled and held in *each* phase before it is applied to the network, then the output of the network is

$$V_{\text{out},SH} = W V_{\text{out,digital}} \tag{12}$$

Things are more complicated if a full, unsampled signal is used for the input. We must solve the systems

$$\mathbf{A}_k\mathbf{p}_k = \mathbf{g}_k \tag{13}$$

in all phases. The result expresses feedthrough to the output; denote it $p_{k,\text{out}}$. In addition, define another frequency-dependent constant

$$R = \frac{1}{N} - Wz^{1/N} \tag{14}$$

The complete solution, including feedthrough, is

$$V_{\text{out,fullsignal}} = Wz^{1/N}\sum_{k=1}^{N} a_k V_{k,\text{out}} + R\sum_{k=1}^{N} p_{k,\text{out}} \tag{15}$$

As an example, consider a two-phase system. It reduces Equation 8 to

$$\begin{bmatrix} \mathbf{A}_1 & -\mathbf{B}_1 \\ -\mathbf{B}_2 & z\mathbf{A}_2 \end{bmatrix}\begin{bmatrix} \mathbf{V}_1 \\ \mathbf{V}_2 \end{bmatrix} = \begin{bmatrix} \mathbf{g}_1 \\ \mathbf{g}_2 z^{1/2} \end{bmatrix} \tag{16}$$

and the constants are

$$\begin{aligned} a_1 &= 1 \\ a_2 &= z^{+1/2} \\ W &= \frac{1 - z^{-1/2}}{\ln z} \end{aligned} \tag{17}$$

In each phase, the output of the *digital* system will be

$$V_{\text{out,digital}} = \sum_{k=1}^{2} a_k V_{k,\text{out}} \tag{18}$$

If the signal is sampled and held in *each* phase *before* it is applied to the network, then the output of the network is given by Equation 12. If the input signal is not sampled, solve Equation 13 and select in each phase the output $p_{k,\text{out}}$. Calculate

$$R = \frac{1}{2} - Wz^{1/2} \tag{19}$$

and the complete solution, including feedthrough, is

$$V_{\text{out,fullsignal}} = Wz^{1/2}\sum_{k=1}^{2} a_k V_{k,\text{out}} + R\sum_{k=1}^{2} p_{k,\text{out}} \tag{20}$$

5. SOLUTION OF THE FREQUENCY-DOMAIN SYSTEM

Solution of the general system (Equation 8) can be done very efficiently, and the theory was developed in References 1 and 4. Here we give the steps for a two-phase system.

In the preprocessing step, done only once, prepare the matrices

$$\mathbf{C}_1 = \mathbf{A}_1^{-1}\mathbf{B}_1, \qquad \mathbf{E}_2 = \mathbf{B}_2\mathbf{C}_1 \tag{21}$$

and vectors

$$\mathbf{p}_1 = \mathbf{A}_1^{-1}\mathbf{g}_1, \qquad \mathbf{g}_1^P = \mathbf{B}_2\mathbf{p}_1 \tag{22}$$

For each frequency solve

$$\left(z\mathbf{A}_2 - \mathbf{E}_2\right)\mathbf{V}_2 = \mathbf{g}_1^P + \mathbf{g}_2 z^{1/2} \tag{23}$$

and evaluate

$$\mathbf{V}_1 = \mathbf{C}_1\mathbf{V}_2 + \mathbf{p}_1 \tag{24}$$

The constants $a_1 = 1$ and $a_2 = z^{1/2}$ are needed to get the digital system output, as given in Equation 18.

We will demonstrate the solution of Equation 16 on the network in Figure 2. The matrices were derived earlier:

$$\mathbf{A}_1 = \begin{bmatrix} 0 & -C_2 \\ 1 & 0 \end{bmatrix} \quad \mathbf{A}_2 = \begin{bmatrix} 0 & -C_2 \\ 1 & 0 \end{bmatrix} \quad \mathbf{B}_1 = \begin{bmatrix} 0 & -C_2 \\ 0 & 0 \end{bmatrix} \quad \mathbf{B}_2 = \begin{bmatrix} C_1 & -C_2 \\ 0 & 0 \end{bmatrix}$$

The following preprocessing steps are done only once:

$$\mathbf{A}_1^{-1} = \begin{bmatrix} 0 & 1 \\ -1/C_2 & 0 \end{bmatrix} \quad \mathbf{C}_1 = \mathbf{A}_1^{-1}\mathbf{B}_1 = \begin{bmatrix} 0 & 0 \\ 0 & 1 \end{bmatrix} \quad \mathbf{p}_1 = \mathbf{A}_1^{-1}\mathbf{g}_1 = \begin{bmatrix} 1 \\ 0 \end{bmatrix}$$

$$\mathbf{E}_2 = \mathbf{B}_2\mathbf{C}_1 = \begin{bmatrix} 0 & -C_2 \\ 0 & 0 \end{bmatrix} \quad \mathbf{g}_1^P = \mathbf{B}_2\mathbf{p}_1 = \begin{bmatrix} C_1 \\ 0 \end{bmatrix}$$

For each frequency prepare Equation 23:

$$\begin{bmatrix} 0 & C_2(1-z) \\ z & 0 \end{bmatrix} \begin{bmatrix} V_{2,(1)} \\ V_{2,(2)} \end{bmatrix} = \begin{bmatrix} C_1 \\ z^{1/2} \end{bmatrix}$$

The solution is

$$\begin{bmatrix} V_{2,(1)} \\ V_{2,(2)} \end{bmatrix} = \begin{bmatrix} z^{-1/2} \\ C_1/[C_2(1-z)] \end{bmatrix}$$

The voltages of the first phase are obtained using Equation 24:

$$\mathbf{V}_1 = \mathbf{C}_1\mathbf{V}_2 + \mathbf{p}_1 = \begin{bmatrix} V_{1,(1)} \\ V_{1,(2)} \end{bmatrix} = \begin{bmatrix} 1 \\ V_{2,(2)} \end{bmatrix}$$

The results must still be multiplied by the coefficients $a_1 = 1$, $a_2 = z^{1/2}$ to get the response of the digital system.

6. CONCLUSION

The chapter described an analysis method for general SC networks with equal phase durations. Readers should be able to prepare the necessary matrices and use packages like MATLAB to get solutions of their problems. General theory, on which this chapter was based, was published in the accompanying references. A program for analysis and design of SC networks with an arbitrary number of phases and arbitrary phase durations, WATSCAD,[5] was also written and is available for SUN workstations and IBM PC equivalents. For information, contact the author.

REFERENCES

1. **J. Vlach, K. Singhal, and M. Vlach:** Computer oriented formulation of equations and analysis of switched capacitor networks. *IEEE Trans. Circuits Syst.*, CAS-31(9), 753–765, 1984.
2. **J. Vlach:** Hand analysis of switched-capacitor networks. *IEEE Circuits Devices Mag.*, 11–16, November 1986.
3. **J. Vlach and E. Christen:** Poles, zeros and their sensitivities in switched capacitor networks. *IEEE Transactions Circuits Syst.*, CAS-32, 279–284, March 1985.
4. **J. Vlach and K. Singhal:** *Computer Methods for Circuit Analysis and Design.* Van Nostrand Reinhold, New York, 1994.
5. WATSCAD: Program for analysis of switched-capacitor networks. *Manual,* University of Waterloo, Waterloo, Ontario, Canada.

Chapter 8

Electromechanical Filters

Introduction to Electromechanical Filters

Robert A. Johnson

CONTENTS

1. INTRODUCTION

What does a quartz watch crystal with a pair of electrodes in each tyne have in common with a 20-crystal cascade of lattice sections? Both are electromechanical filters. More specifically, both are bandpass electromechanical filters. In this chapter we will try to take some of the mystery out of these common devices. By common, we mean filters that are built in the billions for frequency selection or frequency and time control in radios, televisions, telephone systems, watches, computers, and, for telemetry, navigation, games, and many more applications.

0-8493-8951-8/97/$0.00+$.50
© 1997 by CRC Press LLC

2. GENERAL DESCRIPTIONS

2.1 COMMON CHARACTERISTICS

Let us look at the important characteristics of electromechanical filters. They are passive devices; therefore, they require no power. They are generally small compared with other passive filters. They are frequency stable with regard to both time and temperature. Finally, their high resonator Qs and stability allow narrow bandwidths to be realized.

What all electromechanical filters have in common are

1. An electrical input and an electrical output;
2. Electrical-to-mechanical and mechanical-to-electrical energy conversion;
3. Mechanical (acoustic) vibrations in a solid material such as quartz, ceramic, or metal; and
4. Bandpass filter frequency responses.

2.2 DIFFERENTIATING ONE TYPE OF ELECTROMECHANICAL FILTER FROM ANOTHER

Electromechanical bandpass filters have been given names, or at least have retained their names, to differentiate one type from another. Starting with crystal filters:

Crystal Filters: An electromechanical filter that uses quartz as the solid medium in which mechanical vibrations take place. The quartz elements are piezoelectric, and the acoustic energy is distributed throughout the resonator or resonators in so-called bulk waves (as opposed to surface waves).

Ceramic Filters: Ceramic filters have many of the same characteristics as crystal filters except piezoelectric ceramic, usually lead zirconate titanate (PZT) materials, is used in the place of quartz.

Mechanical Filters: An electromechanical filter that uses metal alloys instead of quartz or ceramic as the vibrating medium. Although the resonators are made of materials that are magnetostrictive, the electromechanical energy conversion is usually performed by composite ceramic/metal transducer resonators on the input and output of the filter. In the case of multiresonator filters, the coupling between resonators is usually mechanical.

SAW Filters: Surface acoustic wave (SAW) filters use various solid mediums such as quartz, lithium tantalate, lithium niobate, zinc oxide, and PZT ceramic. The vibration energy is near the surface of the medium, thus using surface rather than bulk waves. SAW filters are of two types. These are FIR transversal (nonrecursive) filters having delay and summing elements like FIR electrical digital filters and IIR (recursive) ladder or lattice filters formed from SAW resonators.

It is important for the filter user when studying various filter types to differentiate electromechanical filters from electromagnetic-field filters (for example, microwave filters). In the latter case the waves within the resonator are electromagnetic rather than acoustic. For example, the term *ceramic filter* is sometimes used to describe a filter composed of resonators that are one quarter wavelength (electromagnetic) lines using ceramic material. These resonators are roughly 0.75 cm long at 1.0 GHz. The frequency of a corresponding half-wavelength (acoustic) ceramic resonator of the same length is 200 kHz.

3. BANDWIDTH VS. CENTER FREQUENCY

Figure 1 shows the maximum and minimum limits of center frequency and bandwidth of each filter type. The bandwidth is expressed as a percentage of the center frequency. These limits are based on the following physical and dimensional criteria:

Minimum Bandwidth: Frequency shift due to temperature and aging.

Maximum Bandwidth: Electromechanical coupling.

Minimum Frequency: Excessive resonator dimensions and number of finger pairs in SAW filters.

Maximum Frequency: Minimum manufacturable resonator dimensions or finger-pair spacing.

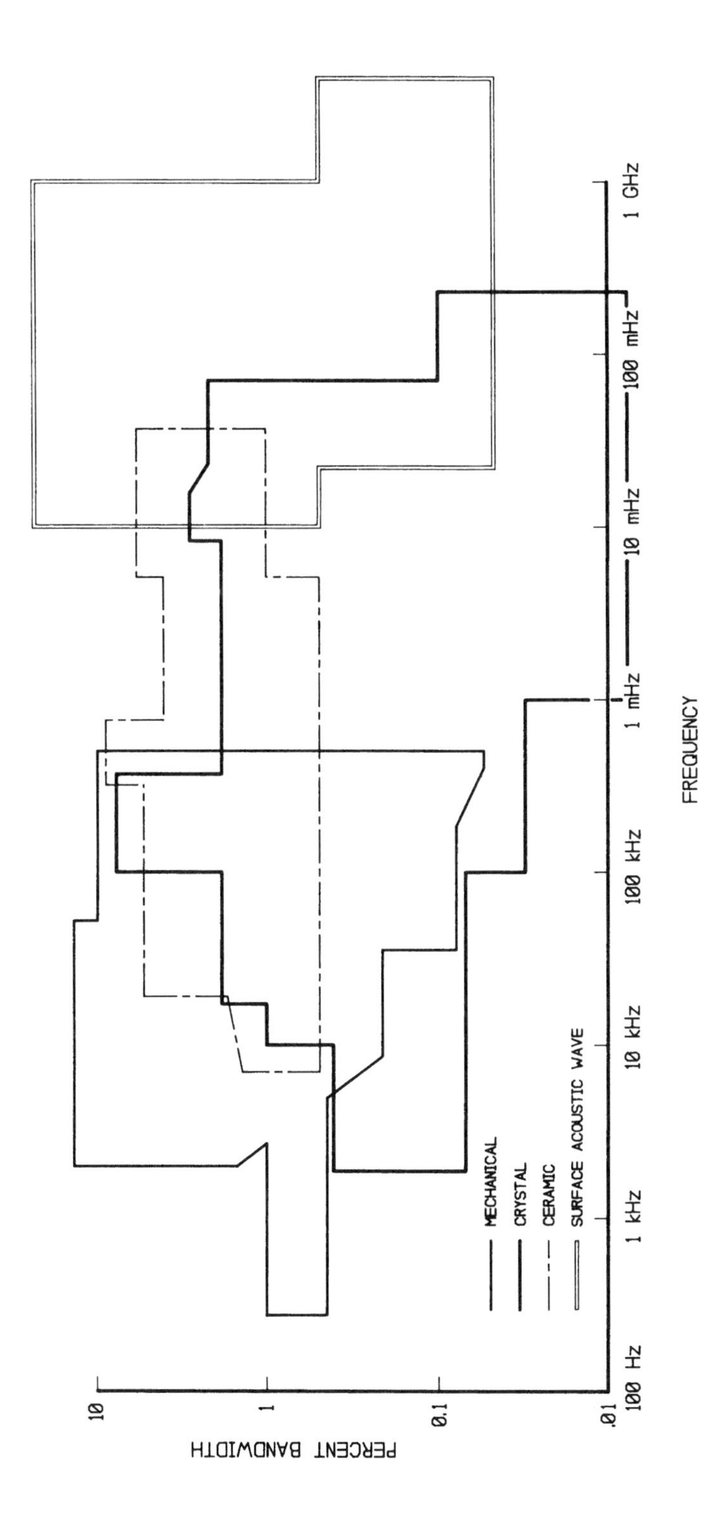

Figure 1 Practical frequency and bandwidth application range of electromechanical filters.

Let us look at each of these in more detail. The specific rules for setting the limits are somewhat subjective, but, if understood, they provide a starting point for determining the value of filter type for a specific application.

3.1 MINIMUM BANDWIDTH

For each of the filter types the major limitation on minimum bandwidth is the frequency shift of the filter with temperature. Therefore, the minimum filter bandwidth shown in the figure corresponds to a frequency shift of 0.2 times the bandwidth over the temperature range of –20 to +70°C. Although this was the criterion chosen, it may be necessary for the user to also consider the following: (1) frequency shift due to aging, (2) insertion loss due to finite resonator Q, and (3) passband "rounding" also due to the resonator Q effects. In addition to the fact that a very narrow filter may have limited use as a result of the above, it is usually more difficult to build very narrow filters because of tighter resonator frequency tolerances, greater impedance ratios, and more finger pairs in the case of SAW filters.

3.2 MAXIMUM BANDWIDTH

The maximum filter bandwidth is limited by the electromechanical coupling coefficient, k_{em}, of the piezoelectric resonators. The resonators are quartz and PZTs in the case of crystal and ceramic filters and ceramic/metal composites in the case of mechanical filters. SAW filters are also limited in bandwidth, in this case, by the k_{em} of the substrate material.

As an example, in inductorless ladder networks the maximum fractional bandwidth is

$$B_{max} / F_0 = q_1 \left(k_{em}^2\right) / 2 \tag{1}$$

where B_{max} is the maximum 3-dB filter bandwidth, F_0 is the center frequency, and q_1 is the normalized Q of the terminating section.[1]

When coils are used to tune the capacitance of the transducer resonators in a ladder network, such as a monolithic crystal or ceramic filter or a mechanical filter, the maximum fractional bandwidth achievable is roughly

$$B_{max} / F_0 = k_{em} / k_{12} \tag{2}$$

where k_{12} is the normalized coupling coefficient between the first and second resonators.[1] The first resonator is the inductor and capacitor; the second resonator is the first mechanical resonator (transducer resonator).

3.3 LOW-FREQUENCY LIMITS

The filter center frequency is usually determined by a single length or thickness dimension of the resonator. An exception is the flexure resonator where both length and thickness determine the frequency. The low-frequency limits are determined by the allowable length of the resonator or its minimum allowable thickness, in the case of flexure modes of vibration. The lower-frequency limits in Figure 1 correspond to a maximum resonator length of 2.5 cm. In the case of SAW filters, the size-determining factor other than frequency is the percentage bandwidth. The narrower the bandwidth, the greater the number of fingers and, therefore, the greater the length of the filter.

3.4 HIGH-FREQUENCY LIMITS

The upper-frequency limits for each filter type are usually proscribed by the minimum manufacturable or usable length, thickness, or in the case of SAW filters by the minimum manufacturable spacing between fingers.

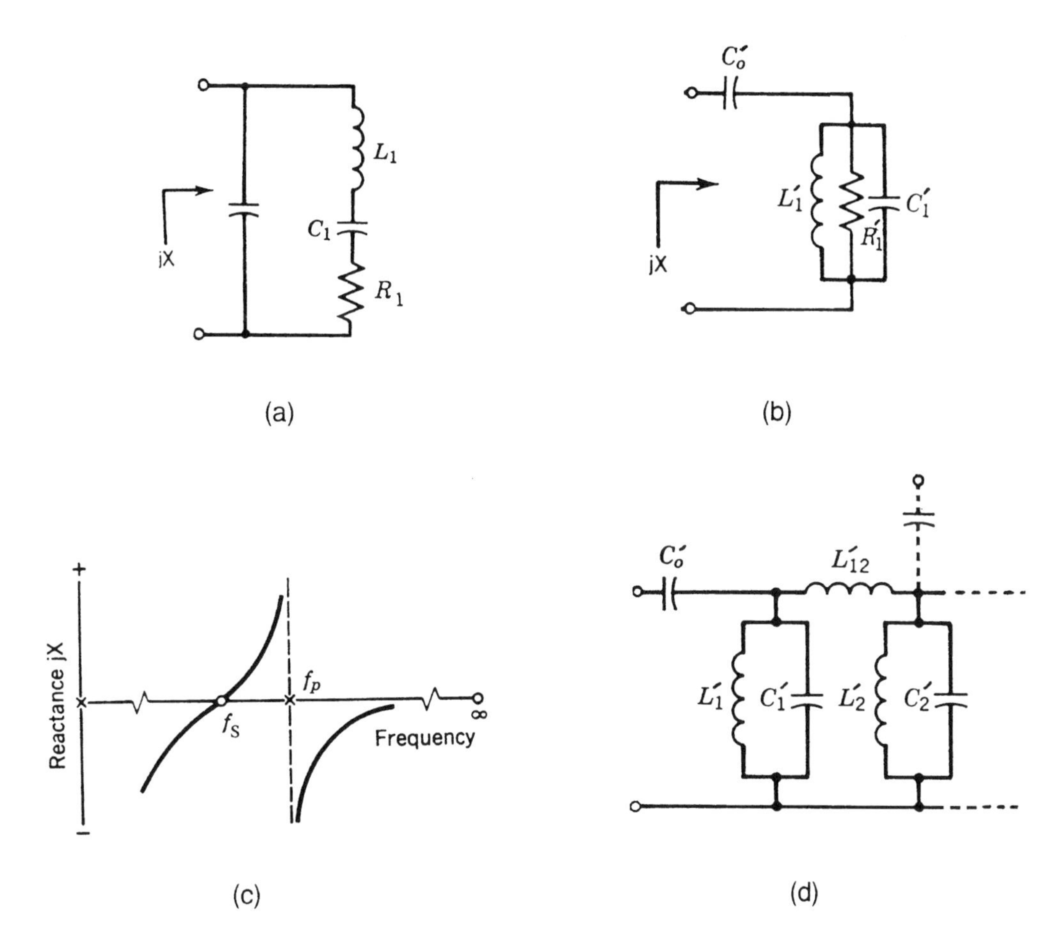

Figure 2 Piezoelectric resonator equivalent circuits: (a) classical analogy, (b) mobility analogy, (c) infinite Q driving-point reactance, and (d) mechanically coupled transducer and second resonator based on the mobility analogy.

4. EQUIVALENT CIRCUITS

The term *equivalent circuit* refers to an electrical equivalent circuit of an electromechanical filter or filter element. The equivalent circuit can contain elements that correspond to both electrical and mechanical portions of the filter. Some elements are actually electrical, and others correspond to the electrical analogy of the mechanical parts. The analogies are of two types, the classical, where voltage corresponds to force (inductance to mass, etc.), and the mobility analogy, where voltage corresponds to velocity (capacitance to mass, etc.). The classical analogy is most commonly used to describe crystal and ceramic filters. The mobility analogy is helpful for describing mechanically coupled mechanical circuits, for example, mechanical filters, but is sometimes used to describe monolithic crystal and ceramic filters.[2] Equivalent circuits for SAW filters are unique to those devices; for example, they can be described in terms of delays and weighting factors, as is done in Chapter 8.5 on SAW filters.

4.1 TRANSDUCER EQUIVALENT CIRCUITS

Since most electromechanical filters, which include SAW filters, use piezoelectric transducers, we will concentrate on filters of this type. A piezoelectric transducer is composed of a material that changes dimensions in the presence of an electric field and, conversely, a material that generates an electric field when mechanically stressed. The most commonly used piezoelectric materials are quartz and PZT ceramics. In most cases, electrodes are plated on opposite surfaces of the transducer, like a capacitor. In converting from electrical to mechanical energy, a voltage is applied across the plated surfaces producing an electric field and a resultant deformation of the material. Since we are using the transducer in a bandpass filter, we are also dealing with a resonant device. The transducer therefore acts as a resonator, a capacitor, and an electromechanical energy conversion device. Equivalent circuits of piezoelectric transducers are shown in Figure 2a (classical analogy) and Figure 2b (mobility analogy).

The equivalent circuit of Figure 2a is most commonly used to represent a two-terminal crystal or ceramic resonator used in a filter or oscillator circuit. C_0 is the capacitance between the transducer electrodes. L_1 and C_1 represent the mechanical portion of the transducer that is physically shaped in such a way that it is resonant at frequency $F_S = 1/\left(2\pi\sqrt{L_1C_1}\right)$. Ignoring the series resistance, when viewed through the input terminals, the circuit of Figure 2a has the reactance characteristics shown in Figure 2c. The equivalent circuit of Figure 2b has an identical frequency response characteristic. The frequency difference between the parallel resonance, F_P, and the series resonance, F_S, is called the pole–zero spacing. The relationships between the pole–zero spacing, $(F_P–F_S)$, the electromechanical coupling coefficient, k_{em}, mentioned earlier, and the equivalent circuit element value ratios, C_1/C_0 and C_0'/C_1', are shown in Equation 3 below.[1]

$$k_{em}^2 \approx 2\left(F_P - F_S\right)/F_S \approx C_1/C_0 \approx C_0'/C_1' \tag{3}$$

The ratio C_0/C_1 is commonly called the capacitance ratio by users of quartz crystals.

By measuring F_S, F_P, and the capacitance C_0, of an actual circuit, it is easy to calculate the remaining terms in Equation 3. In making the calculations, we usually are dealing with low-coupling transducers, where there is only a small error in making C_0' equal to C_0, and the approximately equal signs can be replaced by equal signs. Also, by making a measurement of the resonator Q, and inserting resistance into the tuned circuits, accurate transducer models can be constructed.

4.2 COUPLING BETWEEN TRANSDUCERS AND RESONATORS

Electromechanical filters are either electrically or mechanically coupled. In this section we will study the case of mechanical coupling.

The transducer equivalent circuit shown in Figure 2b is useful in modeling filters with mechanical coupling between the resonators, for example, mechanical filters or even monolithic crystal filters.[2,3] Because the circuit is based on the mobility analogy, the mechanical and electrical circuit topologies are the same. This eliminates the difficulty of constructing dual circuits for complex coupling networks.

The transducer in Figure 2b is shown as a two-port network. If the transducer is mechanically coupled to another resonator through a wire or some other coupling medium, then the output terminals of the transducer are connected to the input terminals of the coupling medium and the output of that element to the input terminals of the next resonator. In other words, the elements are simply cascaded. The resonators and coupling elements are distributed elements, for example, transmission lines. Because most electromechanical filters are narrow-bandwidth devices, the distributed element can be represented by lumped elements. The inverse value of the spring constant of a coupling wire is represented by an inductor. Wire coupling between a transducer and a second resonator is shown as an electrical equivalent circuit in Figure 2d. If the second resonator is a piezoelectric device or an electroded region of a monolithic filter, the dashed-line capacitor is added to the equivalent circuit. This capacitor can be shorted to ground, or, in the case of a two-resonator filter section, external terminal of the capacitors becomes the output terminal. Because the mobility analogy preserves the topologies of mechanical and electrical equivalent circuits, wire bridging to a nonadjacent resonator can be taken into account and is represented by connecting an inductor between the nodes to which the resonators are attached.

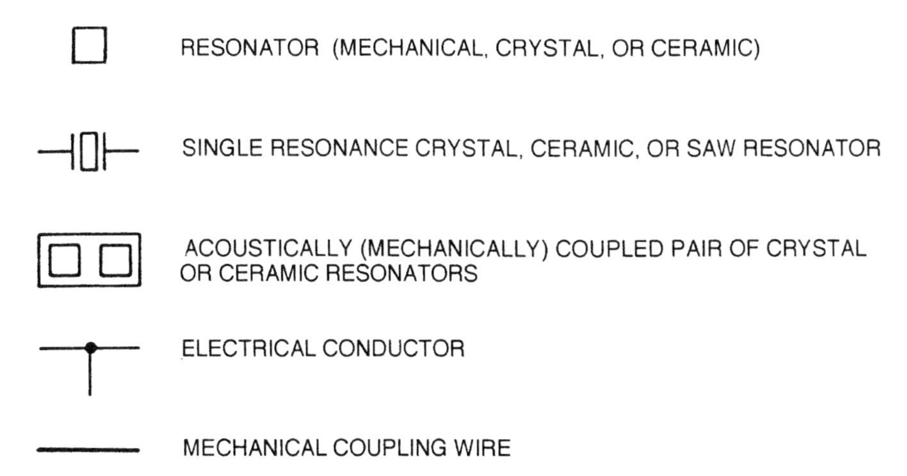

Figure 3 Symbols used in the topology diagrams of Figures 4, 5, and 6. (Adapted from Sheahan, D. F. and Johnson, R. A., *IEEE Trans. Circuits Syst.*, CAS-22, 69, 1975. With permission.)

5. FILTER CONFIGURATIONS

Electromechanical filters are composed of discrete (lumped or distributed) or monolithic elements or combinations of both.[4] By the term *discrete* is meant a resonator, transducer, coupling element, or an inductor or capacitor. By *monolithic*, we mean a single body, which contains both resonant regions and coupling regions or, in the case of SAW filters, finger pairs. The monolithic filters are planer devices where electroded areas differentiate resonant and coupling regions. These various networks are shown below:

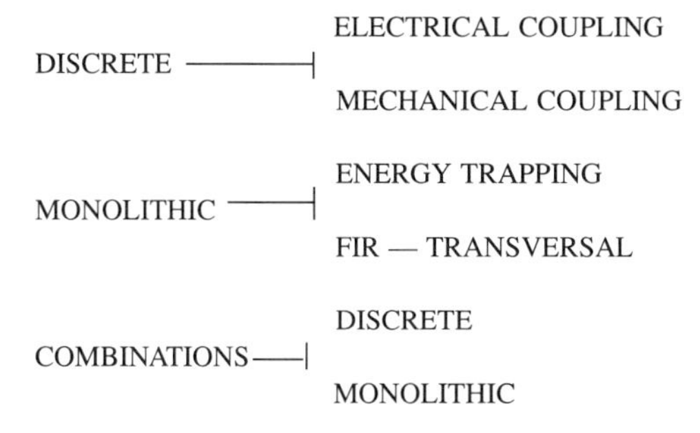

Figure 3 shows various symbols used in schematic and topology diagrams to describe the various types of electromechanical filters. The simple square corresponds to a resonator, whether it be a discrete mechanical filter element or a resonant region of a monolithic crystal or ceramic filter. The discrete single-resonance crystal, ceramic, or SAW resonator element corresponds to the electrical equivalent circuit of Figure 2a. Electrical conductors are recognized by their being connected by nodes, whereas the mechanical filter coupling wires are denoted by simple lines.

5.1 DISCRETE-ELEMENT FILTERS ELECTRICALLY COUPLED

Figure 4 shows network topologies of various crystal, ceramic, and SAW resonator filters. In each case, the resonators are the simple two terminal networks shown in Figure 2a. Not included are more complex three-terminal resonators such as those described in Chapter 8.4 on ceramic filters. In Topologies 1 and 2, attenuation poles are realized by parallel resonances in the series element arms and series resonances in the shunt arms. Topologies 3 and 4 are hybrid lattice networks whose stopband behavior is controlled by the resonance frequencies of the crystals, the motional inductances (L_1) of the crystal, and the total shunt capacitance across each of the lattice arms. The difference between Topologies 3 and 4 is that circuit 4 uses inductors outside the lattice networks to make it possible to build intermediate-bandwidth filters. The inductors

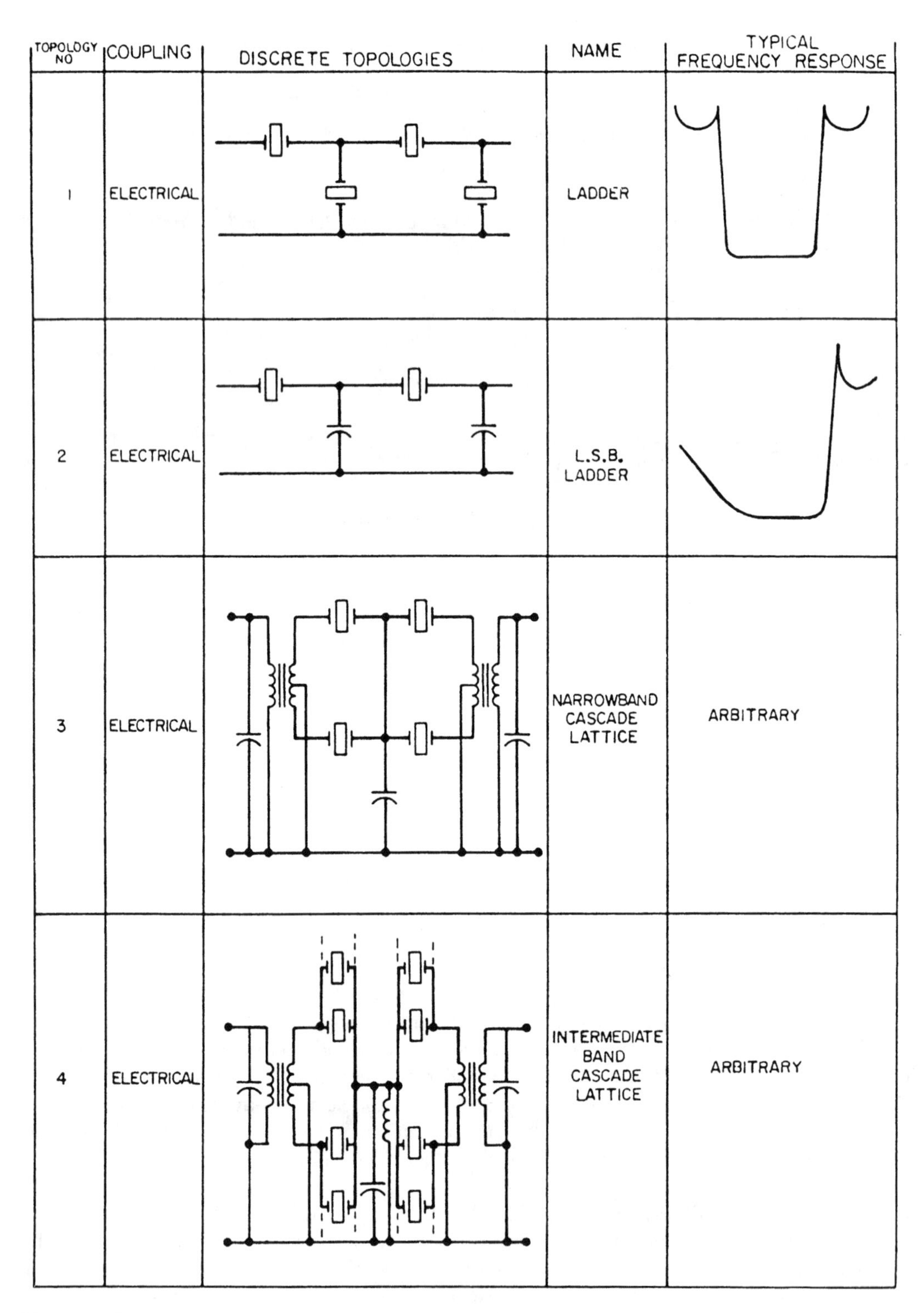

Figure 4 Topologies of discrete crystal, ceramic, and SAW filters and their typical frequency responses. (Adapted from Sheahan, D. F. and Johnson, R. A., *IEEE Trans. Circuits Syst.*, CAS-22, 69, 1975. With permission.)

are reflected into the lattice structure and simply eliminate the narrowing effect of the shunt capacitance across the crystals, but add little or no selectivity. In the case where the electrical tuned circuits add selectivity, these filters are designated as wide-bandwidth filters, and an

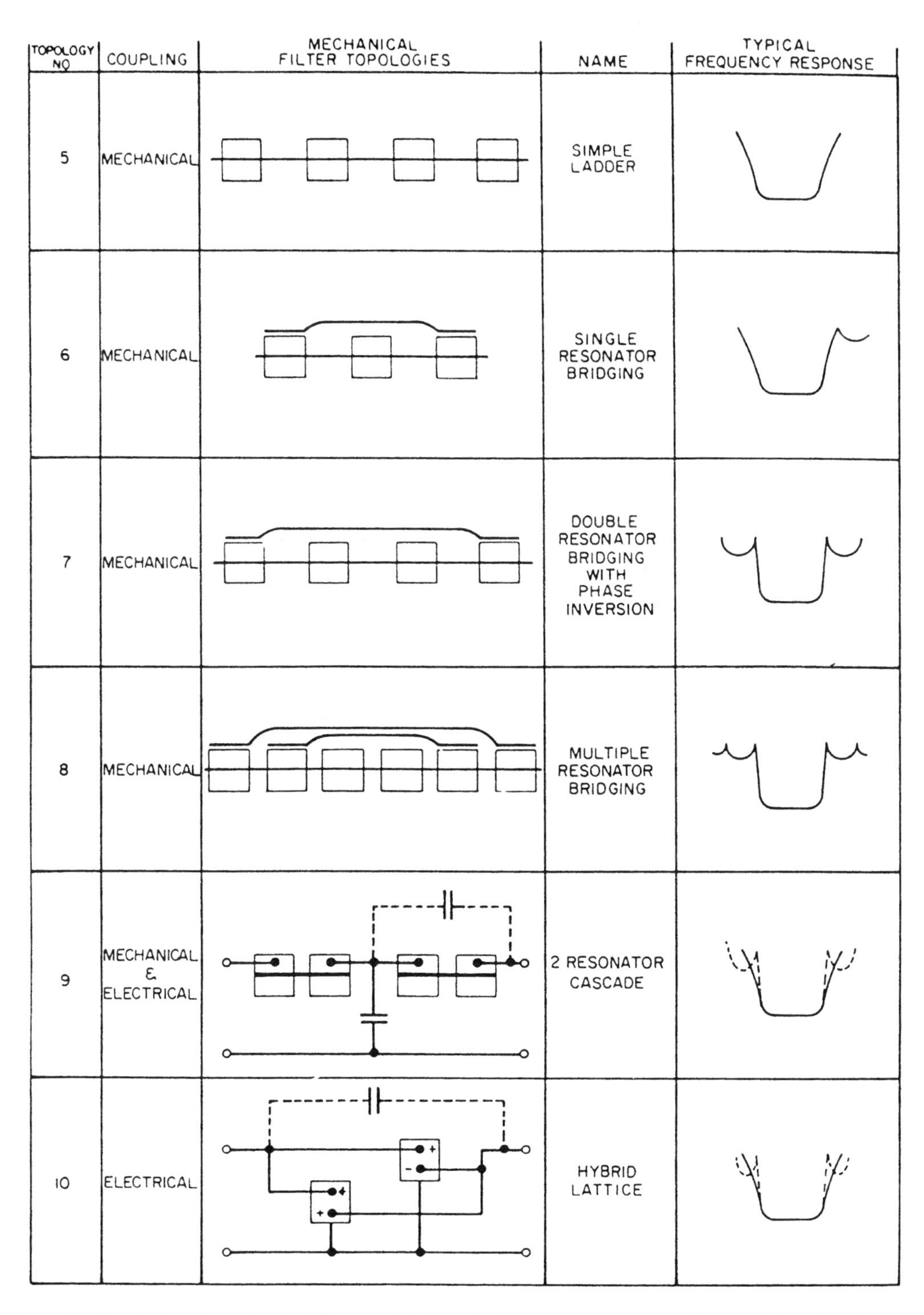

Figure 5 Topologies of mechanical filters and their typical frequency responses. (Adapted from Sheahan, D. F. and Johnson, R. A., *IEEE Trans. Circuits Syst.*, CAS-22, 69, 1975. With permission.)

equation equivalent to Equation 2, but for a lattice, applies to determining the maximum achievable bandwidth.

5.2 DISCRETE-ELEMENT FILTERS MECHANICALLY COUPLED

Figure 5 shows various mechanical filter topologies. If no wire bridging is used in the ladder network, a simple monotonic frequency response results, as shown in Topology 5. Attenuation poles are realized by use of wire bridging across one or two resonators as shown in the next three topologies. In Topology 7, attenuation poles are produced both above and below the filter passband when the bridging wire is between a half-wavelength and a full wavelength long. If the bridging wire is less than a half-wavelength long, delay equalization is obtained. In the case of bridging across a single resonator, the attenuation pole position is a function of the bridging wire length. A short wire will produce an attenuation pole above the filter passband as shown in Topology 6. Means of realizing poles other than by varying the bridging wire length are described in Reference 1.

5.3 MONOLITHIC FILTERS EMPLOYING ENERGY TRAPPING

Topology 11 of Figure 6 shows a model of an eight-resonator monolithic filter. The energy-trapped electrode-pair resonator areas are separated by nonelectroded coupling regions. The coupling regions can be modeled by a simple shunt capacitor (classical analogy) or a series inductor (mobility analogy).

5.4 COMBINATIONS OF VARIOUS DISCRETE AND MONOLITHIC SECTIONS

Topology 12 shows a cascade of four two-resonator monolithic energy-trapped sections coupled with capacitors. A term used for a crystal filter having this type of configuration, where each pair of resonators is on a separate plate, is a polylithic crystal filter. A ceramic filter used in FM radios that is composed of two energy-trapped pairs of resonators and a capacitor on a single wafer is described in Chapter 8.4. Topologies 13 and 14 show ways of combining discrete and monolithic elements to obtain attenuation poles. Returning to Figure 5, Topologies 9 and 10 show methods of cascading mechanical filter sections with bridging and coupling capacitors. The three-terminal resonators of Topology 10 could be also realized by the ring-dot ceramic resonator described in Chapter 8.4.

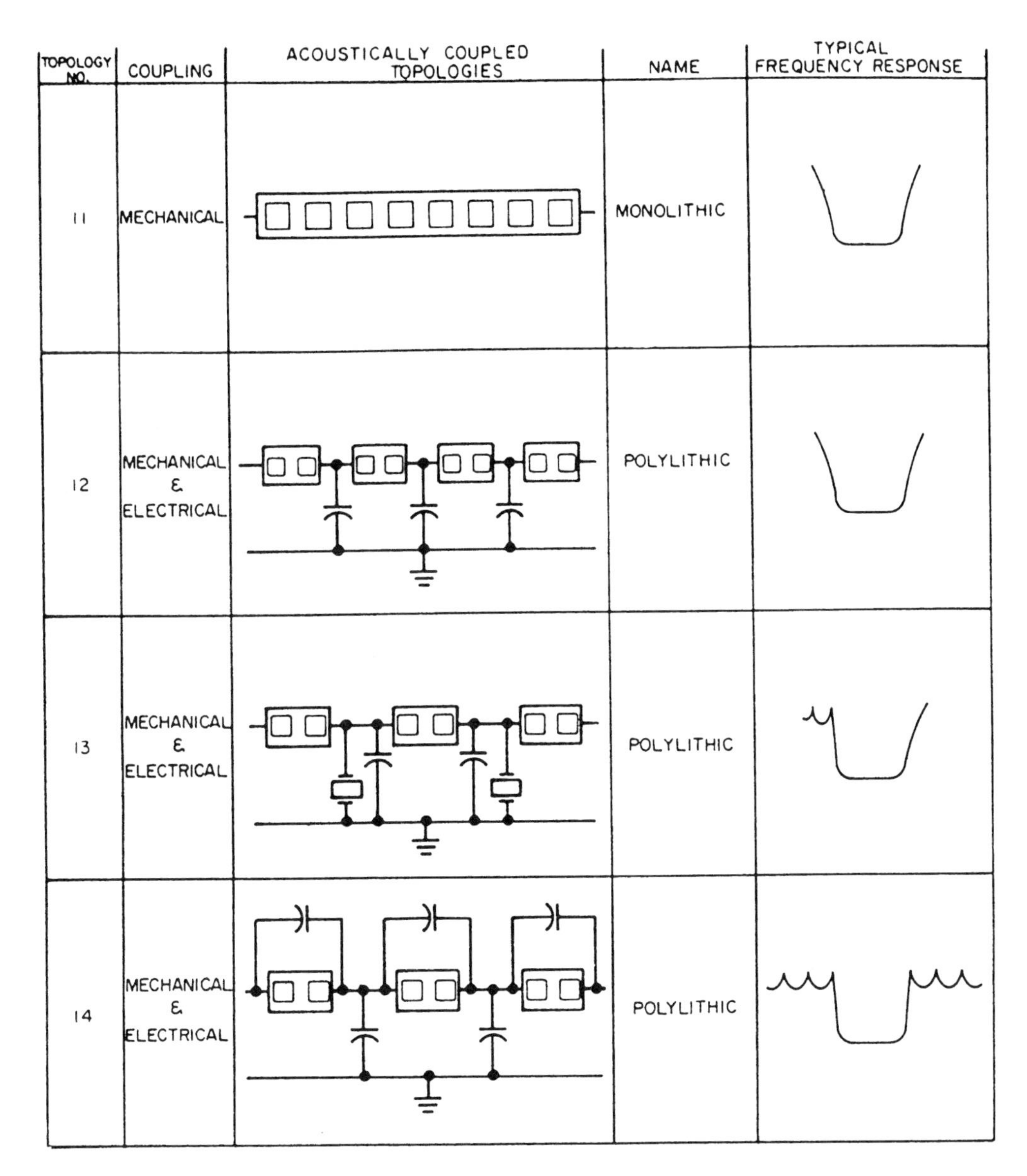

Figure 6 Topologies of monolithic crystal and ceramic filters and their typical frequency responses. (Adapted from Sheahan, D. F. and Johnson, R. A., *IEEE Trans. Circuits Syst.,* CAS-22, 69, 1975. With permission.)

REFERENCES

1. **Johnson, R. A.,** *Mechanical Filters in Electronics,* Wiley-Interscience, New York, 1983, 30, 156, 194.
2. **Rennick, R. C.,** An equivalent circuit approach to the design and analysis of monolithic crystal filters, *IEEE Trans. Sonics Ultrason.,* SU-20, 321, 1973.
3. **Johnson, R. A. and Günther, A. E.,** Mechanical filters and resonators, *IEEE Trans. Sonics Ultrason.,* SU-21, 244, 1974.
4. **Sheahan, D. F. and Johnson, R. A.,** Crystal and mechanical filters, *IEEE Trans. Circuits Syst.,* CAS-22, 69, 1975.

CRYSTAL FILTERS

Robert Kinsman

CONTENTS

1. OVERVIEW

Crystal filters have the distinction of being the oldest of the electromechanical-type filters. The first use of quartz crystals as filter elements was suggested by Professor Walter Cady in his paper "The Piezo-Electric Resonator" in 1922.[1] In 1934 W. P. Mason of Bell Telephone published the results of his work on crystal filters which were developed for use in frequency-division-multiplex (FDM) telephone equipment.[2] This work resulted in the development of the 60 to 108 kHz basic group-band filters used to frequency multiplex 12 telephone voice channels. These filters went into production in 1938 and with some evolutionary changes are still in use today. Production volume for these filters was reportedly as high as 100,000 units per year by Bell Telephone.

The resonance frequency of a quartz crystal, as with any bulk mechanical resonator, is determined primarily by its external dimensions. Quartz crystals have been successfully manufactured over a broad frequency range extending from as low as 1 kHz to well over 100 MHz. A variety of crystal designs are used at frequencies below 1 MHz. These designs essentially all utilize a rectangular or square quartz wafer with the resonance frequency determined by the length and width dimensions. All filter crystals above 1 MHz operate in the "thickness-shear" mode where the resonance frequency is determined by the wafer thickness. Quartz crystals have the highest Q, best temperature stability, and best long-term stability of all the electromechanical resonators, particularly in the frequency region above 1 MHz. For this reason, they are widely used as oscillator elements, as well as for filters.

Because of the excellent temperature and aging properties and high Q of quartz, filters employing quartz crystals provide the ultimate in performance for very narrow bandwidth applications. In

0-8493-8951-8/97/$0.00+$.50
© 1997 by CRC Press LLC

extreme cases, filters can be placed in an oven to provide a very narrow, stable passband over a wide temperature range. The lowest-cost filter designs, the so-called narrowband designs, can be used for bandwidths up to a maximum of 0.3 to 0.4% of center frequency with the limit depending on the particular crystal design used. Wider bandwidths can be achieved by adding inductors to essentially tune out part or all of the shunt capacitance (C_o) of the crystals. These "intermediate-bandwidth" designs are useful up to about 2% of the center frequency, although most applications are in the 0.3 to 1.0% range.[3] In very restricted frequency bands, "wideband" designs can be used to provide bandwidths as wide as 6% by combining crystal resonators with L/C resonators. Only certain crystal types have characteristics that are suitable for use in wideband designs.

The traditional "discrete" crystal filter is typically a hybrid-lattice-type design and consists of an assembly of crystals, capacitors, and RF transformers mounted on a PC board. Each crystal is individually manufactured to a specific frequency and mounted in a hermetically sealed container. Because the crystals can be built to any desired frequency, a wide variety of filter designs is possible. Designs have been developed which can provide filters with Butterworth, Chebyshev, Bessel, or elliptic function responses as well as skewed (single-sideband) characteristics. In addition, a variety of techniques is available for building filters with linear-phase or phase-compensated characteristics. This exceptional design flexibility makes the discrete crystal filter very attractive for specialized requirements. Also, because the filter is essentially a PC-board assembly of individual components, the cost of tooling is minimal, an important factor for small-quantity orders. However, the cost to fabricate crystals is relatively high, and discrete crystal filters are typically the most expensive of the various electromechanical filters. For frequencies below 1 MHz there is extensive overlap between the capabilities of crystal and mechanical filters. In general, the cost of mechanical filters is lower particularly for larger-quantity requirements where the cost of tooling can be amortized.

The newest type of crystal filter is the acoustically coupled or "monolithic" filter.[4,5] A monolithic crystal filter incorporates at least two thickness-shear resonators on a single quartz wafer which are acoustically coupled to each other to provide a multipole filter response. The monolithic device provides a complete filter and requires no additional components except for terminating networks. Monolithic filters are typically manufactured at frequencies of 5 MHz and higher with bandwidths comparable with narrowband discrete filter designs. These filters are most commonly fabricated as two-pole devices, but a few specialized designs have been built with as many as eight resonators on a single wafer. Two-pole elements can be easily cascaded to provide higher-order filter designs. Monolithic crystal filters are considerably less costly to manufacture than comparable discrete filter designs and have been widely used to provide IF selectivity in narrowband communications receivers.

2. CRYSTAL PROPERTIES

The quartz crystal resonator consists essentially of a thin quartz wafer (commonly called a quartz blank in the crystal industry) with coincident metal electrodes placed on opposing surfaces. Typical construction for a high-frequency, thickness-shear crystal is illustrated in Figure 1. When a voltage is applied between the electrodes, the surfaces of the wafer move in opposite directions, and, at the appropriate frequencies, mechanical resonances can be excited in the wafer. The mounting of the quartz blank is extremely critical so as to prevent the mounting system from dampening out (or lowering the Q) of the desired resonance. In the thickness-shear resonator the motion occurs under the electrodes, and with proper design the wafer can be mounted at the edges, as shown in Figure 1, with negligible dampening effects. Lower-frequency crystals are active at the edges of the wafer and are typically mounted by wires which are attached at mechanical nodes on the surface of the wafer.

Crystalline quartz is only moderately piezoelectric; i.e., it has a very low electromechanical coupling coefficient. However, quartz as a material has some extremely attractive properties; it has

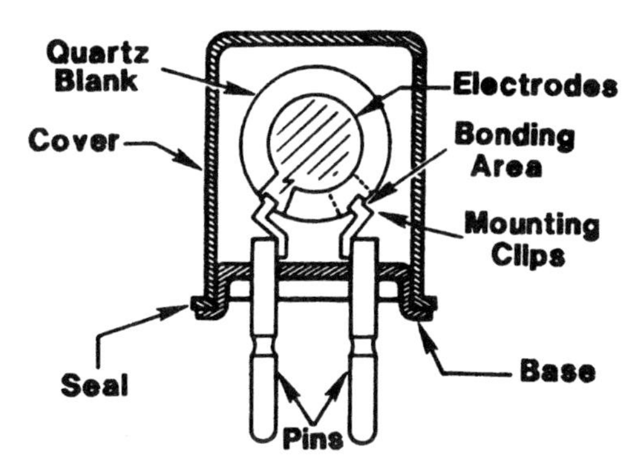

Figure 1 Typical high-frequency quartz crystal resonator construction.

very high mechanical Q, is mechanically very strong, is a very stable inert chemical compound (SiO_2), can withstand very high temperatures (in excess of 500°C), and is a relatively inexpensive material. In addition, quartz can be grown in an extremely pure form. Quartz crystals provide the best frequency stability with temperature and time of all known mechanical and piezoelectric resonators.

All filter crystals used at frequencies of 1 MHz and above operate in the thickness-shear mode and are commonly known as "AT-cut" crystals. Their frequency as a function of temperature is described by a linear plus a third-order function with an inflection temperature near room temperature. The value of the linear coefficient is determined by the exact angle of the wafer plane with reference to the crystallographic axes of the quartz material. A family of curves known as the "Bechmann" curves has been published which describe the temperature performance of AT-cut crystals.[6] Several of these curves are shown in Figure 2. By selecting the proper wafer angle the temperature performance of the crystal can be optimized for minimum frequency shift over a particular temperature range.

The equivalent circuit for a crystal is shown in Chapter 8.1. The static electrode capacitance C_o is determined by the electrode size and the thickness of the quartz wafer. The elements C_1, L_1, and R_1 are called motional parameters and represent the electrical equivalent of the mechanical resonance. The ratio C_o/C_1 is a function of the electromechanical coupling coefficient of quartz and ranges from about 250 to 300 for fundamental mode filter crystals. Practical capacitance values for $C_{o,}$ and correspondingly C_1, are limited by the size of standard production quartz blanks and by electrode sizes which provide resonators with good Q and good suppression of unwanted spurious resonances. Typical values for C_1 are shown in Figure 3. The capacitance ratio for overtone, mode crystals increases by the square of the overtone, and as a result the motional capacitance values are much smaller for those types of crystals.

A variety of crystals at frequencies above 1 MHz are manufactured for use in crystal oscillators. In general, these crystals are unsuitable for use in filters. A typical oscillator crystal will have strong spurious modes which may be suppressed by as little as 3 dB, fine for use in oscillators, but totally unsuitable for filter applications. Filter crystals typically have a higher impedance than oscillator crystals at the same frequency. Also, during the manufacturing process tighter tolerances are required on the wafer flatness and parallelism and on the thickness of the electrode metallization.

Crystals are typically packaged in a metal container with glass-to-metal feedthroughs as shown in Figure 1. The largest package, used at the low end of the frequency range (1 to 5 MHz), measures approximately 20 × 20 × 9 mm. As the frequency increases the size is reduced in several steps with the smallest commercial package measuring approximately 8 × 8 × 3 mm. The package must maintain a true hermetic seal under all conditions as the crystal is extremely sensitive to contamination or moisture. A variety of surface mount type packages have been developed for oscillator applications, but these are rarely used for filter crystals.

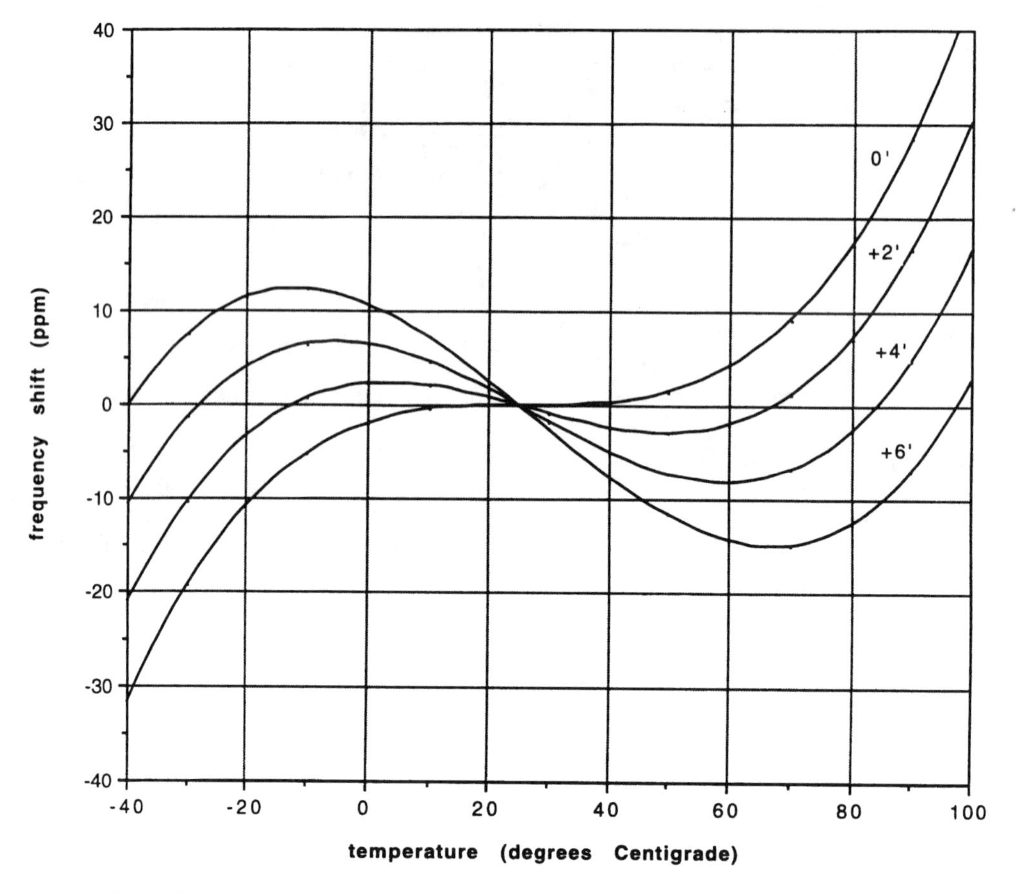

Figure 2 Frequency/temperature characteristic for AT-cut crystals (Bechmann curves).

3. DISCRETE FILTER CIRCUIT DERIVATION

The design of filters using discrete crystals is accomplished by developing a filter network that can absorb the equivalent circuit of the crystal resonators. The relatively large shunt capacitance of the crystal prevents its use in traditional ladder filters for most applications. In addition, crystals are only available over a very limited range of impedance values. Another direct result of the crystal shunt capacitance is to limit the maximum bandwidth which can be achieved with the various filter design types.

Crystals filter networks for all-pole designs with Butterworth, Chebyshev, or Bessel responses can be derived from handbook tables using the L-C narrowband ladder configuration shown in Figure 4. For very narrow bandwidths, crystals can be substituted directly for the series resonant circuits in the series arms. The filter response will be skewed slightly to the low-frequency side, and there will be a transmission zero on the high-frequency side. However, these effects are negligible for bandwidths up to about 0.025% of center frequency for filters using fundamental mode crystals. For wider bandwidths, a bridge-type circuit such as a symmetrical lattice or its hybrid lattice equivalent is typically used as it provides the capability to "balance out" the shunt capacitance of the resonators.

Two circuit transformations are useful in developing the desired filter configuration from the handbook circuit shown in Figure 4. The first transformation known as "Bartlett's bisection theorem," as illustrated in Figure 5, shows how to derive a symmetrical lattice circuit from a symmetrical ladder circuit. The second transformation shown in Figure 6 provides unbalanced equivalent circuits for the lattice section. The full-lattice-type filter was used extensively in the past for FDM filter banks for telephone applications. However, in most applications today the unbalanced

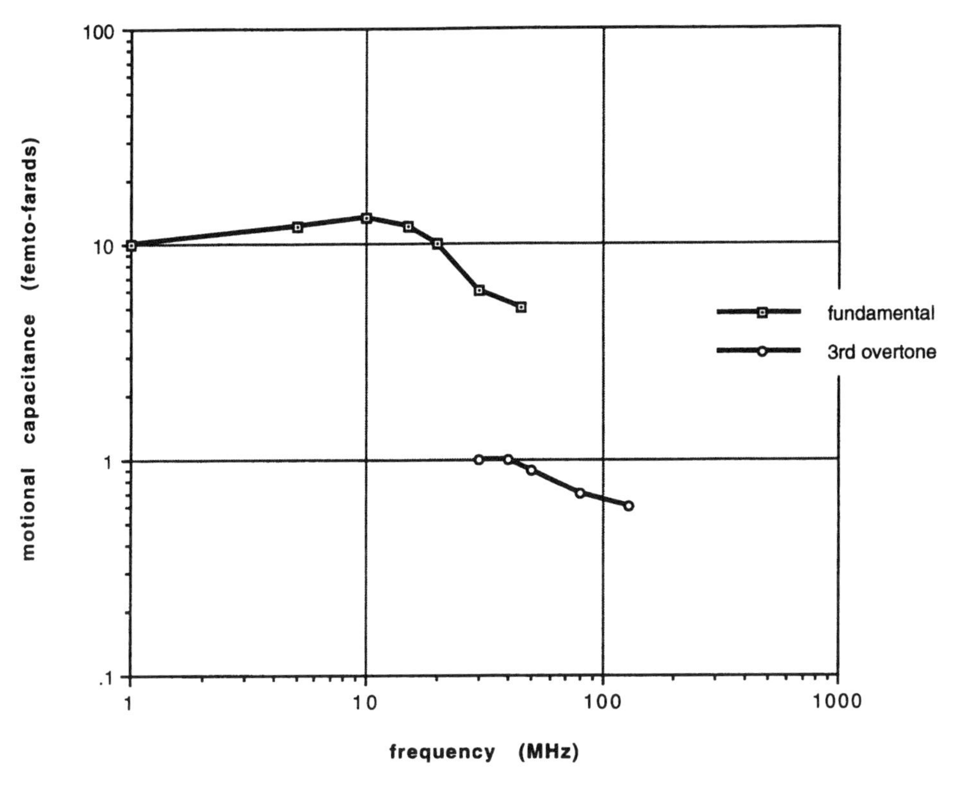

Figure 3 Motional capacitance for typical filter crystals.

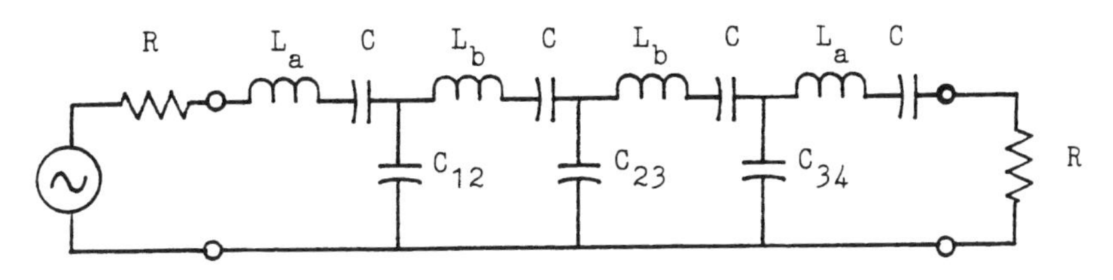

Figure 4 Narrowband L/C ladder filter circuit.

or common ground-type configuration is required. The unbalanced or so-called hybrid-lattice-type filter has the additional feature of using only half as many resonators as the full lattice, although an RF transformer is required.

The most common discrete crystal filter configuration consists of a cascade of two-pole sections. The basic two-pole (two-crystal) section has two primary advantages from a manufacturing standpoint: the crystals are identical except for frequency and the C_o values of the two crystals are equal, thus providing a natural balance between the two lattice arms. The derivation of this type of filter from a handbook design is illustrated in Figure 7. In Figure 7a a typical ladder network is shown for a four-pole filter with capacitive coupling between sections and all resonators having equal inductance. This circuit is then modified as in Figure 7b by adjusting the terminations to force the input and output resonators to have the same resonance frequency as the two center resonators. This circuit is then converted to a cascade of two identical lattice sections in Figure 7c with part of the capacitance of C_t and C_{23} appearing as shunt capacitors across the resonators. In Figure 7d the lattices are replaced with their equivalent hybrid-lattice sections with $C_1 = C/2$ and $C_o = C_p/2$, and in Figure 7e crystals are substituted in the lattice arms. The cascaded two-crystal design has been widely used to provide filters with four-, six-, and eight-pole responses and accounts for a large percentage of all discrete-crystal filters manufactured.

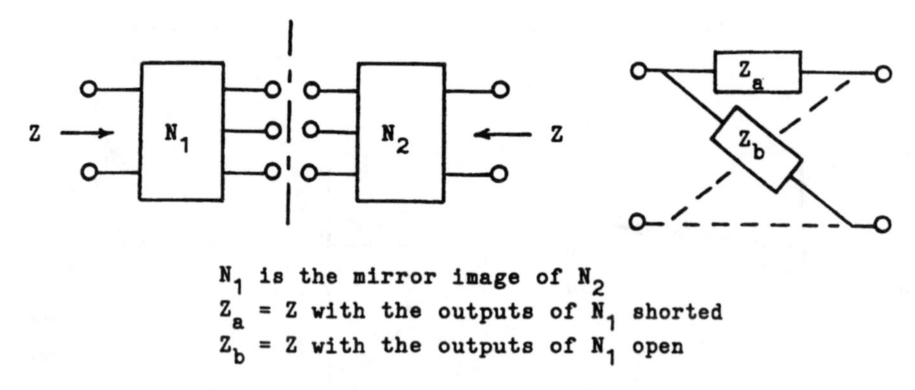

Figure 5 Bartlett's bisection process.

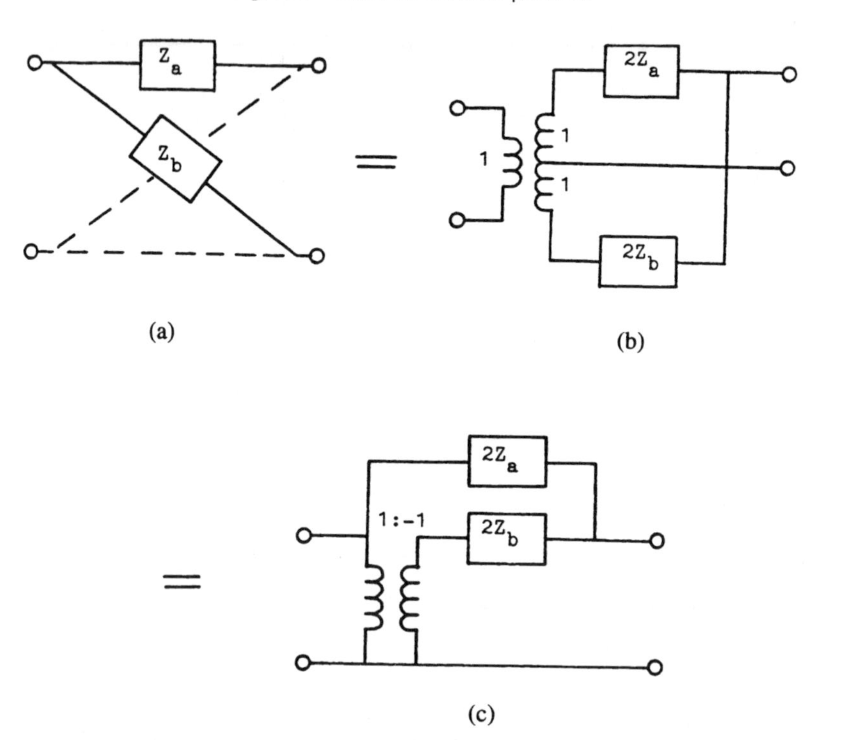

Figure 6 Full lattice (a) and equivalent hybrid-lattice (b, c) filter configurations.

From Figure 7c it can be seen that crystals may be placed in the full-lattice structure to form a four-pole filter using only crystals and capacitors. By convention, this configuration has been defined as a narrowband crystal filter. As the filter design bandwidth is increased, the impedance level increases and the capacitance value of C_{23} decreases. The maximum bandwidth for the narrowband design is reached when the value of $C_{23} = 2C_p$. For filters using fundamental-mode AT-cut crystals the maximum bandwidth is approximately 0.3% of center frequency. The limits of the narrowband design are illustrated in the bandwidth-frequency chart for electromechanical filters at the beginning of this chapter. The bandwidth limit can be increased by adding an inductor across C_{23} to tune out the excess crystal capacitance. However, this is not a satisfactory solution for most applications as the temperature sensitivity of the inductor will cause the filter bandwidth to vary with temperature. It is desirable to keep the number of inductive elements to a minimum in crystal filters because of their cost and temperature-stability problems. Also, because crystal filters are typically rather high-impedance circuits, the tuning of the inductive elements is very sensitive.

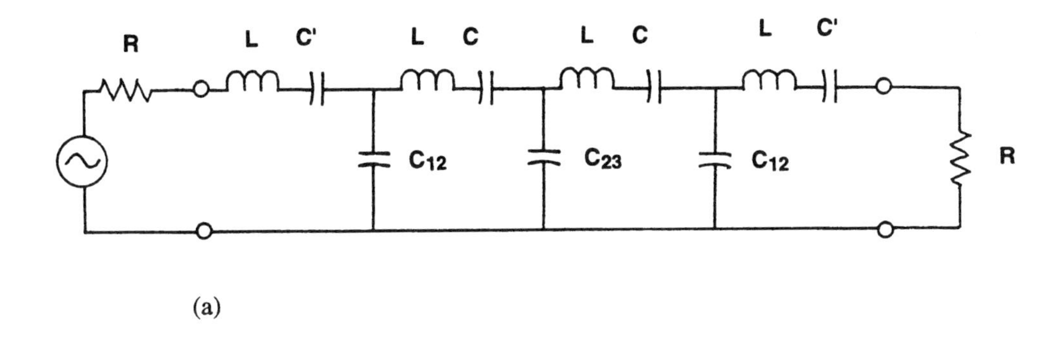

(a)

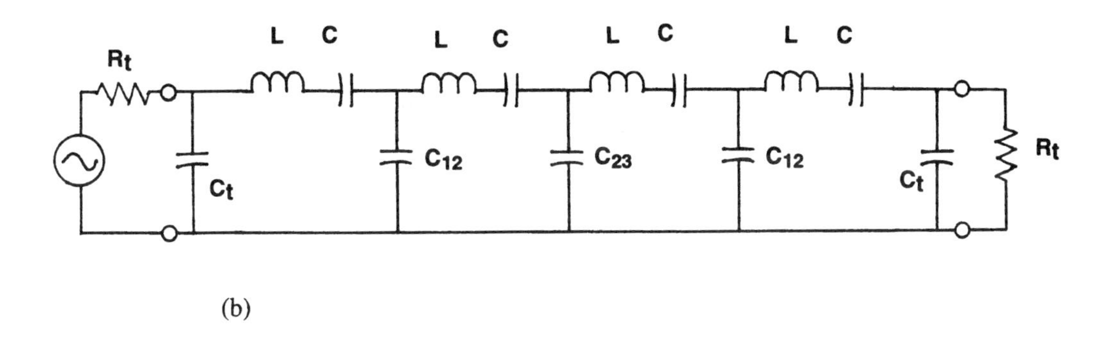

(b)

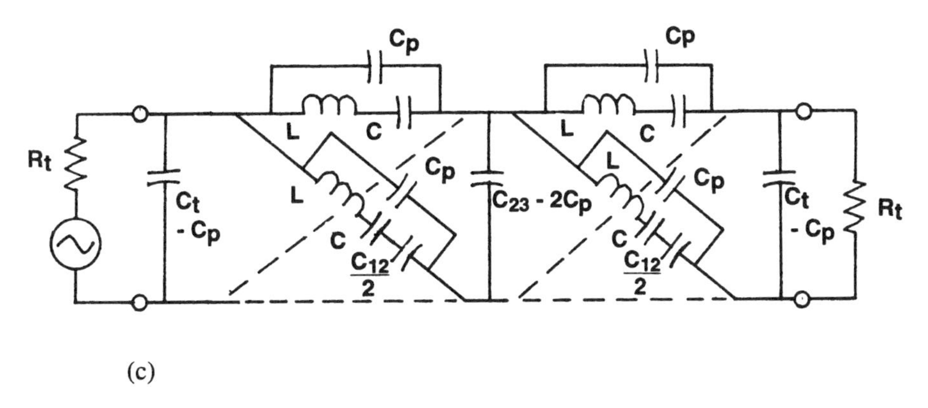

(c)

Figure 7 Four-pole, narrowband, crystal filter circuit derivation. (a) "Handbook" ladder filter circuit; (b) modified ladder filter circuit; (c) equivalent full lattice circuit derived using Bartlett's bisection theorem; (d) hybrid lattice equivalent of the full lattice, $L_1 = 2L$, $C_1 = C/2$, $C_2 = (CC_{12})/2(2C + C_{12})$, $C_o = C_p/2$; (e) crystal filter circuit.

For wider-bandwidth requirements the intermediate-bandwidth design process may be used. In this process tuned circuits are placed across the junction capacitors and the terminations to effectively tune out all of the capacitance reflected from the shunt capacitance of the crystals. For example, in the four-pole design of Figure 7e, a tuned circuit would be added across the center capacitor C_{23}. Tuned circuits would not be required at the input and output because the transformers can be tuned to absorb the extra capacitance. This design is relatively stable with temperature because the resonance frequency of the crystals is symmetrically located on either side of the center frequency. As a result, the bandwidth is defined by the crystal frequency stability, and slight mistuning of the inductors only causes a change in the passband ripple. However, there is a loss in selectivity for this type of design. In effect, the coupling capacitor C_{23} is set equal to zero. By examining the circuit of Figure 7a it can be seen that with $C_{23} = 0$, the two center resonators

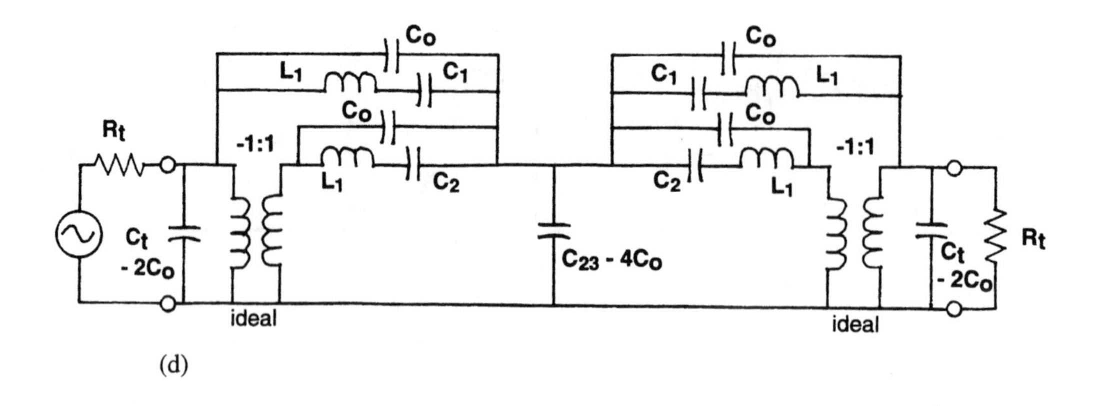

(d)

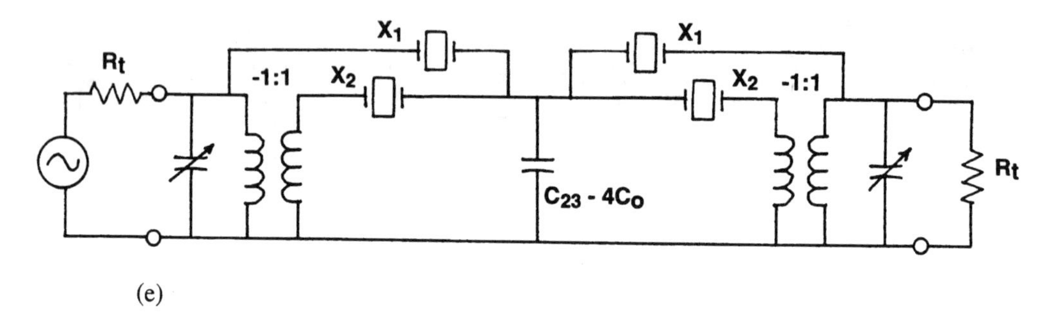

(e)

Figure 7 (continued)

combine to form a single resonator resulting in a three-pole filter design. Effectively, a pole is lost for each internal tuned circuit. To minimize this effect, intermediate filters are often designed with more than two crystals per section. For example, an eight-crystal design using two cascaded four-pole sections provides a seven-pole response. Filter sections with more than two crystals are derived by following the same procedure as outlined in Figure 7. For example, the circuit of Figure 7a is symmetrical about the center capacitor C_{23} and this circuit could be converted into a single four-crystal filter section using Bartlett's bisection theorem.

A third filter design type, the so-called wideband design, combines crystals with L/C resonators. This type of design is useful for bandwidths of approximately 2% of center frequency or greater, but is limited to a very small frequency range where crystals with very low capacitance ratios are available. A major application for this design was in carrier telephone FDM filters operating in the vicinity of 100 kHz. This technique is rarely used with AT-cut-type crystals.

A useful design procedure for generating transmission zeros is illustrated in Figure 8. In this process, a symmetrical two-pole filter section is bridged with a three-element capacitor section containing one real capacitor and two negative capacitive elements. In the final design realization, the negative capacitances are absorbed into coupling capacitors or into capacitance provided at the terminations by a process similar to that shown in Figure 7b. This procedure can be used in a cascaded design to provide any number of symmetrically located zeros up to a maximum equal to the number of poles. The addition of the capacitive elements causes some distortion in the passband ripple, but this can be compensated by small adjustments of the termination impedance and coupling element values.

A critical component in discrete-crystal filters is the RF transformer which is required for the hybrid lattice circuit. This "hybrid transformer" must be linear and stable and tightly balanced. At frequencies above 1 MHz these transformers are typically bifilar wound on powdered iron, toroid cores. Care must be taken to reduce the distributed capacitance as much as possible by carefully spreading the windings on the core. Because the filters are typically fairly high impedance circuits,

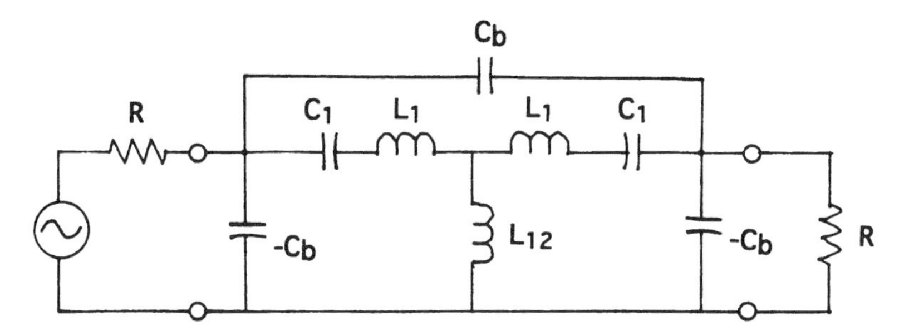

Figure 8 Two-pole, two-zero filter section.

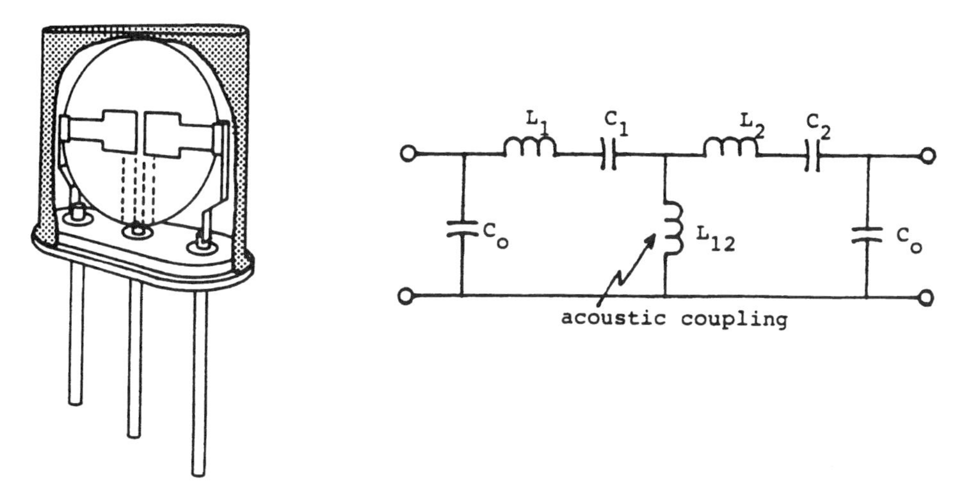

Figure 9 Typical mechanical construction for a two-pole monolithic crystal filter and its electrical equivalent circuit.

the losses in the transformers must be kept to an absolute minimum. Most discrete-crystal filters are packaged in hermetically sealed cases primarily to protect the transformers from moisture and dirt.

Many techniques are available for meeting specialized filter requirements which are beyond the scope of this discussion. For example, lattice designs can be skewed to one side or the other to provide single-sideband-type responses. Also, a variety of methods has been developed for providing linear-phase filters including the incorporation of extra resonators in the lattice structure and the use of crystal, all-pass, equalizer sections which can be cascaded with standard filters. Most of these procedures are unique to crystal filter design because of the requirement to incorporate the crystal equivalent circuit into the filter configuration in an efficient manner.

4. MONOLITHIC CRYSTAL FILTERS

An example of a coupled-resonator, or monolithic crystal filter, is illustrated in Figure 9. A monolithic crystal filter contains two or more resonators on a single quartz wafer which are acoustically coupled to each other. A typical two-pole device with its equivalent circuit is shown in the illustration. From the equivalent circuit it can be seen that the monolithic filter provides a complete two-pole filter with no additional components. As a result, the monolithic filter is considerably less costly and smaller than an equivalent discrete filter, which would require two individual crystals plus an RF transformer with tuning capacitors. Conceptually, the monolithic filter could be built over a wide-bandwidth range. However, there are physical limits on the maximum amount of coupling which can be achieved between resonators and, in general, the bandwidth limits for monolithic filters fall in the same range as for narrowband discrete filters. The range of bandwidths for typical monolithic filters is shown in the bandwidth-frequency chart at the beginning of this chapter.

Two-pole monolithic filters can be easily cascaded to produce higher-order filters. For example, if the circuit of Figure 7a is modified by changing C_{12} and C_{34} to their equivalent inductance values and then modified as in Figure 7b, two identical two-pole devices can be substituted directly. The only extra components which may be required are shunt capacitors at the center junction and at the terminations. This process can be carried out for six- and eight-pole filters also, but more than one type of two-pole device will be required. Because of their low cost and small size, monolithic crystal filters are widely used to provide IF selectivity in communications receivers.

Special monolithic filters have been developed which include up to eight resonators on a single wafer. However, these designs require a substantial tooling effort and are typically manufactured on dedicated equipment because of the complexities in tuning a large number of coupled resonators. In general, the capital investment required for this type of device is very high and can only be justified for high-volume specific applications.

5. APPLICATIONS AND USE

The major application for crystal filters is to provide narrowband IF selectivity in various communications receivers. Typical receivers use anywhere from six to ten poles of selectivity which is supplied with all crystal filters or with a combination of crystal and ceramic filters. Probably the largest single usage is in cellular telephones, which use a deviation of ±12 kHz with typical bandwidths of 30 kHz. Common IF frequencies are 45 MHz and higher frequencies in the 80 to 90 MHz region. Conventional two-way FM radios use a deviation of ±5 kHz with bandwidths of 12 to 13 kHz and one half of those values for split-channel applications. Common IF frequencies are 10.7 and 21.4 MHz, although newer radios are using higher IF frequencies in the 70 to 112 MHz region. Most of these radios incorporate monolithic crystal filters. Many paging receivers also use crystal IF filters with requirements similar to those for conventional two-way FM equipment. Both U.S. cellular and two-way FM systems are now changing from analog to digital modulation using multichannel, time-division-multiplex (TDM) techniques. However, the same bandwidths are being used so the IF filtering requirements have stayed essentially the same. Other newer digital cellular telephone systems, such as the European GSM and CT-2 personal phones, use a much wider channel bandwidth and selectivity is being provided by SAW-type filters.

Crystal filters are commonly used in single-sideband radio exciter applications for selecting the desired sideband. These filters typically are 2 to 3.5 kHz wide and have been built at center frequencies ranging from 1 to 10 MHz. Most designs provide six to eight poles of selectivity.

Another major application for crystal filters has been in channel filter banks for FDM carrier telephone equipment. These filter needs provided the incentive for the early work by Mason which resulted in the development of the narrowband and wideband crystal filter design concepts in the mid 1930s. The original filters were wideband designs with center frequencies in the 60 to 108 kHz frequency range. In the 1970s most of these systems moved higher in frequency to the 5 to 8 MHz range. A unique monolithic filter was developed for this application by Bell Telephone which provided eight poles of selectivity on a single quartz wafer. Today, most of these systems have moved to digital modulation schemes using TDM techniques.

Other applications for crystal filters include narrowband filtering and filter banks for use in a variety of test equipment such as spectrum analyzers and vector voltmeters. Some specialized filters are used in data transmission systems where very flat time delays are required. Most of these applications are relatively low quantity and are ideally suited to discrete-crystal filter designs with their low up-front tooling costs.

Crystal filters are typically high-impedance networks and are sensitive to the termination circuits in which they operate. If the filters are matched directly at their natural impedance, capacitance adjustments may be required for optimum performance. In packaged filter assemblies, matching networks can be included which allow termination at lower impedance levels.

Crystal resonators are sensitive to shock and vibration. A variety of crystal mounting arrangements can be used, and it is essential that the crystal design be optimized for the specific requirement in high shock and vibration applications. Also, because crystals are piezoelectric, they can generate electrical signals when they are vibrated. This situation can cause on-channel signals to be created which can cause problems if the filter is in a very sensitive part of the receiver. A special problem for lower frequency filters (below 1 MHz) is a sensitivity to signal drive level. If too strong a signal is applied, the crystals will shatter. For filters using the higher frequency AT-cut crystals, this is not a problem.

6. FOR FURTHER READING

With the exception of Mason's landmark paper, very little information was published on crystal filter design until the late 1950s. Since that time a primary source of information in this area has been the *Proceedings of the Annual Symposium on Frequency Control.* An excellent source for information on quartz crystals is Salt's *Handbook of Quartz Crystal Devices.*[7] For more-detailed information on the design of crystal filters, please refer to Kinsman's *Crystal Filters.*[8]

REFERENCES

1. **Cady, W. G.,** The piezo-electric resonator, *Proc. IRE,* 10, 83, 1922.
2. **Mason, W. P.,** Electrical wave filters employing quartz crystals as elements, *Bell Syst. Tech. J.,* 13, 405, 1934.
3. **Storch, L.,** U.S. Patent 2,980,872, 1961, *Bandpass Filters.*
4. **Sykes, R. A. and Beaver, W. D.,** High frequency monolithic crystal filters with application to single frequency and single sideband use, *Proc. 20th Annual Frequency Control Symposium,* Atlantic City, N.J., 288, 1966.
5. **Onoe, M., Jumonji, H., and Kobori, N.,** High frequency crystal filters employing multiple mode resonators vibrating in trapped energy modes, *Proc. 20th Annual Frequency Control Symposium,* Atlantic City, N.J., 266, 1966.
6. **Bechmann, R.,** Frequency-temperature-angle characteristics of AT-type resonators made of natural and synthetic quartz, *Proc. IRE,* 44, 1600, 1956.
7. **Salt, D.,** *Handbook of Quartz Crystal Devices,* Von Nostrand Reinhold, Berkshire, England, 1987.
8. **Kinsman, R. G.,** *Crystal Filters,* Wiley-Interscience, New York, 1987.

Mechanical Filters

Robert A. Johnson

CONTENTS

1. INTRODUCTION

Mechanical filters have been manufactured since the early 1950s primarily for use in high-performance HF radios and frequency-division multiplex (FDM) telephone equipment. Other uses have been in applications such as train control, navigation, sonar, direction finding, test equipment, and

0-8493-8951-8/97/$0.00+$.50
© 1997 by CRC Press LLC

more recently as antialiasing or final IF filters in digital and analog radios operating in the VHF/UHF range. The first volume production of mechanical filters for radios was carried out by Collins Radio Company in the early 1950s followed by companies in Japan and Germany in the late 1950s. The earliest voice-channel filters for telephone systems were built by Collins Radio in the late 1950s.[1] A decade and a half later, most of the world major manufacturers of telephone equipment began building mechanical filters for FDM systems.[2] The primary exception was Western Electric which continued to manufacture voice-channel monolithic crystal filters. Why were mechanical filters so popular during the final decade of the life of FDM telephone systems?

The popularity of FDM mechanical filters was because of their high performance, small size, and reliability. The monolithic crystal filter could match the mechanical filter for size, cost, and reliability, but in the case of realizing filters that meet international specifications, only the mechanical filter and lumped element crystal filters could meet those specifications. The most popular of the FDM mechanical filters were the torsional resonator types that had a carrier frequency of 128 kHz. Torsional resonator filters were first manufactured in Germany in the late 1950s. These filters, which operate at frequencies from 100 to 500 kHz, have continued to be popular for use in HF and other high-performance radios, where they compete with crystal and ceramic filters. In radio applications how do these filters compare? Generally speaking, for equivalent performance, the cost to manufacture a mechanical filter is roughly the same as that of a crystal filter. Therefore, the choice of technology is often made on the basis of the desired center frequency and allowable filter dimensions. Because of the higher Q and greater stability of mechanical filter resonators, the performance of mechanical filters is usually better than that of ceramic filters, but the manufacturing cost is greater. The wider the bandwidth, the less performance advantage the mechanical filter has over the ceramic filter.

2. HOW THE MECHANICAL FILTER WORKS

Mechanical filters are designed with various transducers, resonators, means of mechanical coupling, and spatial configurations. Rather than attempt to describe all of these varieties of mechanical filters, we will concentrate on the most widely used configuration, the torsional ladder filter.[3]

Figure 1 shows a seven-resonator torsional mechanical filter. At each end is a transducer resonator and inbetween are five metal resonators. Each resonator rod is supported by small-diameter wires at nodes, i.e., at points of no motion. The transducer resonators and the interior resonators are connected by resistance welding the coupling wires to the rods. The filter is then assembled on the printed circuit board by soldering the support pins to the through-plated holes of the board. The board is then attached with solder to the terminals of the metal base thus completing the assembly. To understand how the mechanical filter operates, let us trace a signal from the input to the output.

An electrical signal is applied to the input pair of terminals shown on the right side of Figure 1. This produces an electric field across the torsional piezoelectric ceramic, which causes the ceramic, as well as the entire transducer resonator, to vibrate in torsion. The mechanical vibration is coupled from resonator to resonator by means of the coupling wires. Torsional vibration within the output transducer resonator produces a voltage across the output terminals of the filter. The steady state vibrations of the filter are quite complex, as in the case of an LC bandpass filter, and involve acoustic waves flowing back and forth between the transducers.

2.1 ELECTRICAL EQUIVALENT CIRCUIT

A second way to understand the operation of a mechanical filter is through the use of an electrical equivalent circuit. The circuit shown in Figure 2 is based on the mobility analogy discussed in Chapter 8.1. This is a symmetrical five-resonator filter (rather than the seven-resonator filter of Figure 1), where the first and last resonators correspond to the torsional transducer resonators of the mechanical filter. The center three resonators correspond to the interior metal alloy resonators.

The external source and load resistors of Figure 2 are shunted by fixed or parasitic values of electrical capacitance C_p. The mechanical filter itself begins with C_0 which is the capacitance between the transducer electrodes, as shown in Figure 2b of Chapter 8.1. Next is the parallel-tuned

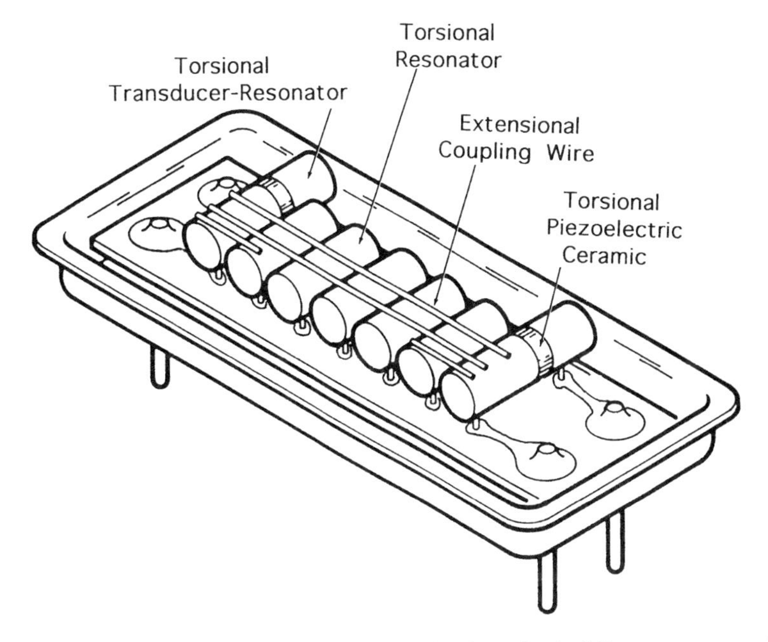

Figure 1 Seven-resonator torsional mechanical filter.

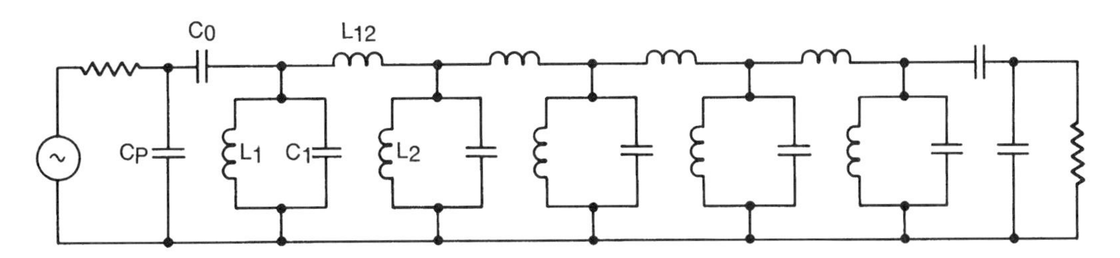

Figure 2 Electrical equivalent circuit of a five-resonator torsional mechanical filter based on the mobility analogy.

circuit representing the transducer resonator followed by the series-arm inductance, L_{12}, which is analogous to the compliance (inverse of stiffness) of the coupling wires between the first two resonators. Next is the first interior resonator, and so on.

The electrical equivalent circuit is directly related to the lumped element representation of the mechanical circuit. The mechanical resonators and coupling wires are actually mechanical transmission lines, but these transmission lines can be represented by lumped springs (inductors) and masses (capacitors) in the case of filter bandwidths that are less than 10% of the center frequency.

The electrical equivalent circuit is helpful in both understanding the operation of the mechanical filter and as an analysis tool for calculating the frequency- and time-domain responses of an actual filter.

2.2 MECHANICAL FILTER COMPONENTS

In this section we will discuss the principles of operation of the elements that compose a mechanical filter. The basic theory of resonators, transducers, and coupling wires is described in greater detail in Reference 4.

2.2.1 Mechanical Filter Resonators

Torsional-type filter elements will be used to illustrate various mechanical resonator principles. Figure 3 shows two mechanical resonators; (a) is a half-wavelength line (first-mode); and (b) is a full-wavelength line (second-mode). The first-mode resonator has a single nodal plane bisecting the resonator. Moving from the nodal plane in either direction, the amplitude of vibration is sinusoidal. A support wire is welded at the node. At that contact point, the wire has no effect on

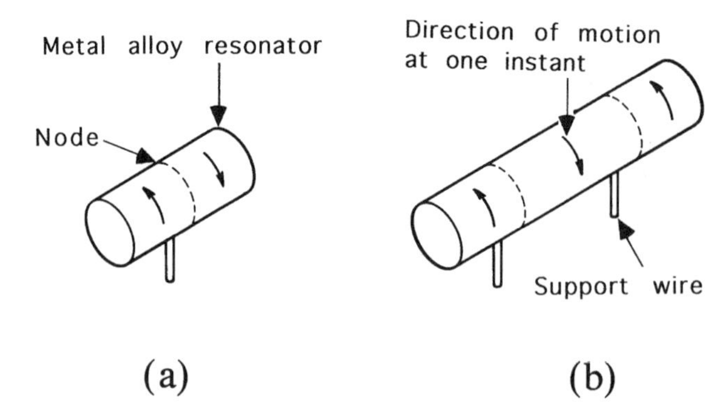

Figure 3 Torsional mechanical resonators. (a) First (fundamental) mode, (b) second mode (full wavelength).

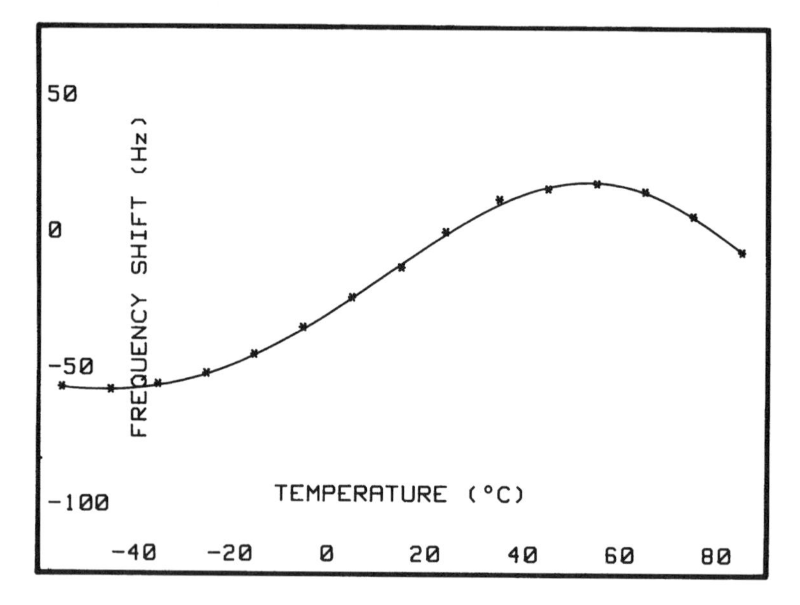

Figure 4 Resonance frequency shift vs. temperature. The frequency at 20°C is 455 kHz. Metal is Thermelast 4290 Fe–Ni–Be–Mo alloy.

the Q or resonance frequency of the resonator. This wire acts as a support during the resonator tuning process, before the filter assembly. The most common form of tuning is by laser, where material is removed from the end of the resonator increasing its frequency. After welding the resonators together with coupling wires, the support wire or wires act as supports for the filter structure as a whole.

Figure 3b shows a full-wavelength resonator that has two nodal planes and two support wires. An advantage of this configuration is that the two support wires provide a more rigid physical structure to enable the filter to withstand high shock levels of 100 *g* or more for 11 millisec.

The mechanical resonators are made of iron–nickel alloys.[5] These various alloys can be heat treated to obtain the frequency vs. operating temperature behavior shown in Figure 4. The resonance frequency is near 455 kHz. Varying the heat treatment temperature causes the cubic-function-shaped curve to rotate, thus, if necessary, providing optimum frequency shift characteristics at either high or low temperatures. Resonators made from this particular material have Q values of almost 20,000.

The length of the resonator is fixed by its resonance frequency and the velocity of torsional waves in the metal. The diameter is usually chosen to avoid flexural resonances near the filter passband and to obtain a minimum package height.

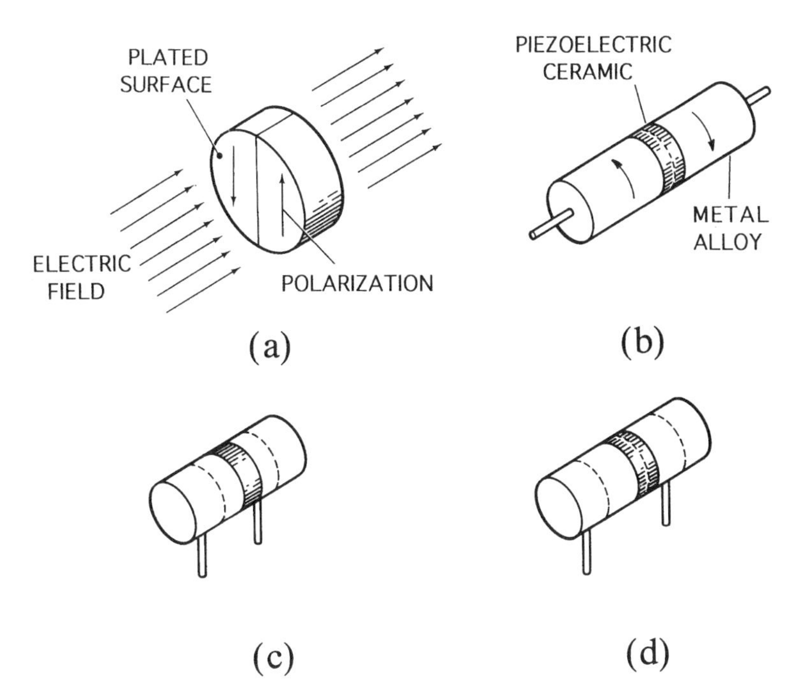

Figure 5 Torsional-mode transducers. (a) Piezoelectric ceramic, (b) first-mode composite transducer resonator with end supports, (c) second-mode resonator, and (d) third-mode resonator. (Adapted from Domino, W. J., *Proc. 1988 Ultrason. Symp.*, 245, 1988. With permission.)

We have only considered the torsional resonator, although resonators with various other modes of vibration are possible. For example, flexure-mode bars are used at the low frequencies, typically below 200 kHz and as low as 2 to 3 kHz. Fork resonators are realizable at frequencies as low as a few hundred Hz. At popular frequencies, such as 450, 455, and 500 kHz, early mechanical filters were designed with flexure-mode discs and some filters with extension-mode rods. Some of the resonator modes and equations that are shown in Figures 6 and 7 of Chapter 8.4 on ceramic filters also apply to mechanical filters.

2.2.2 Transducer Resonators

Transducer resonators employ various modes of vibration and structural configurations. In this section we will concentrate on the torsional-mode design of Figure 1 that employs a piezoelectric disc transducer. The disc is built out of two half-discs polarized in opposite directions, as shown in Figure 5a. The two major surfaces of the epoxy-bonded ceramics are gold or silver plated to provide electrodes and a means to solder bond to the metal rods on either side of the disc. When an electric field is applied in a direction perpendicular to the polarization field, a thickness-shear mode is introduced. This causes the two major surfaces to rotate in opposite directions relative to one another, which in turn generates a torsional vibration of the composite metal-ceramic transducer resonator.

Maximum electromechanical coupling is obtained when the ceramic disc is located at a nodal point of the composite resonator. Locating the ceramic disc in the center of the first-mode resonator shown in Figure 5b creates the problem of not being able to support the resonator with a support wire at the nodal point since it is not practical to attach a metal wire to the ceramic disc. One solution is to support the resonator at the center of the end surface where there is little motion. Another solution is to move the ceramic off of the node in the direction of one of the ends. Although this lowers the electromechanical coupling, k_{em}, the frequency stability with time and temperature is improved. For very narrow filters, where k_{em} can be low and a high degree of stability is needed, the ceramic can be attached to the end of the resonator. Figure 5c shows a second-mode resonator

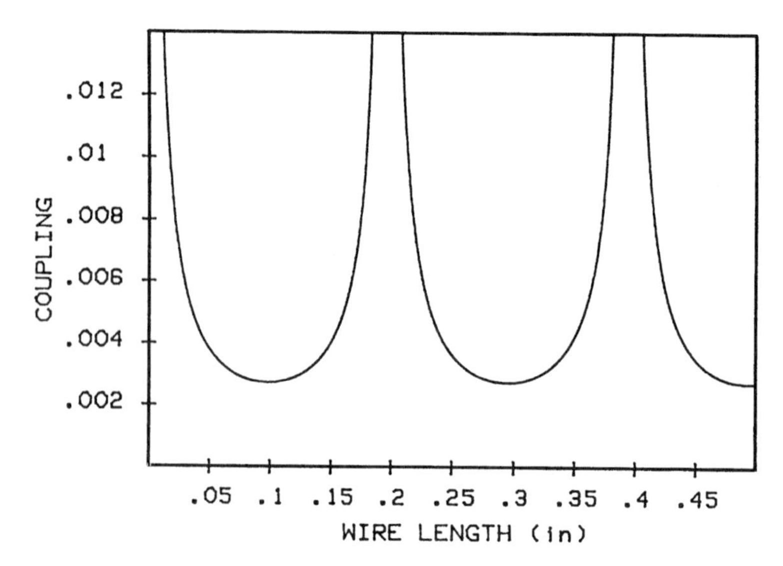

Figure 6 Typical extensional-mode coupling vs. wire length curve at 455 kHz. (From Domino, W. J., *Proc. 1988 Ultrason. Symp.*, 245, 1988. With permission.)

with the ceramic in the center. This is a fine configuration for supporting the resonator, but the k_{em} is zero. To increase the k_{em}, the ceramic must be moved toward one of the nodes.

Figure 5d shows a transducer resonator of the type shown in Figure 1. The third mode of vibration is used in a configuration where the resonator is supported at the two outside nodes and the ceramic is placed at the center node for maximum coupling. If greater stability is needed, the ceramic is moved off of the node. Placing the ceramic at the center of the resonator makes it possible to realize filter bandwidths as wide as 6 kHz, at a center frequency of 455 kHz without having to use coils on the input and output of the filter. Wider bandwidths can be achieved by using a disc polarized in the thickness direction sandwiched between two metal rods. This configuration, which is called a Langevin transducer, vibrates in an extensional (sometimes called longitudinal) mode. Another high-coupling transducer is composed of two plates or half-discs which are polarized in opposite thickness directions causing the resonator to vibrate in a flexure mode. Low-frequency flexure-mode transducer resonators of the bar-and-fork types usually employ a piezoelectric ceramic that is bonded to one of the major surfaces of the metal bar or tyne.

2.2.3 Coupling Wires

The extensional mode of vibration is used to couple the resonators of the filter of Figure 1. The coupling wire or wires are mechanical transmission lines like the torsional or extensional-mode resonators. A typical curve of coupling (between resonators) vs. coupling wire length is shown in Figure 6. This curve is for a typical iron–nickel alloy and a frequency of 455 kHz. When the wire length is short or when the wire length is equal to an integer multiple of half-wavelengths, the coupling between resonators is very high. When the wire length is an odd number of quarter-wavelengths the coupling is a minimum and has zero sensitivity to variations of wire length.

Not shown in the curves are the effects of flexural modes of vibration. Although these modes are not strongly driven, they will affect the filter response if their resonance frequencies are in or near the filter passband. Therefore, the wire length and diameter must be chosen so that these modes are equidistant above and below the filter passband.

Besides extensional-mode coupling, resonators can be coupled by wires that vibrate in torsional and flexural modes. Most low-frequency flexural-mode resonators are coupled in torsion, although fork filter resonators have been coupled by wires vibrating in flexure as have extensional-mode resonators at higher frequencies.

2.2.4 Tasks Performed by the Mechanical Filter Components

In this section, the filtering function of each of the components of the mechanical filter will be briefly described.

The task of the resonators is to

1. Establish the center frequency;
2. Determine the shape factor (the ratio of the 60-dB bandwidth to the 3-dB bandwidth), which is a function of the number of resonators.

The transducers perform the following tasks:

1. Conversion of electrical to mechanical energy and mechanical to electrical energy;
2. Bringing the electrical terminating resistances into the mechanical circuit providing termination for the filter;
3. As the first and last resonators, providing selectivity; and
4. When electrically tuned, the electrical elements provide additional selectivity.

The coupling elements, which are usually wires, determine

1. The 3-dB bandwidth, which is defined by the coupling between the center resonators; and
2. The passband shape (maximally flat, equal ripple, etc.), which is determined by the relative couplings between resonators.

3. MECHANICAL FILTER CHARACTERISTICS

As discussed earlier, the mechanical filter can be modeled as a simple lumped element LC ladder network with resistance in each tuned circuit to account for the finite resonator Q. The model can be used to show frequency-domain or time-domain responses for ideal conditions, temperature extremes, and for Monte Carlo analyses. In this sense, the mechanical filter is no different from crystal, ceramic, and LC filters. But, there are second-order effects that must be accounted for, and these will be discussed in the following sections.

3.1 FREQUENCY RESPONSE CHARACTERISTICS

Figure 7 shows the actual amplitude and group delay response of a 500-Hz-wide mechanical filter with a center frequency of 455 kHz. This is a seven-resonator torsional filter like that shown in Figure 1. The frequency response is monotonic with the response in the stopband reaching an attenuation of roughly 100 dB below the passband level (when test equipment noise is eliminated). Figure 8 shows the frequency response of a 12-resonator torsional mechanical filter. Note that there is a "ledge" starting at approximately 75 dB. This is due to some of the input signal in the input transducer resonator being fed mechanically through the printed circuit board to the output transducer. In this case, the passband-to-stopband attenuation difference ultimately was still greater than 90 dB. A typical frequency response of an eight-resonator filter over a wide frequency range is shown in Figure 9. Care was taken to reduce the test circuit noise level so the ultimate attenuation of close to 100 dB can be shown.

3.1.1 Spurious Responses

The frequency response of a torsional single-sideband (SSB) filter shown in Figure 9 shows the spurious responses due to modes of vibration of the filter resonators and coupling wires. The first strong mode is near 150 kHz. This is the fundamental-mode resonance of the input and output transducer resonators, which at 455 kHz vibrate in the third mode shown in Figure 5d. In this case, because the piezoelectric ceramic is centered in the resonator, the second-mode response at 300 kHz is not excited and therefore not seen. In the case of radio designs, the spurious responses shown in Figure 9 are usually not a problem because other filtering takes place in preceding IF stages. But, in some systems this is not the case. Spurious responses are not unique to mechanical

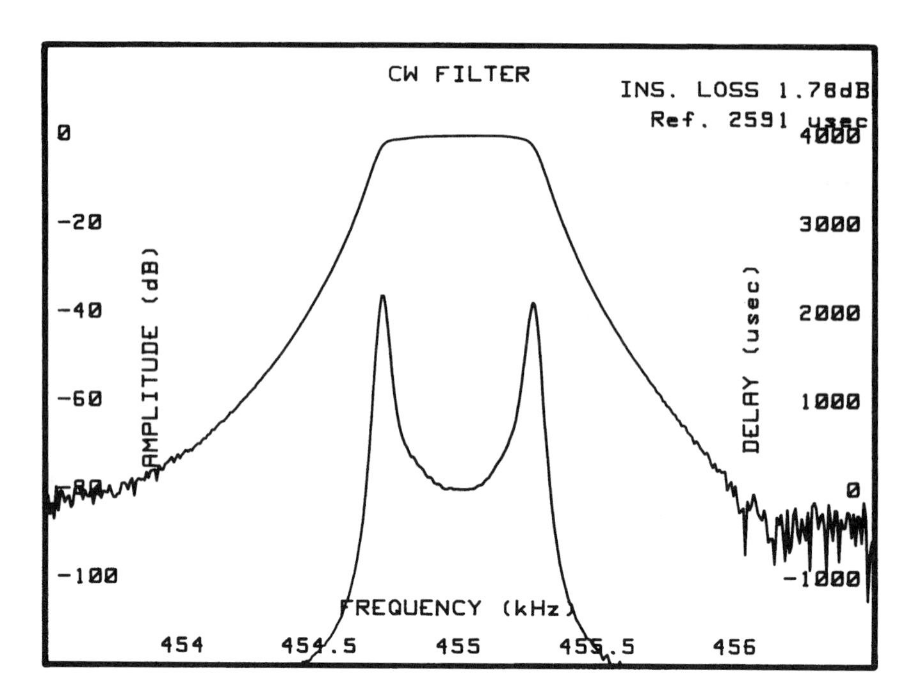

Figure 7 Amplitude and group delay response of a 500-Hz bandwidth torsional mechanical filter, of the type shown in Figure 1. (From Domino, W. J. and Johnson, R. A., *RF Design,* 10, 49, 1991. With permission.)

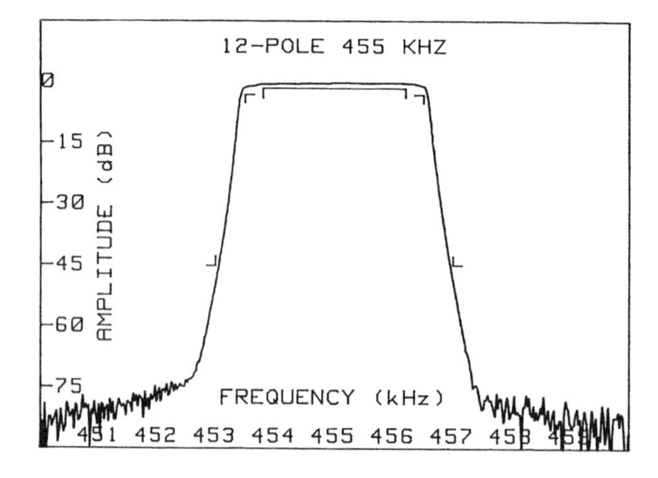

Figure 8 Amplitude vs. frequency response of a 12-resonator torsional mechanical filter. (Courtesy of Rockwell International.)

filters but are present in all filters that utilize distributed resonator and coupling elements. Elimination and control of spurious responses is one of the principal tasks of mechanical filter design.

3.1.2 Group Delay

Returning to Figure 7, we see that the group delay curve in the passband has a typical bandpass filter parabolic shape. Since this is a simple monotonic ladder filter, the delay is directly related to the amplitude response and therefore related to the coupling between resonators. Any attempt to flatten the delay response results in a rounded amplitude response. A means of getting around this dilemma is to add a bridging wire across two resonators (for example, from resonator one to resonator four), which adds a delay peak in the center of the filter passband. The result is as if a delay equalizer section were added to the frequency response. The bridging wire must be less than one-half-wavelength long or a length (or positioned on the resonators) such that the bridging is "in

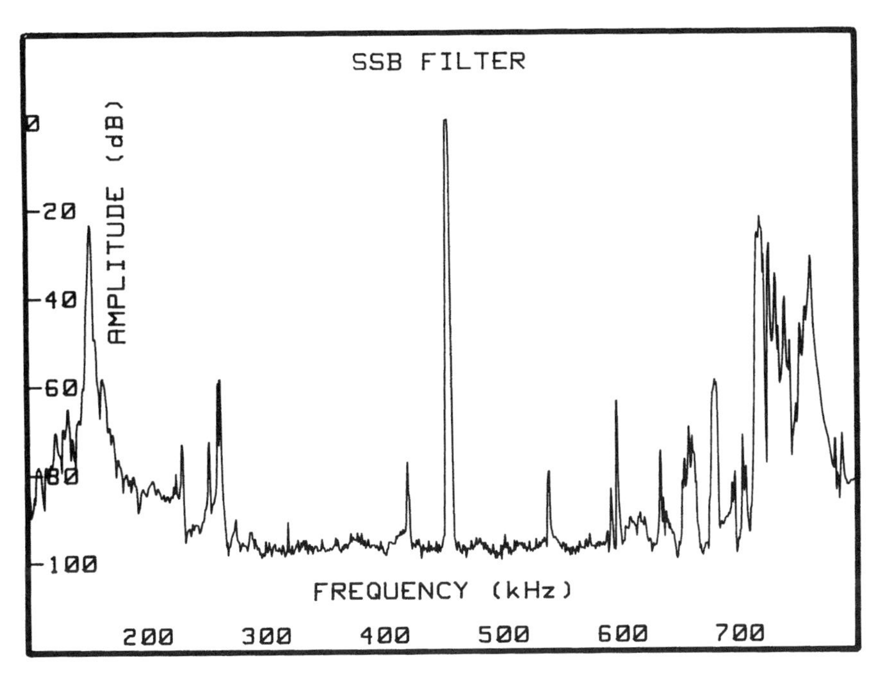

Figure 9 Typical spurious response characteristics of a 2.5-kHz bandwidth torsional mechanical filter. (From Domino, W. J. and Johnson, R. A., *RF Design,* 10, 49, 1991. With permission.)

phase." Out-of-phase bridging results in attenuation poles both above and below the filter passband, as shown in Topology 7 of Figure 5 of Chapter 8.1.

3.1.3 Attenuation Poles

Figure 5 in Chapter 8.1 shows various configurations for realizing attenuation poles (transmission zeros) in mechanical filters. The purpose of attenuation poles in the filter stopband is to increase the filter selectivity near the passband without having to increase the number of resonators and the resulting group delay. Using out-of-phase bridging across a single resonator results in an attenuation on the low side of the filter passband. In-phase bridging across a single resonator realizes an attenuation pole above the filter passband, as is shown in Topology 6 of Figure 5 in Chapter 8.1. The subject of attenuation-pole realizations is described in detail in Reference 4.

3.2 CHARACTERISTICS DEPENDENT ON THE SIGNAL LEVEL

3.2.1 Dynamic Range

To this point, we have considered the relative passband and stopband amplitudes of a mechanical filter response. Next, let us consider the subject of dynamic range. With regard to the high-voltage end of the dynamic range, the issues are the possible damage to the filter as well as issues of a reversible nature, such as nonlinearities in amplitude, phase, and the related issue of intermodulation distortion. Because all filter designs are different, only rough generalizations can be provided. Damage to the filter turns out to not be an issue. Short-term signal levels of up to 20 V will not result in permanent damage to most mechanical filters. With regard to problems of a reversible nature, let us start with amplitude. Most mechanical filters will withstand signal levels of 1.0 V without substantial amplitude distortion. What distortion there is takes the form of a flattening of the passband amplitude response and a frequency shift of the passband due to resonator frequency shifts. Phase shift due to amplitude changes is a problem in some systems and must be considered when operating in the 0.1 to 1.0 V range of input signals. At signals below 0.1 V the only noticeable effects are related to intermodulation distortion.

The low end of the dynamic range is some value smaller than 130 dB below 1.0 V. Since the mechanical filter is a passive device, there is little noise generated internally, so noise is not a

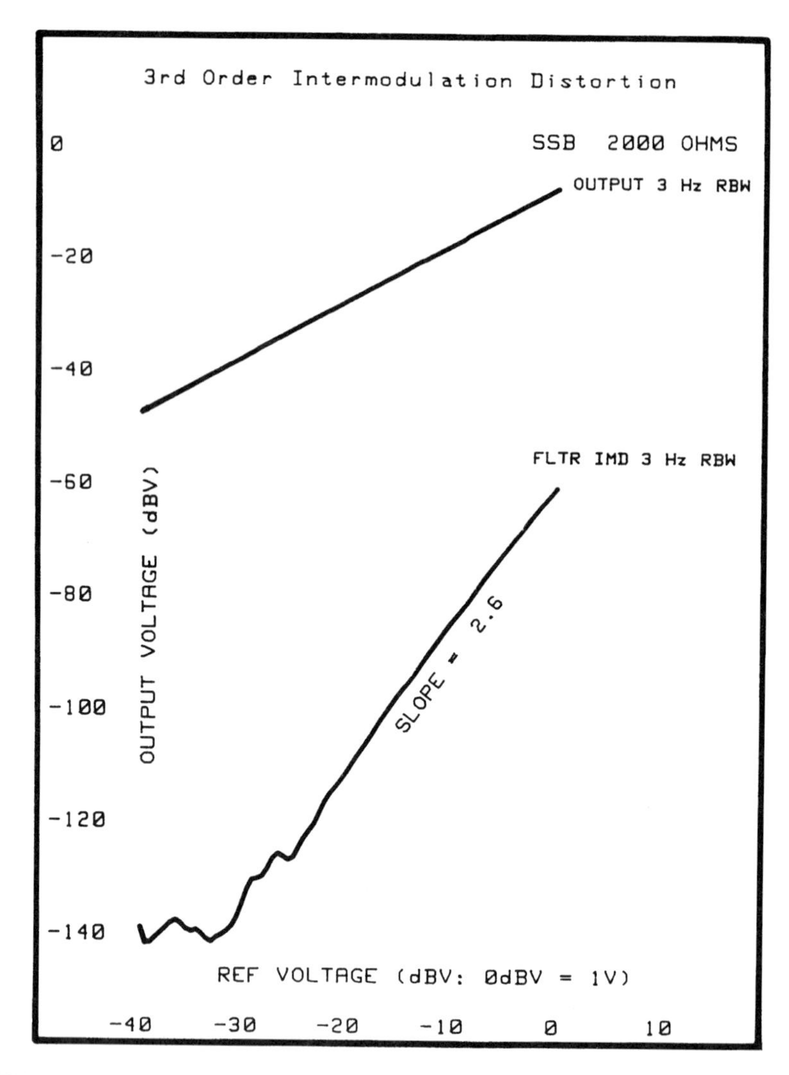

Figure 10 Third-order intermodulation distortion of a 2.5-kHz SSB torsional mechanical filter. (From Domino, W. J. and Johnson, R. A., *RF Design,* 10, 49, 1991. With permission.)

problem. Torsional mechanical filters operate well at low signal levels, but it is difficult to make measurements below –130 dB with standard test equipment in a factory or even in a normal laboratory environment. Because of the higher slope of the third-order intermodulation distortion products, this is not an issue at very low signal levels.

3.2.2 Intermodulation Distortion

Because a mechanical filter, like all filters, is not a perfectly linear device, two signals present at the input of the filter create a third signal, the so-called third-order intermodulation distortion product, or, in short, third-order IMD. Figure 10 shows typical in-band IMD curves for a 3-kHz bandwidth torsional mechanical filter. By *in-band* is meant that both of the input signals are inside the filter passband. The product of signals out of the passband is considerably smaller. The terminating impedance of the filter is 2 kΩ, and the reference voltage is the voltage on the generator side of the source resistor. The slope of the third-order product is 2.6, not 3.0, as would be expected in an LC bandpass filter. This slope, with a value below 3.0, is typical of all electromechanical filters. Until meeting with the noise at the lower end of the curve, the slope of the third-order curve is quite linear. This is not always the case. All electromechanical filters have the potential of erratic slopes over regions of the third-order curve. The higher the frequency, the more common is this effect. At 455 kHz these anomalies are the exception and are usually due to a particle attached to a resonator after tuning. The output level difference between the first- and third-order curves is directly related to the filter bandwidth. The wider the bandwidth, the greater is the spread between the two curves.

4. USING MECHANICAL FILTERS

In this section we will concentrate on the unique features of mechanical filters regarding their use in electronic equipment. For example, because the mechanical filter is a passive device and can operate at very low levels, the filter is usually located as close to the front end of the equipment circuitry as possible. In some low-frequency systems, this may be at the antenna or just after a low gain preamp lifter. In an IF circuit, the mechanical filter usually follows the mixer.

4.1 THE FILTER IN ITS ELECTRICAL ENVIRONMENT

In this section we will look at the mechanical filter and its relationship with surrounding electrical circuits.

4.1.1 Termination

As with all bandpass filters, proper termination is of primary importance. The filter manufacturer specifies a terminating resistance and possibly a value of capacitance or inductance. This is the value of resistance with which the filter is to be terminated, not, for example, the value of the input impedance of the filter. Terminating a mechanical filter is usually not difficult, because small changes in parasitics have little effect on the response. In this way, the mechanical filter is closer to a ceramic filter than most crystal and SAW filters. A typical torsional mechanical filter terminates in a resistance on the order of 2.0 kΩ and requires a shunt capacitor of 30 pF. Although the resistance should be held to 5% of its specified value, the capacitance can vary 10 to 20% when it is not being used to tune any of the mechanical filter elements. For wideband mechanical filters that use or require coils, the tolerance on the capacitance is often tighter. Older disc-resonator filters often operated into impedances as high as 100 kΩ and required capacitors to tune internal coils. These filters were sensitive to capacitance variations. Low-frequency mechanical filters of the bar-and-fork types operate at high impedance levels of 10 kΩ or more, but do not require electrical tuning.

4.1.2 Shielding and Grounding

The effects of poor shielding or poor grounding are the same; some of the signal gets around the filter and causes the stopband rejection to be reduced. The effect is much like the ledge shown in Figure 8, which is caused by signal energy bypassing all but the end resonators. Because the mechanical filter is not a high-frequency device, normal care in both shielding and grounding should prevent the signal from bypassing the filter. When operating at very low signal levels, care should be taken to guarantee that the output of the filter is not fed back to the input, or large variations in the passband frequency response will occur. In the case of some of the older mechanical filters that employed internal coils and magnets, it was necessary to keep the filters from large magnetic fields or from one another. This is not the case with the torsional types that use piezoelectric ceramic transducers.

4.2 THE FILTER IN ITS PHYSICAL ENVIRONMENT

In this section we will look at the mechanical filter in its thermal and shock and vibration environment.

4.2.1 Temperature Effects

The two major temperature effects are the reversible and irreversible changes in filter performance. In almost all cases, mechanical filters will withstand storage temperatures ranging from –55 to

+100°C without damage. An exception is when the filter is not hermetically sealed and is subjected to extreme humidity. In that case, the filter must be protected from the moisture by post coating.

Reversible effects on the filter response are primarily due to frequency shifts of the resonators, changes in capacitance C_0, that cause an apparent shift in the transducer resonator frequency, and changes in k_{em}, that cause terminating resistance-like changes. Each of these changes affects the frequency response. Since each filter design is different, all of these effects are considered by the filter designer in order to meet an agreed-upon specification.

4.2.2 Shock and Vibration

The two major effects of shock and vibration are damage to the part and microphonic responses. In general, mechanical filters are quite rugged and will meet a 50 to 100 *g* shock level without damage. But, there are many exceptions, and the filter user should discuss this issue with the filter designer. Regarding microphonic outputs due to vibration, this is usually not an issue for filters that operate at frequencies above 100 kHz. Low-frequency flexure-mode bar or fork filters are microphonic and care must be taken to either (1) reduce the vibration level the resonator sees or (2) use a highpass filter following the mechanical filter when the external vibration frequency is lower than the filter center frequency.

5. FOR FURTHER READING

In the brief space allowed in this chapter, we have concentrated on only one mechanical filter type, namely the torsional mechanical filter. A broader range of filter types can be found in References 4, 5, and 6. These references, as well as Reference 1, also provide a wide-ranging history of the mechanical filter. A more detailed commentary on the use of torsional mechanical filters in analog and digital radios is found in Reference 7.

REFERENCES

1. **Hathaway, J. C. and Babcock, D. F.,** Survey of mechanical filters and their applications, *Proc. IRE,* 45, 5, 1957.
2. **Yano, T., Futami, T. and Kanazawa, S.,** New torsional mode electromechanical channel filter, *1974 European Conf. on Circuit Theory and Design,* London, 121, 1974.
3. **Domino, W. J.,** Torsional mechanical filters at 500 kHz, *Proc. 1988 Ultrason. Symp.,* Chicago, Vol. 1, 245, 1988.
4. **Johnson, R. A.,** *Mechanical Filters in Electronics,* Wiley-Interscience, New York, 1983.
5. **Johnson, R. A., Börner, M., and Konno, M.,** Mechanical filters — a review of progress, *IEEE Trans. Sonics Ultrason.,* SU-18, 155, 1971.
6. **Sheahan, D. F. and Johnson, R. A.,** *Modern Crystal and Mechanical Filters,* IEEE Press, New York, 1977.
7. **Domino, W. J. and Johnson, R. A.,** Miniature precision bandpass filters solve IF design problems, *RF Design,* 10, 49, 1991.

8.4 Ceramic Filters

Satoru Fujishima

CONTENTS

1. INTRODUCTION

Ceramic filters are filter components that make use of piezoelectric ceramics. The design principles for ceramic filters are almost the same as for crystal filters. However, ceramic filters have many merits that are different from crystal filters. The differences are based on the piezoelectric materials used for these filters. Let us consider the piezoelectric materials first.

Materials, the whole body of which is a crystal, are called single crystals, while those composed of many crystal bodies are called ceramics.

Piezoelectric ceramics are composed of many crystals. Single crystals, like a quartz crystal, in which the internal polar axes are all oriented in one direction, can be used as a piezoelectric material just as they are.

However, ceramics, which are composed of many crystals unevenly directed, do not show the piezoelectric effect since the polar axes of the crystals orient in random directions. In this case the ceramics act as dielectric materials. In order to create the piezoelectric effect in ceramics, it is necessary to orient the polar axis of every crystal in one direction.

This is accomplished by applying a direct electric field of high intensity for a considerable duration. The process is referred to as the polarization treatment. Figure 1A shows the crystal polar axes oriented in random directions before the treatment, and Figure 1B shows all axes oriented in one direction after the treatment.

0-8493-8951-8/97/$0.00+$.50
© 1997 by CRC Press LLC

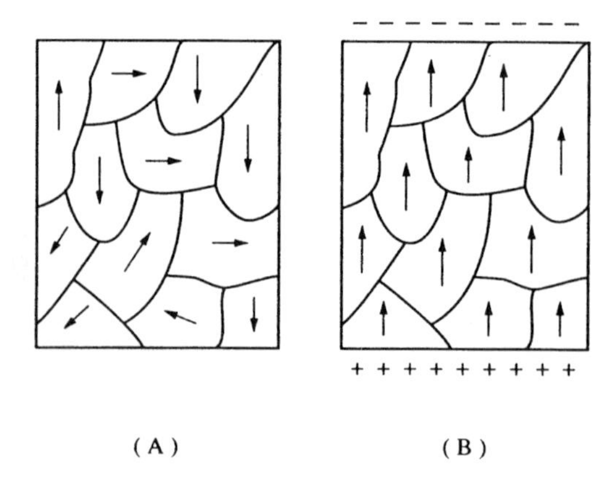

Figure 1 Polar axes directions in ceramics before and after polarization treatment.

Table 1 Material Constants of Piezoelectric Ceramics

Material	Frequency Constant (m/s)	Coupling Coefficient (%)	Permittivity ε	Q_m
PZT-6E	2040	31	820	1100
PCM-18[a]	2520	39	1200	1800

[a] Trade name of Matsushita's material.

The most common piezoelectric ceramic filter materials are lead zirconate/lead titanate. The temperature coefficient of the center frequency can be adjusted by changing the lead zirconate ratio to lead titanate for each vibration mode of the resonator used.

Physical dimensions of the resonator are tightly controlled since they determine the resonance frequency. The resonance frequencies are inversely proportional to physical dimensions in the case of typical vibration modes.

Typical data of piezoelectric ceramics used for filters are listed in Table 1. In Table 1, frequency constants are the resonance frequency for the radial vibration mode of a ceramic disc times diameter. Coupling coefficients are also for the radial vibration mode. From these data, we can calculate the diameter of a disc to make a ceramic filter of a desired frequency. All data in Table 1 are mean values with about ±10% error.

2. FUNDAMENTALS

Ceramic filters are composed of many kinds of ceramic resonators. A piezoelectric ceramic resonator is a small, thin piece of piezoelectric ceramic with metallized electrodes on the opposite surfaces.

Figure 2 shows the symbol for a piezoelectric ceramic resonator in electronic circuits. The impedance and phase characteristics measured between the two terminals in Figure 2 are quite the same as a dielectric ceramic capacitor above and below the resonance frequency range, but the characteristics change drastically as shown in Figure 3 near the resonance frequency.

Figure 4 shows the vector admittance curve by frequency around the resonance. This means that the mechanical vibration of a two-terminal vibrator can be replaced equivalently with a combination of series and parallel resonance circuits as shown in Figure 5A.

Figure 5B shows an equivalent circuit modified from Figure 5A connecting the real part and the imaginary part in series, and Figure 5C shows another equivalent circuit where real and imaginary parts are connected in parallel.

The equivalent circuit constants L_1, C_1, and R_1 in Figure 5A can be determined from the following formulae:

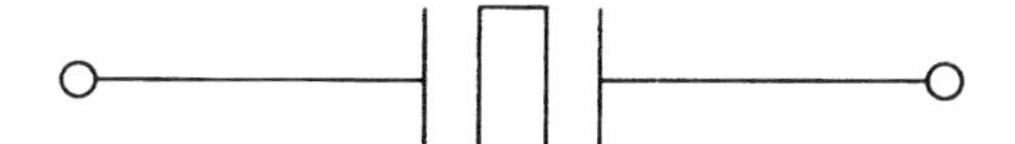

Figure 2 Symbol of a ceramic resonator.

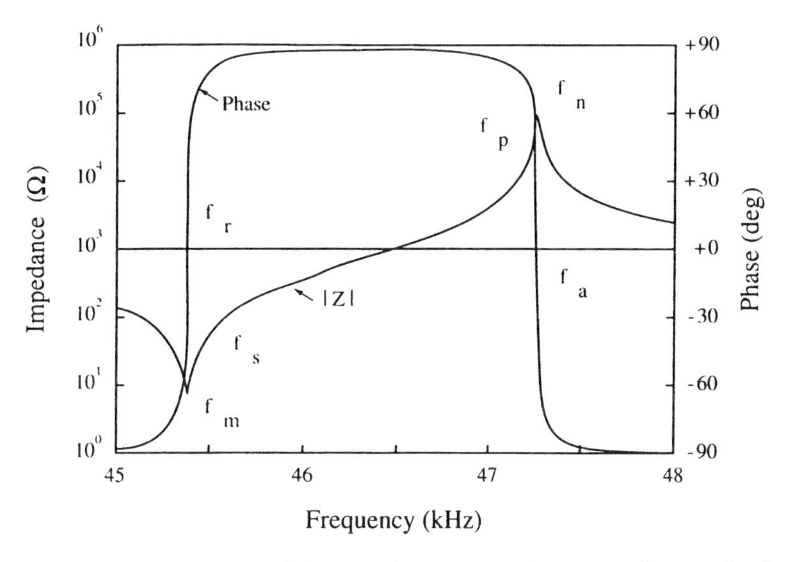

Figure 3 Impedance and phase characteristics near the resonance frequency of a ceramic vibrator.

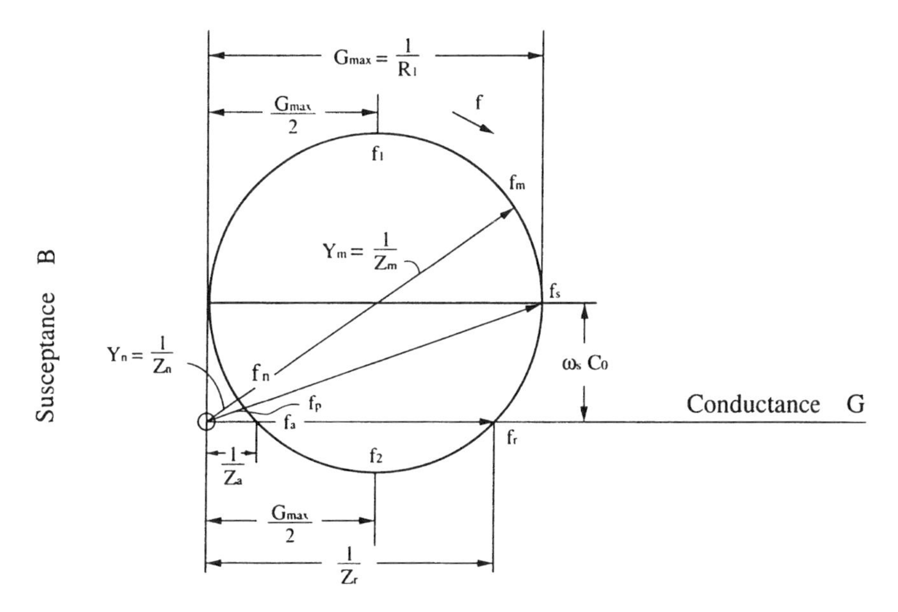

Figure 4 Motional admittance circle near the resonance frequency of a ceramic resonator.

$$f_s = \frac{1}{2\pi\sqrt{L_1 C_1}} \tag{1}$$

$$f_p = f_s\sqrt{1 + C_1 / C_0} \tag{2}$$

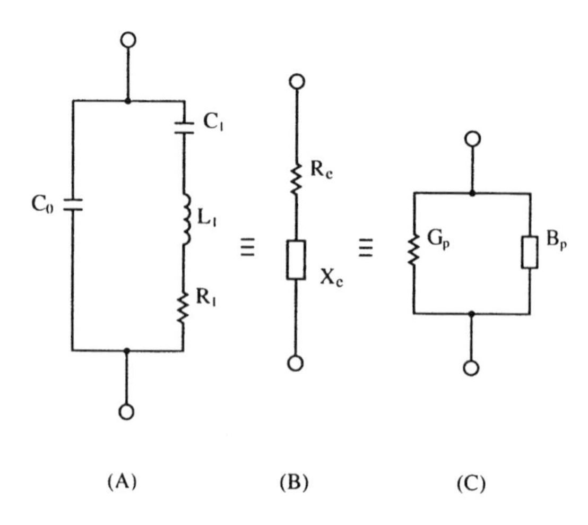

Figure 5 Equivalent circuit of a ceramic vibrator.

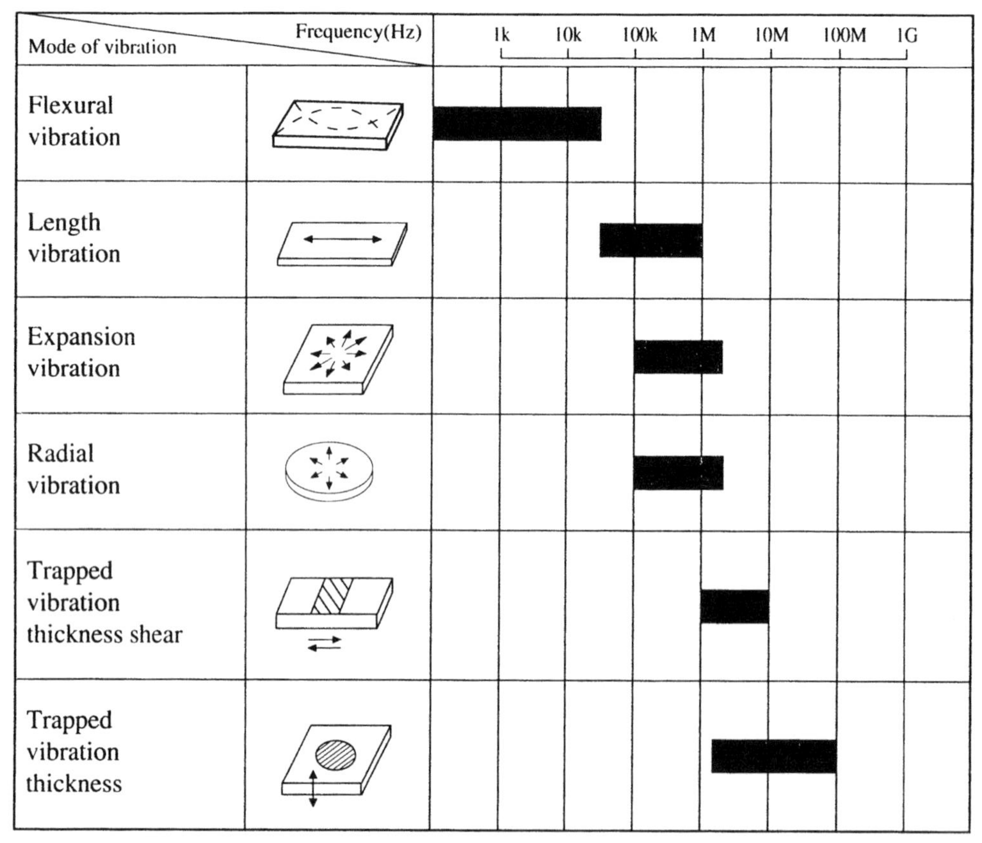

Figure 6 Usable vibration mode by frequency range of piezoelectric ceramic resonators.

$$Q_m = f_s/(f_2 - f_1) = 2\pi f_s L_1 / R_1 \tag{3}$$

Compared with quartz crystals, piezoelectric ceramics have very low Q_m values, which are suitable for wideband filters, as shown in Table 1.

The frequency range covered commercially by piezoelectric ceramic filters is generally from 10 to 30 MHz. The mechanical resonance is classified according to the vibration direction and the

Mode of vibration	Notation	Resonance frequency
Width flexural vibration	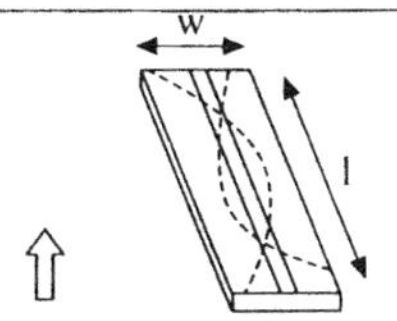 w, l	$\frac{\pi}{4\sqrt{3}} \cdot \frac{w}{l^2} \sqrt{\frac{2}{1+\sigma}} \cdot \sqrt{\frac{1}{\rho s_{11}^E}}$
Length vibration	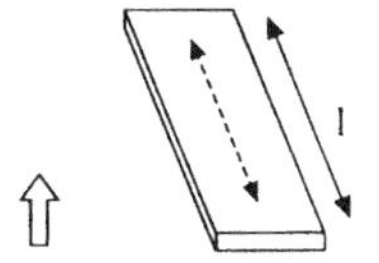 l	$\frac{1}{2l} \cdot \sqrt{\frac{1}{\rho s_{11}^E}}$
Expansion vibration	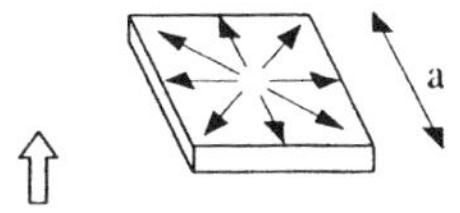 a	$\frac{1}{2a} \sqrt{\frac{1}{\rho (1-\sigma) s_{11}^E}}$
Radial vibration	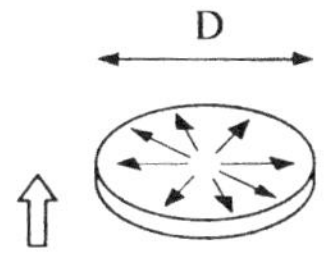 D	$\frac{2}{\pi D} \sqrt{\frac{1}{\rho (1-\alpha^2) s_{11}^E}}$
Trapped vibration in thickness	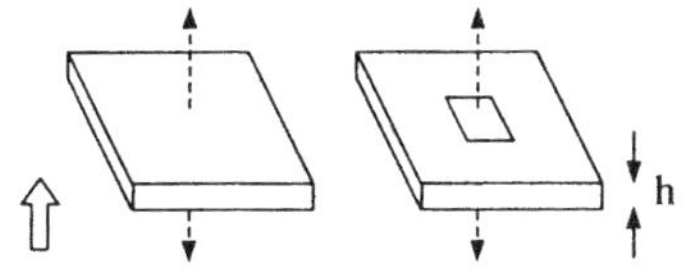 h	$\frac{1}{2h} \sqrt{\frac{c_{33}^D}{\rho} \left(1 - \frac{4k_t^2}{\pi^2}\right)}$
Trapped vibration in thickness shear	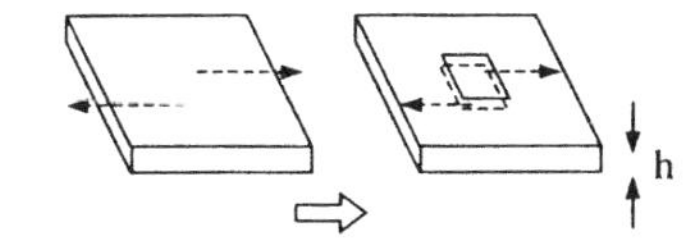 h	$\frac{1}{2h} \sqrt{\frac{c_{44}^D}{\rho} \left(1 - \frac{4k_{15}^2}{\pi^2}\right)}$

⇨ : Poling direction σ : Poisson's ratio s : Compliance c : Stiffness

Figure 7 Resonance frequency of each vibration mode.

type of waves generated. This is referred to as the vibration mode, which depends on the position of the electrodes, direction of polarization, and driving electric field direction. Various modes result in various frequency ranges of the resonator. Figure 6 shows typical vibration modes and their frequency range. Figure 7 shows the fundamental resonance frequency equation of each vibration mode of Figure 6.

The fractional bandwidth of a ceramic filter is mainly determind by the electromechanical coupling coefficient of the vibration mode. Figure 8 shows the maximum pass bandwidth at each center frequency of the ceramic filter. Ceramic filters with pass bandwidths greatly exeeding the upper boundaries shown in Figure 8 cannot be obtained without adding inductors.

3. IF FILTERS FOR AM RADIO RECEIVERS

Ceramic filters are widely used for AM radio receivers and communications equipment in the frequency range from 400 to 500 kHz. Typical IF filters are ladder-type filters composed of many resonators. The ladder-type filter is capable of both high selectivity and linear phase characteristics. Many fundamental sections with series arm and parallel arm resonators are cascade connected as shown in Figure 9.

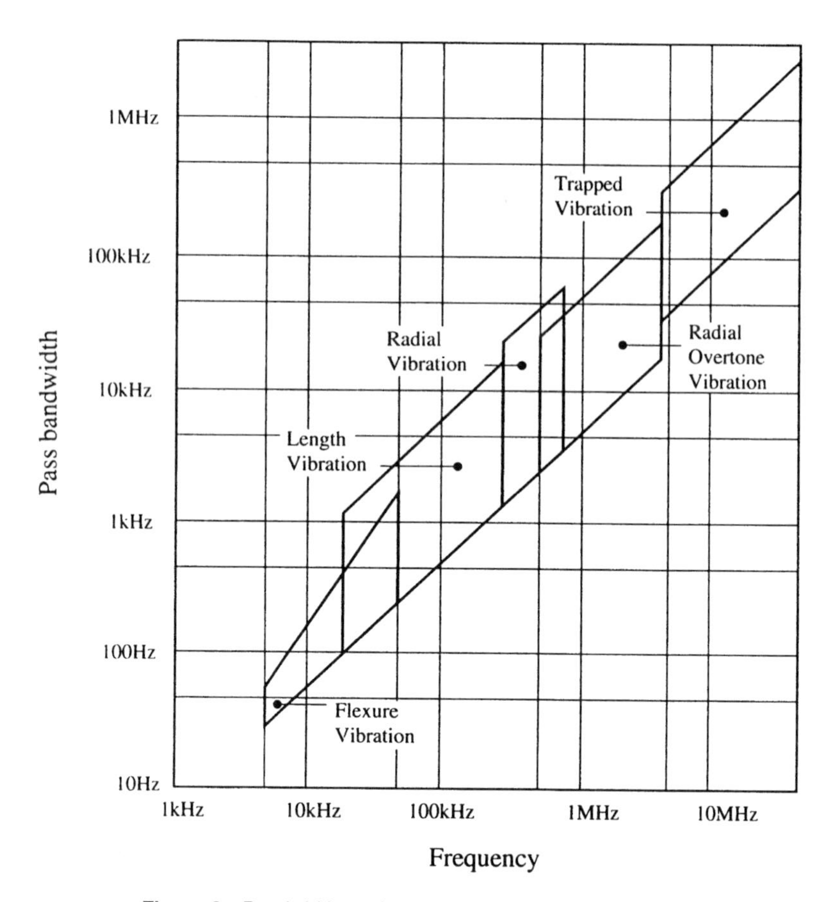

Figure 8 Bandwidth vs. frequency range of ceramic filters.

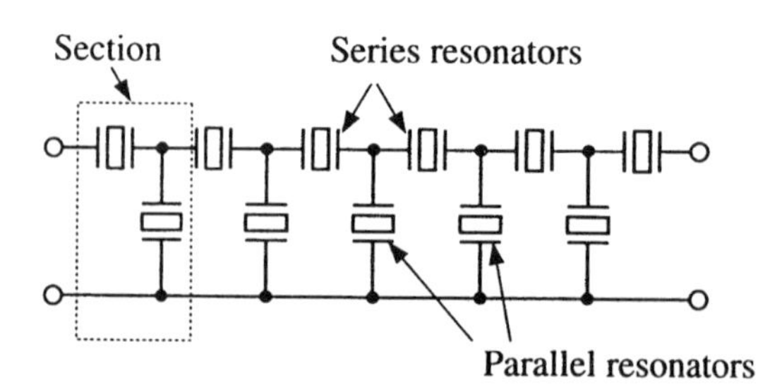

Figure 9 Structure of a ladder-type ceramic filter.

When the resonance frequency of series resonators and the antiresonance frequency of parallel resonators are matched, as shown in Figure 10A, the passband, stopband, and attenuation poles of the filter are determined as shown in Figure 10B.

The attenuation in the stopband increases with increasing number of sections. Figure 11 shows examples of frequency characteristics of ladder filters using two, three, and four sections.

Another typical IF filter is the ring-and-dot electrode type. Figure 12 shows a sketch of the electrodes on the resonator. One side of the piezoelectric ceramic plate has a common ground electrode, while the other side has split electrodes, a ring, and a dot.

The electrical input signal is converted to a mechanical vibration in the form of a square expansion mode by the dot electrode. The mechanical vibration is converted again to the output electrical signal by the ring electrode. Figure 13 shows the equivalent circuit of the vibrator. C_{12} and C_{34} in Figure 13 are the input and output capacitance, respectively. L, C, and R show equivalent

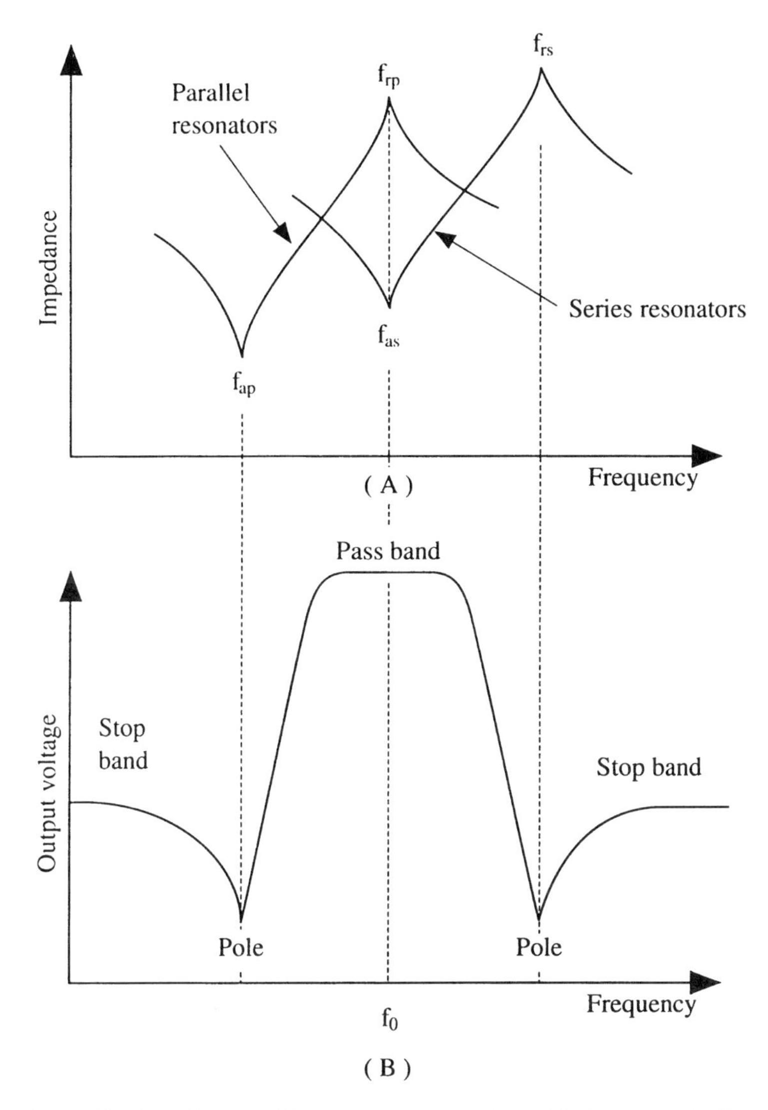

Figure 10 Impedance and frequency response of a ladder-type ceramic filter.

inductance, capacitance, and resistance at the resonance frequency of the expansion mode of the resonator.

One major advantage of the filter is that its selectivity can easily be enhanced by connecting it in cascade as shown in Figure 14. C_c in Figure 14 is the coupling capacitor which determines the bandwidth of the cascade-connected filter.

Figure 15 shows a typical frequency characteristic of a single resonator filter (A) and a two-resonator cascade-connected filter (B). The larger the capacitance of C_c, the wider is the bandwidth and the larger the ripple in the passband, as shown in Figure 16. Note that the frequency response has no attenuation poles.

4. IF FILTERS FOR FM RADIO RECEIVERS

The ceramic filter is widely used for FM radio receivers, television sound IF, and communications equipment in the frequency range between 3 and 30 MHz. The vibration energy of thickness-expansion or thickness-shear modes can be trapped under the partial electrodes of a piezoelectric ceramic vibrator in this frequency range. Distribution of the two modes of the vibration amplitude

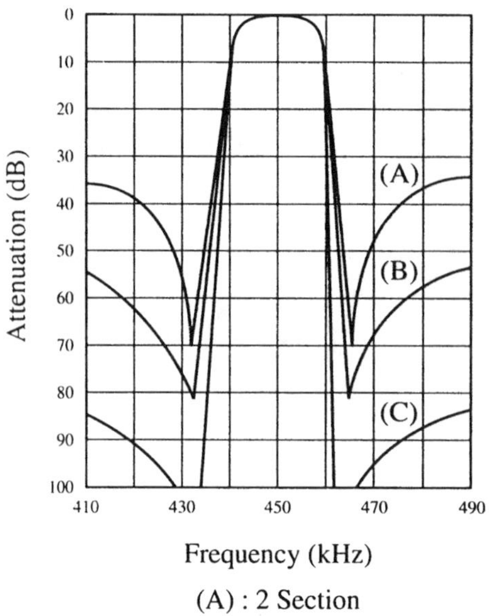

Figure 11 Frequency characteristics of two-, three-, and four-section ladder-type ceramic filters.

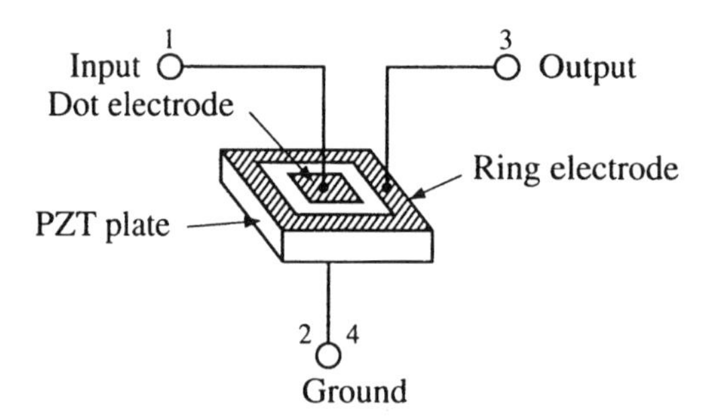

Figure 12 The sketch of electrodes on a ceramic filter.

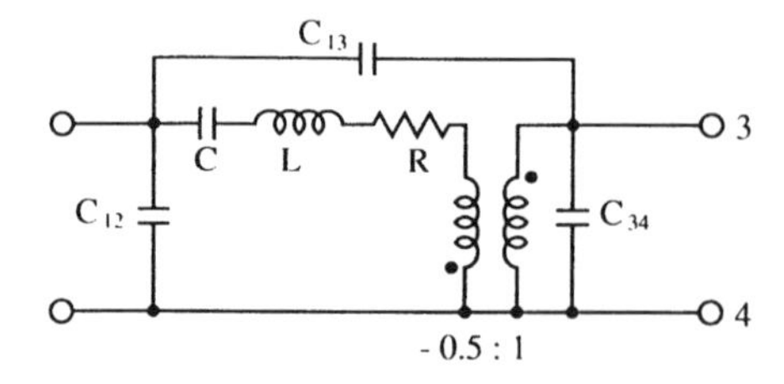

Figure 13 Equivalent circuit of a ring-dot ceramic filter.

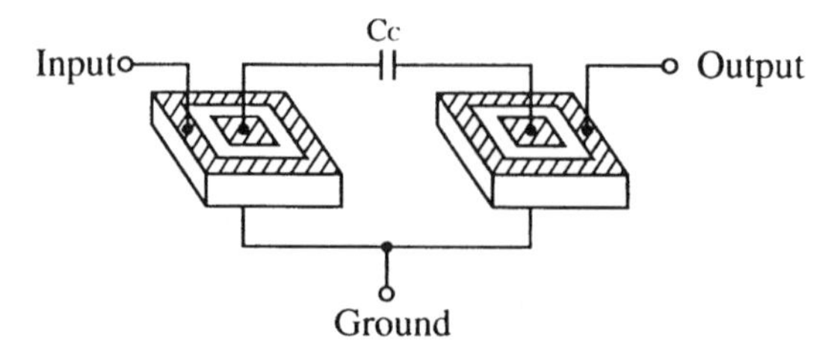

Figure 14 Cascade-connected ceramic filter using a coupling capacitor Cc.

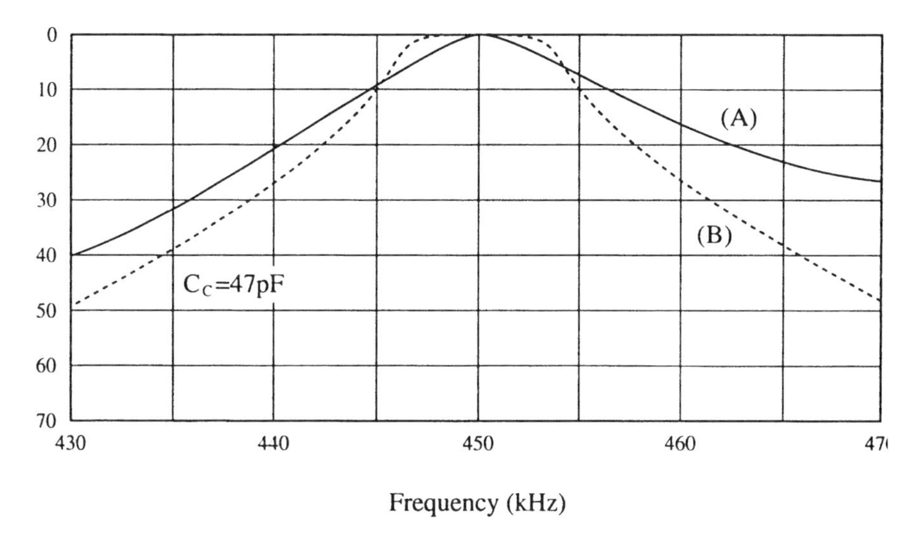

Figure 15 Frequency characteristics of ceramic filters. (A) Single resonator; (B) cascade-connected resonators.

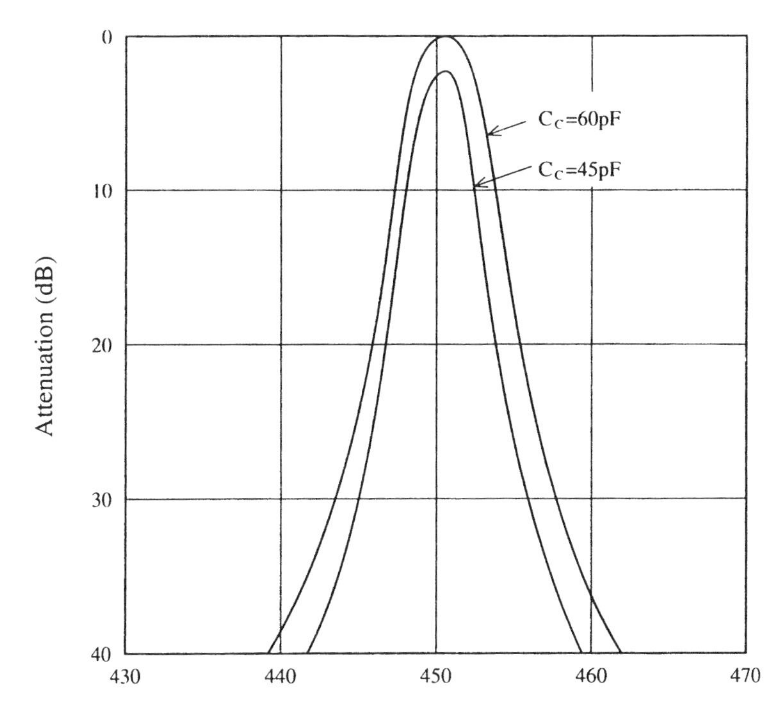

Figure 16 Frequency characteristics of two resonators cascade-connected by different coupling capacitors.

is illustrated in Figure 17, in which the solid line is called the symmetrical mode and the dotted line is called the antisymmetrical mode.

Figure 18 shows the equivalent circuit of Figure 17. It consists of two series resonance circuits, a phase inversion transformer, stray capacitance between split electrodes, and the input and output capacitance.

A piezoelectric ceramic filter can be realized by utilizing the integration of two trapped-mode resonators and a nonpiezoelectric coupling capacitor on one wafer as shown in Figure 19A. Thus, a complete filter can be obtained in monolithic or integrated form. Figure 19B shows the electrical circuit with the coupling capacitor C_c.

The coupling capacitor is made by a depoling process with either a reverse electric field or heat applied to the electrode.

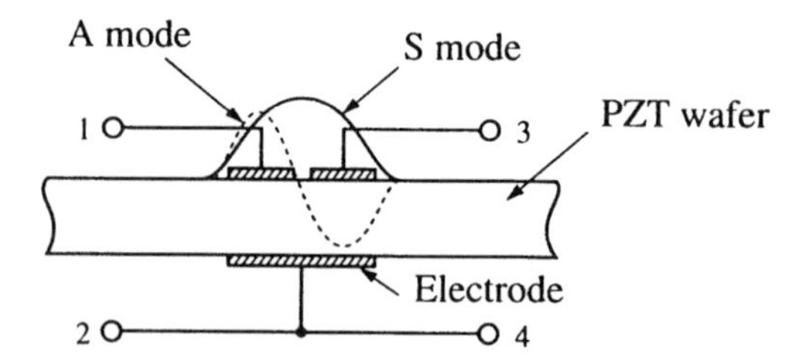

Figure 17 Vibration amplitude of a multimode energy-trapped ceramic filter.

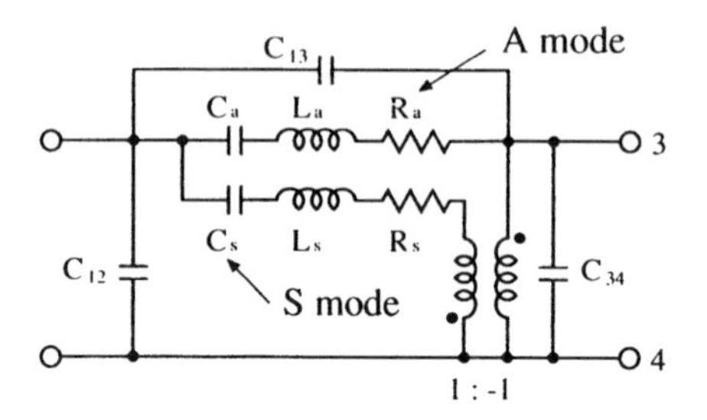

Figure 18 Equivalent circuit of a multimode energy trapped-ceramic filter.

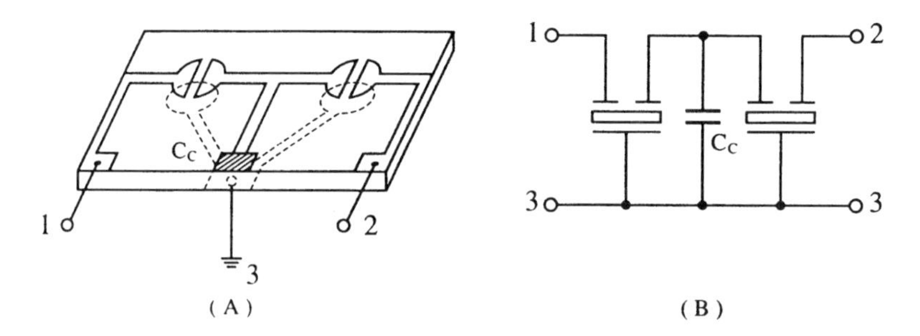

Figure 19 Monolithic ceramic filter. (A) Structure; (B) electrical circuit.

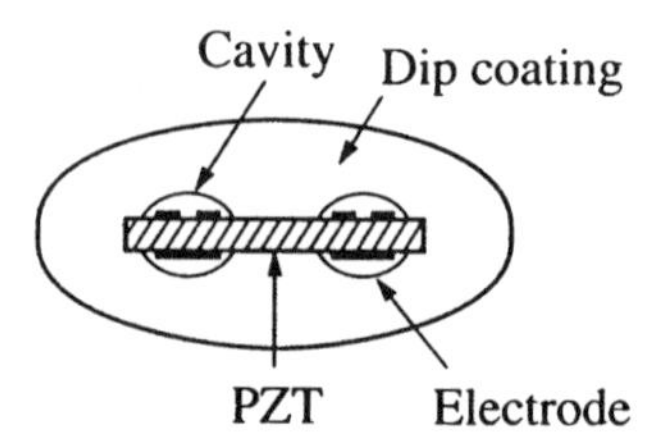

Figure 20 Section of a monolithic ceramic filter.

Figure 20 shows an example of the package for the energy-trapped ceramic filter. Small cavities are made on the partial electrodes using a special coating which is absorbed into the dip coating in the curing process. Figure 21 shows an example of the frequency characteristic of the ceramic filter for FM radio receivers.

5. UTILIZATION AND LIMITATIONS

5.1 EFFECT OF BANDWIDTH ON AVAILABILITY OF CERAMIC FILTERS

The minimum bandwidth of ceramic filters results from temperature and time instability of ceramic vibrators. Filters with a less than 0.5% bandwidth-to-center-frequency ratio are difficult to imple-

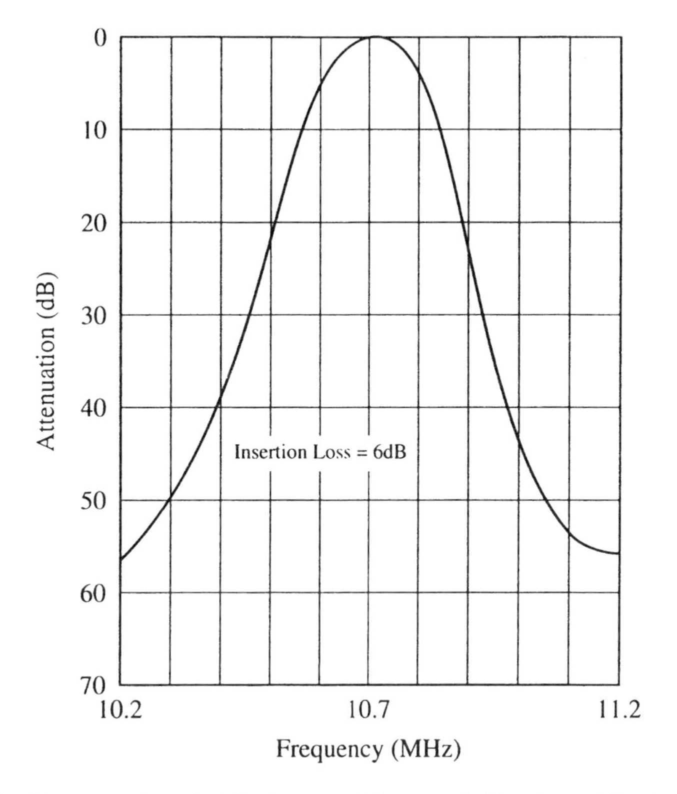

Figure 21 Frequency characteristic of a monolithic ceramic filter for an FM radio receiver.

ment, because the passband is comparable with the relative frequency shift due to the temperature of 0.2% plus aging 0.2% (overall instability of 0.4%).

Filters with bandwidths greatly exeeding the upper boundaries of Figure 8 cannot be made available without adding inductors. In principle, the maximum bandwidth is decided by the electromechanical coupling coefficient of the piezoelectric ceramics.

5.2 LIMITATIONS OF TEMPERATURE PERFORMANCE

Since most resonators do not have a linear temperature coefficient, measurement at a substantial number of temperatures may be necessary to obtain accurate data. For narrow bandwidths, the filter response may show variations of ripple and insertion loss with temperature.

5.3 LIMITATIONS OF ENVIRONMENTAL PERFORMANCE

Ceramic filters are widely used for consumer markets. Resonators are enclosed in a plastic case or resin coating instead of a hermetic seal. They are available in temperature ranges from –20° to +85°C, in 100% humidity at 25°C for 100 h, and 10 years of aging at room temperature. Exceeding these limits, ceramic filters may fall out of the specification.

5.4 IMPEDANCE MATCHING

The input (R_1) and output (R_2) impedances of ceramic filters should be matched correctly to the specified values. If a lower input or output resistance is connected to a ceramic filter, the insertion loss increases, the center frequency shifts toward the lower-frequency side, and ripples increase. On the other hand, if a higher input or output resistance is connected to a ceramic filter, the insertion loss and ripples again increase, but the center frequency shifts toward the higher side.

Figure 22 shows an example of the characteristic, in which Figure 22A shows the case of under-mismatching and Figure 22B shows the case of over-mismatching from the correct value of 3 kΩ. The mismatching range from the correct impedance should be less than ±20% of the specified value.

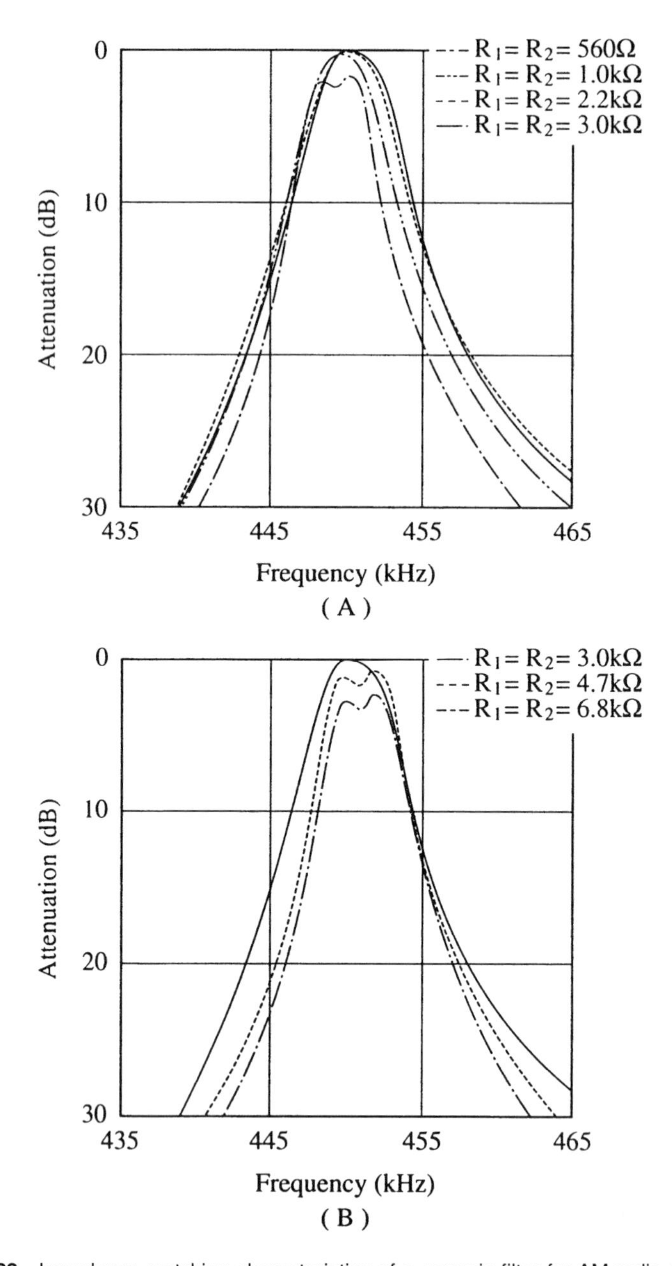

Figure 22 Impedance matching characteristics of a ceramic filter for AM radio receivers.

REFERENCES

1. **B. Jaffe, R. S. Roth, and S. Marzullo,** Properties of piezoelectric ceramics in the solid-solution series lead titanate-lead zirconate-lead oxide, tin oxide and lead titanate-lead hafnate, *J. Res. Natl. Bur. Stand.*, 55, 239, 1955.
2. **H. Ouchi, M. Nagano, and S. Hayakawa,** Piezoelectric properties of Pb ($Mg_{1/3}$ $Nb_{2/3}$) O_3-$PbTiO_3$-$PbZrO_3$ solid solution ceramics, *J. Am. Ceram. Soc.*, 48, 630, 1965.
3. **S. Fujishima, J. Merlina, and J. Miyazaki,** Piezoelectric ceramic resonators and filters, *Proc. 38th Frequency Control Symp.*, 184, 1984.
4. **W. P. Mason,** *Physical Acoustics*, Vol. 1, Part A, Academic Press, New York, 234, 1964.
5. **IEEE Standard:** *Definitions and Methods of Measurement for Piezoelectric Vibrators,* New York, 1966.

Surface Acoustic Wave Filters

Jacqueline H. Hines

CONTENTS

1. INTRODUCTION

Surface acoustic wave (SAW) components have been commercially available for over 20 years. Common SAW device types include bandpass filters, delay lines, dispersive devices, resonators, and resonator filters. Following intensive research and development in the late 1960s and early 1970s, SAW devices became widely accepted for used in a diverse group of military and commercial applications ranging from radar to mobile communications. The status of SAW devices as preferred components for numerous applications is based on the inherent advantages of the technology.[1,2]

0-8493-8951-8/97/$0.00+$.50
© 1997 by CRC Press LLC

SAW devices can provide complex signal processing functions in a single compact device. One example of this is the outstanding bandpass filter characteristics which can routinely be achieved using SAW devices. Comparable performance utilizing LC-filter technology would require numerous components and could occupy many square inches of board space. Since surface acoustic waves propagate with velocities on the order of 10^{-5} times the speed of light, the realization of electronically long delay times on substrates of limited dimensions is possible. Additional performance advantages of SAW technology, which vary based on the application, include small size, linear phase, low shape factor, excellent rejection, and temperature stability. The ruggedness and reliability of SAW devices are characteristic of the physical device structure. Since device operating frequencies are set by photolithographic processes, SAW devices do not require complicated tuning procedures, nor do they become detuned in the field. The semiconductor microfabrication techniques used in manufacturing SAW components allow for the volume production of economical, reproducible devices. The outstanding reproducibility of these devices makes them ideal for applications such as channelized filter banks for spectral analysis. Small size and ruggedness make SAW devices useful for mobile communications and related applications. The relative radiation hardness of conventional SAW devices makes them ideal for space-based applications.[3,4]

2. BASIC PRINCIPLES OF SAW OPERATION

The basic principles of operation of most SAW device types are best introduced by considering the concrete example of a transversal SAW filter. In its simplest form, a transversal SAW filter consists of two transducers with interdigital arrays of thin metal electrodes deposited on a highly polished piezoelectric substrate. Figure 1 illustrates this generic configuration. When an RF signal of the proper frequency is applied across the interdigital transducers (IDTs), the alternating electrode polarities cause the surface of the crystal to expand and contract. This causes the generation of a mechanical (or acoustic) wave. Reciprocally, a propagating acoustic wave generates an electrostatic wave with potentials at the surface of the substrate which can be detected by an IDT. The configuration of the IDTs, in combination with the substrate, determines the signal processing function and response characteristics of the device. The transducer electrodes function as sampling points for both electrical-to-mechanical and mechanical-to-electrical transduction. For fundamental operation, the IDT electrodes are generally spaced at $^1/_4$ or $^1/_2$ of the acoustic wavelength at the operating frequency based on sampling theory considerations. This relationship places physical limitations on the frequency of operation of practical SAW devices. At low frequencies, the acoustic wavelength becomes too large for the devices to be of practical size. SAW devices are seldom used at frequencies below 10 MHz. At high frequencies, the electrodes and spaces become too narrow to fabricate with conventional photolithographic techniques, limiting typical commercially available SAW devices to operating frequencies below 3 GHz. Since SAW filters can be designed to operate efficiently at specific harmonic frequencies, gigahertz frequency operation can be achieved with manufacturable linewidths using harmonic operation.

The basic SAW IDT illustrated in Figure 1 is a bidirectional acoustic radiator. This causes half of the power radiated by the input transducer to be directed away from the output transducer and, therefore, lost and only half of the intercepted acoustic energy at the output transducer to be reconverted into electrical energy. This results in a characteristic minimum insertion loss of 6 dB due to bidirectionality. In addition to the loss due to bidirectionality, a number of other factors introduce loss in a surface wave device. Second-order effects such as resistive losses due to finite electrode resistance, apodization loss due to varying emitter and receiver electrode widths, impedance mismatch, and propagation and diffraction losses all contribute to the overall device insertion loss. For practical bidirectional filters, the insertion loss is generally in the range of 10 to 30 dB.

Recent advances in transducer structures and device configuration allow for the fabrication of unidirectional transducers, eliminating the bidirectional loss. A number of other low-loss filter structures and techniques allow for the production of transversal SAW filters with losses in the 2 to 10 dB range. Infinite impulse response (IIR) devices such as SAW resonators and SAW resonator

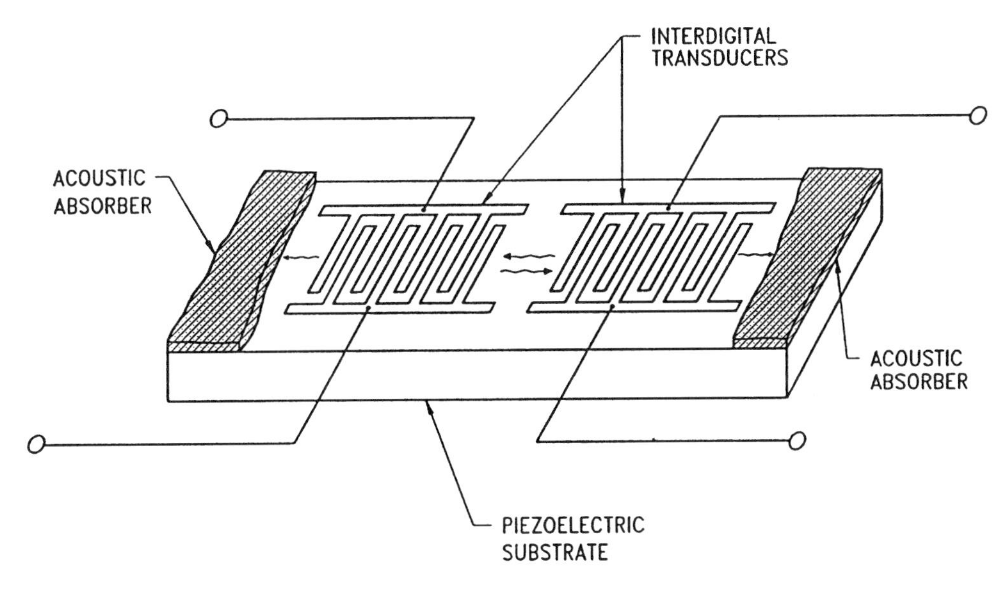

Figure 1 Configuration of a basic SAW transversal filter.

filters also demonstrate very low loss, typically below 5 dB. It should be noted, however, that not every SAW requirement can or should be implemented as a low-loss filter. Various performance trade-offs exist, such as the limited selectivity and restricted bandwidth capability exhibited by specific low-loss filter techniques.

3. APPLICATIONS AND DEVICE EXAMPLES

SAW devices are currently utilized for numerous commercial, consumer, and military applications. The type of device commonly used for each application can be categorized (after Campbell[2]) based on the generic components that make up the SAW device. These components include bidirectional and unidirectional transducers, dispersive or "chirp" transducers, and reflectors. Certain groups of devices which do not fit conveniently into this type of categorization can be grouped by function or by specific device features, such as taking advantage of substrate acoustic nonlinearity under large-signal conditions.

3.1 BIDIRECTIONAL FILTERS

Bidirectional transversal filters constitute the most widely used application of SAW technology. SAW filters are indispensable for applications requiring narrow transition bandwidths, linear phase, and low amplitude ripple. Devices which utilize bidirectional transducers can be designed with fractional bandwidths (defined as bandwidth divided by device center frequency in percentage) ranging from approximately 1 to 100%, with insertion losses typically in excess of 10 dB and out-of-band rejection ranging from 26 dB to in excess of 60 dB. The unique ability to independently specify amplitude and group delay predistortion has been widely used for cable television (CATV) products to correct for various system distortions and in digital data transmission systems to perform bit shaping. The inherent reproducibility of these devices makes them ideal for use in phase coherent systems that require two or more channels to closely track in amplitude and phase.

Applications of such devices include delay lines for radar systems, tapped delay lines for fuses, variable delay lines for reduction of multipath interference, linear and nonlinear FM chirp filters for radar and real-time spectral analysis, reflective array compressors (RACs) for very wide time-bandwidth radar and high-resolution microscan receivers, adaptive filters for spread-spectrum communications, quadraphase code generators for reducing radar spectral splatter, and minimum-shift-keying (MSK) code generators for spread-spectrum links. Other applications of bidirectional SAW filters include clock recovery filters for fiber optic repeaters, Nyquist filters for digital radio, comb filters for multiplexer filter banks, phase shift keying (PSK) filters for binary code generation and detection, Barker code filters and pulse compressors, Fourier-transform processors for spectral identification, microscan (compressive) receivers, sampling FM demodulators, bandpass filters with nonsymmetric amplitude and/or phase for equalizers, and acousto-optic spectrum analyzers. Consumer applications of bidirectional SAW filters include intermediate frequency (IF) filters for TV receivers and for interference suppression in TV video game systems.

3.2 LOW-LOSS FILTERS

Low-loss SAW filters are required for certain applications where system noise figure or dynamic range cannot accommodate the higher loss typical of bidirectional SAW filters. Low-loss filters can be designed that retain the inherent advantages of bidirectional filters, while providing low loss and excellent triple transit suppression. Low-loss SAW devices using unidirectional transducers can be designed with fractional bandwidths of 0.1% up to approximately 15%, and with losses typically ranging from 4 to 12 dB. Such filters can be used in low-power communications receiver front-end or preselector circuit stages and in duplexers for cellular radio. Additional techniques for achieving low-loss performance have been demonstrated, including multitrack transducer structures such as the interdigitated IDT (IIDT), ladder-type resonator filters, and waveguide-coupled resonator filters.[5-7] These techniques are capable of producing losses in the 2 to 13 dB range.

Some examples of the outstanding performance achievable using various types of commercially available SAW bandpass filters are shown in Figures 2 through 6. Figure 2 shows a low-frequency SAW bandpass filter with a low shape factor, the ratio of the filter 40-dB bandwidth to 3-dB bandwidth. In this example the shape factor is less than 1.2, passband ripple is less than ±0.2 dB, and group delay ripple is below ±15 ns. The passband of this filter covers the frequency range 33.9 to 39.65 MHz, the insertion loss in a 50-Ω system is 29.4 dB, and ultimate out-of-band rejection exceeds 55 dBc. Figure 3 illustrates a typical SAW low-loss bandpass filter which utilizes unidirectional transducers, exhibiting a loss of approximately 7.6 dB and a shape factor of slightly more than 2. Figure 4 shows an example of the sculptured amplitude responses possible using SAW devices. Figure 5 illustrates the capability of producing bandpass filters with outstanding amplitude characteristics and independently sculptured group delay. This device was developed for use as a vestigial sideband filter in CATV applications. Figure 6 illustrates the wide bandwidths possible using dispersive SAW transducers. This device consists of dispersive transducers in a structure called a slanted array compressor[2,11,13] and has a center frequency of 1.3 GHz and a fractional bandwidth of 50%.

3.3 RESONATOR FILTERS

Low loss, narrow fractional bandwidth, and small size are highly desirable characteristics for UHF filters. These specifications can be satisfied by multipole SAW resonator filters. SAW resonators and resonator filters utilize the properties of a reflective cavity to support a standing surface wave. Single-pole resonators can be designed with fractional bandwidths of up to roughly 0.1%, and multipole resonator filters can typically be designed with fractional bandwidths of up to 0.4%. These devices generally have insertion losses of less than 6 dB, often less than 3 dB.

Resonators can be used as CATV filters, frequency-control elements in fixed frequency oscillators, in output stages of video cassette recorders (VCRs), in garage-door transmitter control circuits, and in medical-alert transmitter control circuits and similar applications. Resonator filters can be used as preselector filters, local oscillator suppression, clock recovery, and narrowband IF filtering for cellular phones, and as frequency-control elements in voltage-controlled oscillators, to name but a few common applications. Figures 7 and 8 illustrate two responses typical of SAW resonator filters, using either in-line or waveguide coupling.

3.4 NONLINEAR DEVICES

SAW devices such as correlators and convolvers take advantage of the nonlinear properties of specific piezoelectric substrates under large-signal conditions. Convolvers are used for code detection in radar, spread-spectrum communications, electronic counter measures (ECM), air-traffic

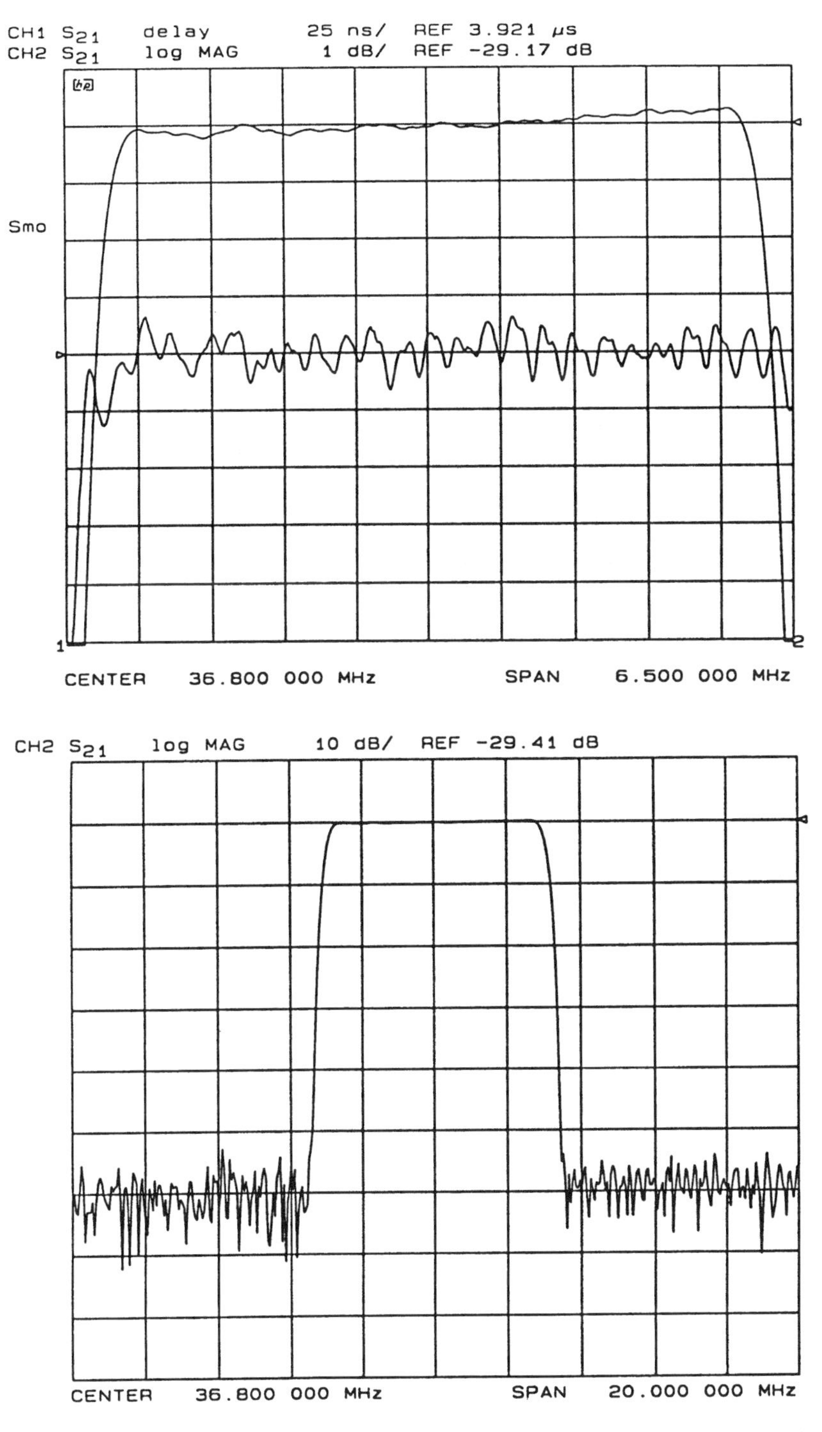

Figure 2 Examples of a high-performance SAW bandpass filter with low shape factor and small group delay ripple (< ±15 ns).

control, and data-handling systems. Multiple-port SAW/silicon convolvers are utilized in spread-spectrum communications systems with MSK or differential PSK (DPSK) modulation.

3.5 IDENTIFICATION TAG DEVICES

SAW technology can be used to generate passive, uniquely coded identification devices. This type of device can be utilized in applications which include stolen vehicle identification, wireless data

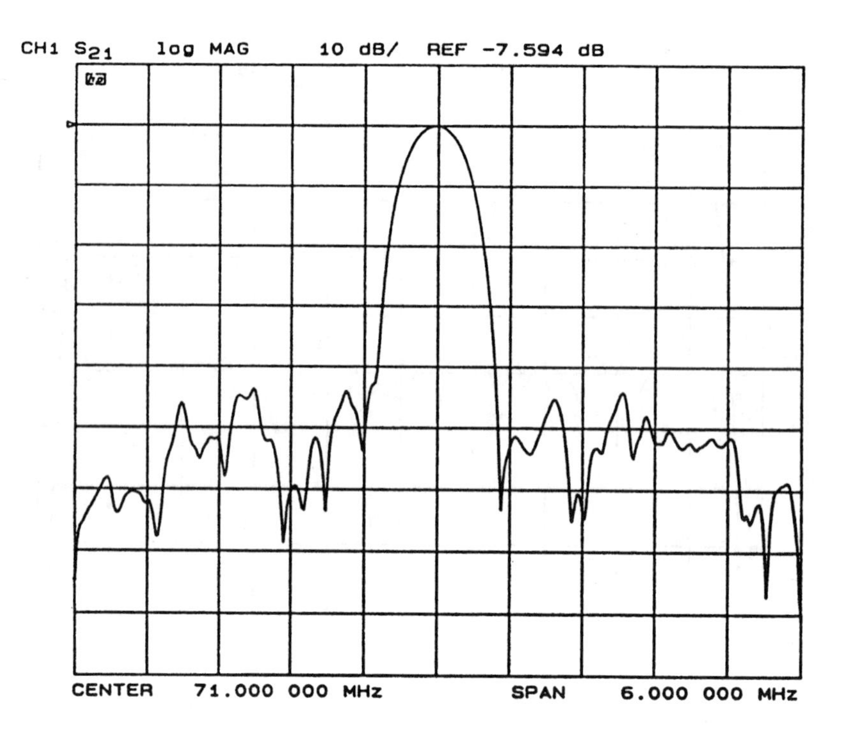

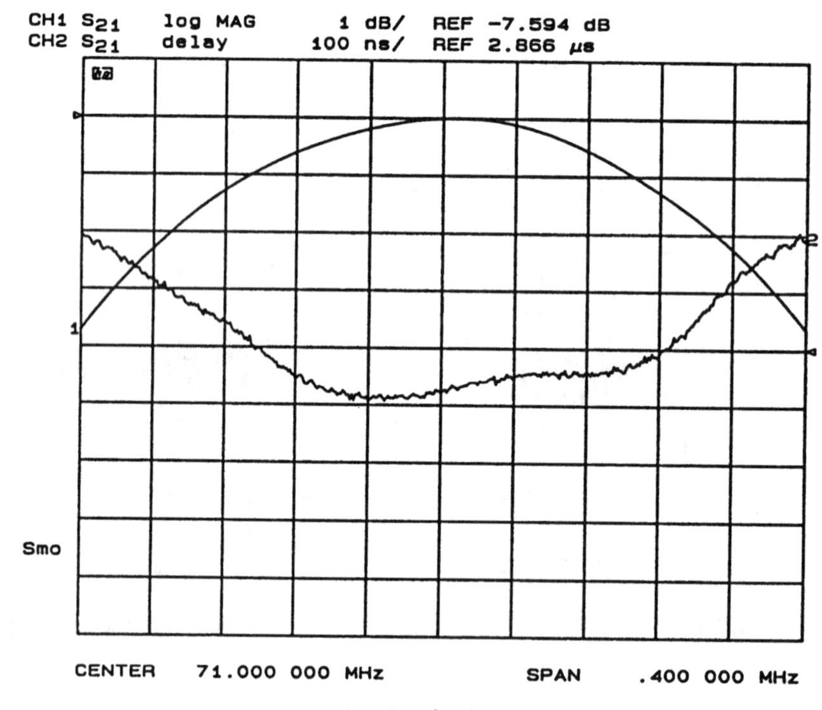

Figure 3 Typical responses for a SAW low-loss filter.

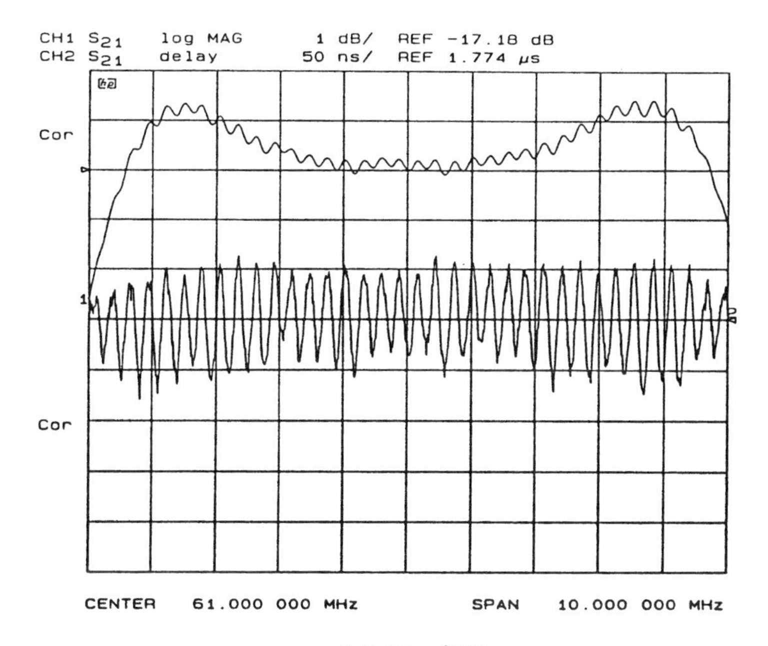

1 MHz / Div
1 dB / Div 50 nsec / Div

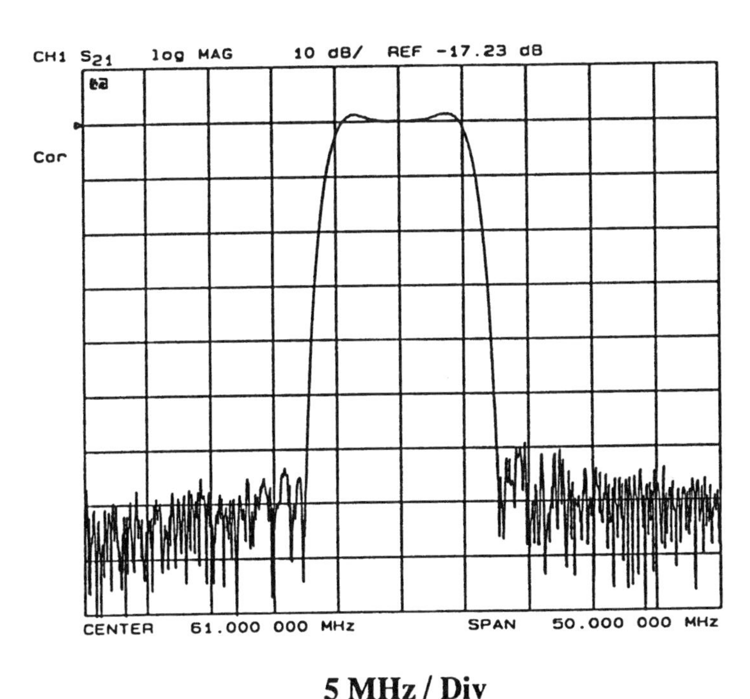

5 MHz / Div
10 dB / Div

Figure 4 Typical responses for a SAW signal-shaping filter.

terminals, automobile toll rings, and personnel and livestock monitoring. Figure 9a shows a typical SAW-coded delay line identification tag device. The amplitude of the reflector responses can be adjusted to yield roughly equal "on" tap responses, as shown in the corresponding time-domain response (Figure 9b). Alternative implementations of this type of device can be used for spread-spectrum PSK identification applications.

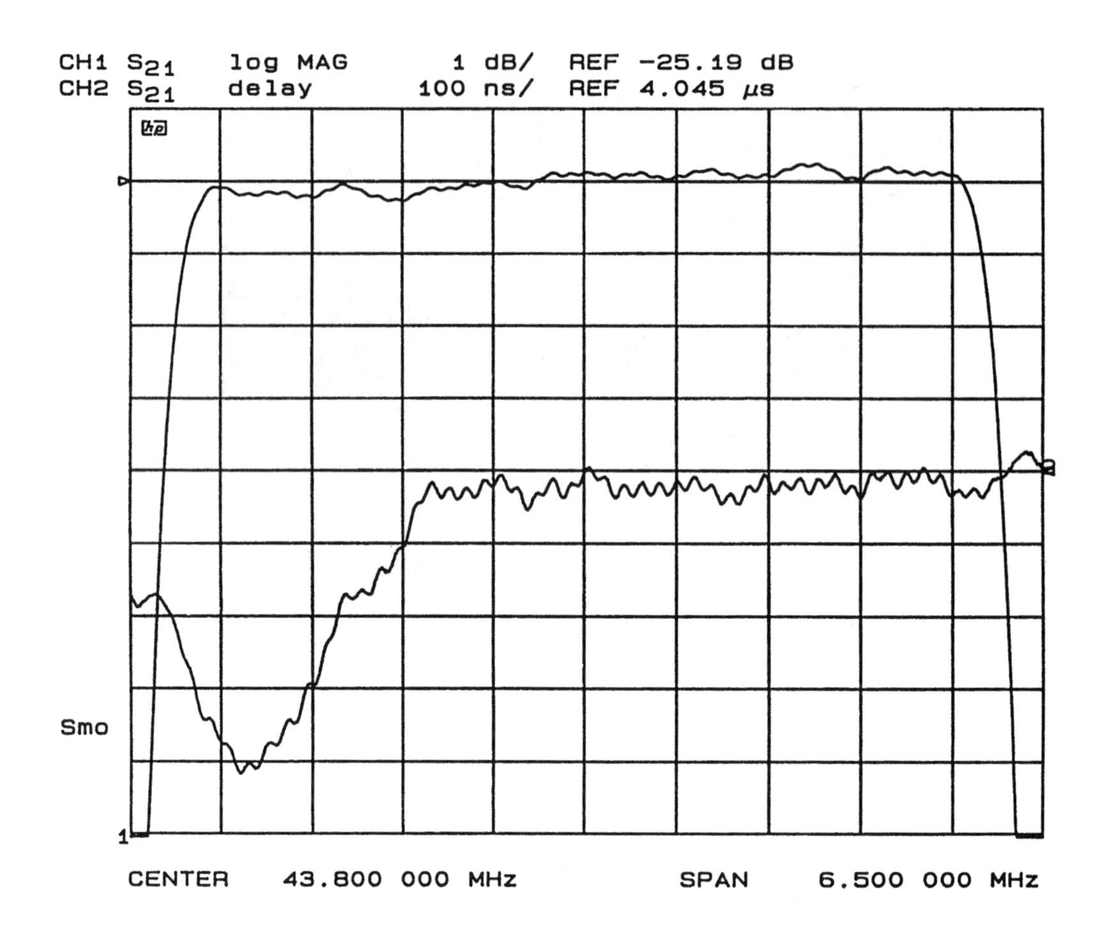

Figure 5 Typical response of a SAW vestigial sideband filter.

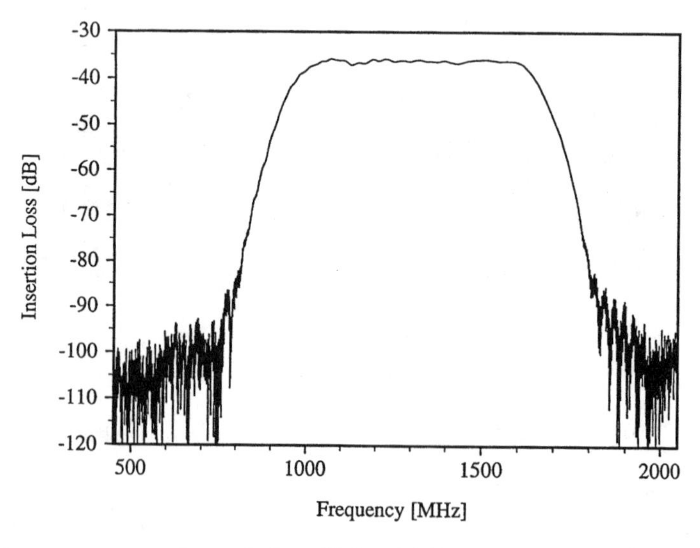

Figure 6 Slanted array compressor demonstrating wide bandwidths possible using dispersive SAW transducers.

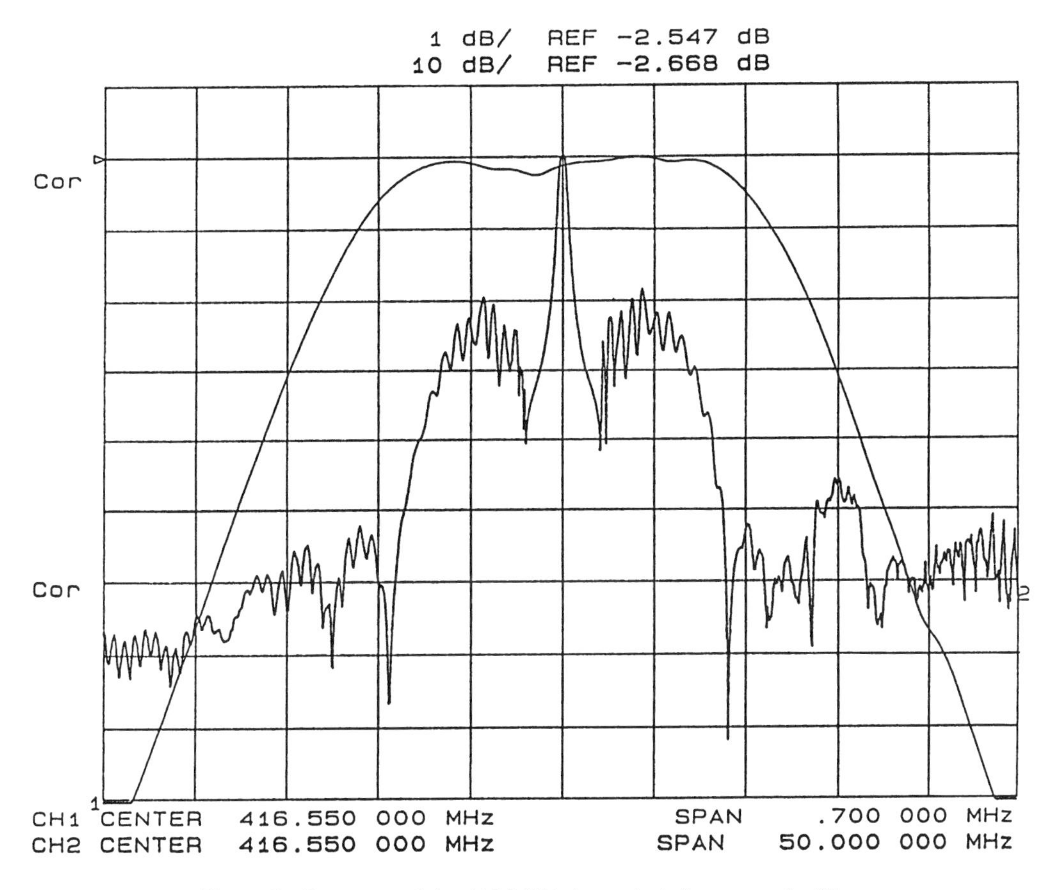

Figure 7 Response of the 416.5-MHz two-pole in-line resonator filter.

3.6 THEORY

The mechanical equation of motion for solids subjected to stress is derived from Newton's third law. Combined with the definition of strain, the constitutive relationships for the elastic and electromagnetic behavior, and constraints on the electric displacement from Maxwell's equations, this yields a systm of four coupled wave equations for the electrical potential and the three components of elastic displacement in a charge-free piezoelectric crystal.[8] The details of this system of equations and the boundary conditions and method of solution for various propagating acoustic waves in piezoelectric substrates may be found in a number of publications.[8-14] This discussion will concentrate on describing basic SAW filter design techniques, rather than on the theoretical wave equation solutions.

4. DESIGN TECHNIQUES AND CONSIDERATIONS

4.1 BIDIRECTIONAL FILTERS

Bidirectional transversal SAW filters use finite impulse response (FIR) design techniques very similar to those used for digital filter design. The SAW designer can use any one of a number of techniques to derive a pair of time-domain impulse responses which, when convolved, yield an overall filter impulse response. The Fourier transform of the impulse response is the ideal filter transfer function or frequency response. Computer-aided design tools can be used to simulate the performance of an actual device. Such tools must take into account characteristics of the substrate, the device configuration, and numerous second-order effects in order to accurately predict device performance.

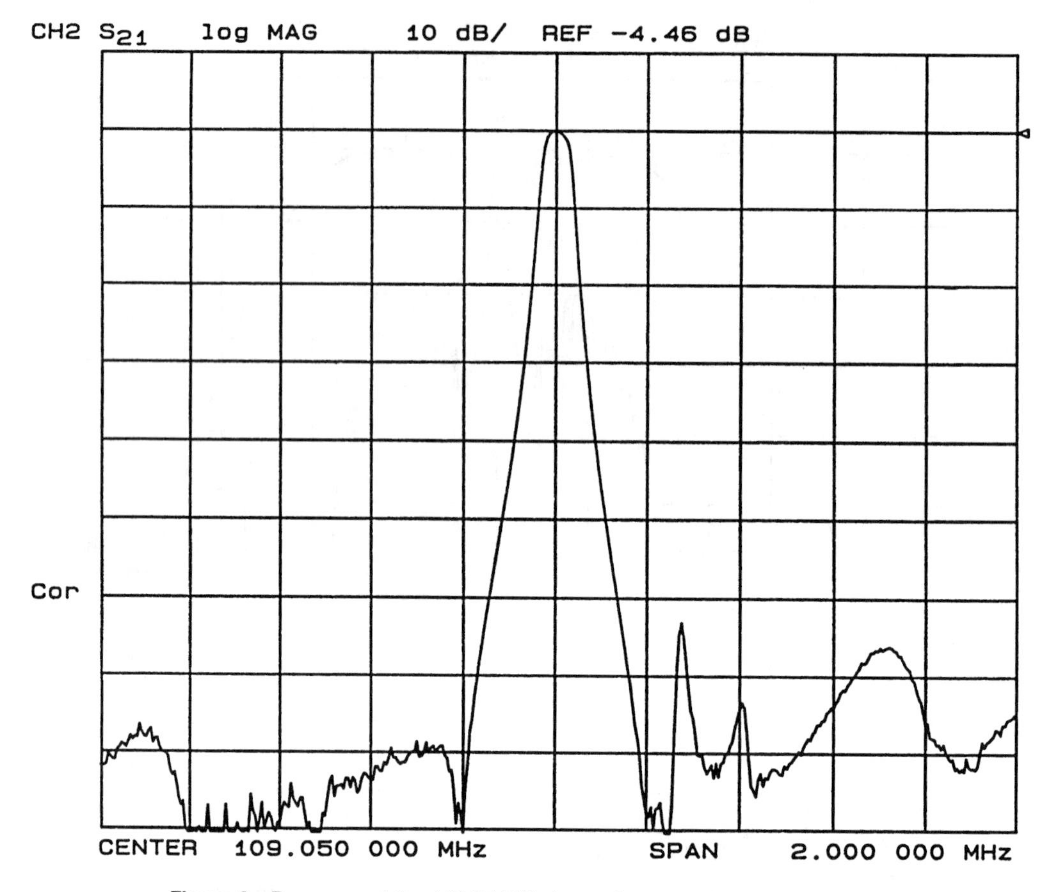

Figure 8 Response of the 109.05-MHz four-pole waveguide resonator filter.

A conventional linear phase filter utilizing bidirectional transducers is nondispersive. The overall filter transfer function in the general case where the IDTs have arbitrary finger separation can be written as follows:

$$H(f) = \frac{V_{\text{output}}}{V_{\text{input}}} = H_1(f)H_2^*(f)e^{-j\beta x(f)} \tag{1}$$

where $H_1(f)$ and $H_2(f)$ are the transfer functions of the individual transducers, the acoustic wave-number is $\beta = 2\pi/\lambda = 2\pi f/v$, and $x(f)$ is the separation distance between segments of input and output transducers that are excited at signal frequency f. When the IDTs both have uniform finger periodicity, so that the same frequency is excited in all transducer segments, the separation $x(f)$ reduces to the transducer electrode center-to-center separation, d. Thus, the filter phase angle is linear with frequency, as demonstrated by the relationship

$$\beta d = \left(\frac{2\pi d}{v}\right) f \tag{2}$$

where transducer separation, d, and acoustic velocity, v, are both constants. This relationship holds for transducers with uniform finger overlaps (unapodized transducers), as well as for transducers with apodization patterns which are symmetric about the transducer center.[2] Numerous practical factors and second-order effects cause the actual device response to vary from this ideal. The SAW

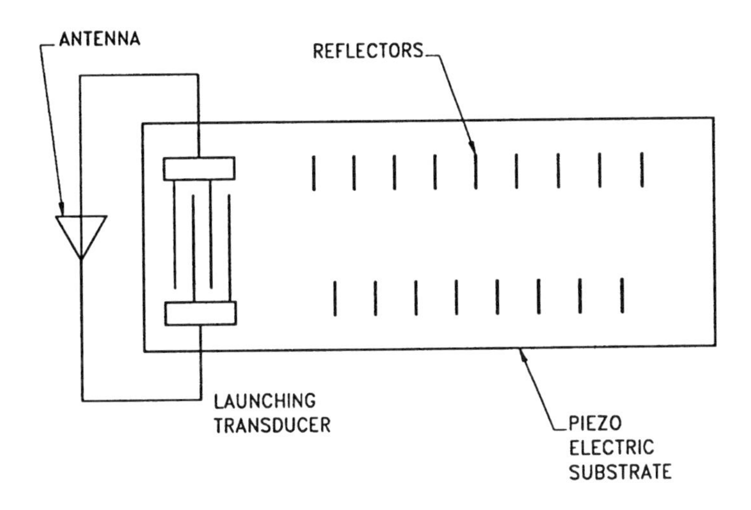

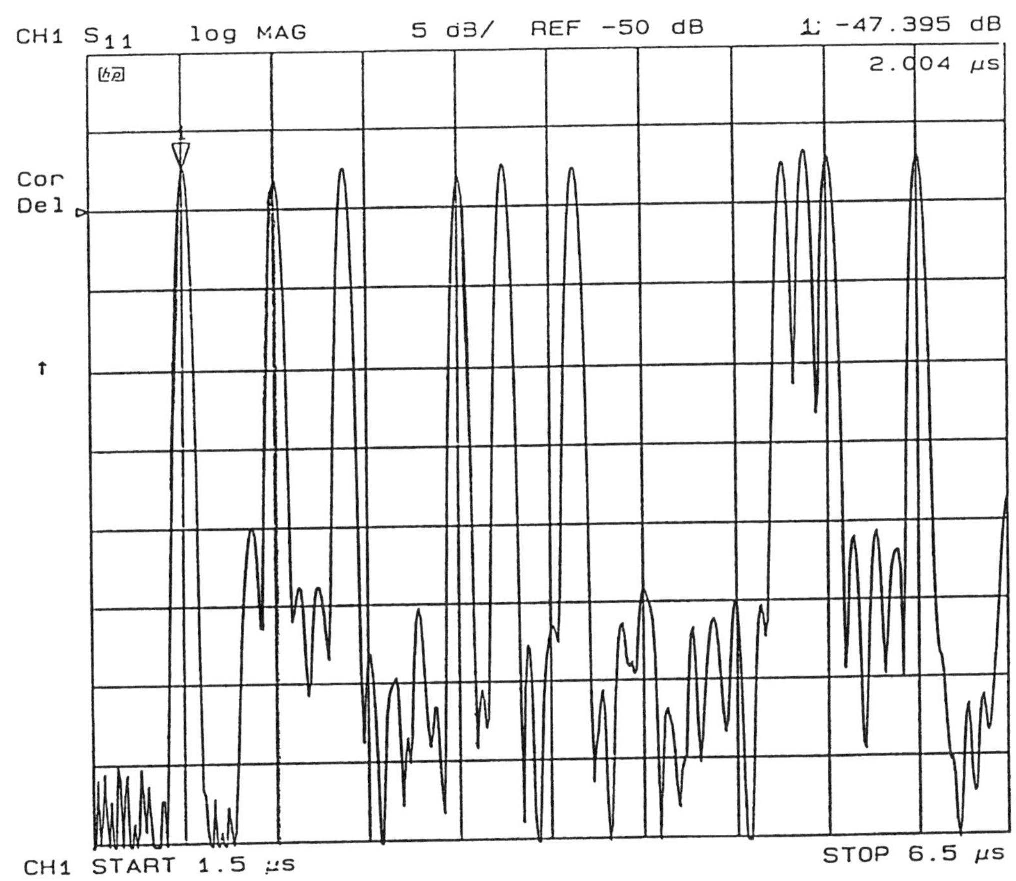

Figure 9 (a) Example of a SAW-coded delay line identification tag device; (b) time-domain response of an 856-MHz coded delay line.

designer must try to devise transfer functions for the individual transducers in a manner so as to reduce or compensate for these undesirable effects, resulting in the desired overall frequency response. Overall device layout and configuration must also be selected to minimize such effects.

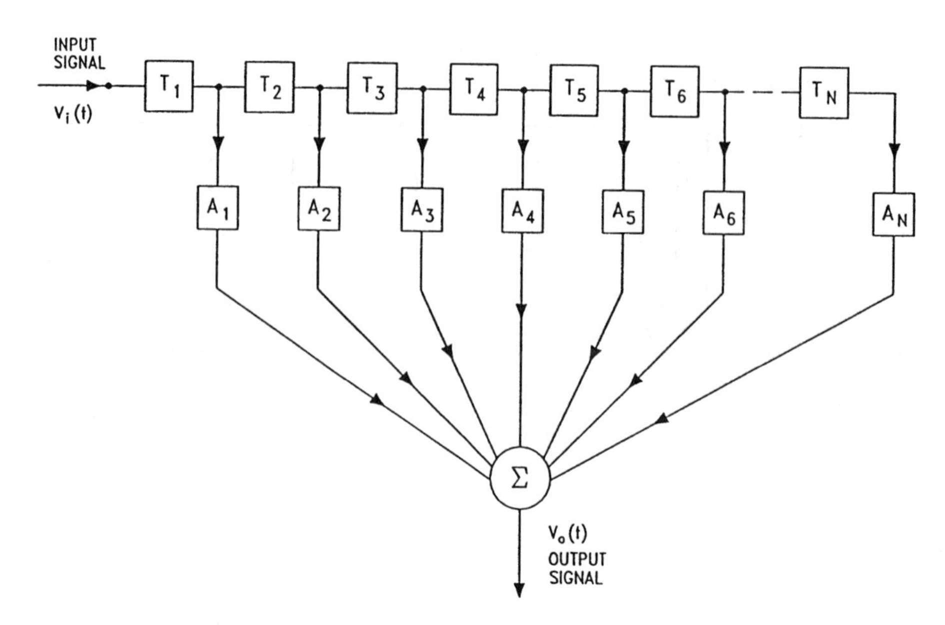

Figure 10 SAW IDT structures function as transversal filters automatically summing a series of delayed, weighted versions of the input signal. The weighting factors (A_N) are implemented by varying electrode overlaps (apodizing the transducer), while the delays (T_N) are dependent on the propagation time of the SAW between electrodes in the transducer. IDT bus bars perform the summation.

Most SAW designers work with computer-based design tools and models which help predict device performance, taking into account numerous factors which influence device operation.

The simplest model commonly used for SAW devices is the delta function model. While this model cannot take into account circuit effects, harmonic operation, bulk wave interference, or diffraction, it can be used to model the effects of electromagnetic feedthrough and triple transit. A major limitation of the delta function model is that it cannot be used to predict absolute insertion loss levels. However, this model yields information on the relative insertion loss as a function of frequency for the transducer and device transfer functions, and it is quite useful for a consideration of basic SAW design concepts. The delta function model approximates the electric field distribution between adjacent electrodes in an IDT as a discrete number of delta function sources normal to the piezoelectric surface.[2] The time-dependent electric field intensity will be proportional to the instantaneous charge accumulated on adjacent electrodes because of the time-varying input voltage. For IDTs with adjacent electrodes attached to opposite polarity bus bars, the relative field polarities will alternate from electrode to electrode. Superposition implies that a summation of these impulse response sources, taking into account their relative phases with reference to a common (arbitrary) position will yield the overall electric field intensity (and therefore SAW potential and wave amplitude) excited by one IDT. The SAW IDT structure automatically performs the summation of a series of weighted, delayed versions of the input stimulus characteristic of a transversal filter (see Figure 10).[2,12,15] This corresponds to a time-domain impulse response given by

$$h_i(t) = \sum_{n=1}^{N} (-1)^n A_n \delta(t - T_n) \tag{3}$$

where A_n is an amplitude-weighting factor proportional to the electrode apodization overlap and T_n is the cumulative delay time required for the SAW to travel through the transducer to the nth finger pair, $T_n = x_n/v$.

Taking the reference position $x = 0$ in the center of a transducer with an odd number of electrodes, N, and recognizing that for linear causal systems the frequency response is given by the Fourier transform of the time-domain impulse response, the transducer frequency response can be written as

$$\begin{aligned} H_i(f) &= \sum_{n=\frac{-(N-1)}{2}}^{\frac{(N-1)}{2}} (-1)^n A_n e^{-j\beta x_n} \\ &= \sum_{n=\frac{-(N-1)}{2}}^{\frac{(N-1)}{2}} (-1)^n A_n \left[\cos(\beta x_n) - j\sin(\beta x_n)\right] \end{aligned} \tag{4}$$

When uniform electrode overlaps are assumed, the delta function sources will all have equal amplitude, which can be normalized to $A_n = 1$. By taking into account symmetry relationships of the trigonometric functions, it becomes apparent that, for symmetric apodization patterns, where $A_{-i} = A_i$, the imaginary terms cancel out in pairs, leaving only the summation of real quantities. For frequencies close to the center frequency f_0, this summation can be approximated by a sinc function (or $\sin(x)/x$), such that

$$H_i(f) = N\left|\frac{\sin N_p n\left(\dfrac{f-f_0}{f_0}\right)}{N_p n\left(\dfrac{f-f_0}{f_0}\right)}\right| \tag{5}$$

This transducer transfer function has several properties characteristic of sinc functions. First, the null-to-null fractional bandwidth is $BW_{n-n} = 2/N_p$, the 1.5 dB fractional bandwidth is $BW_{1.5\text{dB}} = 0.638/N_p$, where N_p is the number of periods (or finger pairs) in the transducer and the first sidelobes are approximately 13.5 dB below the main response at the center frequency f_0. The overall filter frequency response can be obtained by considering the transfer function of both transducers separately and then applying Equation 1. Two unweighted transducers provide a simple bandpass filter with a sinc^2 frequency response, commonly used as a delay line.

A steeper, more "brick-wall" filter amplitude vs. frequency response is desirable for numerous applications. In order to obtain a specified overall device frequency response, the SAW designer must generate individual transducer frequency responses which multiply to yield the overall desired response. This corresponds to generating transducer designs with impulse responses which convolve in the time domain to yield the inverse Fourier transform of the desired frequency response. The transducer impulse responses can be sampled and converted directly into the spatial domain to yield the transducer electrode geometries necessary to generate the desired ideal SAW response. Sampling theory requires these electrodes to be spaced at a frequency corresponding to (at least) twice the response center frequency ($2f_0$) for a symmetric filter response and to be spaced at a frequency corresponding to at least twice the highest frequency of interest for nonsymmetric responses (generally satisfied by $4f_0$ sampling). It should be noted that the time-domain impulse response of an unweighted transducer is simply a rect function, sampled at the appropriate multiple of the carrier frequency. A pair of unweighted transducers thus yields a sinc^2 frequency response, as previously noted. Utilizing the inherent symmetry of Fourier transform pairs, it becomes evident that in order to generate a true brick-wall filter in the frequency domain, a response with a sinc^2 weighting is required in the time domain. Such a time-domain response would be infinite in extent,

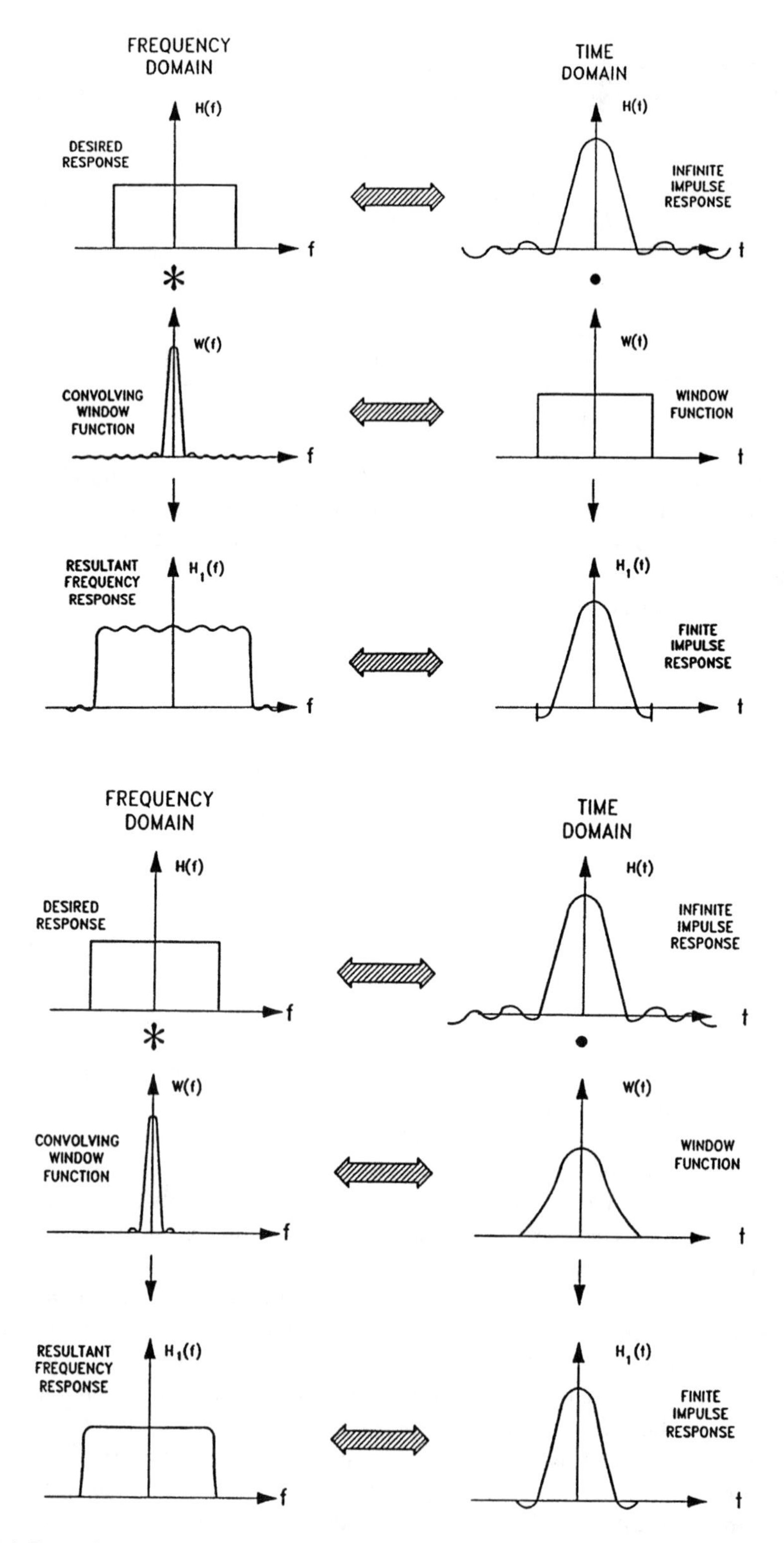

Figure 11 (a) Truncation of infinite impulse response with a rectangular window function results in Gibb's phenomena ripples throughout the frequency response. (b) Using a modified window function to provide less-abrupt truncation of the infinite impulse response results in a frequency response with substantially reduced passband ripple.

a situation which is impossible to implement in real devices. Truncation of the infinite time-domain response in practical devices introduces undesirable amplitude and phase ripples in the passband. These ripples are known as Gibb's phenomena and are introduced since truncation with a rect function in the time domain is equivalent to convolution with a sinc function in the frequency domain. The convolution of the ideal bandpass rect function frequency response with a sinc function results in ripples throughout the frequency response; however, these ripples are generally only of concern in the passband region. This convolution also results in less-steep transition bandwidths. This relationship is shown schematically in Figure 11a. The bandwidth of the frequency response is dependent on the time extent of the main lobe of the transducer apodization, rather than the overall transducer length. Increasing the time length of the transducer (i.e., maintaining more side lobes) while maintaining the transducer electrode periodicity will allow for a better approximation of the desired rectangular frequency response, although it will not reduce the effect of Gibb's phenomena other than to push the ripples closer to the passband edge.[2] A way to achieve a lower passband ripple response is to modify the transducer apodization pattern using window functions. Window functions can provide a more gradual truncation than an abrupt rect function truncation, reducing Gibb's pehnomena, as illustrated in Figure 11b. Various window functions that may be multiplied by the transducer time-domain weighting to provide improved frequency response are known, including cosine window, Hamming, Kaiser, Kaiser–Bessel, Taylor, and Dolph–Chebyshev functions.[2,12,13]

Weighting of the transducer impulse response is generally implemented through adjustment of the transverse electrode overlaps (apodization) or through selective removal of individual transducer electrodes (withdrawal weighting). A less commonly used method for amplitude weighting, capacitive weighting of transducers, has also been demonstrated. In overlap apodization, each sample in the desired sampled time-domain impulse response is normalized relative to the impulse response maximum and assigned an electrode overlap corresponding to this normalized weight. The maximum weight corresponds to an overlap extending across the entire transducer beam width. The impulse response of an apodized transducer can be written using Equation 3 above, with A_n taking on values corresponding to the electrode overlaps. Overlap apodized transducers generate acoustic beam profiles which are nonuniform, with more energy in specific portions of the beam width at any specific point in time. Since the precise geometry of the apodization determines the location of energy within the beam, using two apodized transducers acoustically in series can result in portions of the individual transducer responses not interacting with active regions of the opposite transducer apodization pattern. Thus, the frequency response of two-in-line apodized transducers will not be the product of their individual frequency responses. One common way to compensate for this situation is to use a multistrip coupler (MSC) between the two transducers to collect the spatially weighted wave emitted from the input transducer, convert it to an amplitude-weighted wave, and to transmit this wave with uniform phase fronts over the entire aperture of the second transducer. In this configuration, provided the MSC is wideband compared with the transducer frequency responses, the overall device response will be the product of the individual transducer frequency responses. One additional reason to utilize MSCs is the potential for improvement of filter rejection through bulk mode suppression or frequency selectivity of the MSC. In instances where a MSC is impractical because of low substrate electromechanical coupling (such as quartz) or because of the die surface area required, filters are generally designed with one apodized and one unweighted transducer, or with one apodized and one withdrawal weighted transducer. In either case, the nonapodized transducer has a uniform beam width and will interact with the entire wave generated by the apodized transducer. Alternatively, if two in-line apodized transducers are used, the impulse response of each transducer can be represented as the summation of the responses of a series of tracks, each of which is essentially an unweighted transducer. The impulse and frequency response of each track will be given by Equations 3 and 4, respectively, where the transducer aperture W is interpreted as the individual track width, and the transducer length N_p is interpreted as the number of contiguous active electrode pairs in the track. The overall impulse and frequency responses for transducer $i(i = 1,2)$ can be written

$$h_i(t) = \sum_{j=1}^{m} h_j(t) \qquad H_i(f) = \sum_{j=1}^{m} H_j(f) \tag{6}$$

where $h_j(t)$ is the track impulse response, $H_j(f)$ is the track frequency response, and m is the total number of tracks used for the analysis. Additionally, the acoustic conductance, susceptance, and static capacitance must be evaluated for each track separately and summed to obtain the overall transducer equivalent circuit parameter values. The number and time lengths of these tracks will vary based on the individual apodization pattern and the transducer impulse response accuracy desired, as well as the computational complexity involved. The response of two cascaded apodized transducers is found by taking the products of the individual transducer frequency responses in each track, then summing these product for all tracks. This can be written

$$H(f) = \sum_{j=1}^{m} H_{1j}(f) H_{2j}(f) \neq H_1(f) H_2(f) \tag{7}$$

For each apodized transducer in a device, a small increase in insertion loss will be incurred, since not all of the energy in a uniform acoustic beam can be extracted by an apodized transducer and vice versa. This apodization loss is generally less than 1 dB per transducer and can be calculated as[15]

$$IL_{\text{apodization}} = 10 \log \frac{\left| \sum_{j=1}^{m} H_{ij}(f_0) \right|^2}{m \sum_{j=1}^{m} \left(H_{ij}(f_0) \right)^2} \tag{8}$$

where the track frequency response for track j in transducer i can be found as above.

A range of techniques can be used by the SAW designer to generate or select desirable pairs of transducers. First, the designer can use intuition or trial and error to select pairs of window functions with varying parameters for the two transducers. While this approach is quite simple, it is highly dependent on the expertise of the designer and rarely results in the optimal design solution. Second, eigensynthesis or optimization algorithms such as the Remez exchange algorithm (and others) can be used to generate transducers selectively with specified performance characteristics.[2] In either of these two approaches, the overall filter response would be the product (in frequency) or convolution (in time) of the two individual transducer responses. Another design approach is to use an optimization routine to generate the ideal overall filter response, the use some means to separate out or deconvolve two transducer responses from this overall response. Zero separation is one technique that can be used to generate two individual transducer impulse responses which will convolve (in time) to yield the overall desired device response.[13]

4.2 LOW-LOSS FILTERS

Low-loss SAW filters can be designed with techniques similar to those used for bidirectional filter design, using a number of transducer structures and acoustic element configurations. The three-phase unidirectional transducer structure[2,11,13,15] has historically offered the highest-performance low-loss filter available. The trade-offs required to obtain this performance are a more complex multilevel fabrication process, combined with the requirement for more-complicated phasing and matching networks to achieve unidirectionality. Recent advances in low-loss SAW design techniques have led to the development of numerous single-level, low-loss filter configurations which have performance that rivals that of three-phase unidirectional structures.[2,13,15] These filters are easier to fabricate and can be matched with networks much simpler than those required for three-phase devices. A popular example is the single-phase unidirectional transducer, or SPUDT. This structure takes advantage of the fact that the effective centers of electroacoustic transduction and mechanical reflection are not necessarily geometrically coincident in an IDT. By properly specifying specific IDT parameters (such as finger width, thickness, location, etc. — which parameters are specified depends on the technique used to generate SPUDT action) and electrical matching conditions, the centers of transduction and reflection can be placed such that the waves launched by transduction and those generated by reflection will add in phase in one direction, while adding destructively in the other direction. This yields the desired unidirectional transducer response.[2,13,15] Computer-based techniques similar to those used for bidirectional filter design can be used to design low-loss SAW filters.

4.3 DISPERSIVE DEVICES

SAW devices can be designed with nonuniformly spaced electrodes, resulting in dispersive transducers and either dispersive or nondispersive device performance. If the interelectrode spacing in an IDT monotonically increases toward the output transducer, then the frequency excited by the transducer decreases correspondingly. This configuration is known as an "up-chirp," since the low-frequency signal will be recieved first, with the recieved signal frequency increasing with time. Similarly, if the electrode spacing monotonically decreases, the higher-frequency components of the SAW will be received first, resulting in a "down-chirp." A device with characteristics which are nondispersive can be formed by using two dispersive transducers oriented in the same direction. Such devices are capable of very broadband performance. Dispersive transducers can be used in a number of other configurations, and find wide application in spread-spectrum secure communications and radar applications.[2,13,15]

4.4 RESONATOR FILTERS

Loss loss, narrow fractional bandwidth, and small size can be achieved by multipole SAW resonator filters. Multipole devices utilize multiple coupled resonant cavities. Coupling controls the flow of energy from one resonant region to another in a multipole device, and the strength of the coupling determines the filter bandwidth. Coupling in multipole resonator filters can be realized in a number of ways, two of the most common of which are in-line coupling and transverse acoustic coupling. A discussion of the principles of resonator and resonator filter design is beyond the scope of this publication and can be found in several reference texts, including Reference 2.

5. IMPEDANCE MATCHING AND TRIPLE TRANSIT

Time-spurious responses can arise from a number of sources, depending on the wave mode being utilized for signal processing, the substrate characteristics, and the specific device configuration. If not properly accounted for or avoided, these effects can substantially impact device performance.

Triple transit is the signal resulting from the acoustic wave traversing the delay path three times rather than once. This is caused by part of the acoustic wave energy being reflected or regenerated by the output transducer back toward the input transducer, where it is once again reflected or regenerated, sending a wave back to the output. The spurious signal due to triple transit is delayed by twice the overall device delay, τ, relative to the main response. This results in a passband ripple in the frequency domain, with a period which is the reciprocal of 2τ. This ripple appears in the magnitude, phase, and delay responses.

The triple-transit signal is attenuated during its propagation and reflection/regeneration process, resulting in a delayed signal with attenuation of roughly twice the insertion loss relative to the main output signal. Thus, tight electrical impedance matching which reduces the device insertion loss also aggravates the triple-transit reflection. Many engineers choose to use SAW devices without impedance matching to minimize passband ripple if the insertion loss is acceptable. An alternative approach is simply to utilize a series inductor to cancel the SAW capacitance and reduce insertion loss without increasing triple transit beyond acceptable limits. For specific applications, special SAW devices can be designed with very low levels of triple transit. Triple-transit rejection of over 65 dB has been demonstrated on commercially available devices.

Table 1 Substrate Parameters for Common Single-Crystal Piezoelectrics

Crystal	Cut	k^2 (%)	C_s (pF/cm)	v_0 (m/s)	α_T (ppm/°C)	% BW
Quartz	ST	0.11	0.5	3159	0 (1st order)	0.1–5
$LiNbO_3$	YZ	4.6	4.6	3488	–94	10–30
	128°	5.6	5.0	3992	–72	15–65
$LiTa_2O_3$	XY	0.75	4.4	3295	–23	5–10

6. SURFACE WAVE SUBSTRATES

The principal materials commonly used for SAW devices can be conveniently divided into two categories: single crystals and thin films. Single crystals such as quartz, lithium niobate ($LiNbO_3$), and lithium tantalate ($LiTaO_5$) are widely used. Thin films such as aluminum nitride (AIN) and zinc oxide (ZnO) are routinely deposited on substrates such as glass or silicon for low-cost applications. Sapphire is also used as a substrate for piezoelectric thin films in order to achieve higher SAW velocities, necessary for high-frequency operation with reduced lithographic demands. ZnO on thin film diamond has also been under development for the past few years for high-frequency device applications. An additional potential benefit of thin film piezoelectrics is the possibility of integrating SAW devices with other circuits. The single-crystal materials mentioned above are all anisotropic, indicating that the acoustic properties of the crystal vary with propagation direction (or crystal cut angle). The pertinent physical properties for each of the most commonly used crystal cuts are given in Table 1. Since the electromechanical coupling coefficient k^2 determines the fractional bandwidths that can be realized on a given substrate, rough guidelines for design are included. Also, the temperature coefficient for each substrate indicates the frequency shift which will be induced by a change in temperature.

ACKNOWLEDGMENT

All device plots courtesy of Sawtek, Inc., Orlando, Florida.

REFERENCES

1. **Hartmann, C. S.,** Systems impact of modern Rayleigh wave technology, in Ash, E. A. and Paige, E. G. S., Eds., *Rayleigh-Wave Theory and Application,* from Springer Series in Wave Phenomena, Springer-Verlag, New York, 1985, 238–253.
2. **Campbell, C.,** *Surface Acoustic Wave Devices and Their Signal Processing Applications,* Academic Press, New York, 1989.
3. **Hines, J. H. and Stapor, W. J.,** The effects of ionizing radiation on SAW resonators, *IEEE Ultrasonics Symp. Proc.,* 1990, 471–476.
4. **Stapor, W. J., Hines, J. H., and Dale, C. J.,** Ionizing space radiation effects on surface acoustic wave resonators, *SPIE Symp. Proc.,* 1992.
5. **Yatsuda, H., Inaoka, T., and Horishima, T.,** IIDT type low-loss SAW filters with improved stopband rejection in the range of 1 to 2 GHz, *IEEE Ultrasonics Symp. Proc.,* 1992, 67–70.
6. **Hikita, W., Shibagaki, N., Akagi, T., and Sakiyama, K.,** Design methodology and synthesis techniques for ladder-type SAW resonator coupled filters, *IEEE Ultrasonics Symp. Proc.,* 1993, 15–24.
7. **Martin, G., Wall, B., Kunze, R., and Weihnacht, M.,** Four modes waveguide resonator filters, *IEEE Ultrasonics Symp. Proc.,* 1993, 35–39.
8. **Farnell, G. W.,** Elastic surface waves, in Matthews, H., Ed., *Surface Wave Filters, Design, Construction, and Use,* John Wiley & Sons, New York, 1977, 1–53.
9. **Auld, B. A.,** *Acoustic Fields and Waves in Solids,* 2nd ed., Robert E. Krieger Publishing, Florida, 1990.
10. **Lewis, M. F.,** On Rayleigh waves and related propagating acoustic waves, in Ash, E. A. and Paige, E. G. S., Eds., *Rayleigh-Wave Theory and Application,* from Springer Series in Wave Phenomena, Springer-Verlag, New York, 1985, 37–58.
11. **Morgan, D. P.,** *Surface-Wave Devices for Signal Processing,* Elsevier, New York, 1985.
12. **Datta, S.,** *Surface Acoustic Wave Devices,* Prentice-Hall, Englewood Cliffs, NJ, 1986.
13. **Feldmann, M. and Hénaff, J.,** *Surface Acoustic Waves for Signal Processing,* Artech House, Boston, 1989.
14. **Oliner, A. A.,** *Acoustic Surface Waves,* Springer-Verlag, New York, 1978.
15. **Malocha, D. C.,** Surface acoustic wave technology, in Dorf, R. C., *The Electrical Engineering Handbook,* CRC Press, Boca Raton, FL, 1993, 1062–1076.

Index

A

B

C

D

E

F

G

M

N

O

P

Q

R

S

T

U

V

W

X

Z